Lecture Notes in Mathematics 1938

Editors:
J.-M. Morel, Cachan
F. Takens, Groningen
B. Teissier, Paris

**FONDAZIONE
CIME
ROBERTO CONTI**

CENTRO INTERNAZIONALE MATEMATICO ESTIVO
INTERNATIONAL MATHEMATICAL SUMMER CENTER

C.I.M.E. means Centro Internazionale Matematico Estivo, that is, International Mathematical Summer Center. Conceived in the early fifties, it was born in 1954 and made welcome by the world mathematical community where it remains in good health and spirit. Many mathematicians from all over the world have been involved in a way or another in C.I.M.E.'s activities during the past years.

So they already know what the C.I.M.E. is all about. For the benefit of future potential users and co-operators the main purposes and the functioning of the Centre may be summarized as follows: every year, during the summer, Sessions (three or four as a rule) on different themes from pure and applied mathematics are offered by application to mathematicians from all countries. Each session is generally based on three or four main courses (24–30 hours over a period of 6-8 working days) held from specialists of international renown, plus a certain number of seminars.

A C.I.M.E. Session, therefore, is neither a Symposium, nor just a School, but maybe a blend of both. The aim is that of bringing to the attention of younger researchers the origins, later developments, and perspectives of some branch of live mathematics.

The topics of the courses are generally of international resonance and the participation of the courses cover the expertise of different countries and continents. Such combination, gave an excellent opportunity to young participants to be acquainted with the most advance research in the topics of the courses and the possibility of an interchange with the world famous specialists. The full immersion atmosphere of the courses and the daily exchange among participants are a first building brick in the edifice of international collaboration in mathematical research.

C.I.M.E. Director
Pietro ZECCA
Dipartimento di Energetica "S. Stecco"
Università di Firenze
Via S. Marta, 3
50139 Florence
Italy
e-mail: zecca@unifi.it

C.I.M.E. Secretary
Elvira MASCOLO
Dipartimento di Matematica
Università di Firenze
viale G.B. Morgagni 67/A
50134 Florence
Italy
e-mail: mascolo@math.unifi.it

For more information see CIME's homepage: http://www.cime.unifi.it

CIME's activity is supported by:

– Istituto Nationale di Alta Mathematica "F. Severi"
– Ministero dell'Istruzione, dell'Università e delle Ricerca
– Ministero degli Affari Esteri, Direzione Generale per la Promozione e la Cooperazione, Ufficio V

Denis Auroux · Fabrizio Catanese
Marco Manetti · Paul Seidel
Bernd Siebert · Ivan Smith
Gang Tian

Symplectic 4-Manifolds and Algebraic Surfaces

Lectures given at the
C.I.M.E. Summer School
held in Cetraro, Italy
September 2–10, 2003

Editors:
Fabrizio Catanese
Gang Tian

 Springer

FONDAZIONE
CIME
ROBERTO CONTI

Denis Auroux
Department of Mathematics, Room 2-248
Massachusetts Institute of Technology
77 Massachusetts Avenue
Cambridge, MA 02139-4307, USA
auroux@math.mit.edu

Bernd Siebert
Mathematisches Institut
Albert-Ludwigs-Universität Freiburg
Eckerstr. 1
79104 Freiburg, Germany
bernd.siebert@math.uni-freiburg.de

Fabrizio Catanese
Mathematisches Institut
Lehrstuhl Mathematik VIII
Universität Bayreuth
95447 Bayreuth, Germany
Fabrizio.Catanese@uni-bayreuth.de

Ivan Smith
Centre for Mathematical Sciences
University of Cambridge
Wilberforce Road
Cambridge, CB3 0WB, UK
i.smith@dpmms.cam.ac.uk

Marco Manetti
Dipartimento di Matematica
"G. Castelnuovo"
Università di Roma "La Sapienza"
Piazzale Aldo Moro 5
00185 Roma, Italy
manetti@mat.uniroma1.it

Gang Tian
Department of Mathematics
Princeton University
Fine Hall Room 702, Washington Road
Princeton, NJ 08544-1000, USA
tian@math.princeton.edu

Paul Seidel
Department of Mathematics, Room 2-270
Massachusetts Institute of Technology
77 Massachusetts Avenue
Cambridge, MA 02139-4307, USA
seidel@math.mit.edu

ISBN: 978-3-540-78278-0   e-ISBN: 978-3-540-78279-7
DOI: 10.1007/978-3-540-78279-7

Lecture Notes in Mathematics ISSN print edition: 0075-8434
                    ISSN electronic edition: 1617-9692

Library of Congress Control Number: 2008921369

Mathematics Subject Classification (2000): 14J29, 14D66, 53D99, 32G05, 32G15, 20F36, 37R50

Cover design: *design & production* GmbH, Heidelberg

Cover art: The Togliatti Quintic by Oliver Labs, produced with Stephan Endrass' visualization tool surf
(http://surf.sourceforge.net)

Printed on acid-free paper

9 8 7 6 5 4 3 2 1

springer.com

Preface

The third C.I.M.E. Session 'Symplectic 4-Manifolds and Algebraic Surfaces' took place from September 2 to September 10, 2003 in the customary beautiful location of the Grand Hotel San Michele, Cetraro, Cosenza.

The present volume contains the text of the five series of lectures, which were delivered during the course.

There were also some very interesting seminar lectures during the course. They are of a more specialized nature and are not reproduced here.

The lectures survey recent and important advances in symplectic and differential topology of 4-manifolds and algebraic surfaces.

Relations with real algebraic geometry have been treated only in part in the course by Catanese, and much more in the seminars by Frediani and Welschinger. Indeed, this and other very interesting topics of current research could only be treated rather quickly, in view of the vastness of the central theme of the School, the study of differential, symplectic and complex structures on even dimensional, and especially four-dimensional, manifolds.

The course had at least a double valency: on the one hand the introduction of new methods, for instance symplectic geometry, for the study of moduli spaces of complex structures on algebraic surfaces. On the other hand, the use of algebraic surfaces as concrete models for the investigation of symplectic topology in dimension 4, and for laying down a research plan based on the analogies with the surface classification.

One concrete example of the synergy of these two viewpoints is given for instance by the study of partial compactifications of moduli spaces of singular surfaces, which led to the construction of symplectomorphisms through surgeries associated with smoothings of singularities.

Let us now try to describe the contents of the courses and the interwoven thread which relates them to one another, thus making this volume a coherent exposition of an active field of current research.

As it is well known, every projective variety inherits from the ambient space the Fubini-Study Kähler form, and in this way one obtains the most natural examples of symplectic structures. Even if one wants to consider more general symplectic manifolds, for use in the theory of dynamical systems, or just for the sake of classification, the relation with complex manifolds theory is always present.

In fact, a symplectic manifold always admits almost-complex structures, and the Kähler condition has as an analogue the condition that the almost-complex structure be compatible with the given symplectic structure. Even if almost-complex structures are not integrable (i.e., there are no local holomorphic coordinates), nevertheless one can still consider maps from a complex curve to the given almost-complex manifold whose derivative is complex linear.

One of the fundamental ideas, due to Gromov, is to use such maps to study the topology of symplectic manifolds.

These maps are called pseudoholomorphic curves, and the key point is that the corresponding generalized Cauchy–Riemann equations in the nonintegrable case do not substantially differ from the classical case (analyticity, removable singularities,...). The great advantage is, however, that while complex structures may remain 'nongeneric' even after deformation, a generic almost-complex structure really has an apt behaviour for transversality questions.

The study of pseudoholomorphic curves leads to important invariants, the so-called Gromov–Witten invariants, which have become an active research subject since late 1993 and were also treated in the seminar by Pandharipande.

If we start from a complex manifold, there remains, however, a basic question: if we take symplectic curves, how much do these differ from the holomorphic curves pertaining to the initial complex structure? Can they be deformed isotopically to holomorphic curves? Its analogue in algebraic geometry is to classify holomorphic curves under algebraic equivalence. In general, there are complex manifolds which contain symplectic curves, not isotopic to holomorphic ones. However, if the underlying symplectic manifold is an algebraic surface with positive first Chern class, it is expected that any symplectic curve be isotopic to a holomorphic curve.

This is the so-called symplectic isotopy problem, one of the fundamental problems in the study of symplectic 4-manifolds. It has many topological consequences in symplectic geometry. For instance, the solution to this symplectic isotopy problem provides a very effective way of classifying simply-connected symplectic 4-manifolds. Significant progress has been made. The central topic of the course by Siebert and Tian was the study of the symplectic isotopy problem, describing main tools and recent progress. The course has a pretty strong analytic flavour because of the nonlinearity of the Cauchy–Riemann equations in the nonintegrable case. Some applications to symplectic 4-manifolds were also discussed, in particular, that any genus two symplectic Lefschetz fibration under some mild non-degeneracy conditions is equivalent to a holomorphic surface. This result ties in with what we might call the 'dual' approach.

This is the approach taken by Donaldson for the study of symplectic 4-manifolds. Donaldson was able to extend the algebro-geometric concept of Lefschetz pencils to the case of symplectic manifolds. Even if for any generic almost-complex structure one cannot find holomorphic functions (even locally); one can nevertheless find smooth functions and sections of line bundles whose antiholomorphic derivative $(\bar{\partial}f)$, even if not zero, is still much smaller than the holomorphic derivative ∂f. This condition produces the same type of topological behaviour as the one possessed by holomorphic functions, and the functions satisfying it are called approximately holomorphic functions (respectively, sections).

In this way, Donaldson was able to extend the algebro-geometric concept of Lefschetz pencils to the case of symplectic manifolds.

The topic of Lefschetz pencils has occupied a central role in several courses, one by Auroux and Smith, one by Seidel, and one by Catanese.

In fact, one main use of Lefschetz pencils is, from the results of Kas and Gompf, to encode the differential topology and the symplectic topology of the fibred manifold into a factorization inside the mapping class group, with factors which are (positive) Dehn twists.

While the course of Auroux and Smith used symplectic Lefschetz pencils to study topological invariants of symplectic 4-manifolds and the differences between the world of symplectic 4-manifolds and that of complex surfaces, in the course by Catanese, Lefschetz fibrations were used to describe recent work done in collaboration with Wajnryb to prove explicit diffeomorphisms of certain simply connected algebraic surfaces which are not deformation equivalent.

An important ingredient here is a detailed knowledge of the mapping class group of the fibres, which are compact complex curves of genus at least two.

Applications to higher dimensional symplectic varieties, and to Mirror symmetry, were discussed in the course by Seidel, dedicated to the symplectic mapping class group of 4-manifolds, and to Dehn twists in dimension 4. The resulting picture of symplectic monodromy is surprising. In fact, Seidel shows that the natural homomorphism of the symplectic to the differential mapping class group may not be injective, and moreover reveals a delicate deformation behaviour: there are symplectomorphisms which are not isotopic to the identity for some special symplectic structure, but become isotopic after a small deformation of the symplectic structure.

Lefschetz pencils are the 'generic' maps with target the complex projective line: the course by Auroux and Smith went all the way to consider a generalization of the classical algebro-geometric concept of 'generic multiple planes'.

This notion was extended to the symplectic case, again via approximately holomorphic sections, by Auroux and Katzarkov, and the course discusses the geometric invariants of a symplectic structure (which are deformation invariant) that can be extracted in this way.

This research interest is related to an old problem, posed by Boris Moishezon, namely, whether one can distinguish connected components of moduli spaces of surfaces of general type via these invariants (essentially, fundamental groups of complements of branch curves).

The courses by Catanese and Manetti are devoted instead to a similar question, a conjecture raised by Friedman and Morgan in the 1980s on the grounds of gauge theoretic speculations.

This conjecture is summarized by the acronym def=diff, and stated that diffeomorphic algebraic surfaces should be deformation equivalent. The course by Catanese reports briefly on the first nontrivial counterexamples, obtained by Manetti in the interesting case of surfaces of general type, and on the 'trivial' counterexamples, obtained independently by several authors, where a surface S is not deformation equivalent to the complex conjugate surface.

The focus in both courses is set on the simplest and strongest counterexamples, the so-called 'abc' surfaces, which are simply connected, and for which Catanese and Wajnryb showed that the diffeomorphism type is determined by the integers (a+c) and b. The course by Manetti focuses on the deformation theoretic and degeneration aspects (especially, smoothings of singularities), which up to now were scattered in a long series of articles by Catanese and Manetti. Catanese's course has a broader content and includes also other introductory facts on surfaces of general type and on singularities.

We would like to point out that, in spite of tremendous progress in 4-manifold topology, starting with Michael Freedman's solution of the problem of understanding the topology of simply connected 4-manifolds up to homeomorphism, and Simon Donaldson's gauge theoretic discoveries that smooth and topological structures differ drastically in dimension 4, the differential and symplectic topology even of simply connected symplectic 4-manifolds and algebraic surfaces is still a deep mystery.

Modern approaches to the study of symplectic 4-manifolds and algebraic surfaces combine a wide range of techniques and sources of inspiration. Gauge theory, symplectic geometry, pseudoholomorphic curves, singularity theory, moduli spaces, braid groups, monodromy, in addition to classical topology and algebraic geometry, combine to make this one of the most vibrant and active areas of research in mathematics.

Some keywords for the present volume are therefore pseudoholomorphic curves, algebraic and symplectic Lefschetz pencils, Dehn twists and monodromy, symplectic invariants, deformation theory and singularities, classification and moduli spaces of algebraic surfaces of general type, applications to mirror symmetry.

It is our hope that these texts will be useful to people working in related areas of mathematics and will become standard references on these topics.

We take this opportunity to thank the C.I.M.E. foundation for making the event possible, the authors for their hard work, the other lecturers for their interesting contributions, and the participants of the conference for their lively interest and enthusiastic collaboration.

We would also like to take this opportunity to thank once more the other authors for their work and apologize for the delay in publication of the volume.

May, 2007

Fabrizio Catanese
Gang Tian

Contents

Lefschetz Pencils, Branched Covers and Symplectic Invariants

Denis Auroux[1] and Ivan Smith[2]

[1] Department of Mathematics, M.I.T., Cambridge, MA 02139, USA
   auroux@math.mit.edu
[2] Centre for Mathematical Sciences, University of Cambridge, Wilberforce Road,
   Cambridge CB3 0WB, UK
   i.smith@dpmms.cam.ac.uk

> Two symplectic fibrations are never *exactly* the same. When you
> have two fibrations, they might be canonically isomorphic, but when
> you look closely, the points of one might be numbers while the
> points of the other are bananas.
>
> P. Seidel, 9/9/03

1 Introduction and Background

This set of lectures aims to give an overview of Donaldson's theory of linear
systems on symplectic manifolds and the algebraic and geometric invariants
to which they give rise. After collecting some of the relevant background, we
discuss topological, algebraic and symplectic viewpoints on Lefschetz pencils
and branched covers of the projective plane. The later lectures discuss invariants obtained by combining this theory with pseudo-holomorphic curve
methods.

1.1 Symplectic Manifolds

Definition 1.1 *A symplectic structure on a smooth manifold M is a closed
non-degenerate 2-form ω, i.e., an element $\omega \in \Omega^2(M)$ such that $d\omega = 0$ and
$\forall v \in TM - \{0\}$, $\iota_v \omega \neq 0$.*

For example, $\mathbb{R}^{2n}$ carries a standard symplectic structure, given by the
2-form $\omega_0 = \sum dx_i \wedge dy_i$. Similarly, every orientable surface is symplectic,
taking for ω any non-vanishing volume form.

Since ω induces a non-degenerate antisymmetric bilinear pairing on the tangent spaces to M, it is clear that every symplectic manifold is even-dimensional and orientable (if $\dim M = 2n$, then $\frac{1}{n!}\omega^n$ defines a volume form on M).

Two important features of symplectic structures that set them apart from most other geometric structures are the existence of a large number of symplectic automorphisms, and the absence of local geometric invariants.

The first point is illustrated by the following construction. Consider a smooth function $H : M \to \mathbb{R}$ (a *Hamiltonian*), and define X_H to be the vector field on M such that $\omega(X_H, \cdot) = dH$. Let $\phi_t : M \to M$ be the family of diifeomorphisms generated by the flow of X_H, i.e., $\phi_0 = \mathrm{Id}$ and $\frac{d}{dt}\phi_t(x) = X_H(\phi_t(x))$. Then ϕ_t is a *symplectomorphism*, i.e., $\phi_t^*\omega = \omega$. Indeed, we have $\phi_0^*\omega = \omega$, and

$$\frac{d}{dt}\phi_t^*\omega = \phi_t^*(L_{X_H}\omega) = \phi_t^*(d\iota_{X_H}\omega + \iota_{X_H}d\omega) = \phi_t^*(d(dH) + 0) = 0.$$

Therefore, the group of symplectomorphisms $\mathrm{Symp}(M, \omega)$ is infinite-dimensional, and its Lie algebra contains all Hamiltonian vector fields. An easy consequence is that $\mathrm{Symp}(M, \omega)$ acts transitively on points of M. This is in contrast with the case of Riemannian metrics, where isometry groups are much smaller.

The lack of local geometric invariants of symplectic structures is illustrated by two classical results of fundamental importance, which show that the study of symplectic manifolds largely reduces to *topology* (i.e., to discrete invariants): Darboux's theorem, and Moser's stability theorem. The first one shows that all symplectic forms are locally equivalent, in sharp contrast to the case of Riemannian metrics where curvature provides a local invariant, and the second one shows that exact deformations of symplectic structures are trivial.

Theorem 1.2 (Darboux) *Every point in a symplectic manifold (M^{2n}, ω) admits a neighborhood that is symplectomorphic to a neighborhood of the origin in $(\mathbb{R}^{2n}, \omega_0)$.*

Proof. We first use local coordinates to map a neighborhood of a given point in M diffeomorphically onto a neighborhood V of the origin in $\mathbb{R}^{2n}$. Composing this diffeomorphism f with a suitable linear transformation of $\mathbb{R}^{2n}$, we can ensure that the symplectic form $\omega_1 = (f^{-1})^*\omega$ coincides with ω_0 at the origin. This implies that, restricting to a smaller neighborhood if necessary, the closed 2-forms $\omega_t = t\omega_1 + (1-t)\omega_0$ are non-degenerate over V for all $t \in [0, 1]$.

Using the Poincaré lemma, consider a family of 1-forms α_t on V such that $\frac{d}{dt}\omega_t = -d\alpha_t$. Subtracting a constant 1-form from α_t if necessary, we can assume that α_t vanishes at the origin for all t. Using the non-degeneracy of ω_t we can find vector fields X_t such that $\iota_{X_t}\omega_t = \alpha_t$. Let $(\phi_t)_{t\in[0,1]}$ be the flow generated by X_t, i.e., the family of diffeomorphisms defined by $\phi_0 = \mathrm{Id}$, $\frac{d}{dt}\phi_t(x) = X_t(\phi_t(x))$; we may need to restrict to a smaller neighborhood $V' \subset V$ of the origin in order to make the flow ϕ_t well-defined for all t. We then have

$$\frac{d}{dt}\phi_t^*\omega_t = \phi_t^*(L_{X_t}\omega_t) + \phi_t^*\left(\frac{d\omega_t}{dt}\right) = \phi_t^*(d(\iota_{X_t}\omega_t) - d\alpha_t) = 0,$$

and therefore $\phi_1^*\omega_1 = \omega_0$. Therefore, $\phi_1^{-1} \circ f$ induces a symplectomorphism from a neighborhood of x in (M,ω) to a neighborhood of the origin in $(\mathbb{R}^{2n}, \omega_0)$. □

Theorem 1.3 (Moser) *Let $(\omega_t)_{t\in[0,1]}$ be a continuous family of symplectic forms on a compact manifold M. Assume that the cohomology class $[\omega_t] \in H^2(M, \mathbb{R})$ does not depend on t. Then (M, ω_0) is symplectomorphic to (M, ω_1).*

Proof. We use the same argument as above: since $[\omega_t]$ is constant there exist 1-forms α_t such that $\frac{d}{dt}\omega_t = -d\alpha_t$. Define vector fields X_t such that $\iota_{X_t}\omega_t = \alpha_t$ and the corresponding flow ϕ_t. By the same calculation as above, we conclude that $\phi_1^*\omega_1 = \omega_0$. □

Definition 1.4 *A submanifold $W \subset (M^{2n}, \omega)$ is called* symplectic *if $\omega_{|W}$ is non-degenerate at every point of W (it is then a symplectic form on W);* isotropic *if $\omega_{|W} = 0$; and* Lagrangian *if it is isotropic of maximal dimension $\dim W = n = \frac{1}{2}\dim M$.*

An important example is the following: given any smooth manifold N, the cotangent bundle T^*N admits a canonical symplectic structure that can be expressed locally as $\omega = \sum dp_i \wedge dq_i$ (where (q_i) are local coordinates on N and (p_i) are the dual coordinates on the cotangent spaces). Then the zero section is a Lagrangian submanifold of T^*N.

Since the symplectic form induces a non-degenerate pairing between tangent and normal spaces to a Lagrangian submanifold, the normal bundle to a Lagrangian submanifold is always isomorphic to its cotangent bundle. The fact that this isomorphism extends beyond the infinitesimal level is a classical result of Weinstein:

Theorem 1.5 (Weinstein) *For any Lagrangian submanifold $L \subset (M, \omega)$, there exists a neighborhood of L which is symplectomorphic to a neighborhood of the zero section in the cotangent bundle $(T^*L, \sum dp_i \wedge dq_i)$.*

There is also a neighborhood theorem for symplectic submanifolds; in that case, the local model for a neighborhood of the submanifold $W \subset M$ is a neighborhood of the zero section in the symplectic vector bundle NW over W (since $Sp(2n)$ retracts onto $U(n)$, the classification of symplectic vector bundles is the same as that of complex vector bundles).

1.2 Almost-Complex Structures

Definition 1.6 *An* almost-complex structure *on a manifold M is an endomorphism J of the tangent bundle TM such that $J^2 = -\mathrm{Id}$. An almost-complex structure J is said to be* tamed *by a symplectic form ω if for every* *non-zero tangent vector u we have $\omega(u, Ju) > 0$; it is compatible with ω if it is ω-tame and $\omega(u, Jv) = -\omega(Ju, v)$; equivalently, J is ω-compatible if and only if $g(u, v) = \omega(u, Jv)$ is a Riemannian metric.*

Proposition 1.7 *Every symplectic manifold (M, ω) admits a compatible almost-complex structure. Moreover, the space of ω-compatible (resp. ω-tame) almost-complex structures is contractible.*

This result follows from the fact that the space of compatible (or tame) complex structures on a symplectic vector space is non-empty and contractible (this can be seen by constructing explicit retractions); it is then enough to observe that a compatible (resp. tame) almost-complex structure on a symplectic manifold is simply a section of the bundle $End(TM)$ that defines a compatible (resp. tame) complex structure on each tangent space.

An almost-complex structure induces a splitting of the complexified tangent and cotangent bundles: $TM \otimes \mathbb{C} = TM^{1,0} \oplus TM^{0,1}$, where $TM^{1,0}$ and $TM^{0,1}$ are respectively the $+i$ and $-i$ eigenspaces of J (i.e., $TM^{1,0} = \{v - iJv, \ v \in TM\}$, and similarly for $TM^{0,1}$; for example, on $\mathbb{C}^n$ equipped with its standard complex structure, the $(1, 0)$ tangent space is generated by $\partial/\partial z_i$ and the $(0, 1)$ tangent space by $\partial/\partial \bar{z}_i$. Similarly, J induces a complex structure on the cotangent bundle, and $T^*M \otimes \mathbb{C} = T^*M^{1,0} \oplus T^*M^{0,1}$ (by definition $(1, 0)$-forms are those which pair trivially with $(0, 1)$-vectors, and vice versa). This splitting of the cotangent bundle induces a splitting of differential forms into *types*: $\bigwedge^r T^*M \otimes \mathbb{C} = \bigoplus_{p+q=r} \bigwedge^p T^*M^{1,0} \otimes \bigwedge^q T^*M^{0,1}$. Moreover, given a function $f : M \to \mathbb{C}$ we can write $df = \partial f + \bar{\partial} f$, where $\partial f = \frac{1}{2}(df - i \, df \circ J)$ and $\bar{\partial} f = \frac{1}{2}(df + i \, df \circ J)$ are the $(1, 0)$ and $(0, 1)$ parts of df respectively. Similarly, given a complex vector bundle E over M equipped with a connection, the covariant derivative ∇ can be split into operators $\partial^\nabla : \Gamma(E) \to \Omega^{1,0}(E)$ and $\bar{\partial}^\nabla : \Gamma(E) \to \Omega^{0,1}(E)$.

Although the tangent space to a symplectic manifold (M, ω) equipped with a compatible almost-complex structure J can be pointwise identified with $(\mathbb{C}^n, \omega_0, i)$, there is an important difference between a symplectic manifold equipped with a compatible almost-complex structure and a complex Kähler manifold: the possible lack of *integrability* of the almost-complex structure, namely the fact that the Lie bracket of two $(1, 0)$ vector fields is not necessarily of type $(1, 0)$.

Definition 1.8 *The* Nijenhuis tensor *of an almost-complex manifold (M, J) is the quantity defined by $N_J(X, Y) = \frac{1}{4}([X, Y] + J[X, JY] + J[JX, Y] - [JX, JY])$. The almost-complex structure J is said to be* integrable *if $N_J = 0$.*

It can be checked that N_J is a tensor (i.e., only depends on the values of the vector fields X and Y at a given point), and that $N_J(X, Y) = 2 \operatorname{Re}([X^{1,0}, Y^{1,0}]^{(0,1)})$. The non-vanishing of N_J is therefore an obstruction to the integrability of a local frame of $(1, 0)$ tangent vectors, i.e., to the existence of local holomorphic coordinates. The Nijenhuis tensor is also related to the fact that the exterior differential of a $(1,0)$-form may have a non-zero component of type $(0,2)$, so that the ∂ and $\bar{\partial}$ operators on differential forms do not have square zero ($\bar{\partial}^2$ can be expressed in terms of ∂ and the Nijenhuis tensor).

Theorem 1.9 (Newlander–Nirenberg) *Given an almost-complex manifold (M, J), the following properties are equivalent: (i) $N_J = 0$; (ii) $[T^{1,0}M, T^{1,0}M] \subset T^{1,0}M$; (iii) $\bar{\partial}^2 = 0$; (iv) (M, J) is a complex manifold, i.e., admits complex analytic coordinate charts.*

1.3 Pseudo-Holomorphic Curves and Gromov–Witten Invariants

Pseudo-holomorphic curves, first introduced by Gromov in 1985 [Gr], have since then become the most important tool in modern symplectic topology. In the same way as the study of complex curves in complex manifolds plays a central role in algebraic geometry, the study of pseudo-holomorphic curves has revolutionized our understanding of symplectic manifolds.

The equation for holomorphic maps between two almost-complex manifolds becomes overdetermined as soon as the complex dimension of the domain exceeds 1, so we cannot expect the presence of any almost-complex submanifolds of complex dimension ≥ 2 in a symplectic manifold equipped with a generic almost-complex structure. On the other hand, J-holomorphic curves, i.e., maps from a Riemann surface (Σ, j) to the manifold (M, J) such that $J \circ df = df \circ j$ (or in the usual notation, $\bar{\partial}_J f = 0$), are governed by an elliptic PDE, and their study makes sense even in non-Kähler symplectic manifolds. The questions that we would like to answer are of the following type:

Given a compact symplectic manifold (M, ω) equipped with a generic compatible almost-complex structure J and a homology class $\beta \in H_2(M, \mathbb{Z})$, what is the number of pseudo-holomorphic curves of given genus g, representing the homology class β and passing through r given points in M (or through r given submanifolds in M)?

The answer to this question is given by *Gromov–Witten invariants*, which count such curves (in a sense that is not always obvious, as the result can, e.g., be negative, and need not even be an integer). We will only consider a simple instance of the theory, in which we count holomorphic spheres which are sections of a fibration.

To start with, one must study deformations of pseudo-holomorphic curves, by linearizing the equation $\bar{\partial}_J f = 0$ near a solution. The linearized Cauchy–Riemann operator $D_{\bar{\partial}}$, whose kernel describes infinitesimal deformations of a given curve $f : S^2 \to M$, is a Fredholm operator of (real) index

$$2d := \text{ind}\, D_{\bar{\partial}} = (\dim_{\mathbb{R}} M - 6) + 2\, c_1(TM) \cdot [f(S^2)].$$

When the considered curve is *regular*, i.e., when the linearized operator $D_{\bar{\partial}}$ is surjective, the deformation theory is unobstructed, and we expect the moduli space $\mathcal{M}(\beta) = \{f : S^2 \to M, \ \bar{\partial}_J f = 0, \ [f(S^2)] = \beta\}$ to be locally a smooth manifold of real dimension $2d$.

The main result underlying the theory of pseudo-holomorphic curves is Gromov's compactness theorem (see [Gr, McS], ...):

Theorem 1.10 (Gromov) *Let $f_n : (\Sigma_n, j_n) \to (M, \omega, J)$ be a sequence of pseudo-holomorphic curves in a compact symplectic manifold, representing a fixed homology class. Then a subsequence of $\{f_n\}$ converges (in the "Gromov–Hausdorff topology") to a limiting map f_∞, possibly singular.*

The limiting curve f_∞ can have a very complicated structure, and in particular its domain may be a nodal Riemann surface with more than one component, due to the phenomenon of *bubbling*. For example, the sequence of degree 2 holomorphic curves $f_n : \mathbb{CP}^1 \to \mathbb{CP}^2$ defined by $f_n(u : v) = (u^2 : uv : \frac{1}{n} v^2)$ converges to a singular curve with two degree 1 components: for $(u : v) \neq (0 : 1)$, we have $\lim f_n(u : v) = (u : v : 0)$, so that the sequence apparently converges to a line in $\mathbb{CP}^2$. However the derivatives of f_n become unbounded near $(0 : 1)$, and composing f_n with the coordinate change $\phi_n(u : v) = (\frac{1}{n} u : v)$ we obtain $f_n \circ \phi_n(u : v) = (\frac{1}{n^2} u^2 : \frac{1}{n} uv : \frac{1}{n} v^2) = (\frac{1}{n} u^2 : uv : v^2)$, which converges to $(0 : u : v)$ everywhere except at $(1 : 0)$, giving the other component (the "bubble") in the limiting curve. Therefore, it can happen that the moduli space $\mathcal{M}(\beta)$ is not compact, and needs to be compactified by adding maps with singular (nodal, possibly reducible) domains.

In the simplest case where the dimension of the moduli space is $2d = 0$, and assuming regularity, we can obtain an invariant by counting the number of points of the compactified moduli space $\overline{\mathcal{M}}(\beta)$ (up to sign). In the situations we consider, the moduli space will always be smooth and compact, but may have the wrong (excess) dimension, consisting of curves whose deformation theory is obstructed. In this case there is an *obstruction bundle* Obs $\to \mathcal{M}(\beta)$, whose fiber at $(f : S^2 \to M)$ is Coker $D_{\bar{\partial}}$. In this case the invariant may be recovered as the Euler class of this bundle, viewed as an integer (the degree of Obs).

1.4 Lagrangian Floer Homology

Roughly speaking, Floer homology is a refinement of intersection theory for Lagrangian submanifolds, in which we can only cancel intersection points by Whitney moves along pseudo-holomorphic Whitney discs. Formally the construction proceeds as follows.

Consider two compact orientable (relatively spin) Lagrangian submanifolds L_0 and L_1 in a symplectic manifold (M, ω) equipped with a compatible almost-complex structure J. Lagrangian Floer homology corresponds to the Morse theory of a functional on (a covering of) the space of arcs joining L_0 to L_1, whose critical points are constant paths.

For simplicity, we will only consider situations where it is not necessary to keep track of relative homology classes (e.g., by working over a Novikov ring), and where no bubbling can occur. For example, if we assume that $\pi_2(M) = \pi_2(M, L_i) = 0$, then Floer homology is well-defined; to get well-defined product structures we will only work with exact Lagrangian submanifolds of exact symplectic manifolds (see the final section).

By definition, the *Floer complex* $CF^*(L_0, L_1)$ is the free module with one generator for each intersection point $p \in L_0 \cap L_1$, and grading given by the *Maslov index*.

Given two points $p_\pm \in L_0 \cap L_1$, we can define a moduli space $\mathcal{M}(p_-, p_+)$ of pseudo-holomorphic maps $f : \mathbb{R} \times [0, 1] \to M$ such that $f(\cdot, 0) \in L_0$, $f(\cdot, 1) \in L_1$, and $\lim_{t \to \pm\infty} f(t, \tau) = p_\pm \ \forall \tau \in [0, 1]$; the expected dimension of this moduli space is the difference of Maslov indices. Assuming regularity and compactness of $\mathcal{M}(p_-, p_+)$, we can define an operator ∂ on $CF^*(L_0, L_1)$ by the formula

$$\partial p_- = \sum_{p_+} \#(\mathcal{M}(p_-, p_+)/\mathbb{R}) \ p_+,$$

where the sum runs over all p_+ for which the expected dimension of the moduli space is 1.

In good cases we have $\partial^2 = 0$, which allows us to define the Floer homology $HF^*(L_0, L_1) = \mathrm{Ker} \, \partial / \mathrm{Im} \, \partial$. The assumptions made above on $\pi_2(M)$ and $\pi_2(M, L_i)$ eliminate the serious technical difficulties associated to bubbling (which are more serious than in the compact case, since bubbling can also occur on the boundary of the domain, a real codimension 1 phenomenon which may prevent the compactified moduli space from carrying a fundamental class, see [FO³]).

When Floer homology is well-defined, it has important consequences on the intersection properties of Lagrangian submanifolds. Indeed, for every *Hamiltonian* diffeomorphism ϕ we have $HF^*(L_0, L_1) = HF^*(L_0, \phi(L_1))$; and if L_0 and L_1 intersect transversely, then the total rank of $HF^*(L_0, L_1)$ gives a lower bound on the number of intersection points of L_0 and L_1. A classical consequence, using the definition of Floer homology and the relation between $HF^*(L, L)$ and the usual cohomology $H^*(L)$, is the non-existence of compact simply connected Lagrangian submanifolds in $\mathbb{C}^n$.

Besides a differential, Floer complexes for Lagrangians are also equipped with a product structure, i.e., a morphism $CF^*(L_0, L_1) \otimes CF^*(L_1, L_2) \to CF^*(L_0, L_2)$ (well-defined in the cases that we will consider). This product structure is defined as follows: consider three points $p_1 \in L_0 \cap L_1$, $p_2 \in L_1 \cap L_2$, $p_3 \in L_0 \cap L_2$, and the moduli space $\mathcal{M}(p_1, p_2, p_3)$ of all pseudo-holomorphic maps f from a disc with three marked points q_1, q_2, q_3 on its boundary to M, taking q_i to p_i and the three portions of boundary delimited by the marked points to L_0, L_1, L_2 respectively. We compactify this moduli space and complete it if necessary in order to obtain a well-defined fundamental cycle. The virtual dimension of this moduli space is the difference between the Maslov index of p_3 and the sum of those of p_1 and p_2 . The product of p_1 and p_2 is then defined as

$$p_1 \cdot p_2 = \sum_{p_3} \#\mathcal{M}(p_1, p_2, p_3)\, p_3, \qquad (1)$$

where the sum runs over all p_3 for which the expected dimension of the moduli space is zero.

While the product structure on CF^* defined by (1) satisfies the Leibniz rule with respect to the differential ∂ (and hence descends to a product structure on Floer homology), it differs from usual products by the fact that it is only associative *up to homotopy*. In fact, Floer complexes come equipped with a full set of *higher-order products*

$$\mu^n : CF^*(L_0, L_1) \otimes \cdots \otimes CF^*(L_{n-1}, L_n) \to CF^*(L_0, L_n) \quad \text{for all } n \geq 1,$$

with each μ^n shifting degree by $2 - n$. The first two maps μ^1 and μ^2 are respectively the Floer differential ∂ and the product described above. The definition of μ^n is similar to those of ∂ and of the product structure: given generators $p_i \in CF^*(L_{i-1}, L_i)$ for $1 \leq i \leq n$ and $p_{n+1} \in CF^*(L_0, L_n)$ such that $\deg p_{n+1} = \sum_{i=1}^n \deg p_i + 2 - n$, the coefficient of p_{n+1} in $\mu^n(p_1, \ldots, p_n)$ is obtained by counting (in a suitable sense) pseudo-holomorphic maps f from a disc with $n+1$ marked points $q_1, \ldots, q_{n+1}$ on its boundary to M, such that $f(q_i) = p_i$ and the portions of boundary delimited by the marked points are mapped to $L_0, \ldots, L_n$ respectively.

The maps $(\mu^n)_{n \geq 1}$ define an A_∞-*structure* on Floer complexes, i.e., they satisfy an infinite sequence of algebraic relations:

$$\begin{cases} \mu^1(\mu^1(a)) = 0, \\ \mu^1(\mu^2(a,b)) = \mu^2(\mu^1(a), b) + (-1)^{\deg a}\mu^2(a, \mu^1(b)), \\ \mu^1(\mu^3(a,b,c)) = \mu^2(\mu^2(a,b), c) - \mu^2(a, \mu^2(b,c)) \\ \qquad\qquad \pm \mu^3(\mu^1(a), b, c) \pm \mu^3(a, \mu^1(b), c) \pm \mu^3(a, b, \mu^1(c)), \\ \cdots \end{cases}$$

This leads to the concept of "Fukaya category" of a symplectic manifold. Conjecturally, for every symplectic manifold (M, ω) one should be able to define an A_∞-category $\mathcal{F}(M)$ whose objects are Lagrangian submanifolds (compact, orientable, relatively spin, "twisted" by a flat unitary vector bundle); the space of morphisms between two objects L_0 and L_1 is the Floer complex $CF^*(L_0, L_1)$ equipped with its differential $\partial = \mu^1$, with (non-associative) composition given by the product μ^2, and higher order compositions μ^n.

The importance of Fukaya categories in modern symplectic topology is largely due to the *homological mirror symmetry* conjecture, formulated by Kontsevich. Very roughly, this conjecture states that the phenomenon of mirror symmetry, i.e., a conjectural correspondence between symplectic manifolds and complex manifolds ("mirror pairs") arising from a duality among string theories, should be visible at the level of Fukaya categories of symplectic manifolds and categories of coherent sheaves on complex manifolds: given a mirror pair consisting of a symplectic manifold M and a complex manifold X, the derived categories $D\mathcal{F}(M)$ and $D^b Coh(X)$ should be equivalent (in a more precise form of the conjecture, one should actually consider families of manifolds and deformations of categories). However, due to the very incomplete nature of our understanding of Fukaya categories in comparison to the much better understood derived categories of coherent sheaves, this conjecture has so far only been verified on very specific examples.

1.5 The Topology of Symplectic Four-Manifolds

To end our introduction, we mention some of the known results and open questions in the theory of compact symplectic four-manifolds, which motivate the directions taken in the later lectures.

Recall that, in the case of open manifolds, Gromov's h-principle implies that the existence of an almost-complex structure is sufficient. In contrast, the case of compact manifolds is much less understood, except in dimension 4. Whereas the existence of a class $\alpha \in H^2(M, \mathbb{R})$ such that $\alpha^{\cup n} \neq 0$ and of an almost-complex structure already provide elementary obstructions to the existence of a symplectic structure on a given compact manifold, in the case of four-manifolds a much stronger obstruction arises from Seiberg–Witten invariants. We will not define these, but mention some of their key topological consequences for symplectic four-manifolds, which follow from the work of Taubes ([Ta1, Ta2], ...).

Theorem 1.11 (Taubes) *(i) Let M be a compact symplectic four-manifold with $b_2^+ \geq 2$. Then the homology class $c_1(K_M)$ admits a (possibly disconnected) smooth pseudo-holomorphic representative (in particular $c_1(K_M) \cdot [\omega] \geq 0$). Hence, if M is minimal, i.e., contains no (-1)-spheres, then $c_1(K_M)^2 = 2\chi(M) + 3\sigma(M) \geq 0$.*

(ii) If (M^4, ω) splits as a connected sum $M_1 \# M_2$, then one of the M_i has negative definite intersection form.

When $b_2^+(M) = 1$, Seiberg–Witten theory still has some implications. Using Gromov's characterization of the Fubini-Study symplectic structure of $\mathbb{CP}^2$ in terms of the existence of pseudo-holomorphic lines, Taubes has shown that the symplectic structure of $\mathbb{CP}^2$ is unique up to scaling. This result has been extended by Lalonde and McDuff to the case of rational ruled surfaces, where ω is determined by its cohomology class.

Remark. For any smooth connected symplectic curve Σ in a symplectic four-manifold (M, ω), the genus $g(\Sigma)$ is related to the homology class by the classical *adjunction formula*

$$2 - 2g(\Sigma) + [\Sigma] \cdot [\Sigma] = -c_1(K_M) \cdot [\Sigma],$$

a direct consequence of the splitting $TM_{|\Sigma} = T\Sigma \oplus N\Sigma$. For example, every connected component of the pseudo-holomorphic representative of $c_1(K_M)$

constructed by Taubes satisfies $g(\Sigma) = 1 + [\Sigma] \cdot [\Sigma]$ (this is how one derives the inequality $c_1(K_M)^2 \geq 0$ under the minimality assumption). In fact, Seiberg–Witten theory also implies that symplectic curves have minimal genus among all smoothly embedded surfaces in their homology class.

In parallel to the above constraints on symplectic four-manifolds, surgery techniques have led to many interesting new examples of compact symplectic manifolds.

One of the most efficient techniques in this respect is the *symplectic sum* construction, investigated by Gompf [Go1]: if two symplectic manifolds (M_1^{2n}, ω_1) and (M_2^{2n}, ω_2) contain compact symplectic hypersurfaces W_1^{2n-2}, W_2^{2n-2} that are mutually symplectomorphic and whose normal bundles have opposite Chern classes, then we can cut M_1 and M_2 open along the submanifolds W_1 and W_2, and glue them to each other along their common boundary, performing a fiberwise connected sum in the normal bundles to W_1 and W_2, to obtain a new symplectic manifold $M = M_1{}_{W_1} \#_{W_2} M_2$. This construction has in particular allowed Gompf to show that every finitely presented group can be realized as the fundamental group of a compact symplectic four-manifold. This is in sharp contrast to the Kähler case, where Hodge theory shows that the first Betti number is always even.

A large number of questions remain open, even concerning the Chern numbers realized by symplectic four-manifolds. For instance it is unknown to this date whether the Bogomolov–Miyaoka–Yau inequality $c_1^2 \leq 3c_2$, satisfied by all complex surfaces of general type, also holds in the symplectic case. Moreover, very little is known about the symplectic topology of complex surfaces of general type.

Addendum

Since these notes were first prepared, there have inevitably been many interesting further developments, which we do not have time to survey. We simply draw the reader's attention to the rise of *broken pencils* [ADK2, Pe] and to the role of Lefschetz pencils, open books and symplectic fillings in the resolution of Property P for knots [KM].

2 Symplectic Lefschetz Fibrations

This section will provide a theoretical classification of symplectic four-manifolds in algebraic terms, but we begin very humbly.

2.1 Fibrations and Monodromy

Here is an easy way to build symplectic four-manifolds:

Proposition 2.1 (Thurston) *If $\Sigma_g \to X \to \Sigma_h$ is a surface bundle with fiber non-torsion in homology, then X is symplectic.*

Proof. Let $\eta \in \Omega^2(M)$ be a closed 2-form representing a cohomology class which pairs non-trivially with the fiber. Cover the base Σ_h by balls U_i over which the fibration is trivial: we have a diffeomorphism $\phi_i : f^{-1}(U_i) \to U_i \times \Sigma_g$, which determines a projection $p_i : f^{-1}(U_i) \to \Sigma_g$.

Let σ be a symplectic form on the fiber Σ_g, in the same cohomology class as the restriction of η. After restriction to $f^{-1}(U_i) \simeq U_i \times \Sigma_g$, we can write $p_i^*\sigma = \eta + d\alpha_i$ for some 1-form α_i over $f^{-1}(U_i)$. Let $\{\rho_i\}$ be a partition of unity subordinate to the cover $\{U_i\}$ of Σ_h, and let $\tilde{\eta} = \eta + \sum_i d((\rho_i \circ f)\,\alpha_i)$. The 2-form $\tilde{\eta}$ is well-defined since the support of $\rho_i \circ f$ is contained in $f^{-1}(U_i)$, and it is obviously closed. Moreover, over $f^{-1}(p)$, we have $\tilde{\eta}_{|f^{-1}(p)} = \eta_{|f^{-1}(p)} + \sum_i \rho_i(p)d\alpha_{i|f^{-1}(p)} = \sum_i \rho_i(p)\,(\eta + d\alpha_i)_{|f^{-1}(p)} = \sum_i \rho_i(p)\,(p_i^*\sigma)_{|f^{-1}(p)}$. Since a positive linear combination of symplectic forms over a Riemann surface is still symplectic, the form $\tilde{\eta}$ is non-degenerate on every fiber.

At any point $x \in X$, the tangent space $T_x X$ splits into a vertical subspace $V_x = \operatorname{Ker} df_x$ and a horizontal subspace $H_x = \{v \in T_x X, \tilde{\eta}(v, v') = 0 \;\forall v' \in V_x\}$. Since the restriction of $\tilde{\eta}$ to the vertical subspace is non-degenerate, we have $T_x X = H_x \oplus V_x$. Letting κ be a symplectic form on the base Σ_h, the 2-form $f^*\kappa$ is non-degenerate over H_x, and therefore for sufficiently large $C > 0$ the 2-form $\tilde{\eta} + C\,f^*\kappa$ defines a global symplectic form on X. $\square$

The cohomology class of the symplectic form depends on C, but the structure is canonical up to deformation equivalence. The hypothesis on the fiber is satisfied whenever $g \neq 1$, since $c_1(TX^{\text{vert}})$ evaluates non-trivially on the fiber. That some assumption is needed for $g = 1$ is shown by the example of the Hopf fibration $T^2 \to S^1 \times S^3 \to S^2$. Historically, the first example of a non-Kähler symplectic four-manifold, due to Thurston [Th], is a non-trivial T^2-bundle over T^2 (the product of S^1 with the mapping torus of a Dehn twist, which has $b_1 = 3$).

Unfortunately, not many four-manifolds are fibered.

Definition 2.2 *A Lefschetz pencil on a smooth oriented four-manifold X is a map $f : X - \{b_1, \ldots, b_n\} \to S^2$, submersive away from a finite set $\{p_1, \ldots, p_r\}$, conforming to local models (i) $(z_1, z_2) \mapsto z_1/z_2$ near each b_i, (ii) $(z_1, z_2) \mapsto z_1^2 + z_2^2$ near each p_j. Here the z_i are orientation-preserving local complex-valued coordinates.*

We can additionally require that the critical values of f are all distinct (so that each fiber contains at most one singular point).

This definition is motivated by the complex analogue of Morse theory. Global holomorphic functions on a projective surface must be constant, but interesting functions exist on the complement of finitely many points, and the generic such will have only quadratic singularities. The (closures of the) fibers of the map f cut the four-manifold X into a family of real surfaces all passing through the b_i (locally like complex lines through a point in $\mathbb{C}^2$), and with certain fibers having nodal singularities $((z_1 + iz_2)(z_1 - iz_2) = 0)$. If we blow up the b_i, then the map f extends to the entire manifold and we obtain a *Lefschetz fibration*.

A small generalization of the previous argument to the case of Lefschetz fibrations shows that, if X admits a Lefschetz pencil, then it is symplectic (work on the blow-up and choose the constant C so large that the exceptional sections arising from the b_i are all symplectic and can be symplectically blown down). In fact the symplectic form obtained in this way is canonical up to isotopy rather than just deformation equivalence, as shown by Gompf [GS, Go2].

A real Morse function encodes the topology of a manifold: the critical values disconnect $\mathbb{R}$, and the topology of the level sets changes by a handle addition as we cross a critical value. In the complex case, the critical values do not disconnect, but the local model is determined by its *monodromy*, i.e., the diffeomorphism of the smooth fiber obtained by restricting the fibration to a circle enclosing a single critical value.

The fiber F_t of the map $(z_1, z_2) \to z_1^2 + z_2^2$ above $t \in \mathbb{C}$ is given by the equation $(z_1 + iz_2)(z_1 - iz_2) = t$: the fiber F_t is smooth (topologically an annulus) for all $t \neq 0$, while the fiber above the origin presents a transverse double point, and is obtained from the nearby fibers by collapsing an embedded simple closed loop called the *vanishing cycle*. For example, for $t > 0$ the vanishing cycle is the loop $\{(x_1, x_2) \in \mathbb{R}^2, \ x_1^2 + x_2^2 = t\} = F_t \cap \mathbb{R}^2 \subset F_t$.

Proposition 2.3 *For a circle in the base S^2 of a Lefschetz fibration enclosing a single critical value, whose critical fiber has a single node, the monodromy is a Dehn twist about the vanishing cycle.*

Proof (Sketch). By introducing a cutoff function ψ and by identifying the fiber $z_1^2 + z_2^2 = t$ with the set $z_1^2 + z_2^2 = \psi(\|z\|^2) \, t$, we can see that the monodromy is the identity outside a small neighborhood of the vanishing cycle. This reduces the problem to the local model of the annulus, which has mapping class group (relative to the boundary) isomorphic to $\mathbb{Z}$, so we just need to find one integer. One possibility is to study an example, e.g., an elliptic surface, where we can determine the monodromy by considering its action on homology, interpreted as periods. Alternatively, we can think of the annulus as a double cover of the disc branched at two points (the two square roots of t), and watch these move as t follows the unit circle.

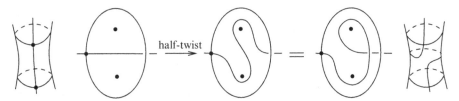

One can also consider higher-dimensional Lefschetz fibrations, which by definition are again submersions away from non-degenerate quadratic singularities. In the local model, the smooth fibers of $f : (z_i) \mapsto \sum z_i^2$ also contain a (Lagrangian) sphere, and the monodromy around the critical fiber $f^{-1}(0)$ is a generalized Dehn twist about this vanishing cycle (see the lectures by Seidel in this volume).

Using this local model, we may now define the monodromy homomorphism. Fix a base point $q_0 \in S^2 - \mathrm{crit}(f)$, and consider a closed loop $\gamma : [0, 1] \to S^2 - \mathrm{crit}(f)$ (starting and ending at q_0). By fixing a horizontal distribution we can perform parallel transport in the fibers of f along γ, which induces a diffeomorphism from $\Sigma = f^{-1}(q_0)$ to itself. (Such a horizontal distribution is canonically defined if we fix a symplectic form on the total space by taking the symplectic orthogonal complements to the fibers.) The isotopy class of this diffeomorphism, which is well-defined independently of the chosen horizontal distribution, is called the *monodromy* of f along γ. Hence, we obtain a monodromy homomorphism characteristic of the Lefschetz fibration f,

$$\psi : \pi_1(S^2 - \mathrm{crit}(f), q_0) \to \pi_0 \mathrm{Diff}^+(\Sigma),$$

which takes a loop encircling one critical value to a Dehn twist as above.

Example. let C_0 and C_1 be generic cubic curves in $\mathbb{CP}^2$, and consider the pencil of cubics $\{C_0 + \lambda C_1 = 0\}_{\lambda \in \mathbb{CP}^1}$. This pencil has nine base points (the intersections of C_0 and C_1), and 12 singular fibers. To see the latter fact, note that the Euler characteristic $\chi(X)$ of the total space of a Lefschetz pencil of genus g is given by $\chi(X) = 4 - 4g - \#b_i + \#p_j$.

After blowing up the base points, we obtain an elliptic Lefschetz fibration, whose monodromy takes values in the genus 1 mapping class group $\mathrm{Map}_1 = \pi_0 \mathrm{Diff}^+(T^2) = SL(2, \mathbb{Z})$. Each local monodromy is conjugate to $\begin{pmatrix} 1 & 1 \\ 0 & 1 \end{pmatrix}$. The monodromy homomorphism is determined by the images of a basis for $\pi_1(S^2 - \mathrm{crit}(f))$ consisting of 12 loops encircling one critical value each. The monodromy around the product of these loops is the identity, as the product loop bounds a disc in S^2 over which the fibration is trivial. In an appropriate basis, the resulting word in 12 Dehn twists in $SL(2, \mathbb{Z})$ can be brought into standard form

$$(A \cdot B)^6 = \left(\begin{pmatrix} 1 & 1 \\ 0 & 1 \end{pmatrix} \cdot \begin{pmatrix} 1 & 0 \\ -1 & 1 \end{pmatrix} \right)^6 = I.$$

Such a word in Dehn twists, called a *positive relation* in the relevant mapping class group, captures the topology of a Lefschetz fibration. In the case where the fibration admits distinguished sections, e.g., coming from the base points of a pencil, we can refine the monodromy by working in the relative mapping class group of the pair $(\Sigma, \{b_i\})$ (see the example in the last lecture).

In fact, a careful analysis of positive relations in $SL(2, \mathbb{Z})$ implies that all elliptic Lefschetz fibrations are Kähler (and have monodromy words of the form $(A \cdot B)^{6n} = 1$), a classical result of Moishezon and Livne [Mo1]. More geometrically, this result can also be deduced from the work of Siebert and Tian [ST] described in their lectures in this volume.

Remark. In the case of Lefschetz fibrations over a disc, the monodromy homomorphism determines the total space of the fibration up to symplectic deformation. When considering fibrations over S^2, the monodromy data determines the fibration over a large disc D containing all critical values, after which we only need to add a trivial fibration over a small disc $D' = S^2 - D$, to be glued in a manner compatible with the fibration structure over the common boundary $\partial D = \partial D' = S^1$. This gluing involves the choice of a map from S^1 to $\mathrm{Diff}^+(\Sigma_g)$, i.e., an element of $\pi_1 \mathrm{Diff}^+(\Sigma_g)$, which is trivial if $g \geq 2$ (hence in this case the positive relation determines the topology completely).

Combining the facts that Lefschetz pencils carry symplectic structures and that they correspond to positive relations has algebraic consequences for the mapping class group.

Proposition 2.4 *There is no positive relation involving only Dehn twists about separating curves.*

Proof (Sketch). (see [Sm1] for a harder proof) – Suppose for contradiction we have such a word of length δ. This defines a four-manifold X with a Lefschetz fibration having this as monodromy. We can compute the signature $\sigma(X)$ by surgery, cutting the manifold open along neighborhoods of the singular fibers; we find (cf. [Oz]) that each local model contributes -1 to the signature, so we obtain $\sigma(X) = -\delta$. This allows us to compute the Betti and Chern numbers of X: $b_1(X) = 2g$, $b_2(X) = \delta + 2$, $c_2(X) = 4 - 4g + \delta$, and $c_1^2(X) = 8 - 8g - \delta < -\delta$. Hence c_1^2 of any minimal model of X is negative, and so X must be rational or ruled by a theorem of Liu. These cases can be excluded by hand.   □

The pencil of cubics on $\mathbb{CP}^2$ is an instance of a much more general construction.

Proposition 2.5 (Lefschetz) *Projective surfaces have Lefschetz pencils.*

Generic hyperplane sections cut out smooth complex curves, and a pencil corresponds to a line of such hyperplanes. Inside the dual projective space $(\mathbb{CP}^N)^*$ pick a line transverse to the dual variety (the locus of hyperplanes which are not transverse to the given surface). A local computation shows that this transversality condition goes over to give exactly the non-degenerate critical points of a Lefschetz pencil. From another point of view, if L is a very ample line bundle on X (so sections generate the fibers of L), we can consider the evaluation map $X \times H^0(L) \to L$. Let $Z \subset X \times H^0(L)$ be the preimage of the zero section, then a regular value of the projection $Z \to H^0(L)$ is a section with smooth zero set. In this way the construction of embedded complex curves in X (and more generally linear systems of such) can be reduced to the existence of regular values, i.e., Sard's theorem.

Certainly the converse to the above cannot be true: not all Lefschetz fibrations are Kähler. The easiest way to see this is to use the (twisted) fiber sum construction. Given a positive relation $\tau_1 \ldots \tau_s = 1$ in Map_g, and some element $\phi \in \mathrm{Map}_g$, we obtain a new positive relation $\tau_1 \ldots \tau_s(\phi^{-1}\tau_1\phi) \ldots (\phi^{-1}\tau_s\phi) = 1$. If $\phi = \mathrm{Id}$ the corresponding four-manifold is the double branched cover of the original manifold over a union of two smooth fibers, but in general the operation has no holomorphic interpretation. The vanishing cycles of the new fibration are the union of the old vanishing cycles and their images under ϕ. Since $H_1(X) = H_1(\Sigma)/\langle\text{vanishing cycles}\rangle$, we can easily construct examples with odd first Betti number, for example starting with a genus 2 pencil on $T^2 \times S^2$ [OS].

More sophisticated examples (for instance with trivial first Betti number) can be obtained by forming infinite families of twisted fiber sums with non-conjugate monodromy groups and invoking the following.

Theorem 2.6 (Arakelov-Paršin) *Only finitely many isotopy classes of Lefschetz fibrations with given fiber genus and number of critical fibers can be Kähler.*

Remark. Twisted fiber sum constructions can often be "untwisted" by subsequently fiber summing with another suitable (e.g., holomorphic) Lefschetz fibration. A consequence of this is a stable isotopy result for genus 2 Lefschetz fibrations [Au6]: any genus 2 Lefschetz fibration becomes isotopic to a holomorphic fibration after repeated fiber sums with the standard holomorphic fibration with 20 singular fibers coming from a genus 2 pencil on $\mathbb{CP}^1 \times \mathbb{CP}^1$. More generally a similar result holds for all Lefschetz fibrations with monodromy contained in the *hyperelliptic* subgroup of the mapping class group. This is a corollary of a recent result of Kharlamov and Kulikov [KK] about braid monodromy factorizations: after repeated (untwisted) fiber sums with copies of a same fixed holomorphic fibration with $8g + 4$ singular fibers, any hyperelliptic genus g Lefschetz fibration eventually becomes holomorphic. Moreover, the fibration obtained in this manner is completely determined by its number of singular fibers of each type (irreducible, reducible with components of given genus), and when the fiber genus is odd by a certain $\mathbb{Z}_2$-valued invariant. (The proof of this result uses the fact that the hyperelliptic mapping class group is an extension by $\mathbb{Z}_2$ of the braid group of $2g + 2$ points on a sphere, which is itself a quotient of B_{2g+2}; this makes it possible to transform the monodromy of a hyperelliptic Lefschetz fibration into a factorization in B_{2g+2}, with different types of factors for the various types of singular fibers and extra contributions belonging to the kernel of the morphism $B_{2g+2} \to B_{2g+2}(S^2)$, and hence reduce the problem to that studied by Kharlamov and Kulikov. This connection between mapping class groups and braid groups will be further studied in later lectures.) It is not clear whether the result should be expected to remain true in the non-hyperelliptic case.

These examples of Lefschetz fibrations differ somewhat in character from those obtained in projective geometry, since the latter always admit exceptional sections of square -1. However, an elementary argument in hyperbolic geometry shows that fiber sums never have this property [Sm3]. To see that not every Lefschetz pencil arises from a holomorphic family of surfaces appears to require some strictly deeper machinery.

To introduce this, let us note that, if we choose a metric on the total space of a genus g Lefschetz pencil or fibration, the fibers become Riemann surfaces and this induces a map $\phi : \mathbb{CP}^1 \to \overline{\mathcal{M}}_g$ to the Deligne–Mumford moduli space of stable genus g curves. There is a line bundle $\lambda \to \overline{\mathcal{M}}_g$ (the Hodge bundle), with fiber $\det H^0(K_\Sigma)$ above $[\Sigma]$, and an index theorem for the family of $\bar{\partial}$-operators on the fibers shows

Proposition 2.7 ([Sm1]) $\sigma(X) = 4\langle c_1(\lambda), [\phi(\mathbb{CP}^1)]\rangle - \delta.$

On the other hand, $c_1(\lambda)$ and the Poincaré duals of the components of the divisor of nodal curves generate $H^2(\overline{\mathcal{M}}_g, \mathbb{Z})$, so the above formula – together with the numbers of singular fibers of different topological types – characterizes the homology class $[\phi(\mathbb{CP}^1)]$.

Clearly, holomorphic Lefschetz fibrations give rise to rational curves in the moduli space, and these have locally positive intersection with all divisors in which they are not contained. This gives another constraint on which Lefschetz pencils and fibrations can be holomorphic. For example, the divisor $\overline{\mathcal{H}}_3$ of hyperelliptic genus 3 curves has homology class $[\overline{\mathcal{H}}_3] = 9c_1(\lambda) - [\Delta_0] - 3[\Delta_1]$, where Δ_0 and Δ_1 are the divisors of irreducible and reducible nodal curves respectively.

Corollary 2.8 *A genus 3 Lefschetz fibration X with irreducible fibers and such that (i) $\chi(X) + 1$ is not divisible by 7, (ii) $9\sigma(X) + 5\chi(X) + 40 < 0$ is not holomorphic.*

Proof (Sketch). For hyperelliptic fibrations of any genus, we have

$$(8g + 4) c_1(\lambda) \cdot [\phi(S^2)] = g [\Delta_0] \cdot [\phi(S^2)] + \sum_{h=1}^{[g/2]} 4h(g - h) [\Delta_h] \cdot [\phi(S^2)].$$

To prove this, we can represent the four-manifold as a double cover of a rational ruled surface (see the lectures of Siebert and Tian in this volume). This gives another expression for $\sigma(X)$ which can be compared to that above. Then assumption (i) and integrality of the signature show that the fibration is not isotopic to a hyperelliptic fibration, while assumption (ii) shows that $[\phi(S^2)] \cdot [\overline{\mathcal{H}}_3] < 0$. $\square$

It is possible to build a Lefschetz fibration (with 74 singular fibers) admitting a (-1)-section and satisfying the conditions of the Corollary [Sm2].

The right correspondence between the geometry and the algebra comes from the work of Donaldson [Do2, Do3]:

Theorem 2.9 (Donaldson) *Any compact symplectic four-manifold admits Lefschetz pencils with symplectic fibers (if $[\omega]$ is integral, Poincaré dual to $k[\omega]$ for any sufficiently large k).*

As explained later, we even get some uniqueness – but only asymptotically with the parameter k. Increasing the parameter k makes the algebraic monodromy descriptions more and more complicated, but in principle, as with surgery theory for high-dimensional smooth manifolds, this gives a complete algebraic encoding. The construction is flexible enough to impose some extra conditions on the pencils, for instance we can assume that all the singular fibers are irreducible. (If the four-manifold has even intersection form, this is completely elementary.)

In order to describe Donaldson's construction of symplectic Lefschetz pencils, we need a digression into *approximately holomorphic geometry*.

2.2 Approximately Holomorphic Geometry

On an almost-complex manifold, the lack of integrability usually prevents the existence of non-trivial holomorphic sections of vector bundles or pseudo-holomorphic maps to other manifolds, but one can work in a similar manner with approximately holomorphic objects.

Let (M^{2n}, ω) be a compact symplectic manifold of dimension $2n$. We will assume throughout this paragraph that $\frac{1}{2\pi}[\omega] \in H^2(M, \mathbb{Z})$; this integrality condition does not restrict the topological type of M, since any symplectic form can be perturbed into another symplectic form ω' whose cohomology class is rational (we can then achieve integrality by multiplication by a constant factor). Moreover, it is easy to check that the submanifolds of M that we will construct are not only ω'-symplectic but also ω-symplectic, hence making the general case of Theorem 2.9 follow from the integral case.

Let J be an almost-complex structure compatible with ω, and let $g(.,.) = \omega(., J.)$ be the corresponding Riemannian metric. We consider a complex line bundle L over M such that $c_1(L) = \frac{1}{2\pi}[\omega]$, endowed with a Hermitian metric and a Hermitian connection ∇^L with curvature 2-form $F(\nabla^L) = -i\omega$. The almost-complex structure induces a splitting of the connection: $\nabla^L = \partial^L + \bar{\partial}^L$, where $\partial^L s(v) = \frac{1}{2}(\nabla^L s(v) - i\nabla^L s(Jv))$ and $\bar{\partial}^L s(v) = \frac{1}{2}(\nabla^L s(v) + i\nabla^L s(Jv))$.

If the almost-complex structure J is integrable, i.e., if M is Kähler, then L is an ample holomorphic line bundle, and for large enough values of k the holomorphic sections of $L^{\otimes k}$ determine an embedding of the manifold M into a projective space (Kodaira's theorem). Generic hyperplane sections of this projective embedding are smooth hypersurfaces in M, and a pencil of hyperplanes through a generic codimension 2 linear subspace defines a Lefschetz pencil.

When the manifold M is only symplectic, the lack of integrability of J prevents the existence of holomorphic sections. Nonetheless, it is possible to find an *approximately holomorphic* local model: a neighborhood of a point

$x \in M$, equipped with the symplectic form ω and the almost-complex structure J, can be identified with a neighborhood of the origin in $\mathbb{C}^n$ equipped with the standard symplectic form ω_0 and an almost-complex structure of the form $i + O(|z|)$. In this local model, the line bundle $L^{\otimes k}$ endowed with the connection $\nabla = (\nabla^L)^{\otimes k}$ of curvature $-ik\omega$ can be identified with the trivial line bundle $\mathbb{C}$ endowed with the connection $d + \frac{k}{4} \sum (z_j \, d\bar{z}_j - \bar{z}_j \, dz_j)$. The section of $L^{\otimes k}$ given in this trivialization by $s_{\mathrm{ref},k,x}(z) = \exp(-\frac{1}{4}k|z|^2)$ is then approximately holomorphic [Do1].

More precisely, a sequence of sections s_k of $L^{\otimes k}$ is said to be approximately holomorphic if, with respect to the rescaled metrics $g_k = kg$, and after normalization of the sections to ensure that $\|s_k\|_{C^r, g_k} \sim C$, an inequality of the form $\|\bar{\partial}s_k\|_{C^{r-1}, g_k} < C'k^{-1/2}$ holds, where C and C' are constants independent of k. The change of metric, which dilates all distances by a factor of $\sqrt{k}$, is required in order to be able to obtain uniform estimates, due to the large curvature of the line bundle $L^{\otimes k}$. The intuitive idea is that, for large k, the sections of the line bundle $L^{\otimes k}$ with curvature $-ik\omega$ probe the geometry of M at small scale ($\sim 1/\sqrt{k}$), which makes the almost-complex structure J almost integrable and allows one to achieve better approximations of the holomorphicity condition $\bar{\partial}s = 0$.

Since the above requirement is an open condition, there is no well-defined "space of approximately holomorphic sections" of $L^{\otimes k}$. Nonetheless, the above local model gives us a large number of approximately holomorphic sections (consider $s_{\mathrm{ref},k,x}$ for a large finite set of $x \in X$), which can be used to embed X as a symplectic submanifold of a (high-dimensional) projective space. However, this embedding by itself is not very useful since it is not clear that any of its hyperplane sections can be used to define a smooth symplectic hypersurface in X.

Hence, in contrast with the complex case, the non-trivial part of the construction is to find, among all the available approximately holomorphic sections, some whose geometric behavior is as generic as possible. That this is at all possible is a subtle observation of Donaldson, which leads to the following result [Do1]:

Theorem 2.10 (Donaldson) *For $k \gg 0$, $L^{\otimes k}$ admits approximately holomorphic sections s_k whose zero sets W_k are smooth symplectic hypersurfaces.*

The proof of this result starts from the observation that, if the section s_k vanishes transversely and if $|\bar{\partial}s_k(x)| \ll |\partial s_k(x)|$ at every point of $W_k = s_k^{-1}(0)$, then the submanifold W_k is symplectic, and even approximately J-holomorphic (i.e., $J(TW_k)$ is close to TW_k). The crucial point is therefore to obtain a lower bound for ∂s_k at every point of W_k, in order to make up for the lack of holomorphicity.

Sections s_k of $L^{\otimes k}$ are said to be *uniformly transverse to* 0 if there exists a constant $\eta > 0$ (independent of k) such that the inequality $|\partial s_k(x)|_{g_k} > \eta$ holds at any point of M where $|s_k(x)| < \eta$. In order to prove Theorem 2.10, it is sufficient to achieve this uniform estimate on the transversality of some approximately holomorphic sections s_k. The idea of the construction of such sections consists of two main steps. The first one is an effective local transversality result for complex-valued functions, for which Donaldson's argument appeals to ideas of Yomdin about the complexity of real semi-algebraic sets (see [Au5] for a simplification of the argument). The second step is a remarkable globalization process, which makes it possible to achieve uniform transversality over larger and larger open subsets by means of successive perturbations of the sections s_k, until transversality holds over the entire manifold M [Do1].

That the interplay between the two steps above is subtle can already be gathered from the delicate statement of the local transversality result. For $\beta > 0$ set $t(\beta) = \beta/(\log \beta^{-1})^d$. Here β represents the maximum size of the allowed perturbation $s_{\text{given}} \mapsto s_{\text{given}} - w s_{\text{ref}}$ with $|w| < \beta$, $t(\beta)$ is the amount of transversality thereby obtained, and $d = d(n)$ is a universal constant that we will mostly ignore. Write $B^+ \subset \mathbb{C}^n$ for a Euclidean ball slightly larger than the unit ball B.

Theorem 2.11 ([Do1, Au2]) *If $f : B^+ \to \mathbb{C}^{n+1}$ satisfies $|f| < 1$ and $|\bar{\partial} f| < t(\beta)$ pointwise, then there is some $w \in \mathbb{C}^{n+1}$ with $|w| < \beta$ such that $|f(z) - w| > t(\beta)$ over $B \subset B^+$.*

To see the relevance of this, for an approximately holomorphic section s_k of $L^{\otimes k}$ we consider the holomorphic 1-jet $(s_k, \partial s_k)$, a section of $L^{\otimes k} \oplus L^{\otimes k} \otimes T^* M^{1,0}$. This is locally a map from the complex n-dimensional ball to $\mathbb{C}^{n+1}$, and by adding reference sections we can explicitly give local perturbations of this. In this context, theorems such as the one above are elementary if we take a polynomial function $t(\beta) = (const)\beta^q$ with $q > 1$, but such perturbations do not patch well.

By way of an example, let $f : \mathbb{R}^4 \to \mathbb{R}^6$ have bounded derivative, so f takes balls of radius ϵ to balls of comparable size. Now $B^4(1)$ is filled by approximately ϵ^{-4} balls, and each is taken to a ball of volume approximately ϵ^6. Hence, the total volume of the image is about ϵ^2 of $B^6(1)$. To change the function $f \mapsto f - w$ by β in order to miss the image of f by $t(\beta) = (const)\beta^3$, say, would be straightforward, because taking $\epsilon = t(\beta)$, the ϵ-neighborhood of $f(B^4)$ has a volume of the order of ϵ^2 and hence cannot contain any ball of radius $\epsilon^{1/3} \sim \beta$.

However, our manifold is covered by $O(k^{2n})$ balls of fixed g_k-radius in which our reference sections $s_{\text{ref},k}$ are concentrated. Perturbing over each ball one by one, all estimates are destroyed and it is impossible to achieve uniform transversality. The solution is to perturb over balls at great distance simultaneously; nonetheless the simultaneous perturbations will *not* be entirely independent. We cover X by a fixed number D^{2n} of collections of balls (the number of balls in each collection, but not the number of collections, will grow with the parameter k), in such a way that any two balls in the same collection are at g_k-distance at least D from each other. To obtain uniformly transverse sections, we (1) start with some approximately holomorphic section (e.g., the zero section $s_0 = 0$); (2) perturb by β_0 over balls of the first collection I_0 to get a section s_1 which is $t(\beta_0)$ transverse over $\cup_{i \in I_0} B_i$; (3) perturb over balls of the second collection I_1 by an amount $\beta_1 \ll t(\beta_0)/2$ to get a section s_2 which is $t(\beta_1)$-transverse over $\cup_{i \in I_0 \cup I_1} B_i$ etc. Continuing, we need the sequence

$$\beta_0, \ \beta_1 \sim t(\beta_0), \ \beta_2 \sim t(\beta_1) \ldots$$

to be chosen so that, at each stage (for all $N < D^{2n}$), $\exp(-D^2 \beta_N) < \beta_{N+1}/2$. Here, the left-hand side $e^{-D^2} \beta_N$ is the effect of the perturbation at a ball B_1 in the collection I_N on another ball B_2 of the *same* collection (which we perturb simultaneously); the right-hand side $\beta_{N+1}/2$ is the transversality obtained at the ball B_2 by virtue of its *own* perturbation. We have $N = D^{2n}$ stages, for some large D, so we need

$$\exp(-N^{1/n}) \beta_N < \beta_{N+1}/2$$

for $N \gg 0$. This inequality fails for any polynomial function $\beta_{N+1} = t(\beta_N) = \beta_N^q/2$ for $q > 1$: one gets $\beta_N \sim (1/2)^{q^N}$ and $\exp(-N^{1/n}) \not< \beta_{N+1}/\beta_N$.

In other words, the estimates coming naively from Sard's theorem do not provide a good enough local theorem to pass to a global one. The remedy is that our functions are approximately holomorphic and not arbitrary, and for holomorphic functions stronger Sard-like theorems are available: the prototype here is that the regular values of a smooth map $B(1) \subset \mathbb{R}^{2n} \to \mathbb{R}^2$ are in general only dense, but the regular values of a holomorphic map $B(1) \subset \mathbb{C}^n \to \mathbb{C}$ form the complement of a finite set. In practice, the proof of Theorem 2.11 proceeds by reduction to the case of polynomial functions, and then appeals either to real algebraic geometry [Do1] or to the classical monotonicity theorem [Au5].

The symplectic submanifolds constructed by Donaldson present several remarkable properties which make them closer to complex submanifolds than to arbitrary symplectic submanifolds. For instance, they satisfy the Lefschetz hyperplane theorem: up to half the dimension of the submanifold, the homology and homotopy groups of W_k are identical to those of M [Do1]. More importantly, these submanifolds are, in a sense, asymptotically unique: for given large enough k, the submanifolds W_k are, up to symplectic isotopy, independent of all the choices made in the construction (including that of the almost-complex structure J) [Au1].

It is worth mentioning that analogues of this construction have been obtained for contact manifolds by Ibort, Martinez-Torres and Presas ([IMP],...); see also recent work of Giroux and Mohsen [Gi].

As an application of this theorem, we mention a symplectic packing result due to Biran [Bi]. Recall that a full symplectic packing of a manifold (M, ω) is an embedding $(\amalg B_i, \omega_{std}) \hookrightarrow (M, \omega)$ of a disjoint union of standard Euclidean symplectic balls of equal volumes whose images fill the entire volume of M. Gromov pointed out that, in contrast to the volume-preserving case, there are obstructions to symplectic packing: for instance, $\mathbb{CP}^2$ cannot be fully packed by two balls.

Theorem 2.12 (Biran) *Let (M^4, ω) be a symplectic four-manifold with integral $[\omega]$. Then M admits a full packing by N balls for all large N.*

The key ingredient in the proof is to reduce to the case of ruled surfaces by decomposing M into a disc bundle over a Donaldson submanifold Σ dual to $k[\omega]$ and an isotropic CW-complex – which takes no volume. It is therefore sufficient to fully pack a ruled surface by balls which remain disjoint from a section at infinity. On the other hand, there is a well-known correspondence between embeddings of symplectic balls of size μ and symplectic forms on a blow-up giving each exceptional curve area μ. Moreover, symplectic forms in a given cohomology class can be constructed by symplectic inflation in the presence of appropriate embedded symplectic surfaces, and for ruled surfaces these are provided by an elementary computation of the Gromov or Seiberg–Witten invariants. The best value of N is not known in general, because it is determined by the best value of k in Donaldson's construction.

We now move on to Donaldson's construction of symplectic Lefschetz pencils [Do2, Do3]. In comparison with Theorem 2.10, the general setup is the same, the main difference being that we consider no longer one, but two sections of $L^{\otimes k}$. A pair of suitably chosen approximately holomorphic sections (s_k^0, s_k^1) of $L^{\otimes k}$ defines a family of symplectic hypersurfaces

$$\Sigma_{k,\alpha} = \{x \in M,\ s_k^0(x) - \alpha s_k^1(x) = 0\}, \quad \alpha \in \mathbb{CP}^1 = \mathbb{C} \cup \{\infty\}.$$

The submanifolds $\Sigma_{k,\alpha}$ are all smooth except for finitely many of them which present an isolated singularity; they intersect transversely along the *base points* of the pencil, which form a smooth symplectic submanifold $Z_k = \{s_k^0 = s_k^1 = 0\}$ of codimension 4.

The two sections s_k^0 and s_k^1 determine a projective map $f_k = (s_k^0 : s_k^1) : M - Z_k \to \mathbb{CP}^1$, whose critical points correspond to the singularities of the fibers $\Sigma_{k,\alpha}$. In the case of a symplectic Lefschetz pencil, the function f_k is a complex Morse function, i.e., near any of its critical points it is given by the local model $f_k(z) = z_1^2 + \cdots + z_n^2$ in approximately holomorphic coordinates. After blowing up M along Z_k, the Lefschetz pencil structure on M gives rise to a well-defined map $\hat{f}_k : \hat{M} \to \mathbb{CP}^1$; this map is a symplectic Lefschetz fibration. Hence, Theorem 2.9 may be reformulated more precisely as follows:

Theorem 2.13 (Donaldson) *For large enough k, the manifold (M^{2n}, ω) admits symplectic Lefschetz pencil structures determined by pairs of suitably chosen approximately holomorphic sections s_k^0, s_k^1 of $L^{\otimes k}$. Moreover, for large enough k these Lefschetz pencil structures are uniquely determined up to isotopy.*

As in the case of submanifolds, Donaldson's argument relies on successive perturbations of given approximately holomorphic sections s_k^0 and s_k^1 in order to achieve uniform transversality properties, not only for the sections (s_k^0, s_k^1) themselves but also for the derivative ∂f_k [Do3].

The precise meaning of the uniqueness statement is the following: assume we are given two sequences of Lefschetz pencil structures on (M, ω), determined by pairs of approximately holomorphic sections of $L^{\otimes k}$ satisfying uniform transversality estimates, but possibly with respect to two different ω-compatible almost-complex structures on M. Then, beyond a certain (non-explicit) value of k, it becomes possible to find one-parameter families of Lefschetz pencil structures interpolating between the given ones. In particular, this implies that for large k the monodromy invariants associated to these Lefschetz pencils only depend on (M, ω, k) and not on the choices made in the construction.

The monodromy invariants associated to a symplectic Lefschetz pencil are essentially those of the symplectic Lefschetz fibration obtained after blow-up along the base points, with only a small refinement. After the blow-up operation, each fiber of $\hat{f}_k : \hat{M} \to \mathbb{CP}^1$ contains a copy of the base locus Z_k embedded as a smooth symplectic hypersurface. This hypersurface lies away from all vanishing cycles, and is preserved by the monodromy. Hence, the monodromy homomorphism can be defined to take values in the group of isotopy classes of symplectomorphisms of the fiber Σ_k whose restriction to the submanifold Z_k is the identity.

3 Symplectic Branched Covers of $\mathbb{CP}^2$

3.1 Symplectic Branched Covers

Definition 3.1 *A smooth map* $f : X^4 \to (Y^4, \omega_Y)$ *from a compact oriented smooth four-manifold to a compact symplectic four-manifold is a (generic) symplectic branched covering if, given any point* $p \in X$, *there exist neighborhoods* $U \ni p$ *and* $V \ni f(p)$ *and orientation-preserving local diffeomorphisms* $\phi : U \to \mathbb{C}^2$ *and* $\psi : V \to \mathbb{C}^2$, *such that* $\psi_* \omega_Y(v, iv) > 0 \ \forall v \neq 0$ *(i.e., the standard complex structure is* $\psi_* \omega_Y$*-tame), and such that* $\psi \circ f \circ \phi^{-1}$ *is one of the following model maps:*

(i) $(u, v) \mapsto (u, v)$ *(local diffeomorphism)*
(ii) $(u, v) \mapsto (u^2, v)$ *(simple branching)*
(iii) $(u, v) \mapsto (u^3 - uv, v)$ *(cusp)*

The three local models appearing in this definition are exactly those describing a generic holomorphic map between complex surfaces, except that the local coordinate systems we consider are not holomorphic.

By computing the Jacobian of f in the given local coordinates, we can see that the *ramification curve* $R \subset X$ is a smooth submanifold (it is given by $\{u = 0\}$ in the second local model and $\{v = 3u^2\}$ in the third one). However, the image $D = f(R) \subset X$ (the *branch curve*, or *discriminant curve*) may be singular. More precisely, in the simple branching model D is given by $\{z_1 = 0\}$, while in the cusp model we have $f(u, 3u^2) = (-2u^3, 3u^2)$, and hence

D is locally identified with the singular curve $\{27z_1^2 = 4z_2^3\} \subset \mathbb{C}^2$. This means that, at the cusp points, D fails to be immersed. Besides the cusps, the branch curve D also generically presents *transverse double points* (or *nodes*), which do not appear in the local models because they correspond to simple branching in two distinct points p_1, p_2 of the same fiber of f. There is no constraint on the orientation of the local intersection between the two branches of D at a node (positive or negative, i.e., complex or anti-complex), because the local models near p_1 and p_2 hold in different coordinate systems on Y.

Generically, the only singularities of the branch curve $D \subset Y$ are transverse double points ("nodes") of either orientation and complex cusps. Moreover, because the local models identify D with a complex curve, the tameness condition on the coordinate systems implies that D is a (singular) symplectic submanifold of Y.

The following result states that a symplectic branched cover of a symplectic four-manifold carries a natural symplectic structure [Au2]:

Proposition 3.2 *If* $f : X^4 \to (Y^4, \omega_Y)$ *is a symplectic branched cover, then* X *carries a symplectic form* ω_X *such that* $[\omega_X] = f^*[\omega_Y]$, *canonically determined up to symplectomorphism.*

Proof. The 2-form $f^*\omega_Y$ is closed, but it is only non-degenerate outside of R. At any point p of R, the 2-plane $K_p = \operatorname{Ker} df_p \subset T_p X$ carries a natural orientation induced by the complex orientation in the local coordinates of Definition 3.1. Using the local models, we can construct an *exact* 2-form α such that, at any point $p \in R$, the restriction of α to K_p is non-degenerate and positive.

More precisely, given $p \in R$ we consider a small ball centered at p and local coordinates (u, v) such that f is given by one of the models of the definition, and we set $\alpha_p = d(\chi_1(|u|)\chi_2(|v|)\, x\, dy)$, where $x = \operatorname{Re}(u)$, $y = \operatorname{Im}(u)$, and χ_1 and χ_2 are suitably chosen smooth cut-off functions. We then define α to be the sum of these α_p when p ranges over a finite subset of R for which the supports of the α_p cover the entire ramification curve R. Since $f^*\omega_Y \wedge \alpha$ is positive at every point of R, it is easy to check that the 2-form $\omega_X = f^*\omega_Y + \epsilon\, \alpha$ is symplectic for a small enough value of the constant $\epsilon > 0$.

The fact that ω_X is canonical up to symplectomorphism follows immediately from Moser's stability theorem and from the observation that the space of exact perturbations α such that $\alpha_{|K_p} > 0 \; \forall p \in R$ is a convex subset of $\Omega^2(X)$ and hence connected. $\square$

Approximately holomorphic techniques make it possible to show that every compact symplectic four-manifold can be realized as a branched cover of $\mathbb{CP}^2$. The general setup is similar to Donaldson's construction of symplectic Lefschetz pencils: we consider a compact symplectic manifold (X, ω), and perturbing the symplectic structure if necessary we may assume that $\frac{1}{2\pi}[\omega] \in H^2(X, \mathbb{Z})$. Introducing an almost-complex structure J and a line bundle L with $c_1(L) = \frac{1}{2\pi}[\omega]$, we consider triples of approximately holomorphic sections (s_k^0, s_k^1, s_k^2) of $L^{\otimes k}$: for $k \gg 0$, it is again possible to achieve a generic behavior for the projective map $f_k = (s_k^0 : s_k^1 : s_k^2) : X \to \mathbb{CP}^2$ associated with the linear system. If the manifold X is four-dimensional, then the linear system generically has no base points, and for a suitable choice of sections the map f_k is a branched covering [Au2].

Theorem 3.3 *For large enough k, three suitably chosen approximately holomorphic sections of $L^{\otimes k}$ over (X^4, ω) determine a symplectic branched covering $f_k : X^4 \to \mathbb{CP}^2$, described in approximately holomorphic local coordinates by the local models of Definition 3.1. Moreover, for $k \gg 0$ these branched covering structures are uniquely determined up to isotopy.*

Because the local models hold in approximately holomorphic (and hence ω-tame) coordinates, the ramification curve R_k of f_k is a symplectic submanifold in X (connected, since the Lefschetz hyperplane theorem applies). Moreover, if we normalize the Fubini-Study symplectic form on $\mathbb{CP}^2$ in such a way that $\frac{1}{2\pi}[\omega_{FS}]$ is the generator of $H^2(\mathbb{CP}^2, \mathbb{Z})$, then we have $[f_k^*\omega_{FS}] = 2\pi c_1(L^{\otimes k}) = k[\omega]$, and it is fairly easy to check that the symplectic form on X obtained by applying Proposition 3.2 to the branched covering f_k coincides up to symplectomorphism with $k\omega$ [Au2]. In fact, the exact 2-form $\alpha = k\omega - f_k^*\omega_{FS}$ is positive over $\mathrm{Ker}\, df_k$ at every point of R_k, and $f_k^*\omega_{FS} + t\alpha$ is a symplectic form for all $t \in (0, 1]$.

The uniqueness statement in Theorem 3.3, which should be interpreted exactly in the same way as that obtained by Donaldson for Lefschetz pencils, implies that for $k \gg 0$ it is possible to define invariants of the symplectic manifold (X, ω) in terms of the monodromy of the branched covering f_k and the topology of its branch curve $D_k \subset \mathbb{CP}^2$. However, the branch curve D_k is only determined up to creation or cancellation of (admissible) pairs of nodes of opposite orientations.

A similar construction can be attempted when $\dim X > 4$; in this case, the set of base points $Z_k = \{s_k^0 = s_k^1 = s_k^2 = 0\}$ is no longer empty; it is generically a smooth codimension six symplectic submanifold. With this understood, Theorem 3.3 admits the following higher-dimensional analogue [Au3]:

Theorem 3.4 *For large enough k, three suitably chosen approximately holomorphic sections of $L^{\otimes k}$ over (X^{2n}, ω) determine a map $f_k : X - Z_k \to \mathbb{CP}^2$ with generic local models, canonically determined up to isotopy.*

The model maps describing the local behavior of f_k in approximately holomorphic local coordinates are now the following:

(0) $(z_1, \ldots, z_n) \mapsto (z_1 : z_2 : z_3)$ near a base point
(i) $(z_1, \ldots, z_n) \mapsto (z_1, z_2)$
(ii) $(z_1, \ldots, z_n) \mapsto (z_1^2 + \cdots + z_{n-1}^2, z_n)$
(iii) $(z_1, \ldots, z_n) \mapsto (z_1^3 - z_1 z_n + z_2^2 + \cdots + z_{n-1}^2, z_n)$

The set of critical points $R_k \subset X$ is again a (connected) smooth symplectic curve, and its image $D_k = f_k(R_k) \subset \mathbb{CP}^2$ is again a singular symplectic curve whose only singularities generically are transverse double points of either orientation and complex cusps. The fibers of f_k are codimension four symplectic submanifolds, intersecting along Z_k; the fiber above a point of $\mathbb{CP}^2 - D_k$ is smooth, while the fiber above a smooth point of D_k presents an ordinary double point, the fiber above a node presents two ordinary double points, and the fiber above a cusp presents an A_2 singularity.

The proof of these two results relies on a careful examination of the various possible local behaviors for the map f_k and on transversality arguments establishing the existence of sections of $L^{\otimes k}$ with generic behavior. Hence, the argument relies on the enumeration of the various special cases, generic or not, that may occur; each one corresponds to the vanishing of a certain quantity that can be expressed in terms of the sections s_k^0, s_k^1, s_k^2 and their derivatives. Therefore, the proof largely reduces to a core ingredient which imitates classical singularity theory and can be thought of as a uniform transversality result for jets of approximately holomorphic sections [Au4].

Given approximately holomorphic sections s_k of very positive bundles E_k (e.g., $E_k = \mathbb{C}^m \otimes L^{\otimes k}$) over the symplectic manifold X, one can consider the r-jets $j^r s_k = (s_k, \partial s_k, (\partial \partial s_k)_{\text{sym}}, \ldots, (\partial^r s_k)_{\text{sym}})$, which are sections of the jet bundles $\mathcal{J}^r E_k = \bigoplus_{j=0}^r (T^* X^{(1,0)})_{\text{sym}}^{\otimes j} \otimes E_k$. Jet bundles can naturally be stratified by approximately holomorphic submanifolds corresponding to the various possible local behaviors at order r for the sections s_k. The generically expected behavior corresponds to the case where the jet $j^r s_k$ is transverse to the submanifolds in the stratification. The result is the following [Au4]:

Theorem 3.5 *Given stratifications $\mathcal{S}_k$ of the jet bundles $\mathcal{J}^r E_k$ by a finite number of approximately holomorphic submanifolds (Whitney-regular, uniformly transverse to fibers, and with curvature bounded independently of k), for large enough k the vector bundles E_k admit approximately holomorphic sections s_k whose r-jets are uniformly transverse to the stratifications $\mathcal{S}_k$. Moreover these sections may be chosen arbitrarily close to given sections.*

A one-parameter version of this result also holds, which makes it possible to obtain results of asymptotic uniqueness up to isotopy for generic sections.

Applied to suitably chosen stratifications, this result provides the main ingredient for the construction of m-tuples of approximately holomorphic sections of $L^{\otimes k}$ (and hence projective maps f_k to $\mathbb{CP}^{m-1}$) with generic behavior. Once uniform transversality of jets has been obtained, the only remaining task is to achieve some control over the antiholomorphic derivative $\bar{\partial} f_k$ near the critical points of f_k (typically its vanishing in some directions), in order to ensure that $\bar{\partial} f_k \ll \partial f_k$ everywhere; for low values of m such as those considered above, this task is comparatively easy.

3.2 Monodromy Invariants for Branched Covers of $\mathbb{CP}^2$

The topological data characterizing a symplectic branched covering $f : X^4 \to \mathbb{CP}^2$ are on one hand the topology of the branch curve $D \subset \mathbb{CP}^2$ (up to isotopy and cancellation of pairs of nodes), and on the other hand a monodromy morphism $\theta : \pi_1(\mathbb{CP}^2 - D) \to S_N$ describing the manner in which the $N = \deg f$ sheets of the covering are arranged above $\mathbb{CP}^2 - D$.

Some simple properties of the monodromy morphism θ can be readily seen by considering the local models of Definition 3.1. For example, the image of a small loop γ bounding a disc that intersects D transversely in a single smooth point (such a loop is called a *geometric generator* of $\pi_1(\mathbb{CP}^2 - D)$) by θ is necessarily a transposition. The smoothness of X above a singular point of D implies some compatibility properties on these transpositions (geometric generators corresponding to the two branches of D at a node must map to disjoint commuting transpositions, while to a cusp must correspond a pair of adjacent transpositions). Finally, the connectedness of X implies the surjectivity of θ (because the subgroup $\mathrm{Im}(\theta)$ is generated by transpositions and acts transitively on the fiber of the covering).

It must be mentioned that the amount of information present in the monodromy morphism θ is fairly small: a classical conjecture in algebraic geometry (Chisini's conjecture, essentially solved by Kulikov [Ku]) asserts that, given an algebraic singular plane curve D with cusps and nodes, a symmetric group-valued monodromy morphism θ compatible with D (in the above sense), if it exists, is unique except for a small list of low-degree counter-examples. Whether Chisini's conjecture also holds for symplectic branch curves is an open question, but in any case the number of possibilities for θ is always finite.

The study of a singular complex curve $D \subset \mathbb{CP}^2$ can be carried out using the braid monodromy techniques developed in complex algebraic geometry by Moishezon and Teicher [Mo2, Te1],...: the idea is to choose a linear projection $\pi : \mathbb{CP}^2 - \{\mathrm{pt}\} \to \mathbb{CP}^1$, for example $\pi(x : y : z) = (x : y)$, in such a way that the curve D lies in general position with respect to the fibers of π, i.e., D is positively transverse to the fibers of π everywhere except at isolated non-degenerate smooth complex tangencies. The restriction $\pi_{|D}$ is then a singular branched covering of degree $d = \deg D$, with *special points* corresponding to the singularities of D (nodes and cusps) and to the tangency points. Moreover, we can assume that all special points lie in distinct fibers of π. A plane curve satisfying these topological requirements is called a *braided* (or *Hurwitz*) curve.

Except for those which contain special points of D, the fibers of π are lines intersecting the curve D in d distinct points. If one chooses a reference point $q_0 \in \mathbb{CP}^1$ (and the corresponding fiber $\ell \simeq \mathbb{C} \subset \mathbb{CP}^2$ of π), and if one restricts to an affine subset in order to be able to trivialize the fibration π, the topology of the branched covering $\pi_{|D}$ can be described by a *braid monodromy* morphism

$$\rho : \pi_1(\mathbb{C} - \{\mathrm{pts}\}, q_0) \to B_d, \tag{2}$$

where B_d is the braid group on d strings. The braid $\rho(\gamma)$ corresponds to the motion of the d points of $\ell \cap D$ inside the fibers of π when moving along the loop γ.

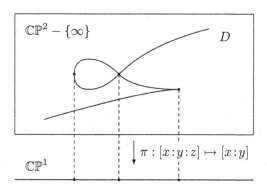

Recall that the braid group B_d is the fundamental group of the configuration space of d distinct points in $\mathbb{R}^2$; it is also the group of isotopy classes of compactly supported orientation-preserving diffeomorphisms of $\mathbb{R}^2$ leaving invariant a set of d given distinct points. It is generated by the standard *half-twists* $X_1, \ldots, X_{d-1}$ (braids which exchange two consecutive points by rotating them counterclockwise by $180°$ around each other), with relations $X_i X_j = X_j X_i$ for $|i - j| \geq 2$ and $X_i X_{i+1} X_i = X_{i+1} X_i X_{i+1}$ (the reader is referred to Birman's book [Bir] for more details).

Another equivalent way to consider the monodromy of a braided curve is to choose an ordered system of generating loops in the free group $\pi_1(\mathbb{C} - \{pts\}, q_0)$. The morphism ρ can then be described by a *factorization* in the braid group B_d, i.e., a decomposition of the monodromy at infinity into the product of the individual monodromies around the various special points of D. By observing that the total space of π is the line bundle $O(1)$ over $\mathbb{CP}^1$, it is easy to see that the monodromy at infinity is given by the central element $\Delta^2 = (X_1 \ldots X_{d-1})^d$ of B_d (called "full twist" because it represents a rotation of a large disc by $360°$). The individual monodromies around the special points are conjugated to powers of half-twists, the exponent being 1 in the case of tangency points, 2 in the case of positive nodes (or -2 for negative nodes), and 3 in the case of cusps.

The braid monodromy ρ and the corresponding factorization depend on trivialization choices, which affect them by *simultaneous conjugation* by an element of B_d (change of trivialization of the fiber ℓ of π), or by *Hurwitz operations* (change of generators of the group $\pi_1(\mathbb{C} - \{pts\}, q_0)$). There is a one-to-one correspondence between braid monodromy morphisms $\rho : \pi_1(\mathbb{C} - \{pts\}) \to B_d$ (mapping generators to suitable powers of half-twists) up to these two algebraic operations and singular (not necessarily complex) braided curves of degree d in $\mathbb{CP}^2$ up to isotopy among such curves (see, e.g., [KK] for a detailed exposition). Moreover, it is easy to check that every braided curve in $\mathbb{CP}^2$ can be deformed into a braided symplectic curve, canonically up to isotopy among symplectic braided curves (this deformation is performed by collapsing the curve D into a neighborhood of a complex line in a way that preserves the fibers of π). However, the curve D is isotopic to a complex curve only for certain specific choices of the morphism ρ.

Unlike the case of complex curves, it is not clear a priori that the symplectic branch curve D_k of one of the covering maps given by Theorem 3.3 can be made compatible with the linear projection π; making the curve D_k braided relies on an improvement of the result in order to control more precisely the behavior of D_k near its special points (tangencies, nodes, cusps). Moreover, one must take into account the possible occurrence of creations or cancellations of admissible pairs of nodes in the branch curve D_k, which affect the braid monodromy morphism $\rho_k : \pi_1(\mathbb{C} - \{\text{pts}\}) \to B_d$ by insertion or deletion of pairs of factors. The uniqueness statement in Theorem 3.3 then leads to the following result [AK]:

Theorem 3.6 (A.-Katzarkov) *For given large enough k, the monodromy morphisms (ρ_k, θ_k) associated to the approximately holomorphic branched covering maps $f_k : X \to \mathbb{CP}^2$ defined by triples of sections of $L^{\otimes k}$ are, up to conjugation, Hurwitz operations, and insertions/deletions, invariants of the symplectic manifold (X, ω). Moreover, these invariants are complete, in the sense that the data (ρ_k, θ_k) are sufficient to reconstruct the manifold (X, ω) up to symplectomorphism.*

It is interesting to mention that the symplectic Lefschetz pencils constructed by Donaldson (Theorem 2.9) can be recovered very easily from the branched covering maps f_k, simply by considering the $\mathbb{CP}^1$-valued maps $\pi \circ f_k$. In other words, the fibers $\Sigma_{k,\alpha}$ of the pencil are the preimages by f_k of the fibers of π, and the singular fibers of the pencil correspond to the fibers of π through the tangency points of D_k.

In fact, the monodromy morphisms ψ_k of the Lefschetz pencils $\pi \circ f_k$ can be recovered very explicitly from θ_k and ρ_k. By restriction to the line $\bar{\ell} = \ell \cup \{\infty\}$, the S_N-valued morphism θ_k describes the topology of a fiber Σ_k of the pencil as an N-fold covering of $\mathbb{CP}^1$ with d branch points; the set of base points Z_k is the preimage of the point at infinity in $\bar{\ell}$. This makes it possible to define a *lifting homomorphism* from a subgroup $B_d^0(\theta_k) \subset B_d$ (*liftable braids*) to the mapping class group $\text{Map}(\Sigma_k, Z_k) = \text{Map}_{g,N}$. The various monodromies are then related by the following formula [AK]:

$$\psi_k = (\theta_k)_* \circ \rho_k. \tag{3}$$

The lifting homomorphism $(\theta_k)_*$ maps liftable half-twists to Dehn twists, so that the tangencies between the branch curve D_k and the fibers of π determine explicitly the vanishing cycles of the Lefschetz pencil $\pi \circ f_k$. On the other hand, the monodromy around a node or cusp of D_k lies in the kernel of $(\theta_k)_*$.

The lifting homomorphism θ_* can be defined more precisely as follows: the space $\tilde{\mathcal{X}}_d$ of configurations of d distinct points in $\mathbb{R}^2$ together with branching data (a transposition in S_N attached to each point, or more accurately an S_N-valued group homomorphism) is a finite covering of the space $\mathcal{X}_d$ of configurations of d distinct points. The morphism θ determines a lift $\tilde{*}$ of the base point in $\mathcal{X}_d$, and the liftable braid subgroup of $B_d = \pi_1(\mathcal{X}_d, *)$ is the stabilizer of θ for the action of B_d by deck transformations of the covering $\tilde{\mathcal{X}}_d \to \mathcal{X}_d$, i.e., $B_d^0(\theta) = \pi_1(\tilde{\mathcal{X}}_d, \tilde{*})$. Moreover, $\tilde{\mathcal{X}}_d$ is naturally equipped with a universal fibration $\mathcal{Y}_d \to \tilde{\mathcal{X}}_d$ by genus g Riemann surfaces with N marked points: the lifting homomorphism $\theta_* : B_d^0(\theta) \to \mathrm{Map}_{g,N}$ is by definition the monodromy of this fibration.

The relation (3) is very useful for explicit calculations of the monodromy of Lefschetz pencils, which is accessible to direct methods only in a few very specific cases. By comparison, the various available techniques for braid monodromy calculations [Mo3, Te1, ADKY] are much more powerful, and make it possible to carry out calculations in a much larger number of cases (see below). In particular, in view of Donaldson's result we are mostly interested in the monodromy of high degree Lefschetz pencils, where the fiber genus and the number of singular fibers are very high, making them inaccessible to direct calculation even for the simplest complex algebraic surfaces.

When considering higher-dimensional manifolds, one may also associate monodromy invariants to a fibration $f_k : X \to \mathbb{CP}^2$; these consist of the braid monodromy of the critical curve $D_k \subset \mathbb{CP}^2$, and the monodromy of the fibration (with values in the mapping class group of the fiber). Furthermore, considering successive hyperplane sections and projections to $\mathbb{CP}^2$, one obtains a complete description of a symplectic manifold (X^{2n}, ω) in terms of $n-1$ braid group-valued monodromy morphisms (describing the critical curves of various $\mathbb{CP}^2$-valued projections) and a single symmetric group-valued homomorphism (see [Au3] for details).

In principle, the above results reduce the classification of compact symplectic manifolds to purely combinatorial questions concerning braid groups and symmetric groups, and symplectic topology seems to largely reduce to the study of certain singular plane curves, or equivalently factorizations in braid groups.

The explicit calculation of these monodromy invariants is hard in the general case, but is made possible for a large number of complex surfaces by the use of "degeneration" techniques and of approximately holomorphic perturbations. These techniques make it possible to compute the braid monodromies of a variety of algebraic branch curves. In most cases, the calculation is only possible for a fixed projection to $\mathbb{CP}^2$ of a given algebraic surface (i.e., fixing $[\omega]$ and considering only $k = 1$); see, e.g., [Mo2, Ro, AGTV],... for various such examples. In a smaller set of examples, the technique applies to projections of arbitrarily large degrees and hence makes it possible to compute explicitly the invariants defined by Theorem 3.6. To this date, the list consists of $\mathbb{CP}^2$ [Mo5, Te3], $\mathbb{CP}^1 \times \mathbb{CP}^1$ [Mo3], certain Del Pezzo and K3 surfaces [Ro, CMT], the Hirzebruch surface $\mathbb{F}_1$ [MRT, ADKY], and all double covers of $\mathbb{CP}^1 \times \mathbb{CP}^1$ branched along connected smooth algebraic curves, which includes an infinite family of surfaces of general type [ADKY].

The degeneration technique, developed by Moishezon and Teicher [Mo3, Te1],..., starts with a projective embedding of the complex surface X, and deforms the image of this embedding to a singular configuration X_0 consisting of a union of planes intersecting along lines. The discriminant curve of a projection of X_0 to $\mathbb{CP}^2$ is therefore a union of lines; the manner in which the smoothing of X_0 affects this curve can be studied explicitly, by considering a certain number of standard local models near the various points of X_0 where three or more planes intersect (see [Te1] and [ADKY] for detailed reviews of the technique). This method makes it possible to handle many examples in low degree, but in the case $k \gg 0$ that we are interested in (very positive linear systems over a fixed manifold), the calculations can only be carried out explicitly for very simple surfaces.

In order to proceed beyond this point, it becomes more efficient to move outside of the algebraic framework and to consider generic approximately holomorphic perturbations of non-generic algebraic maps; the greater flexibility of this setup makes it possible to choose more easily computable local models. For example, the direct calculation of the monodromy invariants becomes possible for all linear systems of the type $\pi^* O(p, q)$ on double covers of $\mathbb{CP}^1 \times \mathbb{CP}^1$ branched along connected smooth algebraic curves of arbitrary degree [ADKY]. It also becomes possible to obtain a general "degree doubling" formula, describing explicitly the monodromy invariants associated to the linear system $L^{\otimes 2k}$ in terms of those associated to the linear system $L^{\otimes k}$ (when $k \gg 0$), both for branched covering maps to $\mathbb{CP}^2$ and for four-dimensional Lefschetz pencils [AK2].

However, in spite of these successes, a serious obstacle restricts the practical applications of monodromy invariants: in general, they cannot be used efficiently to distinguish homeomorphic symplectic manifolds, because no algorithm exists to decide whether two words in a braid group or mapping class group are equivalent to each other via Hurwitz operations. Even if an algorithm could be found, another difficulty is due to the large amount of combinatorial data to be handled: on a typical interesting example, the braid monodromy data can already consist of $\sim 10^4$ factors in a braid group on ~ 100 strings for very small values of the parameter k, and the amount of data grows polynomially with k.

Hence, even when monodromy invariants can be computed, they cannot be *compared*. This theoretical limitation makes it necessary to search for other ways to exploit monodromy data, e.g., by considering invariants that contain less information than braid monodromy but are easier to use in practice.

3.3 Fundamental Groups of Branch Curve Complements

Given a singular plane curve $D \subset \mathbb{CP}^2$, e.g., the branch curve of a covering, it is natural to study the fundamental group $\pi_1(\mathbb{CP}^2 - D)$. The study of this group for various types of algebraic curves is a classical subject going back to the work of Zariski, and has undergone a lot of development in the 1980s and 1990s, in part thanks to the work of Moishezon and Teicher [Mo2,Mo3,Te1],.... The relation to braid monodromy invariants is a very direct one: the Zariski–van Kampen theorem provides an explicit presentation of the group $\pi_1(\mathbb{CP}^2 - D)$ in terms of the braid monodromy morphism $\rho : \pi_1(\mathbb{C} - \{pts\}) \to B_d$. However, if one is interested in the case of symplectic branch curves, it is important to observe that the introduction or the cancellation of pairs of nodes affects the fundamental group of the complement, so that it cannot be used directly to define an invariant associated to a symplectic branched covering. In the symplectic world, the fundamental group of the branch curve complement must be replaced by a suitable quotient, the *stabilized fundamental group* [ADKY].

Using the same notations as above, the inclusion $i : \ell - (\ell \cap D_k) \to \mathbb{CP}^2 - D_k$ of the reference fiber of the linear projection π induces a surjective morphism on fundamental groups; the images of the standard generators of the free group $\pi_1(\ell - (\ell \cap D_k))$ and their conjugates form a subset $\Gamma_k \subset \pi_1(\mathbb{CP}^2 - D_k)$ whose elements are called *geometric generators*. Recall that the images of the geometric generators by the monodromy morphism θ_k are transpositions in S_N. The creation of a pair of nodes in the curve D_k amounts to quotienting $\pi_1(\mathbb{CP}^2 - D_k)$ by a relation of the form $[\gamma_1, \gamma_2] \sim 1$, where $\gamma_1, \gamma_2 \in \Gamma_k$; however, this creation of nodes can be carried out by deforming the branched covering map f_k only if the two transpositions $\theta_k(\gamma_1)$ and $\theta_k(\gamma_2)$ have disjoint supports. Let K_k be the normal subgroup of $\pi_1(\mathbb{CP}^2 - D_k)$ generated by all such commutators $[\gamma_1, \gamma_2]$. Then we have the following result [ADKY]:

Theorem 3.7 (A.-D.-K.-Y.) *For given $k \gg 0$, the stabilized fundamental group $\bar{G}_k = \pi_1(\mathbb{CP}^2 - D_k)/K_k$ is an invariant of the symplectic manifold (X^4, ω).*

This invariant can be calculated explicitly for the various examples where monodromy invariants are computable ($\mathbb{CP}^2$, $\mathbb{CP}^1 \times \mathbb{CP}^1$, some Del Pezzo and K3 surfaces, Hirzebruch surface $\mathbb{F}_1$, double covers of $\mathbb{CP}^1 \times \mathbb{CP}^1$); namely, the extremely complicated presentations given by the Zariski–van Kampen theorem in terms of braid monodromy data can be simplified in order to obtain a manageable description of the fundamental group of the branch curve complement. These examples lead to various observations and conjectures.

A first remark to be made is that, for all known examples, when the parameter k is sufficiently large the stabilization operation becomes trivial, i.e., geometric generators associated to disjoint transpositions already commute in $\pi_1(\mathbb{CP}^2 - D_k)$, so that $K_k = \{1\}$ and $\bar{G}_k = \pi_1(\mathbb{CP}^2 - D_k)$. For example, in the case of $X = \mathbb{CP}^2$ with its standard Kähler form, we have $\bar{G}_k = \pi_1(\mathbb{CP}^2 - D_k)$ for all $k \geq 3$. Therefore, when $k \gg 0$ no information seems to be lost when quotienting by K_k (the situation for small values of k is very different).

The following general structure result can be proved for the groups $\bar{G}_k$ (and hence for $\pi_1(\mathbb{CP}^2 - D_k)$) [ADKY]:

Theorem 3.8 (A.-D.-K.-Y.) *Let $f : (X, \omega) \to \mathbb{CP}^2$ be a symplectic branched covering of degree N, with braided branch curve D of degree d, and let $\bar{G} = \pi_1(\mathbb{CP}^2 - D)/K$ be the stabilized fundamental group of the branch curve complement. Then there exists a natural exact sequence*

$$1 \longrightarrow G^0 \longrightarrow \bar{G} \longrightarrow S_N \times \mathbb{Z}_d \longrightarrow \mathbb{Z}_2 \longrightarrow 1.$$

Moreover, if X is simply connected then there exists a natural surjective homomorphism $\phi : G^0 \twoheadrightarrow (\mathbb{Z}^2/\Lambda)^{n-1}$, where

$$\Lambda = \{(c_1(K_X) \cdot \alpha, [f^{-1}(\bar{\ell})] \cdot \alpha), \ \alpha \in H_2(X, \mathbb{Z})\}.$$

In this statement, the two components of the morphism $\bar{G} \to S_N \times \mathbb{Z}_d$ are respectively the monodromy of the branched covering, $\theta : \pi_1(\mathbb{CP}^2 - D) \to S_N$, and the linking number (or abelianization, when D is irreducible) morphism

$$\delta : \pi_1(\mathbb{CP}^2 - D) \to \mathbb{Z}_d \ (\simeq H_1(\mathbb{CP}^2 - D, \mathbb{Z})).$$

The subgroup Λ of $\mathbb{Z}^2$ is entirely determined by the numerical properties of the canonical class $c_1(K_X)$ and of the hyperplane class (the homology class of the preimage of a line $\bar{\ell} \subset \mathbb{CP}^2$: in the case of the covering maps of Theorem 3.3 we have $[f^{-1}(\bar{\ell})] = c_1(L^{\otimes k}) = \frac{k}{2\pi}[\omega]$). The morphism ϕ is defined by considering the N lifts in X of a closed loop γ belonging to G^0, or more precisely their homology classes (whose sum is trivial) in the complement of a hyperplane section and of the ramification curve in X.

Moreover, in the known examples we have a much stronger result on the structure of the subgroups G_k^0 for the branch curves of large degree covering maps (determined by sufficiently ample linear systems) [ADKY].

Say that the simply connected complex surface (X, ω) belongs to the class $(\mathcal{C})$ if it belongs to the list of computable examples: $\mathbb{CP}^1 \times \mathbb{CP}^1$, $\mathbb{CP}^2$, the Hirzebruch surface $\mathbb{F}_1$ (equipped with any Kähler form), a Del Pezzo or K3 surface (equipped with a Kähler form coming from a specific complete intersection realization), or a double cover of $\mathbb{CP}^1 \times \mathbb{CP}^1$ branched along a connected smooth algebraic curve (equipped with a Kähler form in the class $\pi^* O(p, q)$ for $p, q \geq 1$). Then we have:

Theorem 3.9 (A.-D.-K.-Y.) *If (X, ω) belongs to the class $(\mathcal{C})$, then for all large enough k the homomorphism ϕ_k induces an isomorphism on the abelianized groups, i.e., $\mathrm{Ab}\, G_k^0 \simeq (\mathbb{Z}^2/\Lambda_k)^{N_k - 1}$, while $\mathrm{Ker}\, \phi_k = [G_k^0, G_k^0]$ is a quotient of $\mathbb{Z}_2 \times \mathbb{Z}_2$.*

It is natural to make the following conjecture:

Conjecture 3.10 *If X is a simply connected symplectic four-manifold, then for all large enough k the homomorphism ϕ_k induces an isomorphism on the abelianized groups, i.e., $\mathrm{Ab}\, G_k^0 \simeq (\mathbb{Z}^2/\Lambda_k)^{N_k - 1}$.*

3.4 Symplectic Isotopy and Non-Isotopy

While it has been well-known for many years that compact symplectic four-manifolds do not always admit Kähler structures, it has been discovered more recently that symplectic curves (smooth or singular) in a given manifold can also offer a wider range of possibilities than complex curves. Proposition 3.2 and Theorem 3.3 establish a bridge between these two phenomena: indeed, a covering of $\mathbb{CP}^2$ (or any other complex surface) branched along a complex curve automatically inherits a complex structure. Therefore, starting with a non-Kähler symplectic manifold, Theorem 3.3 always yields branch curves that are not isotopic to any complex curve in $\mathbb{CP}^2$. The study of isotopy and non-isotopy phenomena for curves is therefore of major interest for our understanding of the topology of symplectic four-manifolds.

The *symplectic isotopy problem* asks whether, in a given complex surface, every symplectic submanifold representing a given homology class is isotopic to a complex submanifold. The first positive result in this direction was due to Gromov, who showed using his compactness result for pseudo-holomorphic curves (Theorem 1.10) that, in $\mathbb{CP}^2$, a smooth symplectic curve of degree 1 or 2 is always isotopic to a complex curve. Successive improvements of this technique have made it possible to extend this result to curves of higher degree in $\mathbb{CP}^2$ or $\mathbb{CP}^1 \times \mathbb{CP}^1$; the currently best known result is due to Siebert and Tian, and makes it possible to handle the case of smooth curves in $\mathbb{CP}^2$ up to degree 17 [ST] (see also their lecture notes in this volume). Isotopy results are also known for sufficiently simple singular curves (Barraud, Shevchishin [Sh], . . .).

Contrarily to the above examples, the general answer to the symplectic isotopy problem appears to be negative. The first counterexamples among smooth connected symplectic curves were found by Fintushel and Stern [FS], who constructed by a *braiding* process infinite families of mutually non-isotopic symplectic curves representing a same homology class (a multiple of the fiber) in elliptic surfaces (a similar construction can also be performed in higher genus). However, these two constructions are preceded by a result of Moishezon [Mo4], who established in the early 1990s a result implying the existence in $\mathbb{CP}^2$ of infinite families of pairwise non-isotopic singular symplectic curves of given degree with given numbers of node and cusp singularities. A reformulation of Moishezon's construction makes it possible to see that it also relies on braiding; moreover, the braiding construction can be related to a surgery operation along a Lagrangian torus in a symplectic four-manifold, known as *Luttinger surgery* [ADK1]. This reformulation makes it possible to vastly simplify Moishezon's argument, which was based on lengthy and delicate calculations of fundamental groups of curve complements, while relating it with various constructions developed in four-dimensional topology.

Given an embedded Lagrangian torus T in a symplectic four-manifold (X, ω), a homotopically non-trivial embedded loop $\gamma \subset T$ and an integer k, Luttinger surgery is an operation that consists in cutting out from X a tubular neighborhood of T, foliated by parallel Lagrangian tori, and gluing it back in such a way that the new meridian loop differs from the old one by k twists along the loop γ (while longitudes are not affected), yielding a new symplectic manifold $(\tilde{X}, \tilde{\omega})$. This relatively little-known construction, which, e.g., makes it possible to turn a product $T^2 \times \Sigma$ into any surface bundle over T^2, or to transform an untwisted fiber sum into a twisted one, can be used to described in a unified manner numerous examples of exotic symplectic four-manifolds constructed in the past few years.

Meanwhile, the braiding construction of symplectic curves starts with a (possibly singular) symplectic curve $\Sigma \subset (Y^4, \omega_Y)$ and two symplectic cylinders embedded in Σ, joined by a Lagrangian annulus contained in the complement of Σ, and consists in performing k half-twists between these two cylinders in order to obtain a new symplectic curve $\tilde{\Sigma}$ in Y.

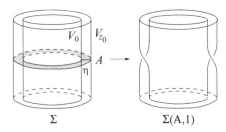

When Σ is the branch curve of a symplectic branched covering $f : X \to Y$, the following result holds [ADK1]:

Proposition 3.11 *The covering of Y branched along the symplectic curve $\tilde{\Sigma}$ obtained by braiding Σ along a Lagrangian annulus $A \subset Y - \Sigma$ is naturally symplectomorphic to the manifold $\tilde{X}$ obtained from the branched cover X by Luttinger surgery along a Lagrangian torus $T \subset X$ formed by the union of two lifts of A.*

Hence, once an infinite family of symplectic curves has been constructed by braiding, it is sufficient to find invariants that distinguish the corresponding branched covers in order to conclude that the curves are not isotopic. In the Fintushel–Stern examples, the branched covers are distinguished by their Seiberg–Witten invariants, whose behavior is well understood in the case of elliptic fibrations and their surgeries.

In the case of Moishezon's examples, a braiding construction makes it possible to construct, starting from complex curves $\Sigma_{p,0} \subset \mathbb{CP}^2$ ($p \geq 2$) of degree $d_p = 9p(p-1)$ with $\kappa_p = 27(p-1)(4p-5)$ cusps and $\nu_p = 27(p-1)(p-2)$ $(3p^2 + 3p - 8)/2$ nodes, symplectic curves $\Sigma_{p,k} \subset \mathbb{CP}^2$ for all $k \in \mathbb{Z}$, with the same degree and numbers of singular points. By Proposition 3.11, these curves can be viewed as the branch curves of symplectic coverings whose total spaces $X_{p,k}$ differ by Luttinger surgeries along a Lagrangian torus $T \subset X_{p,0}$. The effect of these surgeries on the canonical class and on the symplectic form can be described explicitly, which makes it possible to distinguish the manifolds $X_{p,k}$: the canonical class of $(X_{p,k}, \omega_{p,k})$ is given by $p\,c_1(K_{p,k}) = (6p-9)[\omega_{p,k}]+(2p-3)k\,PD([T])$. Moreover, $[T] \in H_2(X_{p,k}, \mathbb{Z})$ is not a torsion class, and if $p \not\equiv 0 \mod 3$ or $k \equiv 0 \mod 3$ then it is a primitive class [ADK1]. This implies that infinitely many of the curves $\Sigma_{p,k}$ are pairwise non-isotopic.

It is to be observed that the argument used by Moishezon to distinguish the curves $\Sigma_{p,k}$, which relies on a calculation of the fundamental groups $\pi_1(\mathbb{CP}^2 - \Sigma_{p,k})$ [Mo4], is related to the one in [ADK1] by means of Conjecture 3.10, of which it can be concluded a posteriori that it is satisfied by the given branched covers $X_{p,k} \to \mathbb{CP}^2$: in particular, the fact that $\pi_1(\mathbb{CP}^2 - \Sigma_{p,k})$ is infinite for $k = 0$ and finite for $k \neq 0$ is consistent with the observation that the canonical class of $X_{p,k}$ is proportional to its symplectic class iff $k = 0$.

4 Symplectic Surfaces from Symmetric Products

This section will describe an approach, developed in [DS, Sm4, Us], to various theorems due to Taubes, but in the context of Lefschetz pencils rather than Seiberg–Witten gauge theory and the equivalence "SW=Gr." The main result on which we will focus is Theorem 1.11 (i) about the existence of embedded symplectic submanifolds representing the canonical class (although we will establish the result only in the case where $b_+ > 1 + b_1$). Note that a Lefschetz pencil is a family of symplectic surfaces of arbitrarily large volume and complexity (depending on k), so somewhat different techniques are needed to find symplectic surfaces in a *prescribed* homology class.

4.1 Symmetric Products

Let (X, ω_X) be a symplectic four-manifold with integral $[\omega_X]$, and fix a Lefschetz pencil, giving rise to a Lefschetz fibration $f : \hat{X} \to S^2$ on the blow-up. We will always equip $\hat{X}$ with the symplectic form $\omega_C = p^*\omega_X + C\,f^*\omega_{S^2}$ for some large $C > 0$.

Definition 4.1 A *standard surface* in $\hat{X}$ is a smooth embedded surface Σ such that $f_{|\Sigma}$ is a simple branched covering over S^2, and at each branch point df gives an oriented isomorphism between the normal bundle to Σ and TS^2.

The condition on df ensures that we have positive branching, that is, near each branch point there are complex-valued coordinates such that Σ looks like $\{(z_1, z_2) \in \mathbb{C}^2,\ z_1^2 = z_2\}$, with f being the projection to z_2.

Lemma 4.2 A *standard surface* in $\hat{X}$, disjoint from the exceptional sections E_i, defines a symplectic surface in X.

This follows from a local calculation showing that $(\hat{X} - \bigcup E_i, \omega_C)$ is symplectomorphic to $(X - \{b_i\}, (1 + kC)\omega_X)$, where k is the degree of the pencil.

By taking C sufficiently large it is clear that the given standard surface is symplectic in $(\hat{X}, \omega_C)$.

Rather than encoding a standard surface algebraically in the monodromy representation, we will take a geometric standpoint and obtain such surfaces from sections of a "relative symmetric product." This construction will in fact yield singular unions of such surfaces, so for later use we therefore provide a smoothing lemma. As a piece of notation, a *positive symplectic divisor* is a union of symplectic surfaces $D = \sum a_i \Sigma_i$ with $a_i > 0$, with the Σ_i pairwise having only isolated positive transverse intersections, and no triple intersections. (In fact, a careful inspection of the later argument will show that, in the positive divisors we construct, if Σ_i and Σ_j intersect, then either one is an exceptional sphere, or one has multiplicity $a_i > 1$.) In any case, we will be able to use the following criterion:

Proposition 4.3 *If a symplectic divisor satisfies $D \cdot \Sigma_j \geq 0$ for all j, then it can be symplectically smoothed.*

Proof (Sketch). Suppose first that all the Σ_i are complex curves in a Kähler surface. The hypothesis says that the line bundle $O(D)$ has non-negative degree on each component. Then we can find a smooth section γ of $O(D)$ which is holomorphic near each intersection point and near its (transverse) zero set. If the section s defines the divisor D, then the zero set of $s - \epsilon\gamma$ is a suitable symplectic smoothing for sufficiently small ϵ.

In general, the Σ_i may not be symplectomorphic to any complex model. Near an intersection point of Σ ad Σ', write Σ' as a graph of a linear map A over the symplectic orthogonal of Σ. The symplecticity of Σ' implies that $\det(A) > -1$; for a complex model (in which A is an element of $\mathbb{C} \subset \mathrm{End}(\mathbb{R}^2)$), we need $\det(A) > 0$ (or $A = 0$). This can be achieved by a small compactly supported perturbation of Σ'. $\square$

We now turn to the main construction. The d-fold symmetric product of a Riemann surface is canonically a smooth complex manifold of dimension d. A holomorphic atlas is provided by noticing that the elementary symmetric functions define a biholomorphism $\mathrm{Sym}^d(\mathbb{C}) = \mathbb{C}^d/S_d \xrightarrow{\sim} \mathbb{C}^d$. For instance, taking a non-zero $(d + 1)$-tuple of numbers to the roots of the degree d polynomial having these as its coefficients gives an identification between $\mathbb{CP}^d$ and $\mathrm{Sym}^d(\mathbb{CP}^1)$.

Note that the symmetric product of a smooth two-manifold has no natural smooth atlas, although it is well-defined up to non-canonical diffeomorphisms. We will always work with almost-complex structures on our Lefschetz fibrations which make the projection map pseudo-holomorphic (call these almost-complex structures *fibered*). In this case the fibers become Riemann surfaces.

Theorem 4.4 *Given a Lefschetz fibration $f : \hat{X} \to \mathbb{CP}^1$ with a fibered almost-complex structure, for each $d \geq 0$ there is a smooth compact symplectic manifold $X_d(f) \to \mathbb{CP}^1$ with fiber at $t \notin \mathrm{crit}(f)$ the symmetric product $\mathrm{Sym}^d(f^{-1}(t))$.*

Away from the singular fibers, this follows by taking charts on X of the form $\theta : D_1 \times D_2 \to X$, such that: for each $x \in D_1$, $\theta_x : \{x\} \times D_2 \to X$ gives a holomorphic disc in a fiber of f; and $f \circ \theta = i \circ \pi_1$, for i an inclusion $D_1 \hookrightarrow \mathbb{CP}^1$. In other words, the fibers of f are Riemann surfaces whose complex structures vary smoothly in the obvious way. Now we restrict further to work with almost complex structures which are integrable near the singular fibers. In this case, we can fill in the fibration by appealing to algebraic geometry. Namely, replace the symmetric product (a moduli space of structure sheaves) by a *Hilbert scheme* (a moduli space of ideal sheaves). We will use a concrete description of the Hilbert scheme of $\mathbb{C}^2$ together with our local model for the singularities of f.

Definition 4.5 $Hilb_d(\mathbb{C}^2)$ *comprises the triples* $\{B_1, B_2, g\}/\sim$ *where* B_1, B_2 *are* $d \times d$ *commuting matrices, and* g *is a vector which is not contained in any subspace* $S \subset \mathbb{C}^d$ *stabilized by both* B_i. $GL_d(\mathbb{C})$ *acts by conjugating the* B_i*'s and on the left of* g.

To unravel this, think of an ideal $J \subset \mathbb{C}[z_1, z_2]$ of codimension d. Identify $\mathbb{C}[z_1, z_2]/J = \mathbb{C}^d$, and define B_1, B_2 to be the action of z_1, z_2, and g the image of 1. Conversely, given a triple, define a map $\mathbb{C}[z_1, z_2] \to \mathbb{C}^d$ by $f(z_1, z_2) \mapsto f(B_1, B_2)g$. The stability condition on g implies that this is surjective, and the kernel gives an ideal of codimension d. Nakajima's notes [Na] show that this correspondence is indeed a holomorphic parametrization of the Hilbert scheme of length d subschemes of $\mathbb{C}^2$ by triples of matrices.

Now consider the relative version of this, for the map $\mathbb{C}^2 \to \mathbb{C}$, $(z_1, z_2) \mapsto z_1 z_2$. The fiber over t is by definition those ideals which contain $z_1 z_2 - t$, giving

$$B_1 B_2 = tI, \quad B_2 B_1 = tI, \quad \text{plus stability.}$$

If the B_i are simultaneously diagonalizable, we get a d-tuple of pairs of eigenvalues, i.e., a point of $Sym^d(f^{-1}(t))$. More generally, the B_i can be represented as a pair of upper triangular matrices, and again there is a map to the symmetric product given by taking eigenvalues. For $t \neq 0$ this map is an isomorphism because of the stability condition; for $t = 0$ this is no longer the case, and suitable off-diagonal matrix entries become coordinates on the fibers. In any case, we now have explicit equations defining $X_d(f)$; one can use these to give a concrete description of its topology, and show smoothness.

Proposition 4.6 *The total space of this relative Hilbert scheme is smooth.*

Proof (Sketch). One can show (for instance from the Abel-Jacobi discussion below) that the $t = 0$ fiber has normal crossings. Given this, at each point of the singular locus there are two line bundles ν_i – the normal bundles to the two branches whose intersection defines the singular locus - and the given deformation of the singular fiber gives a section of $\nu_1^* \otimes \nu_2^*$. The total space is smooth if this section has no zeros. In our case, the ν_i are canonically identified with the tangent spaces at the node of the singular fiber of f, so the bundle $\nu_1^* \otimes \nu_2^*$ is trivial. Hence we only need check smoothness at one point. This one can do by writing down explicit local sections of the map. $\square$

At least when we assume that all the singular fibers of f are irreducible, we get another point of view from the Abel-Jacobi map. Let Σ be a Riemann surface.

Definition 4.7 *The Abel-Jacobi map $\mathrm{Sym}^d \Sigma \to \mathrm{Pic}^d(\Sigma)$ takes a divisor D to the line bundle $\mathcal{O}(D)$.*

The key point is to identify the fibers of this map with linear systems $\mathbb{P}(H^0(L))$, hence projective spaces whose dimension may vary as L moves in Pic.

Example. Let $d = 2g - 2$ (by the adjunction formula, this is the relevant case if a standard surface in $\hat{X}$ is to represent $K_{\hat{X}}$). From the Riemann–Roch theorem $h^0(L) - h^1(L) = d - g + 1 = g - 1$, whilst by Serre duality, $h^1(L) > 0$ only if $L = K$ (since only the trivial degree 0 bundle has a section). Thus $\mathrm{Sym}^{2g-2}(\Sigma) \to \mathrm{Pic}^{2g-2}(\Sigma)$ is a (locally trivial) $\mathbb{CP}^{g-2}$-bundle away from a single exceptional fiber, which is a $\mathbb{CP}^{g-1}$. For instance, $\mathrm{Sym}^2(\Sigma_2) \to T^4$ blows down a single exceptional sphere.

A line bundle on a nodal Riemann surface Σ_0 is given by a line bundle on its normalization $\tilde{\Sigma}_0$ together with a $\mathbb{C}^*$ gluing parameter λ to identify the fibers over the preimages of the node. This $\mathbb{C}^*$-bundle over T^{2g-2} naturally compactifies to a $\mathbb{CP}^1$-bundle; if we glue the 0 and ∞ sections of this over a translation in the base, we compactify the family of Picard varieties for a family of curves with an irreducible nodal fiber. (Think of an elliptic Lefschetz fibration with section, which does precisely this in the $g = 1$ case!) Thus, under our assumptions, the relative Picard fibration $P_d(f)$ is also a smooth compact symplectic manifold.

Given $L \in \mathrm{Pic}^{2g-2}(\Sigma_0)$ in the smooth locus, the line bundle on the normalization has a space of sections of dimension g. Hence the λ-hyperplane of sections which transform correctly over the node gives a $\mathbb{CP}^{g-2}$. The situation is similar along the normal crossing divisor in Pic (just take $\lambda = 0$ or ∞), but changes if L gives rise to the canonical bundle of the normalization, where the dimension jumps. Summing up:

Proposition 4.8 *The symplectic manifold $X_{2g-2}(f)$ is the total space of a family of projective spaces over $P_{2g-2}(f)$, the family being a locally trivial $\mathbb{CP}^{g-1}$ bundle over the section defined by the canonical line bundles of the fibers of f and a locally trivial $\mathbb{CP}^{g-2}$ bundle away from this section.*

There is an analogous description for other values of d, and one sees that the singular locus of the singular fiber of $X_d(f)$ is a copy of $\mathrm{Sym}^{d-1}(\tilde{\Sigma}_0)$. It may be helpful to point out that tuples set-theoretically including the node are *not* the same thing as singular points of $\mathrm{Hilb}(\Sigma_0)$. The relevance is that a smooth family of surfaces in X may well include members passing through critical points of f, but smooth sections of $X_d(f)$ can never pass through the singular loci of fibers. (Roughly speaking, the singular locus of $\mathrm{Hilb}_d(\Sigma_0)$ comprises tuples which contain the node with odd multiplicity.)

To see that $X_d(f)$ admits symplectic structures, one uses an analogue of the argument that applies to Lefschetz fibrations – we have a family of Kähler manifolds with locally holomorphic singularities, so we can patch together local forms to obtain something vertically non-degenerate and then add the pull-back of a form from the base.

4.2 Taubes' Theorem

The construction of symplectic surfaces representing $[K_X]$ proceeds in two stages. Let $r = 2g - 2$. A section s of $X_r(f)$ defines a cycle $C_s \subset \hat{X}$. There is a homotopy class h of sections such that $[C_h] = PD(-c_1(\hat{X}))$. First, we show that the Gromov invariant counting sections of $X_r(f)$ in the class h is non-zero, and then from the pseudoholomorphic sections of $X_r(f)$ thereby provided we construct standard surfaces (which have a suitable component which descends from $\hat{X}$ to X). For the first stage the Abel-Jacobi map is the key, whilst for the second we work with almost complex structures compatible with the "diagonal" strata of the symmetric products (tuples of points not all of which are pairwise distinct).

That the Gromov invariant is well-defined follows easily from a description of the possible bubbles arising from cusp curves in $X_r(f)$. Again, we will always use "fibered" almost complex structures J on $X_r(f)$, i.e., ones making the projection to $\mathbb{CP}^1$ holomorphic.

Lemma 4.9 $\pi_2(Sym^r(\Sigma)) = \mathbb{Z}$, generated by a line l in a fiber of the Abel-Jacobi map. All bubbles in cusp limits of sections of $X_r(f) \to \mathbb{CP}^1$ are homologous to a multiple of l.

Proof (Sketch). The first statement follows from the Lefschetz hyperplane theorem and the fact that $Sym^d(\Sigma)$ is a projective bundle for $d > r$; it is also an easy consequence of the Dold–Thom theorem in algebraic topology. The second statement is then obvious away from singular fibers, but these require special treatment, cf. [DS]. $\square$

Since $\langle c_1(Sym^r(\Sigma)), h \rangle > 0$ this shows that spaces of holomorphic sections can be compactified by high codimension pieces, after which it is easy to define the Gromov invariant.

Proposition 4.10 *For a suitable J, the moduli space of J-holomorphic sections in the class h is a projective space of dimension $(b_+ - b_1 - 1)/2 - 1$.*

Proof. The index of the $\bar{\partial}$-operator on $f_*K \to \mathbb{CP}^1$ is $(b_+ - b_1 - 1)/2$. Choose J on $X_r(f)$ so that the projection τ to $P_r(f)$ is holomorphic, and the latter has the canonical section s_K as an isolated holomorphic section (the index of the normal bundle to this is $-(b_+ - b_1 - 1)/2$). Now all holomorphic sections of $X_r(f)$ lie over s_K. If the degree of the original Lefschetz pencil was sufficiently large, the rank of the bundle f_*K is much larger than its first Chern class.

Then, for a generic connection, the bundle becomes $\mathcal{O} \oplus \mathcal{O} \oplus \cdots \oplus \mathcal{O}(-1) \oplus \cdots \oplus \mathcal{O}(-1)$ (by Grothendieck's theorem). Hence, all nonzero sections are nowhere zero, so yield homotopic sections of the projectivisation $\mathbb{P}(f_*K) = \tau^{-1}(s_K)$, each defining a cycle in $\hat{X}$ in the fixed homology class $-c_1(\hat{X})$. □

We now have a J such that $\mathcal{M}_J(h)$ is compact and smooth – but of the wrong dimension (index arguments show that for a section s of $X_r(f)$ giving a cycle C_s in X, the space of holomorphic sections has virtual dimension $C_s \cdot C_s - C_s \cdot K_X$). The actual invariant is given as the Euler class of an *obstruction bundle*, whose fiber at a curve C is $H^1(\nu_C)$ (from the above, all our curves are embedded so this bundle is well-defined). Let Q be the *quotient bundle* over $\mathbb{CP}^n$ (defined as the cokernel of the inclusion of the tautological bundle into $\underline{\mathbb{C}}^{n+1}$). It has Euler class $(-1)^{n+1}$.

Lemma 4.11 *For J as above, the obstruction bundle $Obs \to \mathcal{M}_J(h)$ is $Q \to \mathbb{CP}^N$ with $N = (b_+ - b_1 - 1)/2 - 1$.*

Proof (Sketch). It is not hard to show that there is a global holomorphic model $M(f_*K)$ for the map $\tau : X_r(f) \to P_r(f)$ in a neighborhood of the fibers of excess dimension, i.e., $\tau^{-1}(s_K) = \mathbb{P}(f_*K)$. In other words, the obstruction computation reduces to algebraic geometry (hence our notation for the bundle $f_*K \to \mathbb{CP}^1$ with fiber $H^0(K_\Sigma)$). We have a sequence of holomorphic vector bundles

$$0 \to T(\mathbb{P}(f_*K)) \to TM(f_*K) \to T(W) \to \operatorname{coker}(D\tau) \to 0$$

where W is fiberwise dual to f_*K, but globally twisted by $\mathcal{O}(-2)$ (this is because $K_{\hat{X}}|f^{-1}(t)$ is not canonically $K_{f^{-1}(t)}$, introducing a twist into all the identifications). If τ was actually a submersion, clearly the obstruction bundle would be trivial (as ν_C would be pulled back from $\tau(C) = s_K \subset P_r(f)$); the deviation from this is measured by $H^1(\operatorname{coker}(D\tau))$. We compute this by splitting the above sequence into two short exact sequences and take the long exact sequences in cohomology. □

Summing up, when $b_+ > 1 + b_1(X)$, we have shown that the Gromov invariant for the class h is ± 1. (One can improve the constraint to $b_+ > 2$, but curiously it seems tricky to exactly reproduce Taubes' sharp bound $b_+ \geq 2$.)

The J-holomorphic curves in $X_r(f)$ give cycles in $\hat{X}$, but there is no reason these should be symplectic. We are trying to build standard surfaces, i.e., we need our cycles to have positive simple tangencies to fibers. This is exactly saying we want sections of $X_r(f)$ *transverse* to the diagonal locus Δ and intersecting it *positively*. To achieve this, we will need to construct almost complex structures which behave well with respect to Δ, which has a natural stratification by the combinatorial type of the tuple. Problems arise because we cannot a priori suppose that our sections do not have image entirely contained inside the diagonals.

As a first delicacy, the diagonal strata are not smoothly embedded, but are all the images of smooth maps which are generically bijective. This leads to "smooth models" of the diagonal strata, and for suitable J any holomorphic section will lift to a unique such model. (In particular, if a given section lies inside Δ because it contains multiple components, then we "throw away" this multiplicity; restoring it later will mean that we construct positive symplectic divisors, in the first instance, and not embedded submanifolds.) As a second delicacy, we have not controlled the index of sections once we lift them to smooth models of the strata. Finally, although we are restricting to the class of almost complex structures which are compatible with Δ, nonetheless we hope to find holomorphic sections transverse to Δ itself. The result we need, then, can loosely be summed up as:

Proposition 4.12 *There are "enough" almost complex structures compatible with the strata.*

A careful proof is given in [DS], Sects. 6–7. The key is that we can obtain deformations of such almost complex structures from vector fields on the fibers of f. Since we can find a vector field on $\mathbb{C}$ taking any prescribed values at a fixed set of points, once we restrict to an open subset of the graph of any given holomorphic section which lies in the top stratum of its associated model there are enough deformations to generate the whole tangent space to the space of J which are both fibered and compatible with Δ.

Remark. There are also strata coming from exceptional sections inside X, each with fiber Sym^{2g-3}. These also vary topologically trivially, and we repeat the above discussion for them. This lets us get standard surfaces in $\hat{X}\backslash \cup E_i \cong X\backslash\{b_j\}$.

The upshot is that the nonvanishing of the Gromov invariant for the class $h \in H_2(X_r(f))$ means we can find a holomorphic section of $X_r(f)$ which is transverse and positive to all strata of the diagonal in which it is not contained. Taking into account the strata coming from exceptional sections as well, and with care, one obtains a positive symplectic divisor $D = \sum a_i \Sigma_i$ in X in the homology class dual to K_X. We must finally check that we can satisfy the conditions of our earlier smoothing result. By adjunction,

$$D \cdot \Sigma_j + \Sigma_j^2 = 2g(\Sigma_j) - 2 = (1 + a_j)\Sigma_j^2 + \sum_{i \neq j} a_i\, \Sigma_i \cdot \Sigma_j$$

For $D \cdot \Sigma_j$ to be negative, it must be that Σ_j is a rational -1 curve; but clearly it suffices to prove the theorem under the assumption that X is minimal. Provided $b_+ > 1 + b_1(X)$, we therefore obtain an embedded symplectic surface in X representing the class K_X. (Strictly we have proven this theorem only when ω_X is rational, but one can remove this assumption, cf. [Sm4].)

This result gives "gauge-theory free" proofs of various standard facts on symplectic four-manifolds: for instance, if $b_+(X) > 1 + b_1(X)$ and X is minimal, then $c_1^2(X) \geq 0$. In particular, manifolds such as $K3\#K3\#K3$ admit no symplectic structure, illustrating the claim (made in the Introduction) that the symplectic condition cannot be reduced to cohomological or homotopical conditions, in contrast to the case of open manifolds covered by Gromov's h-principle.

To close, we point out that we can also see Taubes' remarkable duality $Gr(D) = \pm Gr(K - D)$ very geometrically in this picture. Let $\iota : Pic^{2g-2-r}(\Sigma) \to Pic^r(\Sigma)$ denote the map $O(D) \mapsto O(K - D)$ and write τ_d for the Abel-Jacobi map $Sym^d(\Sigma) \to Pic^d(\Sigma)$. If r is small relative to g, then generically we expect the map τ_r to be an embedding, and its image to be exactly where the map $\iota \circ \tau_{2g-2-r}$ has fibers of excess dimension. If we fix a homology class on X and take pencils of higher and higher degree, we can make r (the intersection number with the fiber) as small as we like relative to g (the genus of the fiber) – the latter grows quadratically, not linearly. Combining this with some deep results in the Brill–Noether theory of Riemann surfaces, one can show the following generalization of Proposition 4.8:

Proposition 4.13 ([Sm4]) *For r small enough relative to g, there exists J on $X_{2g-2-r}(f)$ such that $X_r(f)$ is embedded in $P_r(f) \cong P_{2g-2-r}(f)$, and such that $X_{2g-2-r}(f) \to P_{2g-2-r}(f)$ is holomorphic and locally trivial over $X_r(f)$ and its complement.*

Using this and repeating all the above, we can show there are well-defined integer-valued invariants $\mathcal{I}_f$ counting sections of the $X_r(f)$'s which satisfy the duality

$$\mathcal{I}_f(D) = \pm \mathcal{I}_f(K_X - D).$$

In fact, a smooth fibered J on X defines a canonical fibered $\mathbb{J}$ on $X_r(f)$ but this is only C^0 along the diagonal strata. Suitably interpreted, J-curves in X and $\mathbb{J}$-curves in $X_r(f)$ are tautologically equivalent. Motivated by this, and a (very rough) sketch showing $\mathcal{I}_f = Gr$ mod 2 for symplectic manifolds with $K_X = \lambda[\omega]$ for non-zero λ (more generally whenever there are no embedded square zero symplectic tori), [Sm4] conjectured that the $\mathcal{I}$-invariants and Taubes' Gromov invariants coincide. Recent work of Michael Usher [Us] clarifies and completes this circle of ideas by showing that in full generality $Gr = \mathcal{I}_f$ (in $\mathbb{Z}$) for any pencil f of sufficiently high degree. In this sense, the above equation shows that Taubes' duality can be understood in terms of Serre duality on the fibers of a Lefschetz fibration.

5 Fukaya Categories and Lefschetz Fibrations

Many central questions in symplectic topology revolve around Lagrangian submanifolds, their existence and their intersection properties. From a formal point of view these can be encoded into so-called Fukaya categories, which also play an important role of their own in Kontsevich's homological mirror symmetry conjecture. It turns out that Lefschetz pencils provide some of the most powerful tools available in this context, if one considers vanishing cycles and the "matching paths" between them [Se4]; the most exciting applications to this date are related to Seidel's construction of "directed Fukaya categories" of vanishing cycles [Se1, Se2], and to verifications of the homological mirror conjecture [Se5].

Fukaya categories of symplectic manifolds are intrinsically very hard to compute, because relatively little is known about embedded Lagrangian submanifolds in symplectic manifolds of dimension 4 or more, especially in comparison to the much better understood theory of coherent sheaves over complex varieties, which play the role of their mirror counterparts. The input provided by Lefschetz fibrations is a reduction from the symplectic geometry of the total space to that of the fiber, which in the four-dimensional case provides a crucial simplification.

5.1 Matching Paths and Lagrangian Spheres

Let $f : X \to S^2$ be a symplectic Lefschetz fibration (e.g., obtained by blowing up the base points of a Lefschetz pencil or, allowing the fibers to be open, by simply removing them), and let $\gamma \subset S^2$ be an embedded arc joining a regular value p_0 to a critical value p_1, avoiding all the other critical values of f. Using the horizontal distribution given by the symplectic orthogonal to the fibers, we can transport the vanishing cycle at p_1 along the arc γ to obtain a Lagrangian disc $D_\gamma \subset X$ fibered above γ, whose boundary is an embedded Lagrangian sphere S_γ in the fiber $\Sigma_0 = f^{-1}(p_0)$. The Lagrangian disc D_γ is called the *Lefschetz thimble* over γ, and its boundary S_γ is the vanishing cycle already considered in Sect. 2.

If we now consider an arc γ joining two critical values p_1, p_2 of f and passing through p_0, then the above construction applied to each half of γ yields two Lefschetz thimbles D_1 and D_2, whose boundaries are Lagrangian spheres $S_1, S_2 \subset \Sigma_0$. If S_1 and S_2 coincide exactly, then $D_1 \cup D_2$ is an embedded Lagrangian sphere in X, fibering above the arc γ (see the picture below); more generally, if S_1 and S_2 are Hamiltonian isotopic to each other, then perturbing slightly the symplectic structure we can reduce to the previous case and obtain again a Lagrangian sphere in X. The arc γ is called a *matching path* in the Lefschetz fibration f [Se4].

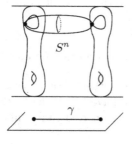

Matching paths are an important source of Lagrangian spheres, and more generally (extending suitably the notion of matching path to embedded arcs passing through several critical values of f) of embedded Lagrangian submanifolds. Conversely, a folklore theorem asserts that any given embedded Lagrangian sphere in a compact symplectic manifold is isotopic to one that fibers above a matching path in a Donaldson-type symplectic Lefschetz pencil of sufficiently high degree. More precisely, the following result holds (see [AMP] for a detailed proof):

Theorem 5.1 *Let L be a compact Lagrangian submanifold in a compact symplectic manifold (X, ω) with integral $[\omega]$, and let $h : L \to [0, 1]$ be any Morse function. Then for large enough k there exist Donaldson pencils $f_k : X - \{base\ points\} \to S^2$, embedded arcs $\gamma_k : [0, 1] \hookrightarrow S^2$, and Morse functions $h_k : L \to [0, 1]$ isotopic to h, such that the restriction of f_k to L is equal to $\gamma_k \circ h_k$.*

The intersection theory of Lagrangian spheres that fiber above matching paths is much nicer than that of arbitrary Lagrangian spheres, because if two Lagrangian spheres $S, S' \subset X$ fiber above matching paths γ, γ', then all intersections of S with S' lie in the fibers above the intersection points of γ with γ'. Hence, the Floer homology of S and S' can be computed by studying intersection theory for Lagrangian spheres in the fibers of f rather than in X. In particular, when X is a four-manifold the vanishing cycles are just closed loops in Riemann surfaces, and the computation of Floer homology essentially reduces to a combinatorial count.

The enumeration of matching paths, if possible, would lead to a complete understanding of isotopy classes of Lagrangian spheres in a given symplectic manifold, with various topological consequences and applications to the definition of new symplectic invariants. However, no finite-time algorithm is currently available for this problem, although some improvements on the "naive" search are possible (e.g., using maps to $\mathbb{CP}^2$ and projective duality to identify certain types of pencil automorphisms). Nonetheless, Theorem 5.1 roughly says that all Lagrangians are built out of Lefschetz thimbles. This implies that, at the formal level of (derived) Fukaya categories, it is sometimes possible to identify a "generating" collection of Lagrangian submanifolds (out of which all others can be built by gluing operations, or, in the language of categories, by passing to bounded complexes). A spectacular illustration is provided by Seidel's recent verification of homological mirror symmetry for quartic K3 surfaces [Se5].

5.2 Fukaya Categories of Vanishing Cycles

The above considerations, together with ideas of Kontsevich about mirror symmetry for Fano varieties, have led Seidel to the following construction of a Fukaya-type A_∞-category associated to a symplectic Lefschetz pencil f on a compact symplectic manifold (X, ω) [Se1]. Let f be a symplectic Lefschetz pencil determined by two sections s_0, s_1 of a sufficiently positive line bundle $L^{\otimes k}$ as in Theorem 2.9. Assume that $\Sigma_\infty = s_1^{-1}(0)$ is a smooth fiber of the pencil, and consider the symplectic manifold with boundary X^0 obtained from X by removing a suitable neighborhood of Σ_∞. The map f induces a Lefschetz fibration $f^0 : X^0 \to D^2$ over a disc, whose fibers are symplectic submanifolds with boundary obtained from the fibers of f by removing a neighborhood of their intersection points with the symplectic hypersurface Σ_∞ (the base points of the pencil). Choose a reference point $p_0 \in \partial D^2$, and consider the fiber $\Sigma_0 = (f^0)^{-1}(p_0) \subset X^0$.

Let $\gamma_1, \ldots, \gamma_r$ be a collection of arcs in D^2 joining the reference point p_0 to the various critical values of f^0, intersecting each other only at p_0, and ordered in the clockwise direction around p_0. As discussed above, each arc γ_i gives rise to a Lefschetz thimble $D_i \subset X^0$, whose boundary is a Lagrangian sphere $L_i \subset \Sigma_0$. To avoid having to discuss the orientation of moduli spaces, we give the following definition using $\mathbb{Z}_2$ (instead of $\mathbb{Z}$) as the coefficient ring [Se1]:

Definition 5.2 (Seidel) *The Fukaya category of vanishing cycles, denoted by $\mathcal{F}_{vc}(f; \{\gamma_i\})$, is a (directed) A_∞-category with r objects $L_1, \ldots, L_r$ (corresponding to the vanishing cycles, or more accurately to the thimbles); the morphisms between the objects are given by*

$$\text{Hom}(L_i, L_j) = \begin{cases} CF^*(L_i, L_j; \mathbb{Z}_2) = \mathbb{Z}_2^{|L_i \cap L_j|} & \text{if } i < j \\ \mathbb{Z}_2 \ id & \text{if } i = j \\ 0 & \text{if } i > j; \end{cases}$$

and the differential μ^1, composition μ^2 and higher order compositions μ^n are given by Lagrangian Floer homology inside Σ_0. More precisely,

$$\mu^n : \text{Hom}(L_{i_0}, L_{i_1}) \otimes \cdots \otimes \text{Hom}(L_{i_{n-1}}, L_{i_n}) \to \text{Hom}(L_{i_0}, L_{i_n})[2 - n]$$

is trivial when the inequality $i_0 < i_1 < \cdots < i_n$ fails to hold (i.e., it is always zero in this case, except for μ^2 where composition with an identity morphism is given by the obvious formula). When $i_0 < \cdots < i_n$, μ^n is defined by counting pseudo-holomorphic maps from the disc to Σ_0, mapping $n + 1$ cyclically ordered marked points on the boundary to the given intersection points between vanishing cycles, and the portions of boundary between them to $L_{i_0}, \ldots, L_{i_n}$.

One of the most attractive features of this definition is that it only involves Floer homology for Lagrangians inside the hypersurface Σ_0; in particular, when X is a symplectic four-manifold, the definition becomes purely combinatorial, since in the case of a Riemann surface the pseudo-holomorphic discs appearing in the definition of Floer homology and product structures are just immersed polygonal regions with convex corners.

From a technical point of view, a property that greatly facilitates the definition of Floer homology for the vanishing cycles L_i is *exactness*. Namely, the symplectic structure on the manifold X^0 is *exact*, i.e., it can be expressed as $\omega = d\theta$ for some 1-form θ (up to a scaling factor, θ is the 1-form describing the connection on $L^{\otimes k}$ in the trivialization of $L^{\otimes k}$ over $X - \Sigma_\infty$ induced by the section $s_1/|s_1|$). With this understood, the submanifolds L_i are all *exact Lagrangian*, i.e., the restriction $\theta_{|L_i}$ is not only closed ($d\theta_{|L_i} = \omega_{|L_i} = 0$) but also exact, $\theta_{|L_i} = d\phi_i$. Exactness has two particularly nice consequences. First, Σ^0 contains no closed pseudo-holomorphic curves (because the cohomology class of $\omega = d\theta$ vanishes). Secondly, there are no non-trivial pseudo-holomorphic discs in Σ_0 with boundary contained in one of the Lagrangian submanifolds L_i. Indeed, for any such disc D, we have $\text{Area}(D) = \int_D \omega = \int_{\partial D} \theta = \int_{\partial D} d\phi_i = 0$. Therefore, bubbling never occurs (neither in the interior nor on the boundary of the domain) in the moduli spaces used to define the Floer homology groups $HF(L_i, L_j)$. Moreover, the exactness of L_i provides a priori estimates on the area of all pseudo-holomorphic discs contributing to the definition of the products μ^n ($n \geq 1$); this implies the finiteness of the number of discs to be considered and solves elegantly the convergence problems that normally make it necessary to define Floer homology over Novikov rings.

Example. We now illustrate the above definition by considering the example of a pencil of degree 2 curves in $\mathbb{CP}^2$ (see also [Se2]). Consider the two sections $s_0 = x_0(x_1 - x_2)$ and $s_1 = x_1(x_2 - x_0)$ of the line bundle $O(2)$ over $\mathbb{CP}^2$: their zero sets are singular conics, in fact the unions of two lines each containing two of the four intersection points $(1:0:0)$, $(0:1:0)$, $(0:0:1)$, $(1:1:1)$. Moreover, the zero set of the linear combination $s_0 + s_1 = x_2(x_1 - x_0)$ is also singular; on the other hand, it is fairly easy to check that all other linear combinations $s_0 + \alpha s_1$ (for $\alpha \in \mathbb{CP}^1 - \{0, 1, \infty\}$) define smooth conics. Removing a neighborhood of a smooth fiber of the pencil generated by s_0 and s_1, we obtain a Lefschetz fibration over the disc, with fiber a sphere with four punctures. The three singular fibers of the pencil are nodal configurations consisting of two transversely intersecting spheres, with each component containing two of the four base points; each of the three different manners in which four points can be split into two groups of two is realized at one of the singular fibers. The following diagram represents the three singular conics of the pencil inside $\mathbb{CP}^2$ (left), and the corresponding vanishing cycles inside a smooth fiber (right):

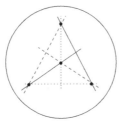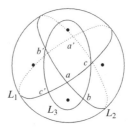

We can describe the monodromy of this Lefschetz pencil by a homomorphism $\psi : \pi_1(\mathbb{C} - \{p_1, p_2, p_3\}) \to \mathrm{Map}_{0,4}$ with values in the mapping class group of a genus 0 surface with 4 boundary components. After choosing a suitable ordered basis of the free group $\pi_1(\mathbb{C} - \{p_1, p_2, p_3\})$, we can make sure that ψ maps the generators to the Dehn twists τ_1, τ_2, τ_3 along the three loops shown on the diagram. On the other hand, because the normal bundles to the exceptional sections of the blown-up Lefschetz fibration have degree -1, the monodromy at infinity is given by the boundary twist $\prod \delta_i$, the product of the four Dehn twists along small loops encircling the four base points in the fiber; on the other hand it is also the product of the monodromies around each of the three singular fibers (τ_1, τ_2, τ_3). Hence, the monodromy of a pencil of conics in $\mathbb{CP}^2$ can be expressed by the relation $\prod \delta_i = \tau_1 \cdot \tau_2 \cdot \tau_3$ in the mapping class group $\mathrm{Map}_{0,4}$ (*lantern relation*).

Any two of the three vanishing cycles intersect transversely in two points, so $\mathrm{Hom}(L_1, L_2) = \mathbb{Z}_2 \, a \oplus \mathbb{Z}_2 \, a'$, $\mathrm{Hom}(L_2, L_3) = \mathbb{Z}_2 \, b \oplus \mathbb{Z}_2 \, b'$, and $\mathrm{Hom}(L_1, L_3) = \mathbb{Z}_2 \, c \oplus \mathbb{Z}_2 \, c'$ are all two-dimensional. There are no immersed 2-sided polygons in the punctured sphere Σ_0 with boundary in $L_i \cup L_j$ for any pair (i, j), since each of the four regions delimited by L_i and L_j contains one of the punctures, so $\mu^1 \equiv 0$. However, there are four triangles with boundary in $L_1 \cup L_2 \cup L_3$ (with vertices abc, $ab'c'$, $a'b'c$, $a'bc'$ respectively), and in each case the cyclic ordering of the boundary is compatible with the ordering of the vanishing cycles. Therefore, the composition of morphisms is given by the formulas $\mu^2(a, b) = \mu^2(a', b') = c$, $\mu^2(a, b') = \mu^2(a', b) = c'$. Finally, the higher compositions μ^n, $n \geq 3$ are all trivial in this category, because the ordering condition $i_0 < \cdots < i_n$ never holds [Se2].

The objects L_i of the category $\mathcal{F}_{vc}(f; \{\gamma_i\})$ actually correspond not only to Lagrangian spheres in Σ_0 (the vanishing cycles), but also to Lagrangian discs in X^0 (the Lefschetz thimbles D_i); and the Floer intersection theory in Σ_0 giving rise to $\mathrm{Hom}(L_i, L_j)$ and to the product structures can also be thought of in terms of intersection theory for the thimbles D_i in X^0 (this is actually the reason of the asymmetry between the cases $i < j$ and $i > j$ in Definition 5.2). In any case, the properties of these objects depend very much on the choice of the ordered collection of arcs $\{\gamma_i\}$. Therefore, $\mathcal{F}_{vc}(f; \{\gamma_i\})$ has little geometric meaning in itself, and should instead be viewed as a collection of *generators* of a much larger category which includes not only the Lefschetz thimbles, but also more general Lagrangian submanifolds of X^0. More precisely, the category naturally associated to the Lefschetz pencil f is not the finite directed A_∞-category defined above, but rather the (split-closed) *derived category* $D\mathcal{F}_{vc}(f)$ obtained from $\mathcal{F}_{vc}(f; \{\gamma_i\})$ by considering (twisted) complexes of formal direct sums of objects (also including idempotent splittings and formal inverses of quasi-isomorphisms). Replacing the ordered collection $\{\gamma_i\}$ by another one $\{\gamma_i'\}$ leads to a different "presentation" of the same derived category. Indeed, we have the following result [Se1]:

Theorem 5.3 (Seidel) *Given any two ordered collections $\{\gamma_i\}$ and $\{\gamma_i'\}$, the categories $\mathcal{F}_{vc}(f; \{\gamma_i\})$ and $\mathcal{F}_{vc}(f; \{\gamma_i'\})$ differ by a sequence of mutations (operations that modify the ordering of the objects while twisting some of them along others). Hence, the derived category $D\mathcal{F}_{vc}(f)$ does not depend on the choice of $\{\gamma_i\}$.*

Roughly speaking, complexes in the derived category correspond to Lagrangian cycles obtained by gluing the corresponding Lefschetz thimbles. For example, assume that the vanishing cycles L_i and L_j ($i < j$) are Hamiltonian isotopic to each other, so that (a smoothing of) $\gamma_i \cup \gamma_j$ is a matching path. Then $\mathrm{Hom}(L_i, L_j)$ has rank two, with one generator in degree 0 and one in degree $n - 1 = \dim L_i$; let a be the degree 0 generator. Then the complex $C = \{0 \to L_i \xrightarrow{a} L_j \to 0\}$, viewed as an object of the derived category, represents the Lagrangian sphere associated to the matching path $\gamma_i \cup \gamma_j$; for example it is easy to check that $\mathrm{Hom}_{D\mathcal{F}_{vc}}(C, C) \simeq H^*(S^n, \mathbb{Z}_2)$.

Building Fukaya-type categories out of vanishing cycles may seem arbitrary, but has a solid geometric underpinning, as suggested by the discussion at the end of Sect. 5.1. In fairly general circumstances, *every* closed Lagrangian submanifold (with well-defined and non-zero Floer homology with itself) of a Kähler manifold X must intersect one of the vanishing cycles of any Lefschetz pencil containing X as a smooth member. For the theory of Biran and Cieliebak [BC] shows that otherwise such a submanifold could be displaced off itself by a Hamiltonian isotopy (first by moving the Lagrangian into an open domain in X admitting a plurisubharmonic function with no top index critical points), a contradiction to the non-triviality of Floer homology. Seidel has pushed this even further: for certain K3 surfaces arising as anticanonical divisors in Fano threefolds, every Lagrangian submanifold L must have non-trivial Floer cohomology with one of the vanishing cycles of the pencil. Otherwise, by repeatedly applying the exact sequence in Floer cohomology, one sees that $HF(L, L)$ is graded isomorphic to a shifted version of itself, which is absurd. In this sense, the vanishing cycles really do "generate" the Fukaya category; if $HF(L, V) = 0$ for all vanishing cycles V, that is if L has no morphisms to any vanishing cycle, then $HF(L, L) = 0$ so L has no identity morphism, and must represent the zero object of the category.

5.3 Applications to Mirror Symmetry

The construction described above has various applications to homological mirror symmetry. In the context of Calabi–Yau manifolds, these have to do with a conjecture of Seidel about the relationship between the derived Fukaya category $D\mathcal{F}_{vc}(f)$ of the Lefschetz pencil f and the derived Fukaya category $D\mathcal{F}(X)$ of the closed symplectic manifold X [Se3]. As seen above, when passing to the derived category $D\mathcal{F}_{vc}(f)$, we hugely increase the number of objects, by considering not only the thimbles D_i but also arbitrary complexes obtained from them; this means that the objects of $D\mathcal{F}_{vc}(f)$ include all sorts of (not necessarily closed) Lagrangian submanifolds in X^0, with boundary in Σ_0. Since Fukaya categories are only concerned with closed Lagrangian submanifolds, it is necessary to consider a subcategory of $D\mathcal{F}_{vc}(f)$ whose objects correspond to the closed Lagrangian submanifolds in X^0 (i.e., combinations of D_i for which the boundaries cancel); it is expected that this can be done in purely categorical terms by considering those objects of $D\mathcal{F}_{vc}(f)$ on which the Serre functor acts simply by a shift. The resulting subcategory should be closely related to the derived Fukaya category of the open manifold X^0. This leaves us with the problem of relating $\mathcal{F}(X^0)$ with $\mathcal{F}(X)$. These two categories have the same objects and morphisms (Lagrangians in X can be made disjoint from Σ_∞), but the differentials and product structures differ. More precisely, the definition of μ^n in $\mathcal{F}(X^0)$ only involves counting pseudo-holomorphic discs contained in X^0, i.e., disjoint from the hypersurface Σ_∞. In order to account for the missing contributions, one should introduce a formal parameter q and count the pseudo-holomorphic discs with boundary in $\bigcup L_i$ that intersect Σ_∞ in m points (with multiplicities) with a coefficient q^m. The introduction of this parameter q leads to a *deformation of A_∞-structures*, i.e., an A_∞-category in which the differentials and products μ^n are defined over a ring of formal power series in the variable q; the limit $q = 0$ corresponds to the (derived) Fukaya category $D\mathcal{F}(X^0)$, while non-zero values of q are expected to yield $D\mathcal{F}(X)$.

These considerations provide a strategy to calculate Fukaya categories (at least for some examples) by induction on dimension [Se3]; an important recent development in this direction is Seidel's proof of homological mirror symmetry for quartic K3 surfaces [Se5].

Returning to more elementary considerations, another context in which the construction of Definition 5.2 is relevant is that of mirror symmetry for Fano manifolds ($c_1(TM) > 0$). Rather than manifolds, mirrors of Fano manifolds are *Landau–Ginzburg models*, i.e., pairs (Y, w), where Y is a non-compact manifold and $w : Y \to \mathbb{C}$, the "superpotential," is a holomorphic function. The complex (resp. symplectic) geometry of a Fano manifold M is then expected to correspond to the symplectic (resp. complex) geometry of the *critical points* of the superpotential w on its mirror. In particular, the homological mirror conjecture of Kontsevich now asserts that the categories $D^b\mathrm{Coh}(M)$ and $D\mathcal{F}_{vc}(w)$ should be equivalent. Following the ideas of various people (Kontsevich, Hori, Vafa, Seidel, etc.), the conjecture can be verified on many examples, at least in those cases where the critical points of w are isolated and non-degenerate.

For example, let a, b, c be three mutually prime positive integers, and consider the weighted projective plane $M = \mathbb{CP}^2(a, b, c) = (\mathbb{C}^3 - \{0\})/\mathbb{C}^*$, where $\mathbb{C}^*$ acts by $t \cdot (x, y, z) = (t^a x, t^b y, t^c z)$. In general M is a Fano orbifold (when $a = b = c = 1$ it is the usual projective plane). Consider the mirror Landau–Ginzburg model (Y, w), where $Y = \{x^a y^b z^c = 1\}$ is a hypersurface in $(\mathbb{C}^*)^3$, equipped with an exact Kähler form, and $w = x + y + z$. The superpotential w has $a + b + c$ isolated non-degenerate critical points, and hence determines an affine Lefschetz fibration to which we can apply the construction of

Definition 5.2. To be more precise, in this context one should actually define the Fukaya category of vanishing cycles over a coefficient ring R larger than $\mathbb{Z}_2$ or $\mathbb{Z}$, for example $R = \mathbb{C}$, counting each pseudo-holomorphic disc $u : D^2 \to \Sigma_0$ with a weight $\pm \exp(-\int_{D^2} u^*\omega)$, or a Novikov ring. Then we have the following result:

Theorem 5.4 ([AKO]) *The derived categories $D^b\mathrm{Coh}(M)$ and $D\mathcal{F}_{vc}(w)$ are equivalent.*

The proof relies on the identification of suitable collections of generators for both categories, and an explicit verification that the morphisms on both sides obey the same composition rules. Moreover, it can also be shown that non-exact deformations of the Kähler structure on Y correspond to non-commutative deformations of M, with an explicit relationship between the deformation parameter on M and the complexified Kähler class $[\omega + iB] \in H^2(Y, \mathbb{C})$ [AKO].

The homological mirror conjecture can be similarly verified for various other examples, and one may reasonably expect that, in the near future, vanishing cycles will play an important role in our understanding of mirror symmetry, not only for Fano and Calabi–Yau varieties, but maybe also for varieties of general type.

References

[ABKP]   J. Amorós, F. Bogomolov, L. Katzarkov, T. Pantev, *Symplectic Lefschetz fibrations with arbitrary fundamental groups*, J. Differ. Geom. **54** (2000), 489–545.

[AGTV]   M. Amram, D. Goldberg, M. Teicher, U. Vishne, *The fundamental group of the Galois cover of the surface $\mathbb{CP}^1 \times T$*, Alg. Geom. Topol. **2** (2002), 403–432.

[Au1]   D. Auroux, *Asymptotically holomorphic families of symplectic submanifolds*, Geom. Funct. Anal. **7** (1997), 971–995.

[Au2]   D. Auroux, *Symplectic 4-manifolds as branched coverings of $\mathbb{CP}^2$*, Invent. Math. **139** (2000), 551–602.

[Au3]   D. Auroux, *Symplectic maps to projective spaces and symplectic invariants*, Turkish J. Math. **25** (2001), 1–42 (`math.GT/0007130`).

[Au4]   D. Auroux, *Estimated transversality in symplectic geometry and projective maps*, "Symplectic Geometry and Mirror Symmetry", Proc. 4th KIAS International Conference, Seoul (2000), World Sci., Singapore, 2001, pp. 1–30 (`math.SG/0010052`).

[Au5]   D. Auroux, *A remark about Donaldson's construction of symplectic submanifolds*, J. Symplectic Geom. **1** (2002), 647–658 (`math.DG/0204286`).

[Au6]   D. Auroux, *Fiber sums of genus 2 Lefschetz fibrations*, Turkish J. Math. **27** (2003), 1–10 (`math.GT/0204285`).

[ADK1]   D. Auroux, S. K. Donaldson, L. Katzarkov, *Luttinger surgery along Lagrangian tori and non-isotopy for singular symplectic plane curves*, Math. Ann. **326** (2003), 185–203.

[ADK2]   D. Auroux, S. K. Donaldson, L. Katzarkov, *Singular Lefschetz pencils*, Geom. Topol. **9** (2005), 1043–1114.

[ADKY]   D. Auroux, S. K. Donaldson, L. Katzarkov, M. Yotov, *Fundamental groups of complements of plane curves and symplectic invariants*, Topology **43** (2004), 1285–1318.

[AK]   D. Auroux, L. Katzarkov, *Branched coverings of $\mathbb{CP}^2$ and invariants of symplectic 4-manifolds*, Invent. Math. **142** (2000), 631–673.

[AK2]   D. Auroux, L. Katzarkov, *The degree doubling formula for braid monodromies and Lefschetz pencils*, Pure Appl. Math. Q. **4** (2008), 237–318 (math.SG/0605001).

[AKO]   D. Auroux, L. Katzarkov, D. Orlov, *Mirror symmetry for weighted projective planes and their noncommutative deformations*, Ann. Math. **167** (2008), in press (math.AG/0404281).

[AMP]   D. Auroux, V. Muñoz, F. Presas, *Lagrangian submanifolds and Lefschetz pencils*, J. Symplectic Geom. **3** (2005), 171–219 (math.SG/0407126).

[Bi]   P. Biran, *A stability property of symplectic packing in dimension 4*, Invent. Math. **136** (1999), 123–155.

[BC]   P. Biran, K. Cieliebak, *Symplectic topology on subcritical manifolds*, Comment. Math. Helv. **76** (2001), 712–753.

[Bir]   J. Birman, *Braids, Links and Mapping class groups*, Annals of Math. Studies **82**, Princeton Univ. Press, Princeton, 1974.

[CMT]   C. Ciliberto, R. Miranda, M. Teicher, *Pillow degenerations of K3 surfaces*, Applications of Algebraic Geometry to Coding Theory, Physics, and Computation, NATO Science Series II, vol. **36**, 2001, pp. 53–63.

[Do1]   S. K. Donaldson, *Symplectic submanifolds and almost-complex geometry*, J. Differ. Geom. **44** (1996), 666–705.

[Do2]   S. K. Donaldson, *Lefschetz fibrations in symplectic geometry*, Documenta Math., Extra Volume ICM 1998, II, 309–314.

[Do3]   S. K. Donaldson, *Lefschetz pencils on symplectic manifolds*, J. Differ. Geom. **53** (1999), 205–236.

[DS]   S. Donaldson, I. Smith, *Lefschetz pencils and the canonical class for symplectic 4-manifolds*, Topology **42** (2003), 743–785.

[FO³]   K. Fukaya, Y.-G. Oh, H. Ohta, K. Ono, *Lagrangian intersection Floer theory: Anomaly and obstruction*, preprint.

[FS]   R. Fintushel, R. Stern, *Symplectic surfaces in a fixed homology class*, J. Differ. Geom. **52** (1999), 203–222.

[Gi]   E. Giroux, *Géométrie de contact: de la dimension trois vers les dimensions supérieures*, Proc. International Congress of Mathematicians, Vol. II (Beijing, 2002), Higher Ed. Press, Beijing, 2002, pp. 405–414.

[Go1]   R. E. Gompf, *A new construction of symplectic manifolds*, Ann. Math. **142** (1995), 527–595.

[Go2]   R. E. Gompf, *Symplectic structures from Lefschetz pencils in high dimensions*, Geom. Topol. Monogr. **7** (2004), 267–290 (math.SG/0210103, math.SG/0409370).

[GS]   R. E. Gompf, A. I. Stipsicz, *4-manifolds and Kirby calculus*, Graduate Studies in Math. **20**, Am. Math. Soc., Providence, 1999.

[Gr]   M. Gromov, *Pseudo-holomorphic curves in symplectic manifolds*, Invent. Math. **82** (1985), 307–347.

[IMP]   A. Ibort, D. Martinez-Torres, F. Presas, *On the construction of contact submanifolds with prescribed topology*, J. Differential Geom. **56** (2000), 235–283.

[KK]   V. Kharlamov, V. Kulikov, *On braid monodromy factorizations*, Izvestia Math. **67** (2003), 79–118 (math.AG/0302113).

[KM]   P. B. Kronheimer, T. S. Mrowka, *Witten's conjecture and property P*, Geom. Topol. **8** (2004), 295–310 (math.GT/0311489).

[Ku]   V. Kulikov, *On a Chisini conjecture*, Izvestia Math. **63** (1999), 1139–1170 (math.AG/9803144).

[McS]   D. McDuff and D. Salamon, *J-holomorphic curves and symplectic topology*, Am. Math. Soc. Colloq. Publ. **52**, Providence, 2004.

[Mo1]   B. Moishezon, *Complex surfaces and connected sums of complex projective planes*, Lecture Notes in Math. **603**, Springer, Heidelberg, 1977.

[Mo2]   B. Moishezon, *Stable branch curves and braid monodromies*, Algebraic Geometry (Chicago, 1980), Lecture Notes in Math. **862**, Springer, Heidelberg, 1981, pp. 107–192.

[Mo3]   B. Moishezon, *On cuspidal branch curves*, J. Algebraic Geom. **2** (1993), 309–384.

[Mo4]   B. Moishezon, *The arithmetic of braids and a statement of Chisini*, Geometric Topology (Haifa, 1992), Contemp. Math. **164**, Am. Math. Soc., Providence, 1994, pp. 151–175.

[Mo5]   B. Moishezon, *Topology of generic polynomial maps in complex dimension two*, preprint.

[MRT]   B. Moishezon, A. Robb, M. Teicher, *On Galois covers of Hirzebruch surfaces*, Math. Ann. **305** (1996), 493–539.

[Na]   H. Nakajima, *Lectures on Hilbert schemes of points on surfaces*, Univ. Lecture Series No. **18**, Am. Math. Soc., Providence, 1999.

[Oz]   B. Ozbagci, *Signatures of Lefschetz fibrations*, Pacific J. Math. **202** (2002), 99–118.

[OS]   B. Ozbagci, A. Stipsicz, *Noncomplex smooth 4-manifolds with genus 2 Lefschetz fibrations*, Proc. Am. Math. Soc. **128** (2000), 3125–3128.

[Pe]   T. Perutz, *Lagrangian matching invariants for fibred four-maniolds I & II*, Geom. Topol. **11** (2007), 759–828 (math.SG/0606061, math.SG/0606062).

[Ro]   A. Robb, *On branch curves of algebraic surfaces*, Singularities and Complex Geometry (Beijing, 1994), Am. Math. Soc./Int. Press Stud. Adv. Math. **5**, Am. Math. Soc., Providence, 1997, pp. 193–221.

[Se1]   P. Seidel, *Vanishing cycles and mutation*, Proc. 3rd European Congress of Mathematics (Barcelona, 2000), Vol. II, Progr. Math. **202**, Birkhäuser, Basel, 2001, pp. 65–85 (math.SG/0007115).

[Se2]   P. Seidel, *More about vanishing cycles and mutation*, "Symplectic Geometry and Mirror Symmetry", Proc. 4th KIAS International Conference, Seoul (2000), World Sci., Singapore, 2001, pp. 429–465 (math.SG/0010032).

[Se3]   P. Seidel, *Fukaya categories and deformations*, Proc. International Congress of Mathematicians, Vol. II (Beijing, 2002), Higher Ed. Press, Beijing, 2002, pp. 351–360.

[Se4]   P. Seidel, *Fukaya categories and Picard-Lefschetz theory*, in preparation.

[Se5]   P. Seidel, *Homological mirror symmetry for the quartic surface*, preprint (math.SG/0310414).

[Sh]   V. Shevchishin, *On the local version of the Severi problem*, Int. Math. Res. Not. (2004), 211–237 (`math.AG/0207048`).

[ST]   B. Siebert, G. Tian, *On the holomorphicity of genus two Lefschetz fibrations*, Ann. Math. **161** (2005), 959–1020 (`math.SG/0305343`).

[Sm1]   I. Smith, *Lefschetz fibrations and the Hodge bundle*, Geom. Topol. **3** (1999), 211–233.

[Sm2]   I. Smith, *Lefschetz pencils and divisors in moduli space*, Geom. Topol. **5** (2001), 579–608.

[Sm3]   I. Smith, *Geometric monodromy and the hyperbolic disc*, Quarterly J. Math. **52** (2001), 217–228 (`math.SG/0011223`).

[Sm4]   I. Smith, *Serre-Taubes duality for pseudoholomorphic curves*, Topology **42** (2003), 931–979.

[Ta1]   C. H. Taubes, *The Seiberg–Witten and the Gromov invariants*, Math. Res. Lett. **2** (1995), 221–238.

[Ta2]   C. H. Taubes, *The geometry of the Seiberg–Witten invariants*, Surveys in Differential Geometry, Vol. III (Cambridge, 1996), Int. Press, Boston, 1998, pp. 299–339.

[Te1]   M. Teicher, *Braid groups, algebraic surfaces and fundamental groups of complements of branch curves*, Algebraic Geometry (Santa Cruz, 1995), Proc. Sympos. Pure Math., **62** (part 1), Amer. Math. Soc., Providence, 1997, pp. 127–150.

[Te2]   M. Teicher, *New invariants for surfaces*, Tel Aviv Topology Conference: Rothenberg Festschrift (1998), Contemp. Math. **231**, Am. Math. Soc., Providence, 1999, pp. 271–281 (`math.AG/9902152`).

[Te3]   M. Teicher, *The fundamental group of a $\mathbb{CP}^2$ complement of a branch curve as an extension of a solvable group by a symmetric group*, Math. Ann. **314** (1999) 19–38.

[Th]   W. Thurston, *Some simple examples of symplectic manifolds*, Proc. Am. Math. Soc. **55** (1976), 467–468.

[Us]   M. Usher, *The Gromov invariant and the Donaldson-Smith standard surface count*, Geom. Topol. **8** (2004), 565–610 (`math.SG/0310450`).

Differentiable and Deformation Type of Algebraic Surfaces, Real and Symplectic Structures

Fabrizio Catanese

Lehrstuhl Mathematik VIII, Universität Bayreuth, NWII D-95440 Bayreuth,
Germany
fabrizio.catanese@uni-bayreuth.de

> Che differenza c'e' fra il palo, il Paolo e la banana?
> *(G. Lastrucci, Firenze, 9/11/62)*

1 Introduction

As already announced in Cetraro at the beginning of the C.I.M.E. course, we deflected from the broader target 'Classification and deformation types of complex and real manifolds', planned and announced originally.

First of all, the lectures actually delivered focused on the intersection of the above vast area with the theme of the School, 'Algebraic surfaces and symplectic 4-manifolds'.

Hence the title of the Lecture Notes has been changed accordingly.

Moreover, the Enriques classification of real algebraic surfaces is not touched upon here, and complex conjugation and real structures appear mostly through their relation to deformation types of complex manifolds, and in particular through their relation with strong and weak rigidity theorems.

In some sense then this course is a continuation of the C.I.M.E. course I held some 20 years ago in Montecatini [Cat88], about 'Moduli of algebraic surfaces'.

But whereas those Lecture Notes had an initial part of considerable length which was meant to be a general introduction to complex deformation theory, here the main results of deformation theory which we need are only stated.

Nevertheless, because the topic can be of interest not only to algebraic geometers, but also to people working in differential or symplectic topology, we decided to start dedicating the first lecture to recalling basic notions concerning projective and Kähler manifolds. Especially, we recall the main principles of classification theory, and state the Enriques classification of algebraic surfaces of special type.

Since surfaces of general type and their moduli spaces are a major theme for us here, it seemed worthwhile to recall in detail in lecture two the structure of their canonical models, in particular of their singularities, the socalled Rational Double Points, or Kleinian quotient singularities. The rest of lecture two is devoted to proving Bombieri's theorem on pluricanonical embeddings, to the analysis of other quotient singularities, and to the deformation equivalence relation (showing that two minimal models are deformation equivalent iff the respective canonical models are). Bombieri's theorem is proven in every detail for the case of an ample canonical divisor, with the hope that some similar result may soon be proven also in the symplectic case.

In lecture three we show first that deformation equivalence implies diffeomorphism, and then, using a result concerning symplectic approximations of projective varieties with isolated singularities and Moser's theorem, we show that a surfaces of general type has a 'canonical symplectic structure', i.e., a symplectic structure whose class is the class of the canonical divisor, and which is unique up to symplectomorphism.

In lecture three and the following ones we thus enter 'in medias res', since one of the main problems that we discuss in these Lecture Notes is the comparison of differentiable and deformation type of minimal surfaces of general type, keeping also in consideration the canonical symplectic structure (unique up to symplectomorphism and invariant for smooth deformation) which these surfaces possess.

We present several counterexamples to the DEF = DIFF speculation of Friedman and Morgan [F-M88] that deformation type and diffeomorphism type should coincide for complex algebraic surfaces. The first ones were obtained by Manetti [Man01], and exhibit non simply connected surfaces which are pairwise not deformation equivalent. We were later able to show that they are canonically symplectomorphic (see [Cat02] and also [Cat06]). An account of these results is to be found in Chap. 6, which is an extra chapter with title 'Epilogue' (we hope however that this title may soon turn out to be inappropriate in view of future further developments).

In lecture 4, after discussing some classical results (like the theorem of Castelnuovo and De Franchis) and some 'semi-classical' results (by the author) concerning the topological characterization of irrational pencils on Kähler manifolds and algebraic surfaces, we discuss orbifold fundamental groups and triangle covers.

We use the above results to describe varieties isogenous to a product. These yield several examples of surfaces not deformation equivalent to their complex conjugate surface. We describe in particular the examples by the present author [Cat03], by Bauer–Catanese–Grunewald [BCG05], and then the ones by Kharlamov–Kulikov [KK02] which yield ball quotients. In this lecture we discuss complex conjugation and real structures, starting from elementary examples and ending with a survey of recent results and with open problems on the theory of 'Beauville surfaces'.

The beginning of lecture 5 is again rather elementary, it discusses connected sums and other surgeries, like fibre sums, and recalls basic definitions and results on braid groups, mapping class groups and Hurwitz equivalence.

After recalling the theory of Lefschetz pencils, especially the differentiable viewpoint introduced by Kas [Kas80], we recall Freedman's basic results on the topology of simply connected compact (oriented) fourmanifolds (see [F-Q90]).

We finally devote ourselves to our main objects of investigation, namely, the socalled '(abc)-surfaces' (introduced in [Cat02]), which are simply connected. We finish Lecture 5 explaining our joint work with Wajnryb [CW04] dedicated to the proof that these last surfaces are diffeomorphic to each other when the two integers b and $a + c$ are fixed.

In Chap. 6 we sketch the proof that these, under suitable numerical conditions, are not deformation equivalent. A result which is only very slightly weaker is explained in the Lecture Notes by Manetti, but with many more details; needless to say, we hope that the combined synergy of the two Lecture Notes may turn out to be very useful for the reader in order to appreciate the long chain of arguments leading to the theorem that the abc-surfaces give us the simply connected counterexamples to a weaker version of the DEF= DIFF question raised by Friedman and Morgan in [F-M88].

An interesting question left open (in spite of previous optimism) concerns the canonical symplectomorphism of the (abc)-surfaces. We discuss this and other problems, related to the connected components of moduli spaces of surfaces of general type, and to the corresponding symplectic structures, again in Chap. 6.

The present text not only expands the contents of the five lectures actually held in Cetraro. Indeed, since otherwise we would not have reached a satisfactory target, we added the extra Chap. 6.

As we already mentioned, since the course by Manetti does not explain the construction of his examples (which are here called Manetti surfaces), we give a very brief overview of the construction, and sketch a proof of the canonical symplectomorphism of these examples.

2 Lecture 1: Projective and Kähler Manifolds, the Enriques Classification, Construction Techniques

2.1 Projective Manifolds, Kähler and Symplectic Structures

The basic interplay between complex algebraic geometry, theory of complex manifolds, and theory of real symplectic manifolds starts with projective manifolds.

We consider a closed connected $\mathbb{C}$-submanifold $X^n \subset \mathbb{P}^N := \mathbb{P}^N_{\mathbb{C}}$.

This means that, around each point $p \in X$, there is a neighbourhood U_p of p and a permutation of the homogeneous coordinates such that, setting

$$x_0 = 1, \ x' := (x_1, \ldots x_n), \ x'' := (x_{n+1}, \ldots x_N),$$

the intersection $X \cap U_p$ coincides with the graph of a holomorphic map Ψ:

$$X \cap U_p = \{(x', x'') \in U_p | x'' = \Psi(x')\}.$$

We can moreover assume, after a linear change of the homogeneous coordinates, that the Taylor development of Ψ starts with a second order term (i.e., p is the point $(1, 0, \ldots 0)$ and the projective tangent space to X at p is the complex subspace $\{x'' = 0\}$.

Definition 2.1 *The Fubini-Study form is the differential 2-form*

$$\omega_{FS} := \frac{i}{2\pi} \partial \bar{\partial} log |z|^2,$$

where z is the homogeneous coordinate vector representing a point of $\mathbb{P}^N$.
   In fact the above 2-form on $\mathbb{C}^{N+1} \backslash \{0\}$ is invariant:
   (1) For the action of $\mathbb{U}(N, \mathbb{C})$ on homogeneous coordinate vectors
   (2) For multiplication of the vector z by a nonzero holomorphic scalar function $f(z)$ (z and $f(z)z$ represent the same point in $\mathbb{P}^N$), hence
   (3) ω_{FS} descends to a differential form on $\mathbb{P}^N$ (being $\mathbb{C}^*$-invariant)

The restriction ω of the Fubini-Study form to a submanifold X of $\mathbb{P}^n$ makes the pair (X, ω) a Kähler manifold according to the following

Definition 2.2 *A pair (X, ω) of a complex manifold X, and a real differential 2-form ω is called a Kähler pair if*
   (i) ω is closed ($d\omega = 0$)
   (ii) ω is of type $(1,1) \Leftrightarrow$ for each pair of tangent vectors v, w one has (J being the operator on complex tangent vectors given by multiplication by $i = \sqrt{-1}$),

$$\omega(Jv, Jw) = \omega(v, w)$$

   (iii) the associated Hermitian form is strictly positive definite $\Leftrightarrow$ the real symmetric bilinear form $\omega(v, Jw)$ is positive definite

The previous definition becomes clearer if one recalls the following easy bilinear algebra lemma.

Lemma 2.3 *Let V be a complex vector space, and H a Hermitian form. Then, decomposing H in real and imaginary part,*

$$H = S + \sqrt{-1}A,$$

we have that S is symmetric, A is alternating, $S(u, v) = A(u, Jv)$ and $A(Ju, Jv) = A(u, v)$.
   Conversely, given a real bilinear and alternating form A, A is the imaginary part of a Hermitian form $H(u, v) = A(u, Jv) + \sqrt{-1}A(u, v)$ if and only if A satisfies the socalled first Riemann bilinear relation:

$$A(Ju, Jv) = A(u, v).$$

Observe that property (iii) implies that ω is nondegenerate (if in the previous lemma S is positive definite, then A is nondegenerate), thus a Kähler pair yields a symplectic manifold according to the standard definition

Definition 2.4 *A pair* (X, ω) *consisting of a real manifold* X, *and a real differential 2-form* ω *is called a* symplectic pair *if*

(i) ω *is a symplectic form, i.e.,* ω *is closed* $(d\omega = 0)$ *and* ω *is nondegenerate at each point (thus* X *has even dimension).*

A symplectic pair (X, ω) *is said to be* integral *iff the De Rham cohomology class of* ω *comes from* $H^2(X, \mathbb{Z})$, *or, equivalently, there is a complex line bundle* L *on* X *such that* ω *is a first Chern form of* L.

An almost complex structure J *on* X *is a differentiable endomorphism of the real tangent bundle of* X *satisfying* $J^2 = -Id$. *It is said to be*

(ii) compatible *with* ω *if*

$$\omega(Jv, Jw) = \omega(v, w),$$

(iii) tame *if the quadratic form* $\omega(v, Jv)$ *is strictly positive definite.*

Finally, a symplectic manifold is a manifold admitting a symplectic form ω.

Observe that compatibility and tameness are the symplectic geometry translation of the two classical Riemann bilinear relations which ensure the existence of a hermitian form, respectively the fact that the latter is positive definite: the point of view changes mainly in the order of the choice for J, resp. ω.

Definition 2.5 *A submanifold* Y *of a symplectic pair* (X, ω) *is a* symplectic submanifold *if* $\omega|_Y$ *is nondegenerate.*

Let (X', ω') *be another symplectic pair. A diffeomorphism* $f : X \to X'$ *is said to be a* symplectomorphism *if* $f^*(\omega') = \omega$.

Thus, unlike the Kähler property for complex submanifolds, the symplectic property is not automatically inherited by submanifolds of even real dimension.

A first intuition about symplectic submanifolds is given by the following result, which holds more generally on any Kähler manifold, and says that a good differentiable approximation of a complex submanifold is a symplectic submanifold.

Lemma 2.6 *Let* $W \subset \mathbb{P}^N$ *be a differentiable submanifold of even dimension* $(dimW = 2n)$, *and assume that the tangent space of* W *is 'close to be complex' in the sense that for each vector* v *tangent to* W *there is another vector* v' *tangent to* W *such that*

$$Jv = v' + u, |u| < |v|.$$

Then the restriction to W *of the Fubini Study form* ω_{FS} *makes* W *a symplectic submanifold of* $\mathbb{P}^N$.

Proof. Let A be the symplectic form on projective space, so that for each vector v tangent to W we have:

$|v|^2 = A(v, Jv) = A(v, v') + A(v, u)$.

Since $|A(v, u)| < |v|^2$, $A(v, v') \neq 0$ and A restricts to a nondegenerate form. □

The above intuition does not hold globally, since it was observed by Thurston [Thur76] that there are symplectic complex manifolds which are not Kähler. The first example of this situation was indeed given by Kodaira [Kod66] who described the socalled Kodaira surfaces $\mathbb{C}^2/\Gamma$, which are principal holomorphic bundles with base and fibre an elliptic curve (they are not Kähler since their first Betti number equals 3). Many more examples have been given later on.

To close the circle between the several notions, there is the following characterization of a Kähler manifold (the full statement is very often referred to as 'folklore', but it follows from the statements contained in Theorem 3.13, page 74 of [Vois02], and Proposition 4.A.8, page 210 of [Huy05]).

Kähler manifolds Theorem *Let (X, ω) be a symplectic pair, and let J be an almost complex structure which is compatible and tame for ω. Let $g(u, v) := \omega(u, Jv)$ be the associated Riemannian metric. Then J is parallel for the Levi Civita connection of g (i.e., its covariant derivative is zero in each direction) if and only if J is integrable (i.e., it yields a complex structure) and ω is a Kähler form.*

Returning to the Fubini-Study form, it has an important normalization property, namely, if we consider a linear subspace $\mathbb{P}^m \subset \mathbb{P}^N$ (it does not matter which one, by the unitary invariance mentioned in (1) above), then integration in pluripolar coordinates yields

$$\int_{\mathbb{P}^m} \frac{1}{m!} \omega_{FS}^m = 1.$$

The above equation, together with Stokes' Lemma, and a multilinear algebra calculation for which we refer for instance to Mumford's book [Mum76] imply

Wirtinger's Theorem *Let $X := X^n$ be a complex submanifold of $\mathbb{P}^N$. Then X is a volume minimizing submanifold for the n-dimensional Riemannian volume function of submanifolds M of real dimension $2n$,*

$$vol(M) := \int dVol_{FS},$$

where $dVol_{FS} = \sqrt{det(g_{ij})(x)} \, |dx|$ is the volume measure of the Riemannian metric $g_{ij}(x)$ associated to the Fubini Study form. Moreover, the global volume of X equals a positive integer, called the degree *of X.*

The previous situation is indeed quite more general:

Let (X, ω) be a symplectic manifold, and let Y be an oriented submanifold of even dimension $= 2m$: then the global symplectic volume of Y

$vol(Y) := \int_Y \frac{1}{n!}\omega^m$ depends only on the homology class of Y, and will be an integer if the pair (X, ω) is integral (i.e., if the De Rham class of ω comes from $H^2(X, \mathbb{Z})$).

If moreover X is Kähler, and Y is a complex submanifold, then Y has a natural orientation, and one has the

Basic principle of Kähler geometry: *Let Y be a compact submanifold of a Kähler manifold X: then $vol(Y) := \int_Y \omega^m > 0$, and in particular the cohomology class of Y in $H^{2m}(X, \mathbb{Z})$ is nontrivial.*

The main point of the basic principle is that the integrand of $vol(Y) := \int_Y \omega^m$ is pointwise positive, because of condition (iii). So we see that a similar principle holds more generally if we have a symplectic manifold X and a compact submanifold Y admitting an almost complex structure compatible and tame for the restriction of ω to Y.

Wirtinger's theorem and the following theorem of Chow provide the link with algebraic geometry mentioned in the beginning.

Chow's Theorem *Let $X := X^n$ be a (connected) complex submanifold of $\mathbb{P}^N$. Then X is an algebraic variety, i.e., X is the locus of zeros of a homogeneous prime ideal $\mathcal{P}$ of the polynomial ring $\mathbb{C}[x_0, \ldots x_N]$.*

We would now like to show how Chow's theorem is a consequence of another result:

Siegel's Theorem *Let $X := X^n$ be a compact (connected) complex manifold of (complex) dimension n. Then the field $\mathbb{C}^{Mer}(X)$ of meromorphic functions on X is finitely generated, and its transcendence degree over $\mathbb{C}$ is at most n.*

The above was proven by Siegel just using the lemma of Schwarz and an appropriate choice of a finite cover of a compact complex manifold made by polycylinder charts (see [Sieg73], or [Corn76]).

Idea of proof of Chow's theorem.

Let $p \in X$ and take coordinates as in 2.1: then we have an injection $\mathbb{C}(x_1, \ldots x_n) \hookrightarrow \mathbb{C}^{Mer}(X)$, thus $\mathbb{C}^{Mer}(X)$ has transcendency degree n by Siegel's theorem.

Let Z be the Zariski closure of X: this means that Z is the set of zeros of the homogeneous ideal $\mathcal{I}_X \subset \mathbb{C}[x_0, \ldots x_N]$ generated by the homogeneous polynomials vanishing on X.

Since X is connected, it follows right away, going to nonhomogeneous coordinates and using that the ring of holomorphic functions on a connected open set is an integral domain, that the ideal $\mathcal{I}_X = \mathcal{I}_Z$ is a prime ideal.

We consider then the homogeneous coordinate ring $\mathbb{C}[Z] := \mathbb{C}[x_0, \ldots x_N]/\mathcal{I}_X$ and the field of rational functions $\mathbb{C}(Z)$, the field of the fractions of the integral domain $\mathbb{C}[Z]$ which are homogeneous of degree 0. We observe that we have an injection $\mathbb{C}(Z) \hookrightarrow \mathbb{C}^{Mer}(X)$.

Therefore $\mathbb{C}(x_1, \ldots x_n) \hookrightarrow \mathbb{C}(Z) \hookrightarrow \mathbb{C}^{Mer}(X)$. Thus the field of rational functions $\mathbb{C}(Z)$ has transcendency degree n and Z is an irreducible algebraic subvariety of $\mathbb{P}^N$ of dimension n. Since the smooth locus $Z^* := Z \backslash Sing(Z)$ is dense in Z for the Hausdorff topology, is connected, and contains X, it follows that $X = Z$. $\square$

The above theorem extends to the singular case: a closed complex analytic subspace of $\mathbb{P}^N$ is also a closed set in the Zariski topology, i.e., a closed algebraic set.

We have seen in the course of the proof that the dimension of an irreducible projective variety is given by the transcendency degree over $\mathbb{C}$ of the field $\mathbb{C}(Z)$ (which, by a further extension of Chow's theorem, equals $\mathbb{C}^{Mer}(Z)$).

The degree of Z is then defined through the

Emmy Noether Normalization Lemma. *Let Z be an irreducible subvariety of $\mathbb{P}^N$ of dimension n: then for general choice of independent linear forms $(x_0, \ldots x_n)$ one has that the homogeneous coordinate ring of Z, $\mathbb{C}[Z] := \mathbb{C}[x_0, \ldots x_N]/\mathcal{I}_Z$ is an integral extension of $\mathbb{C}[x_0, \ldots x_n]$. One can view $\mathbb{C}[Z]$ as a torsion free $\mathbb{C}[x_0, \ldots x_n]$-module, and its rank is called the degree d of Z.*

The geometrical consequences of Noether's normalization are (see [Shaf74]):

- The linear projection with centre $L := \{x | x_0 = \ldots x_n = 0\}$, $\pi_L : \mathbb{P}^N \setminus L \to \mathbb{P}^n$ is defined on Z since $Z \cap L = \emptyset$, and $\pi := \pi|_L : X \to \mathbb{P}^n$ is surjective and finite.
- For $y \in \mathbb{P}^n$, the finite set $\pi^{-1}(y)$ has cardinality at most d, and equality holds for y in a Zariski open set $U \subset \mathbb{P}^n$.

The link between the volume theoretic and the algebraic notion of degree is easily obtained via the Noether projection π_L.

In fact, the formula $(x_0, x', x'') \to (x_0, x', (1-t)x'')$ provides a homotopy between the identity map of Z and a covering of $\mathbb{P}^n$ of degree d, by which it follows that $\int_{Z*} \omega_{FS}^n$ converges and equals precisely d.

We end this subsection by fixing the standard notation: for X a projective variety, and x a point in X we denote by $\mathcal{O}_{X,x}$ the local ring of algebraic functions on X regular in x, i.e.,

$$\mathcal{O}_{X,x} := \{f \in \mathbb{C}(X) | \exists a, b \in \mathbb{C}[X], \text{homogeneous}, s.t. f = a/b \text{ and } b(x) \neq 0\}.$$

This local ring is contained in the local ring of restrictions of local holomorphic functions from $\mathbb{P}^N$, which we denote by $\mathcal{O}_{X,x}^h$.

The pair $\mathcal{O}_{X,x} \subset \mathcal{O}_{X,x}^h$ is a faithfully flat ring extension, according to the standard

Definition 2.7 *A ring extension $A \to B$ is said to be flat, respectively faithfully flat, if the following property holds: a complex of A-modules (M_i, d_i) is exact only if (respectively, if and only if) $(M_i \otimes_A B, d_i \otimes_A B)$ is exact.*

This basic algebraic property underlies the so called (see [Gaga55-6]).

G.A.G.A. Principle. *Given a projective variety, and a coherent (algebraic) $\mathcal{O}_X$-sheaf $\mathcal{F}$, let $\mathcal{F}^h := \mathcal{F} \otimes_{\mathcal{O}_X} \mathcal{O}_X^h$ be the corresponding holomorphic coherent sheaf: then one has a natural isomorphism of cohomology groups*

$$H^i(X_{Zar}, \mathcal{F}) \cong H^i(X_{Haus}, \mathcal{F}^h),$$

where the left hand side stands for Čech cohomology taken in the Zariski topology, the right hand side stands for Čech cohomology taken in the Hausdorff topology. The same holds replacing $\mathcal{F}$ by $\mathcal{O}_X^*$.

Due to the GAGA principle, we shall sometimes make some abuse of notation, and simply write, given a divisor D on X, $H^i(X, D)$ instead of $H^i(X, \mathcal{O}_X(D))$.

2.2 The Birational Equivalence of Algebraic Varieties

A rational map of a (projective) variety $\phi : X \dashrightarrow \mathbb{P}^N$ is given through N rational functions $\phi_1, \ldots \phi_N$.

Taking a common multiple s_0 of the denominators b_j of $\phi_j = a_j/b_j$, we can write $\phi_j = s_j/s_0$, and write $\phi = (s_0, \ldots s_N)$, where the s_j's are all homogeneous of the same degree, whence they define a graded homomorphism $\phi^* : \mathbb{C}[\mathbb{P}^N] \to \mathbb{C}[X]$.

The kernel of ϕ^* is a prime ideal, and its zero locus, denote it by Y, is called the image of ϕ, and we say that X dominates Y.

One says that ϕ is a morphism in p if there is such a representation $\phi = (s_0, \ldots s_N)$ such that some $s_j(p) \neq 0$. One can see that there is a maximal open set $U \subset X$ such that ϕ is a morphism on U, and that $Y = \overline{\phi(U)}$.

If the local rings $\mathcal{O}_{X,x}$ are factorial, in particular if X is smooth, then one can take at each point x relatively prime elements a_j, b_j, let s_0 be the least common multiple of the denominators, and it follows then that the *Indeterminacy Locus* $X \backslash U$ is a closed set of codimension at least 2. In particular, every rational map of a smooth curve is a morphism.

Definition 2.8 *Two algebraic varieties X, Y are said to be* birational *iff their function fields $\mathbb{C}(X), \mathbb{C}(Y)$ are isomorphic, equivalently if there are two dominant rational maps $\phi : X \dashrightarrow Y, \psi : Y \dashrightarrow X$, which are inverse to each other. If $\phi, \psi = \phi^{-1}$ are morphisms, then X and Y are said to be* isomorphic.

By Chow's theorem, biholomorphism and isomorphism is the same notion for projective varieties (this ceases to be true in the non compact case, cf. [Ser59]).

Over the complex numbers, we have [Hir64].

Hironaka's theorem on resolution of singularities. *Every projective variety is birational to a smooth projective variety.*

As we already remarked, two birationally equivalent curves are isomorphic, whereas for a smooth surface S, and a point $p \in S$, one may consider the blow-up of the point p, $\pi : \hat{S} \to S$. $\hat{S}$ is obtained glueing together $S \backslash \{p\}$ with the closure of the projection with centre p, $\pi_p : S \backslash \{p\} \to \mathbb{P}^{N-1}$. One can moreover show that $\hat{S}$ is projective. The result of blow up is that the point p is replaced by the projectivization of the tangent plane to S at p, which is a curve $E \cong \mathbb{P}^1$, with normal sheaf $\mathcal{O}_E(E) \cong \mathcal{O}_{\mathbb{P}^1}(-1)$. In other words, the selfintersection of E, i.e., the degree of the normal bundle of E, is -1, and we simply say that E is an *Exceptional curve of the I Kind*.

Theorem of Castelnuovo and Enriques. *Assume that a smooth projective surface Y contains an irreducible curve $E \cong \mathbb{P}^1$ with selfintersection $E^2 = -1$: then there is a birational morphism $f : Y \to S$ which is isomorphic to the blow up $\pi : \hat{S} \to S$ of a point p (in particular E is the only curve contracted to a point by f).*

The previous theorem justifies the following

Definition 2.9 *A smooth projective surface is said to be* minimal *if it does not contain any exceptional curve of the I kind.*

One shows then that every birational transformation is a composition of blow ups and of inverses of blow ups, and each surface X is birational to a smooth minimal surface S. This surface S is unique, up to isomorphism, if X is not ruled (i.e., not birational to a product $C \times \mathbb{P}^1$), by the classical

Theorem of Castelnuovo. *Two birational minimal models S, S' are isomorphic unless they are birationally* ruled, *i.e., birational to a product $C \times \mathbb{P}^1$, where C is a smooth projective curve. In the ruled case, either $S \cong \mathbb{P}^2$, or S is isomorphic to the projectivization $\mathbb{P}(V)$ of a rank 2 vector bundle V on C.*

Recall now that a variety X is smooth if and only if the sheaf of differential forms Ω_X^1 is locally free, and locally generated by $dx_1, \ldots dx_n$, if $x_1, \ldots x_n$ yield local holomorphic coordinates.

The vector bundle (locally free sheaf) Ω_X^1 and its associated bundles provide birational invariants in view of the classical [B-H75].

Kähler's lemma. *Let $f : X^n \dashrightarrow Y^m$ be a dominant rational map between smooth projective varieties of respective dimensions n, m. Then one has injective pull back linear maps $H^0(Y, \Omega_Y^{1 \otimes r}) \to H^0(X, \Omega_X^{1 \otimes r})$. Hence the vector spaces $H^0(X, \Omega_X^{1 \otimes r_1} \otimes \cdots \otimes \Omega_X^{n \otimes r_n})$ are birational invariants.*

Of particular importance is the top exterior power $\Omega_X^n = \Lambda^n(\Omega_X^1)$, which is locally free of rank 1, thus can be written as $\mathcal{O}_X(K_X)$ for a suitable Cartier divisor K_X, called the canonical divisor, and well defined only up to linear equivalence.

Definition 2.10 *The ith* pluriirregularity *of a smooth projective variety X is the dimension $h^{0,i} := dim(H^i(X, \mathcal{O}_X))$, which by Hodge Theory equals $dim(H^0(X, \Omega_X^i))$. The mth* plurigenus P_m *is instead the dimension $P_m(X) := dim(H^0(X, \Omega_X^{n \otimes m})) = h^0(X, mK_X)$.*

A finer birational invariant is the canonical ring of X.

Definition 2.11 *The* canonical ring *of a smooth projective variety X is the graded ring*

$$\mathcal{R}(X) := \bigoplus_{m=0}^{\infty} H^0(X, mK_X).$$

If $\mathcal{R}(X) = \mathbb{C}$ one defines $Kod(X) = -\infty$, otherwise the Kodaira dimension of X is defined as the transcendence degree over $\mathbb{C}$ of the canonical subfield of $\mathbb{C}(X)$, given by the field $\mathcal{Q}(X)$ of homogeneous fractions of degree zero of $\mathcal{R}(X)$.

X is said to be of general type if its Kodaira dimension is maximal (i.e., equal to the dimension n of X).

As observed in [Andr73] $\mathcal{Q}(X)$ is algebraically closed inside $\mathbb{C}(X)$, thus one obtains that X is of general type if and only if there is a positive integer m such that $H^0(X, mK_X)$ yields a birational map onto its image Σ_m.

One of the more crucial questions in classification theory is whether the canonical ring of a variety of general type is finitely generated, the answer being affirmative [Mum62, Mori88] for dimension $n \leq 3$.[1]

2.3 The Enriques Classification: An Outline

The main discrete invariant of smooth projective curves C is the *genus* $g(C) := h^0(K_C) = h^1(\mathcal{O}_C)$.

It determines easily the Kodaira dimension, and the Enriques classification of curves is the subdivision:

- $Kod(C) = -\infty \Leftrightarrow g(C) = 0 \Leftrightarrow C \cong \mathbb{P}^1$
- $Kod(C) = 0 \Leftrightarrow g(C) = 1 \Leftrightarrow C \cong \mathbb{C}/(\mathbb{Z} + \tau\mathbb{Z})$, with $\tau \in \mathbb{C}, Im(\tau) > 0 \Leftrightarrow C$ is an elliptic curve
- $Kod(C) = 1 \Leftrightarrow g(C) \geq 2 \Leftrightarrow C$ is of general type

Before giving the Enriques classification of projective surfaces over the complex numbers, it is convenient to discuss further the birational invariants of surfaces.

Remark 2.12 *An important birational invariant of smooth varieties X is the fundamental group $\pi_1(X)$.*

For surfaces, the most important invariants are:

- *The* irregularity $q := h^1(\mathcal{O}_X)$
- *The* geometric genus $p_g := P_1 := h^0(X, K_X)$, *which for surfaces combines with the irregularity to give the* holomorphic Euler–Poincaré characteristic $\chi(S) := \chi(\mathcal{O}_S) := 1 - q + p_g$
- *The* bigenus $P_2 := h^0(X, 2K_X)$ *and especially the* twelfth plurigenus $P_{12} := h^0(X, 12K_X)$

If S is a non ruled minimal surface, then also the following are birational invariants:

---

[1] The question seems to have been settled for varieties of general type, and with a positive answer.

- *The selfintersection of a canonical divisor K_S^2, equal to $c_1(S)^2$*
- *The topological Euler number $e(S)$, equal to $c_2(S)$ by the Poincaré Hopf theorem, and which by Noether's theorem can also be expressed as*

$$e(S) = 12\chi(S) - K_S^2 = 12(1 - q + p_g) - K_S^2$$

- *The topological index $\sigma(S)$ (the index of the quadratic form $q_S : H^2(S,\mathbb{Z}) \times H^2(S,\mathbb{Z}) \to \mathbb{Z}$), which, by the Hodge index theorem, satisfies the equality*

$$\sigma(S) = \frac{1}{3}(K_S^2 - 2e(S))$$

- *In particular, all the Betti numbers $b_i(S)$*
- *The positivity $b^+(S)$ and the negativity $b^-(S)$ of q_S (recall that $b^+(S) + b^-(S) = b_2(S)$)*

The Enriques classification of complex algebraic surfaces gives a very simple description of the surfaces with nonpositive Kodaira dimension:

- S is a ruled surface of irregularity $g \iff$:
  $\iff$: S is birational to a product $C_g \times \mathbb{P}^1$, where C_g has genus $g \iff$
  $\iff P_{12}(S) = 0, q(S) = g \iff$
  $\iff Kod(S) = -\infty, q(S) = g$
- S has $Kod(S) = 0 \iff P_{12}(S) = 1$

There are four classes of such surfaces with $Kod(S) = 0$:

- Tori $\iff P_1(S) = 1, q(S) = 2$
- K3 surfaces $\iff P_1(S) = 1, q(S) = 0$
- Enriques surfaces $\iff P_1(S) = 0, q(S) = 0, P_2(S) = 1$
- Hyperelliptic surfaces $\iff P_{12}(S) = 1, q(S) = 1$

Next come the surfaces with strictly positive Kodaira dimension:

- S is a properly elliptic surface $\iff$:
  $\iff$: $P_{12}(S) > 1$, and $H^0(12K_S)$ yields a map to a curve with fibres elliptic curves $\iff$
  $\iff S$ has $Kod(S) = 1 \iff$
  $\iff$ assuming that S is minimal: $P_{12}(S) > 1$ and $K_S^2 = 0$
- S is a surface of general type $\iff$:
  $\iff$: S has $Kod(S) = 2 \iff$
  $\iff P_{12}(S) > 1$, and $H^0(12K_S)$ yields a birational map onto its image $\Sigma_{12} \iff$
  $\iff$ assuming that S is minimal: $P_{12}(S) > 1$ and $K_S^2 \geq 1$

2.4 Some Constructions of Projective Varieties

Goal of this subsection is first of all to illustrate concretely the meaning of the concept 'varieties of general type'. This means, roughly speaking, that if we have a construction of varieties of a fixed dimension involving some integer parameters, most of the time we get varieties of general type when these parameters are all sufficiently large.

[1] *Products.*

Given projective varieties $X \subset \mathbb{P}^n$ and $Y \subset \mathbb{P}^m$, their product $X \times Y$ is also projective. This is an easy consequence of the fact that the product $\mathbb{P}^n \times \mathbb{P}^m$ admits the Segre embedding in $\mathbb{P}^{mn+n+m} \cong \mathbb{P}(Mat(n+1, m+1))$ onto the subspace of rank one matrices, given by the morphism $(x, y) \to x \cdot {}^t y$.

[2] *Complete intersections.*

Given a smooth variety X, and divisors $D_1 = \{f_1 = 0\}, \ldots, D_r = \{f_r = 0\}$ on X, their intersection $Y = D_1 \cap \cdots \cap D_r$ is said to be a complete intersection if Y has codimension r in X. If Y is smooth, or, more generally, reduced, locally its ideal is generated by the local equations of the D_i's ($\mathcal{I}_Y = (f_1, \ldots f_r)$).

Y tends to inherit much from the geometry of X, for instance, if $X = \mathbb{P}^N$ and Y is smooth of dimension $N - r \geq 2$, then Y is simply connected by the theorem of Lefschetz.

[3] *Finite coverings according to Riemann, Grauert and Remmert.*

Assume that Y is a normal variety (this means that each local ring $\mathcal{O}_{X,x}$ is integrally closed in the function field $\mathbb{C}(X)$), and that B is a closed subvariety of Y (the letter B stands for branch locus).

Then there is (cf. [GR58]) a correspondence between

[3a] subgroups $\Gamma \subset \pi_1(Y \backslash B)$ of finite index, and

[3b] pairs (X, f) of a normal variety X and a finite map $f : X \to Y$ which, when restricted to $X \backslash f^{-1}(B)$, is a local biholomorphism and a topological covering space of $Y \backslash B$.

The datum of the covering is equivalent to the datum of the sheaf of $\mathcal{O}_Y$-algebras $f_* \mathcal{O}_X$. As an $\mathcal{O}_Y$-module $f_* \mathcal{O}_X$ is locally free if and only if f is flat (this means that, $\forall x \in X$, $\mathcal{O}_{Y,f(x)} \to \mathcal{O}_{X,x}$ is flat), and this is indeed the case when f is finite and Y is smooth of dimension 2.

[4] *Finite Galois coverings.*

Although this is just a special case of the previous one, namely when Γ is a normal subgroup with factor group $G := \pi_1(Y \backslash B)/\Gamma$, in the more special case (cf. [Par91]) where G is Abelian and Y is smooth, one can give explicit equations for the covering. This is due to the fact that all irreducible representations of an abelian group are 1-dimensional, so we are in the *split case* where $f_* \mathcal{O}_X$ is a direct sum of invertible sheaves.

The easiest example is the one of

[4a] Simple cyclic coverings of degree n.

In this case there is

(i) an invertible sheaf $\mathcal{O}_Y(L)$ such that

$$f_* \mathcal{O}_X = \mathcal{O}_Y \oplus \mathcal{O}_Y(-L) \oplus \cdots \oplus \mathcal{O}_Y(-(n-1)L).$$

(ii) A section $0 \neq \sigma \in H^0(\mathcal{O}_Y(nL))$ such that X is the divisor, in the geometric line bundle $\mathbb{L}$ whose sheaf of regular sections is $\mathcal{O}_Y(L)$, given by the equation $z^n = \sigma(y)$.

Here, z is the never vanishing section of $p^*(\mathcal{O}_Y(L))$ giving a tautological linear form on the fibres of $\mathbb{L}$: in other words, one has an open cover U_α of Y which is trivializing for $\mathcal{O}_Y(L)$, and X is obtained by glueing together the local equations $z_\alpha^n = \sigma_\alpha(y)$, since $z_\alpha = g_{\alpha,\beta}(y)z_\beta$, $\sigma_\alpha(y) = g_{\alpha,\beta}(y)^n\sigma_\beta(y)$.

One has as branch locus $B = \Delta := \{\sigma = 0\}$, at least if one disregards the multiplicity (indeed $B = (n-1)\Delta$). Assume Y is smooth: then X is smooth iff Δ is smooth, and, via direct image, all the calculations of cohomology groups of basic sheaves on X are reduced to calculations for corresponding sheaves on Y. For instance, since $K_X = f^*(K_Y + (n-1)L)$, one has:

$$f_*(\mathcal{O}_X(K_X)) = \mathcal{O}_Y(K_Y) \oplus \mathcal{O}_Y(K_Y + L) \oplus \cdots \oplus \mathcal{O}_Y(K_Y + (n-1)L)$$

(the order is exactly as above according to the characters of the cyclic group).

We see in particular that X is of general type if L is sufficiently positive.

[4b] Simple iterated cyclic coverings.

Suppose that we take a simple cyclic covering $f : Y_1 \to Y$ as above, corresponding to the pair (L, σ), and we want to consider again a simple cyclic covering of Y_1. A small calculation shows that it is not so easy to describe $H^1(\mathcal{O}_{Y_1}^*)$ in terms of the triple (Y, L, σ); but in any case $H^1(\mathcal{O}_{Y_1}^*) \supset H^1(\mathcal{O}_Y^*)$. Thus one defines an *iterated simple cyclic covering* as the composition of a chain of simple cyclic coverings $f_i : Y_{i+1} \to Y_i$, $i = 0, \ldots k-1$ (thus $X := Y_k$, $Y := Y_0$) such that at each step the divisor L_i is the pull back of a divisor on $Y = Y_0$.

In the case of iterated double coverings, considered in [Man97], we have at each step $(z_i)^2 = \sigma_i$ and each σ_i is written as $\sigma_i = b_{i,0} + b_{i,1}z_1 + b_{i,2}z_2 + \cdots + b_{i,1,\ldots i-1}z_1 \ldots z_{i-1}$, where, for $j_1 < j_2 \cdots < j_h$, we are given a section $b_{i,j_1,\ldots j_h} \in H^0(Y, \mathcal{O}_Y(2L_i - L_{j_1} - \cdots - L_{j_h}))$.

In principle, it looks like one could describe the Galois covers with solvable Galois group G by considering iterated cyclic coverings, and then imposing the Galois condition. But this does not work without resorting to more complicated cyclic covers and to special geometry.

[4c] Bidouble covers (Galois with group $(\mathbb{Z}/2)^2$).

The *simple bidouble covers* are simply the fibre product of two double covers, thus here X is the complete intersection of the following two divisors

$$z^2 = \sigma_0, \quad w^2 = s_0$$

in the vector bundle $\mathbb{L} \oplus \mathbb{M}$.

These are the examples we shall mostly consider.

More generally, a bidouble cover of a smooth variety Y occurs [Cat84] as the subvariety X of the direct sum of three line bundles $\mathbb{L}_1 \oplus \mathbb{L}_2 \oplus \mathbb{L}_3$, given by equations

$$Rank \begin{pmatrix} x_1 & w_3 & w_2 \\ w_3 & x_2 & w_1 \\ w_2 & w_1 & x_3 \end{pmatrix} = 1 \qquad (*)$$

Here, we have three Cartier divisors $D_j = div(x_j)$ on Y and three line bundles $\mathbb{L}_i$, with fibre coordinate w_i, such that the following linear equivalences hold on Y,

$$L_i + D_i \equiv L_j + L_k,$$

for each permutation (i, j, k) of $(1, 2, 3)$.

One has: $f_*\mathcal{O}_X = \mathcal{O}_Y \bigoplus (\oplus_i \mathcal{O}_Y(-L_i))$.

Assume in addition that Y is a smooth variety, then:

- X is normal if and only if the divisors D_j are reduced and have no common components.
- X is smooth if and only if the divisors D_j are smooth, they do not have a common intersection and have pairwise transversal intersections.
- X is Cohen–Macaulay and for its dualizing sheaf ω_X (which, if Y is normal, equals the sheaf of Zariski differentials that we shall discuss later) we have $f_*\omega_X = \mathcal{H}om_{\mathcal{O}_Y}(f_*\mathcal{O}_X, \omega_Y) = \omega_Y \bigoplus (\oplus_i \omega_Y(L_i))$.

[5] Natural deformations.

One should in general consider Galois covers as 'special varieties'.

For instance, if we have a line bundle $\mathbb{L}$ on Y, we consider in it the divisor X described by an equation

$$z^n + a_2 z^{n-2} + \ldots a_{n-1} z + a_n = 0, \; for \; a_i \in H^0(Y, \mathcal{O}_Y(iL)).$$

It is clear that we obtain a simple cyclic cover if we set $a_n = -\sigma_0$, and, for $j \neq n$, we set $a_j = 0$.

The family of above divisors (note that we may assume $a_1 = 0$ after performing a Tschirnhausen transformation) is called the family of *natural deformations* of a simple cyclic cover.

One can define more generally a similar concept for any Abelian covering. In particular, for simple bidouble covers, we have the following family of natural deformations

$$z^2 = \sigma_0(y) + w\sigma_1(y), \; w^2 = s_0(y) + zs_1(y),$$

where $\sigma_0 \in H^0(Y, \mathcal{O}_Y(2L)), \sigma_1 \in H^0(Y, \mathcal{O}_Y(2L - M)), s_0 \in H^0(Y, \mathcal{O}_Y(2M))$ $s_1 \in H^0(Y, \mathcal{O}_Y(2M - L))$.

[6] Quotients.

In general, given an action of a finite group G on the function field $\mathbb{C}(X)$ of a variety X, one can always take the birational quotient, corresponding to the invariant subfield $\mathbb{C}(X)^G$.

Assume that $X \subset \mathbb{P}^N$ is a projective variety and that we have a finite group $G \subset \mathbb{P}GL(N + 1, \mathbb{C})$, such that $g(X) = X$, $\forall g \in G$.

We want then to construct a biregular quotient X/G with a projection morphism $\pi : X \to X/G$.

For each point $x \in X$ consider a hyperplane H such that $H \cap Gx = \emptyset$, and let $U := X \backslash (\cup_{g \in G} g(H))$.

U is an invariant affine subset, and we consider on the quotient set U/G the ring of invariant polynomials $\mathbb{C}[U]^G$, which is finitely generated since we are in characteristic zero and we have a projector onto the subspace of invariants.

It follows that if X is normal, then also X/G is normal, and moreover projective since there are very ample g-invariant divisors on X.

If X is smooth, one has that X/G is smooth if

(1) G acts freely or, more generally, if and only if

(2) For each point $p \in X$, the stabilizer subgroup $G_p := \{g|g(p) = p\}$ is generated by pseudoreflections (theorem of Chevalley, cf. for instance [Dolg82]).

To explain the meaning of a *pseudoreflection*, observe that, if $p \in X$ is a smooth point, by a theorem of Cartan [Car57], one can *linearize* the action of G_p, i.e., there exist local holomorphic coordinates $z_1, \dots z_n$ such that the action in these coordinates is linear. Thus, $g \in G_p$ acts by $z \to A(g)z$, and one says that g is a pseudoreflection if $A(g)$ (which is diagonalizable, having finite order) has $(n - 1)$ eigenvalues equal to 1.

[7] *Rational Double Points = Kleinian singularities.*

These are exactly the quotients $Y = \mathbb{C}^2/G$ by the action of a finite group $G \subset SL(2, \mathbb{C})$. Since $A(g) \in SL(2, \mathbb{C})$ it follows that G contains no pseudoreflection, thus Y contains exactly one singular point p, image of the unique point with a nontrivial stabilizer, $0 \in \mathbb{C}^2$.

These singularities (Y, p) will play a prominent role in the next section.

In fact, one of their properties is due to the fact that the differential form $dz_1 \wedge dz_2$ is G-invariant (because $det(A(g)) = 1$), thus the sheaf Ω_Y^2 is trivial on $Y \backslash \{p\}$.

Then the dualizing sheaf $\omega_Y = i_*(\Omega_{Y \backslash \{p\}}^2)$ is also trivial.

3 Lecture 2: Surfaces of General Type and Their Canonical Models: Deformation Equivalence and Singularities

3.1 Rational Double Points

Let us take up again the Kleinian singularities introduced in the previous section

Definition 3.1 *A Kleinian singularity is a singularity (Y, p) analytically isomorphic to a quotient singularity $\mathbb{C}^n/G$ where G is a finite subgroup $G \subset SL(n, \mathbb{C})$.*

Example 3.2 *The surface singularity A_n corresponds to the cyclic group $\mu_n \cong \mathbb{Z}/n$ of nth roots of unity acting with characters 1 and $(n - 1)$.*

I.e., $\zeta \in \mu_n$ acts by $\zeta(u, v) := (\zeta u, \zeta^{n-1}v)$, and the ring of invariants is the ring $\mathbb{C}[x, y, z]/(xy - z^n)$, where

$$x := u^n, \ y := v^n, z := uv.$$

Example 3.3 *One has more generally the cyclic quotient surface singularities corresponds to the cyclic group $\mu_n \cong \mathbb{Z}/n$ of nth roots of unity acting with characters a and b, which are denoted by $\frac{1}{n}(a, b)$.*

Here, $\zeta(u, v) := (\zeta^a u, \zeta^b v)$.

We compute the ring of invariants in the case $n = 4, a = b = 1$: the ring of invariants is generated by

$$y_0 := u^4, y_1 := u^3 v, \ y_2 := u^2 v^2, \ y_3 := u v^3, y_4 := v^4,$$

and the ring is $\mathbb{C}[y_0, \ldots, y_4]/J$, where J is the ideal of 2×2 minors of the matrix $\begin{pmatrix} y_0 & y_1 & y_2 & y_3 \\ y_1 & y_2 & y_3 & y_4 \end{pmatrix}$, or equivalently of the matrix $\begin{pmatrix} y_0 & y_1 & y_2 \\ y_1 & y_2 & y_3 \\ y_2 & y_3 & y_4 \end{pmatrix}$. The first realization of the ideal J corresponds to the identification of the singularity Y as the cone over a rational normal curve of degree 4 (in $\mathbb{P}^4$), while in the second Y is viewed as a linear section of the cone over the Veronese surface.

We observe that $2y_2$ and $y_0 + y_4$ give a map to $\mathbb{C}^2$ which is finite of degree 4. They are invariant for the group of order 16 generated by

$$(u, v) \mapsto (iu, iv), \ (u, v) \mapsto (iu, -iv), \ (u, v) \mapsto (v, u),$$

hence Y is a bidouble cover of $\mathbb{C}^2$ branched on three lines passing through the origin (cf. (*), we set $x_3 := x_1 - x_2$ and we choose as branch divisors $x_1, x_2, x_3 := x_1 - x_2$).

In dimension two, the classification of Kleinian singularities is a nice chapter of geometry ultimately going back to Thaetethus' Platonic solids. Let us briefly recall it.

First of all, by averaging the positive definite Hermitian product in $\mathbb{C}^n$, one finds that a finite subgroup $G \subset SL(n, \mathbb{C})$ is conjugate to a finite subgroup $G \subset SU(n, \mathbb{C})$. Composing the inclusion $G \subset SU(n, \mathbb{C})$ with the surjection $SU(n, \mathbb{C}) \to \mathbb{P}SU(n, \mathbb{C}) \cong SU(n, \mathbb{C})/\mu_n$ yields a finite group G' acting on $\mathbb{P}^{n-1}$.

Thus, for $n = 2$, we get $G' \subset \mathbb{P}SU(2, \mathbb{C}) \cong SO(3)$ acting on the Riemann sphere $\mathbb{P}^1 \cong S^2$.

The consideration of the Hurwitz formula for the quotient morphism $\pi : \mathbb{P}^1 \to \mathbb{P}^1/G'$, and the fact that $\mathbb{P}^1/G'$ is a smooth curve of genus 0, (hence $\mathbb{P}^1/G' \cong \mathbb{P}^1$) allows the classification of such groups G'.

Letting in fact $p_1, \ldots p_k$ be the branch points of π, and $m_1, \ldots m_k$ the respective multiplicities (equal to the order in G' of the element corresponding to the local monodromy), we have *Hurwitz's formula* (expressing the degree of the canonical divisor $K_{\mathbb{P}^1}$ as the sum of the degree of the pull back of $K_{\mathbb{P}^1}$ with the degree of the ramification divisor)

$$-2 = |G'|(-2 + \sum_{i=1}^{k} [1 - \frac{1}{m_i}]).$$

Each term in the square bracket is $\geq \frac{1}{2}$, and the left hand side is negative: hence $k \leq 3$.

The situation to classify is the datum of a ramified covering of $\mathbb{P}^1 \backslash \{p_1, \ldots p_k\}$, Galois with group G'.

By the Riemann existence theorem, and since $\pi_1(\mathbb{P}^1 \backslash \{p_1, \ldots p_k\})$ is the socalled infinite polygonal group $T(\infty^k) = T(\infty, \ldots, \infty)$ generated by simple geometric loops $\alpha_1, \ldots, \alpha_k$, satisfying the relation $\alpha_1 \cdots \alpha_k = 1$, the datum of such a covering amounts to the datum of an epimorphism $\phi : T(\infty, \ldots, \infty) \to G'$ such that, for each $i = 1, \ldots, k$, $a_i := \phi(\alpha_i)$ is an element of order m_i.

The group $T(\infty^k)$ is trivial for $k = 1$, infinite cyclic for $k = 2$, in general a free group of rank $k - 1$.

Since $a_i := \phi(\alpha_i)$ is an element of order m_i, the epimorphism factors through the polygonal group

$$T(m_1, \ldots, m_k) := \langle \alpha_1, \ldots, \alpha_k | \alpha_1 \cdots \alpha_k = \alpha_1^{m_1} = \cdots = \alpha_k^{m_k} = 1 \rangle.$$

If $k = 2$, then we may assume $m_1 = m_2 = m$ and we have a cyclic subgroup G' of order m of $\mathbb{P}SU(2, \mathbb{C})$, which, up to conjugation, is generated by a transformation $\zeta(u, v) := (\zeta u, \zeta^{n-1} v)$, with ζ a primitive mth root of 1 for m odd, and a primitive 2mth root of 1 for m even. Thus, our group G is a cyclic group of order n, with $n = 2m$ for m even, and with $n = 2m$ or $n = m$ for m odd. G is generated by a transformation $\zeta(u, v) := (\zeta u, \zeta^{n-1} v)$ (with ζ a primitive nth root of 1), and we have the singularity A_n previously considered.

If $k = 3$, the only numerical solutions for the Hurwitz' formula are

$$m_1 = 2, m_2 = 2, m_3 = m \geq 2,$$

$$m_1 = 2, m_2 = 3, m_3 = 3, 4, 5.$$

Accordingly the order of the group G' equals $2m, 12, 24, 60$. Since m_3, for $m_3 \geq 3$, is not the least common multiple of m_1, m_2, the group G' is not abelian, and it follows (compare [Klein1884]) that G' is respectively isomorphic to $D_m, \mathcal{A}_4, \mathcal{S}_4, \mathcal{A}_5$.

Accordingly, since as above the lift of an element in G' of even order k has necessarily order $2k$, it follows that G is the full inverse image of G', and G is respectively called the binary dihedral group, the binary tetrahedral group, the binary octahedral group, the binary icosahedral group.

Felix Klein computed explicitly the ring of polynomial invariants for the action of G, showing that $\mathbb{C}[u, v]^G$ is a quotient ring $\mathbb{C}[x, y, z]/(z^2 - f(x, y))$, where

- $f(x, y) = x^2 + y^{n+1}$ for the A_n case
- $f(x, y) = y(x^2 + y^{n-2})$ for the D_n case $(n \geq 4)$
- $f(x, y) = x^3 + y^4$ for the E_6 case, when $G' \cong \mathcal{A}_4$
- $f(x, y) = y(x^2 + y^3)$ for the E_7 case, when $G' \cong \mathcal{S}_4$
- $f(x, y) = x^3 + y^5$ for the E_8 case, when $G' \cong \mathcal{A}_5$

We refer to [Durf79] for several equivalent characterizations of Rational Double points, another name for the Kleinian singularities. An important property (cf. [Reid80] and [Reid87]) is that these singularities may be resolved just by a sequence of point blow ups: in this procedure no points of higher multiplicity than 2 appear, whence it follows once more that the canonical divisor of the minimal resolution is the pull back of the canonical divisor of the singularity.

A simpler way to resolve these singularities (compare [BPV84], pages 86 and following) is to observe that they are expressed as double covers branched over the curve $f(x, y) = 0$. Then the standard method, explained in full generality by Horikawa in [Hor75] is to resolve the branch curve by point blow ups, and keeping as new branch curve at each step $B'' - 2D''$, where B'' is the total transform of the previous branch curve B, and D'' is the maximal effective divisor such that $B'' - 2D''$ is also effective. One obtains the following

Theorem 3.4 *The minimal resolution of a Rational Double Point has as exceptional divisor a finite union of curves $E_i \cong \mathbb{P}^1$, with selfintersection -2, intersecting pairwise transversally in at most one point, and moreover such that no three curves pass through one point. The dual graph of the singularity, whose vertices correspond to the components E_i, and whose edges connect E_i and E_j exactly when $E_i \cdot E_j = 1$, is a tree, which is a linear tree with $n - 1$ vertices exactly in the A_n case. In this way one obtains exactly all the Dynkin diagrams corresponding to the simple Lie algebras.*

Remark 3.5 *(i) See the forthcoming Theorem 3.9 for a list of these Dynkin diagrams.*

(ii) The relation to simple Lie algebras was clarified by Brieskorn in [Briesk71]: these singularities are obtained by intersecting the orbits of the coadjoint action with a three dimensional submanifold in general position.

We end this subsection with an important observation concerning the automorphisms of a Rational Double Point (X, x_0).

Let H be a finite group of automorphisms of the germ $(X, x_0) = (\mathbb{C}^2, 0)/G$.

Then the quotient $(X, x_0)/H$ is a quotient of $(\mathbb{C}^2, 0)$ by a group H' such that $H'/G \cong H$. Moreover, by the usual averaging trick (Cartan's lemma, see [Car57]) we may assume that $H' \subset GL(2, \mathbb{C})$. Therefore H' is contained in the normalizer N_G of G inside $GL(2, \mathbb{C})$. Obviously, N_G contains the centre $\mathbb{C}^*$ of $GL(2, \mathbb{C})$, and $\mathbb{C}^*$ acts on the graded ring $\mathbb{C}[x, y, z]/(z^2 - f(x, y))$ by multiplying homogeneous elements of degree d by t^d. Therefore H is a finite subgroup of the group H^* of graded automorphisms of the ring $\mathbb{C}[x, y, z]/(z^2 - f(x, y))$, which is determined as follows (compare [Cat87])

Theorem 3.6 *The group H^* of graded automorphisms of a RDP is:*
(1) $\mathbb{C}^*$ for E_8, E_7
(2) $\mathbb{C}^* \times \mathbb{Z}/2$ for $E_6, D_n (n \geq 5)$
(3) $\mathbb{C}^* \times \mathcal{S}_3$ for D_4

(4) $(\mathbb{C}^*)^2 \times \mathbb{Z}/2$ *for* $A_n (n \geq 2)$
(5) $GL(2, \mathbb{C})/\{\pm 1\}$ *for* A_1

Idea of proof . The case of A_1 is clear because $G = \{\pm 1\}$ is contained in the centre. In all the other cases, except D_4, y is the generator of smallest degree, therefore it is an eigenvector, and, up to using $\mathbb{C}^*$, we may assume that y is left invariant by an automorphism h. Some calculations allow to conclude that h is the identity in case (1), or the *trivial involution* $z \mapsto -z$ in case of E_6 and of D_n for n odd; while for D_n with n even the extra involution is $y \mapsto -y$.

Finally, for D_4, write the equation as $z^2 = y(x + iy)(x - iy)$ and permute the three lines which are the components of the branch locus. For A_n, one finds that the normalizer is the semidirect product of the diagonal torus with the involution given by $(u, v) \mapsto (v, u)$.

One may also derive the result from the symmetries of the Dynkin diagram.   □

3.2 Canonical Models of Surfaces of General Type

Assume now that S is a smooth minimal (projective) surface of general type.

We have (as an easy consequence of the Riemann Roch theorem) that S is minimal of general type if $K_S^2 > 0$ and K_S is nef (we recall that a divisor D is said to be *nef* if, for each irreducible curve C, we have $D \cdot C \geq 0$).

In fact, S is minimal of general type iff $K_S^2 > 0$ and K_S is nef. Since, if D is nef and, for $m > 0$, we write $|mD| = |M| + \Phi$ as the sum of its movable part and its fixed part, then $M^2 = m^2 D^2 - mD \cdot \Phi - M \cdot \Phi \leq m^2 D^2$. Hence, if $D^2 \leq 0$, the linear system $|mD|$ yields a rational map whose image has dimension at most 1.

Recall further that the Neron-Severi group $NS(S) = Div(S)/\sim$ is the group of divisors modulo numerical equivalence (D is numerically equivalent to 0, $\Leftrightarrow D \cdot C = 0$ for every irreducible curve C on S).

The Neron Severi group is a discrete subgroup of the vector space $H^1(\Omega_S^1)$, and indeed on a projective manifold Y it equals the intersection $(H^2(Y, \mathbb{Z})/Torsion) \cap H^{1,1}(Y)$.

By definition, the intersection form is non degenerate on the Neron Severi group, whose rank ρ is called the *Picard number*. But the Hodge index theorem implies the

Algebraic index theorem *The intersection form on $NS(S)$ has positivity index precisely 1 if S is an algebraic surface.*

The criterion of Nakai-Moishezon says that a divisor L on a surface S is ample if and only if $L^2 > 0$ and $L \cdot C > 0$ for each irreducible curve C on S. Hence:

The canonical divisor K_S of a minimal surface of general type S is ample iff there does not exist an irreducible curve C ($\neq 0$) on S with $K \cdot C = 0$.

Remark 3.7 *Let S be a minimal surface of general type and C an irreducible curve on S with $K \cdot C = 0$. Then, by the index theorem, $C^2 < 0$ and by the adjunction formula we see that $2p(C) - 2 = K \cdot C + C^2 = C^2 < 0$.*

In general $p(C) := 1 - \chi(\mathcal{O}_C)$ is the arithmetic genus of C, which is equal to the sum $p(C) = g(\tilde{C}) + \delta$ of the geometric genus of C, i.e., the genus of the normalization $p : \tilde{C} \to C$ of C, with the number δ of double points of C, defined as $\delta := h^0(p_* \mathcal{O}_{\tilde{C}}/\mathcal{O}_C)$.

Therefore here $p(C) = 0$, so that $C \cong \mathbb{P}^1$, and $C^2 = -2$.

These curves are called (-2)-curves.

Thus K_S is not ample if and only if there exists a (-2)-curve on S. There is an upper bound for the number of these (-2)-curves.

Lemma 3.8 *Let $C_1, \ldots, C_k$ be irreducible (-2)-curves on a minimal surface S of general type. We have:*

$$(\Sigma n_i C_i)^2 \leq 0,$$

and

$$(\Sigma n_i C_i)^2 = 0 \text{ if and only if } n_i = 0 \text{ for all } i.$$

Thus their images in the Neron-Severi group $NS(S)$ are independent and in particular $k \leq \rho - 1$ (ρ is the rank of $NS(S)$), and $k \leq h^1(\Omega_S^1) - 1$.

Proof. Let $\Sigma n_i C_i = C^+ - C^-$, ($C^+$ and C^- being effective divisors without common components) be the (unique) decomposition of $\Sigma n_i C_i$ in its positive and its negative part. Then $K \cdot C^+ = K \cdot C^- = 0$ and $C^+ \cdot C^- \geq 0$, whence $(C^+ - C^-)^2 = (C^+)^2 + (C^-)^2 - 2(C^+ \cdot C^-) \leq (C^+)^2 + (C^-)^2$. By the index theorem $(C^+)^2 + (C^-)^2$ is ≤ 0 and $= 0$ iff $C^+ = C^- = 0$. $\quad\square$

We can classify all possible configurations of (-2)-curves on a minimal surface S of general type by the following argument.

If C_1 and C_2 are two (-2)-curves on S, then:

$$0 > (C_1 + C_2)^2 = -4 + 2C_1 \cdot C_2,$$

hence $C_1.C_2 \leq 1$, i.e., C_1 and C_2 intersect transversally in at most one point.

If C_1, C_2, C_3 are (-2)-curves on S, then again we have

$$0 > (C_1 + C_2 + C_3)^2 = 2(-3 + C_1 \cdot C_2 + C_1 \cdot C_3 + C_2 \cdot C_3).$$

Therefore no three curves meet in one point, nor do they form a triangle.

We associate to a configuration $\cup C_i$ of (-2)-curves on S its *Dynkin graph*: the vertices correspond to the (-2)-curves C_i, and two vertices (corresponding to C_i, C_j) are connected by an edge if and only if $C_i \cdot C_j = 1$.

Obviously the Dynkin graph of a configuration $\cup C_i$ is connected iff $\cup C_i$ is connected. So, let us assume that $\cup C_i$ is connected.

Theorem 3.9 *Let S be a minimal surface of general type and $\cup C_i$ a (connected) configuration of (-2)-curves on S. Then the associated (dual) Dynkin graph of $\cup C_i$ is one of those listed in Fig. 1.*

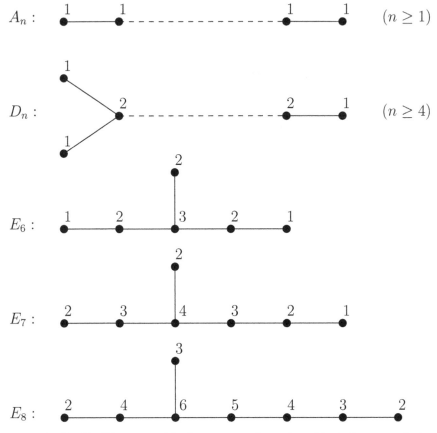

Fig. 1. The Dynkin-Diagrams of (-2)-curves configurations (the index n stands for the number of vertices, i.e. of curves). The labels for the vertices are the coefficients of the fundamental cycle

Remark 3.10 *The figure indicates also the weights n_i of the vertices of the respective trees. These weights correspond to a divisor, called fundamental cycle*

$$Z := \Sigma n_i C_i$$

defined (cf. [ArtM66]) by the properties

$(**)$ $Z \cdot C_i \leq 0$ for all i, $Z^2 = -2$, and $n_i > 0$.

Idea of proof of 3.9. The simplest proof is obtained considering the above set of *Dynkin-Diagrams* $\mathcal{D} := \{A_n, D_n, E_6, E_7, E_8\}$ and the corresponding set of *Extended-Dynkin-Diagrams* $\tilde{\mathcal{D}} := \{\tilde{A}_n, \tilde{D}_n, \tilde{E}_6, \tilde{E}_7, \tilde{E}_8\}$ which classify the *divisors of elliptic type* made of (-2)-curves and are listed in Fig. 2 (note that the divisors of elliptic type classify all the possible nonmultiple fibres

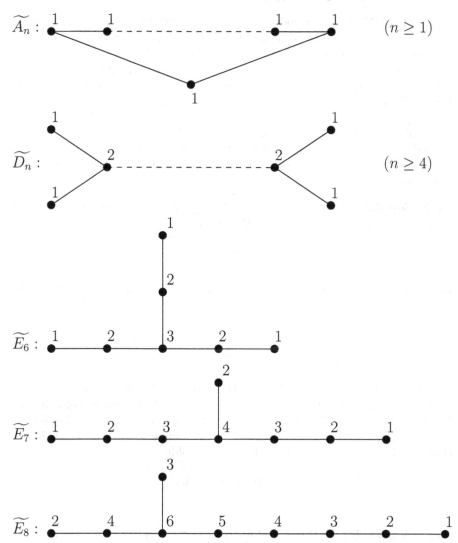

Fig. 2. The extended Dynkin-Diagrams of (-2)-curves configurations. The labels for the vertices are the coefficients of the divisor F of elliptic type

F of elliptic fibrations). Notice that each graph Γ in $\mathcal{D}$ is a subgraph of a corresponding graph $\tilde{\Gamma}$ in $\tilde{\mathcal{D}}$, obtained by adding exactly a (-2)-curve: $\Gamma = \tilde{\Gamma} - C_{\text{end}}$. In this correspondence the fundamental cycle equals $Z = F - C_{\text{end}}$ thus (**) is proven since $F \cdot C_i = 0$ for each i. Moreover, by Zariski's Lemma [BPV84] the intersection form on Γ is negative definite. If moreover Γ is a graph with a negative definite intersection form, then Γ does not contain as a subgraph a graph in $\tilde{\mathcal{D}}$, since $F^2 = 0$. The proof can now be easily concluded. $\square$

Artin [ArtM66] showed indeed that the above configurations can be holo-morphically contracted to Rational Double Points, and that the fundamental cycle is indeed the inverse image of the maximal ideal in the local ring of the singularity. By applying these contractions to the minimal model S of a surface of general type one obtains in this way a normal surface X with Rational Double Points as singularities, called the *canonical model of S*.

We prefer however to sketch briefly how the canonical model is more directly obtained from the pluricanonical maps of S, and ultimately it can be defined as the Projective Spectrum (set of homogeneous prime ideals) of the canonical ring $\mathcal{R}(S)$. We need first of all Franchetta's theory of numerical connectedness.

Definition 3.11 *An effective divisor D is said to be m-connected if, each time we write $D = A + B$, with $A, B > 0$, then*

$$(*) \qquad\qquad\qquad A \cdot B \geq m.$$

Lemma 3.12 *Let D be a nef divisor on a smooth surface S, with $D^2 > 0$. Then, if D is effective, then D is 1-connected.*

Proof. Since D is nef,

$$A^2 + A \cdot B = D \cdot A \geq 0, B^2 + A \cdot B = D \cdot B \geq 0.$$

Assume $A \cdot B \leq 0$: then $A^2, B^2 \geq -(AB) \geq 0 \implies A^2 \cdot B^2 \geq (AB)^2$.

But, by the Index Theorem, $A^2 B^2 \leq (AB)^2$. Thus equality holds in the Index theorem $\iff \exists L$ such that $A \sim aL$, $B \sim bL$, $D \sim (a+b)L$. Moreover, since $D^2 > 0$ we have $L^2 \geq 1$, and we may assume $a, b > 0$ since A, B are effective. Thus $A \cdot B = a \cdot b\, L^2 \geq 1$, equality holding

$$\iff a = b = 1 (\implies D \sim 2L), L^2 = 1.$$

$\square$

Remark 3.13 *Let $A \cdot B = 1$ and assume $A^2 B^2 < (AB)^2 \implies A^2 \cdot B^2 \leq 0$, but $A^2, B^2 \geq -1$. Thus, up to exchanging A and B, either $A^2 = 0$, and then $D \cdot A = 1, A^2 = 0$; or $A^2 > 0$, $B^2 = -1$, and then $D \cdot B = 0$, $B^2 = -1$.*

Hence the following

Corollary 3.14 *Let S be minimal of general type, $D \sim mK$, $m \geq 1$: then D is 2-connected except possibly if $K^2 = 1$, and $m = 2$, or $m = 1$, and $K \sim 2L$, $L^2 = 1$.*

Working a little more one finds

Proposition 3.15 *Let K be nef and big as before, $D \sim mK$ with $m \geq 2$. Then D is 3-connected except possibly if*

- $D = A + B, A^2 = -2, A \cdot K = 0 (\implies A \cdot B = 2)$

- $m = 2, K^2 = 1, 2$
- $m = 3, K^2 = 1$

We use now the *Curve embedding Lemma* of [C-F-96], improved in [CFHR99] to the more general case of any curve C (i.e., a pure 1-dimensional scheme).

Lemma 3.16 *(Curve-embedding lemma) Let C be a curve contained in a smooth algebraic surface S, and let H be a divisor on C. Then H is very ample if, for each length 2 0-dimensional subscheme ζ of C and for each effective divisor $B \leq C$, we have*

$$\mathrm{Hom}(\mathcal{I}_\zeta, \omega_B(-H)) = 0.$$

In particular H is very ample on C if $\forall\ B \leq C$, $H \cdot B > 2p(B) - 2 +$ length $\zeta \cap B$, where length $\zeta \cap B :=$ colength $(\mathcal{I}_\zeta \mathcal{O}_B)$. A fortiori, H is very ample on C if, $\forall\ B \leq C$,

$$H \cdot B \geq 2p(B) + 1. \tag{$*$}$$

Proof. It suffices to show the surjectivity $H^0(\mathcal{O}_C(H)) ->> H^0(\mathcal{O}_\zeta(H))\ \forall$ such ζ. In fact, we can take either $\zeta = \{x, y\}$ [2 diff. points], or $\zeta = (x, \xi)$, ξ a tangent vector at x. The surjectivity is implied by $H^1(\mathcal{I}_\zeta \mathcal{O}_C(H)) = 0$.

By Serre-Grothendieck duality, and since $\omega_C = \mathcal{O}_C(K_S + C)$, we have, in case of nonvanishing, $0 \neq H^1(\mathcal{I}_\zeta \mathcal{O}_C(H))^\vee \cong \mathrm{Hom}(\mathcal{I}_\zeta \mathcal{O}_C(H), \mathcal{O}_C(K_S + C)) \ni \sigma \neq 0$.

Let Z be the maximal subdivisor of C such that σ vanishes on Z (i. e., $Z = \mathrm{div}(z)$, with $z|\sigma$) and let $B = V(\mathrm{Ann}(\sigma))$. Then $B + Z = C$ since, if $C = \{(\beta \cdot z) = 0\}$, $\mathrm{Ann}(\sigma) = (\beta)$.

Indeed, let $f \in \mathcal{I}_\zeta$ be a non zero divisor: then σ is identified with the rational function $\sigma = \frac{\sigma(f)}{f}$; we can lift everything to the local ring $\mathcal{O}_S$, then f is coprime with the equation $\gamma := (\beta z)$ of C, and $z = \mathrm{G.C.D.}(\sigma(f), \gamma)$. Clearly now $\mathrm{Ann}(\sigma) = \{u \mid u\sigma(f) \in (\beta z)\} = (\beta)$.

Hence σ induces

$$\hat{\sigma} := \frac{\sigma}{z} : \mathcal{I}_\zeta \mathcal{O}_B(H) \to \mathcal{O}_B(K_S + C - Z)$$

which is 'good' (i.e., it is injective and with finite cokernel), thus we get

$$0 \to \mathcal{I}_\zeta \mathcal{O}_B \xrightarrow{\hat{\sigma}} \mathcal{O}_B(K_B - H) \to \Delta \to 0$$

where $\mathrm{supp}(\Delta)$ has dim $= 0$.

Then, taking the Euler Poincaré characteristics χ of the sheaves in question, we obtain

$$0 \leq \chi(\Delta) = \chi(\mathcal{O}_B(K_B - H)) - \chi(\mathcal{I}_\zeta \mathcal{O}_B) = -H \cdot B + 2p(B) - 2 + length(\zeta \cap B) < 0,$$

a contradiction. $\square$

The basic-strategy for the study of pluricanonical maps is then to find, for every length 2 subscheme of S, a divisor $C \in |(m-2)K|$ such that $\zeta \subset C$.

Since then, in characteristic $= 0$ we have the vanishing theorem [Ram72-4]

Theorem 3.17 (Kodaira, Mumford, Ramanujam) . *Let $L^2 > 0$ on S, L nef $\Longrightarrow H^i(-L) = 0, i \geq 1$. In particular, if S is minimal of general type, then $H^1(-K_S) = H^1(2K_S) = 0$.*

As shown by Ekedahl [Eke88] this vanishing theorem is false in positive characteristic, but only if char $= 2$, and for 2 very special cases of surfaces!

Corollary 3.18 *If $C \in |(m-2)K|$, then $H^0(\mathcal{O}_S(mK)) \longrightarrow\!\!\!> H^0(\mathcal{O}_C(mK))$. Therefore, $|mK_S|$ is very ample on S if $h^0((m-2)K_S) \geq 3$ and if the hypothesis on $H = mK_S$ in the curve embedding Lemma is verified for any $C \equiv (m-2)K$.*

We shall limit ourselves here to give the proof of a weaker version of Bombieri's theorem [Bom73]

Theorem on Pluricanonical-Embeddings. (Bombieri). (mK) is almost very ample (it embeds ζ except if $\exists B$ with $\zeta \subset B$, and $B \cdot K = 0$) if $m \geq 5$, $m = 4$ and $K^2 \geq 2$, $m = 3$, $p_g \geq 3$, $K^2 \geq 3$.

One first sees when $h^0((m-2)K) \geq 2$.

Lemma 3.19 *For $m \geq 3$ we have $h^0((m-2)K) \geq 3$ except if $m = 3$ $p_g \leq 2$, $m = 4$, $\chi = K^2 = 1$ (then $q = p_g = 0$) and ≥ 2 except if $m = 3$, $p_g \leq 1$.*

Proof. $p_g = H^0(K)$, so let us assume $m \geq 4$.

$$h^0((m-2)K) \geq \chi((m-2)K) \geq \chi + \frac{(m-2)(m-3)}{2} K^2.$$

Now, $\chi \geq 1$ and $K^2 \geq 1$, so we are done unless $m = 4$, $\chi = K^2 = 1$. $\square$

The possibility that K_S may not be ample is contemplated in the following

Lemma 3.20 *Let $H = mK$, $B \leq C \equiv (m-2)K$ and assume $K \cdot B > 0$. Then*

$$H \cdot B \geq 2p(B) + 1 \text{ except possibly if}$$

(A) $m = 4$ and $K^2 = 1$, or $m = 3$ and $K^2 \leq 2$.

Proof. Let $C = B + Z$ as above. Then we want

$$mK \cdot B \geq 2p(B) - 2 + 3 = (K+B) \cdot B + 3 = (K+C-Z) \cdot B + 3 = [(m-1)K - Z] \cdot B + 3,$$

i.e.,

$$K \cdot B + B \cdot Z \geq 3.$$

Since we assumed $K \cdot B \geq 1$, if $Z = 0$ we use $K^2 \geq 2$ if $m \geq 4$, and $K^2 \geq 3$ if $m = 3$, else it suffices to have $B \cdot Z \geq 2$, which is implied by the previous Corollary 3.14 (if $m = 3$, $B \sim Z \sim L$, $L^2 = 1$, then $K \cdot B = 2$). $\square$

Remark Note that then ζ is contracted iff $\exists B$ with $\zeta \subset B$, $K \cdot B = 0$! Thus, if there are no (-2) curves, the theorem says that we have an embedding of S. Else, we have a birational morphism which exactly contracts the fundamental cycles Z of S. To obtain the best technical result one has to replace the subscheme ζ by the subscheme $2Z$, and use that a fundamental cycle Z is 1-connected. We will not do it here, we simply refer to [CFHR99].

The following is the more precise theorem of Bombieri [Bom73]

Theorem 3.21 *Let S be a minimal surface of general type, and consider the linear system $|mK|$ for $m \geq 5$, for $m = 4$ when $K^2 \geq 2$, for $m = 3$ when $p_g \geq 3$, $K^2 \geq 3$.*

Then $|mK|$ yields a birational morphism onto its image, which is a normal surface X with at most Rational Double Points as singularities. For each singular point $p \in X$ the inverse image of the maximal ideal $\mathfrak{M}_p \subset \mathcal{O}_{X,p}$ is a fundamental cycle.

Here we sketch another way to look at the above surface X (called canonical model of S).

Proposition 3.22 *If S is a surface of general type the canonical ring $\mathcal{R}(S)$ is a graded $\mathbb{C}$-algebra of finite type.*

Proof. We choose a natural number such that $|mK|$ is without base points, and consider a pluricanonical morphism which is birational onto its image

$$\phi_m : S \rightarrow \Sigma_m = \Sigma \subset \mathbb{P}^N.$$

For $r = 0, \ldots, m - 1$, we set $\mathcal{F}_r := \phi_*(\mathcal{O}_S(rK))$.
The Serre correspondence (cf. [FAC55]) associates to $\mathcal{F}_r$ the module

$$M_r := \bigoplus_{i=1}^{\infty} H^0(\mathcal{F}_r(i)) = \bigoplus_{i=1}^{\infty} H^0(\phi_*(\mathcal{O}_S(rK))(i)) =$$

$$= \bigoplus_{i=1}^{\infty} H^0(\phi_*(\mathcal{O}_S((r+im)K))) = \bigoplus_{i=1}^{\infty} H^0(\mathcal{O}_S((r+im)K)) = \bigoplus_{i=1}^{\infty} \mathcal{R}_{r+im}.$$

M_r is finitely generated over the ring $\mathcal{A} = \mathbb{C}[y_0, \ldots, y_N]$, hence $\mathcal{R} = \bigoplus_{r=0}^{m-1} M_r$ is a finitely generated $\mathcal{A}$-module.

We consider the natural morphism $\alpha : \mathcal{A} \rightarrow \mathcal{R}$, $y_i \mapsto s_i \in \mathcal{R}_m$, (then the s_i generate a subring B of $\mathcal{R}$ which is a quotient of $\mathcal{A}$). If $v_1, \ldots, v_k$ generate $\mathcal{R}$ as a graded $\mathcal{A}$-module, then $v_1, \ldots, v_k, s_0, \ldots, s_N$ generate $\mathcal{R}$ as a $\mathbb{C}$-algebra. $\square$

The relation between the *canonical ring* $\mathcal{R}(S, K_S)$ and the image of pluri-canonical maps for $m \geq 5$ is then that $X = \mathrm{Proj}(\mathcal{R}(S, K_S))$.

In practice, since $\mathcal{R}$ is a finitely generated graded $\mathbb{C}$-algebra, generated by elements x_i of degree r_i, there is a surjective morphism

$$\lambda : \mathbb{C}[z_0, \dots, z_N] \longrightarrow\!\!\!\!> \mathcal{R}, \ \lambda(z_i) = x_i.$$

If we decree that z_i has degree r_i, then λ is a graded surjective homomorphism of degree zero.

With this grading (where z_i has degree r_i) one defines (see [Dolg82]) the *weighted projective space* $\mathbb{P}(r_0, \dots r_n)$ as $\mathrm{Proj}(\mathbb{C}[z_0, \dots, z_N])$.

$\mathbb{P}(r_0, \dots r_n)$ is simply the quotient $:= \mathbb{C}^{N+1} - \{0\}/\mathbb{C}^*$, where $\mathbb{C}^*$ acts on $\mathbb{C}^{N+1}$ in the following way:

$$t(z) = (z_0 t^{r_0}, \dots, z_N t^{r_N}).$$

The surjective homomorphism λ corresponds to an embedding of X into $\mathbb{P}(r_0, \dots r_n)$.

With the above notation, one can easily explain some classical examples which show that Bombieri's theorem is the best possible result.

Ex. 1: $m \geq 5$ is needed. Take a hypersurface $X_{10} \subset \mathbb{P}(1, 1, 2, 5)$ with Rational Double Points defined by a (weighted) homogeneous polynomial F_{10} of degree 10. Then $\omega_X = \mathcal{O}_X(10 - \Sigma e_i) = \mathcal{O}_X(1)$, $K_X^2 = 10/\prod e_i = 1$, and any m-canonical map with $m \leq 4$ is not birational.

In fact here the quotient ring $\mathbb{C}[y_0, y_1, x_3, z_5]/(F_{10})$, where $\deg y_i = 1, \deg x_3 = 2, \deg z_5 = 5$ is exactly the canonical ring $\mathcal{R}(S)$.

Ex. 2: $m = 3$, $K^2 = 2$ is also an exception.

Take $S = X_8 \subset \mathbb{P}(1, 1, 1, 4)$. Here S was classically described as a double cover $S \to \mathbb{P}^2$ branched on a curve B of degree 8 (since $F_8 = z^2 - f_8(x_0, x_1, x_2)$).

The canonical ring, since also here $\omega_S \cong \mathcal{O}_S(1)$, equals

$$\mathcal{R}(S) = \mathbb{C}[x_0, x_1, x_2, z]/(F_8).$$

Thus $p_g = 3$, $K^2 = 8/4 = 2$ but $|3K|$ factors through the double cover of $\mathbb{P}^2$.

3.3 Deformation Equivalence of Surfaces

The first important consequence of the theorem on pluricanonical embeddings is the finiteness, up to deformation, of the minimal surfaces S of general type with fixed invariants K^2 and χ.

In fact, their 5-canonical models Σ_5 are surfaces with Rational Double Points and of degree $25K^2$ in a fixed projective space $\mathbb{P}^N$, where $N + 1 = P_5 = h^0(5K_S) = \chi + 10K^2$.

In fact, the Hilbert polynomial of Σ_5 equals

$$P(m) := h^0(5mK_S) = \chi + \frac{1}{2}(5m - 1)5mK^2.$$

Grothendieck [Groth60] showed that there is

(i) An integer d and

(ii) A subscheme $\mathcal{H} = \mathcal{H}_P$ of the Grassmannian of codimension $P(d)$-subspaces of $H^0(\mathbb{P}^N, \mathcal{O}(d))$, called Hilbert scheme, such that

(iii) $\mathcal{H}$ parametrizes the degree d pieces $H^0(\mathcal{I}_\Sigma(d))$ of the homogeneous ideals of all the subschemes $\Sigma \subset \mathbb{P}^N$ having the given Hilbert polynomial P.

Inside $\mathcal{H}$ one has the open set

$$\mathcal{H}^0 := \{\Sigma | \Sigma \text{ is reduced with only R.D.P.'s as singularities}\}$$

and one defines

Definition 3.23 *The 5-pseudo moduli space of surfaces of general type with given invariants K^2, χ is the closed subset $\mathcal{H}_0 \subset \mathcal{H}^0$,*

$$\mathcal{H}_0(\chi, K^2) := \{\Sigma \in \mathcal{H}^0 | \omega_\Sigma^{\otimes 5} \cong \mathcal{O}_\Sigma(1)\}$$

Remark 3.24 *The group $\mathbb{P}GL(N+1, \mathbb{C})$ acts on $\mathcal{H}_0$ with finite stabilizers (corresponding to the groups of automorphisms of each surface) and the orbits correspond to the isomorphism classes of minimal surfaces of general type with invariants K^2, χ. A quotient by this action exists as a complex analytic space. Gieseker showed in [Gie77] that if one replaces the 5-canonical embedding by an m-canonical embedding with much higher m, then the corresponding quotient exists as a quasi-projective scheme.*

Since $\mathcal{H}_0$ is a quasi-projective scheme, it has a finite number of irreducible components (to be precise, these are the irreducible components of $(\mathcal{H}_0)_{red}$).

Definition 3.25 *The connected components of $\mathcal{H}_0(\chi, K^2)$ are called the deformation types of the surfaces of general type with given invariants K^2, χ.*

The above deformation types coincide with the equivalence classes for the relation of deformation equivalence (a more general definition introduced by Kodaira and Spencer), in view of the following:

Definition 3.26 *(1) A* deformation *of a compact complex space X is a pair consisting of*

(1.1) A flat morphism $\pi : \mathcal{X} \to T$ between connected complex spaces (i.e., $\pi^* : \mathcal{O}_{T,t} \to \mathcal{O}_{\mathcal{X},x}$ is a flat ring extension for each x with $\pi(x) = t$)

(1.2) An isomorphism $\psi : X \cong \pi^{-1}(t_0) := X_0$ of X with a fibre X_0 of π

(2) Two compact complex manifolds X, Y are said to be direct deformation equivalent *if there are a deformation $\pi : \mathcal{X} \to T$ of X with T irreducible and where all the fibres are smooth, and an isomorphism $\psi' : Y \cong \pi^{-1}(t_1) := X_1$ of Y with a fibre X_1 of π.*

(3) Two canonical models X, Y of surfaces of general type are said to be direct deformation equivalent if there are a deformation $\pi : \mathcal{X} \to T$ of X where T is irreducible and where all the fibres have at most Rational Double

Points as singularities , and an isomorphism $\psi' : Y \cong \pi^{-1}(t_1) := X_1$ *of* Y
with a fibre X_1 *of* π.

(4) Deformation equivalence *is the equivalence relation generated by direct deformation equivalence.*

(5) A small deformation *is the germ* $\pi : (\mathcal{X}, X_0) \to (T, t_0)$ *of a deformation.*

(6) Given a deformation $\pi : \mathcal{X} \to T$ *and a morphism* $f : T' \to T$ *with* $f(t'_0) = t_0$, *the* pull-back $f^*(\mathcal{X})$ *is the fibre product* $\mathcal{X}' := \mathcal{X} \times_T T'$ *endowed with the projection onto the second factor* T' *(then* $X \cong X'_0$*).*

The two definitions (2) and (3) introduced above do not conflict with each other in view of the following

Theorem 3.27 *Given two minimal surfaces of general type* S, S' *and their respective canonical models* X, X', *then*

S and S' are deformation equivalent (resp.: direct deformation equivalent)
$\Leftrightarrow X$ *and* X' *are deformation equivalent (resp.: direct deformation equivalent).*

We shall highlight the idea of proof of the above proposition in the next subsection: we observe here that the proposition implies that the deformation equivalence classes of surfaces of general type correspond to the deformation types introduced above (the connected components of $\mathcal{H}_0$), since over $\mathcal{H}$ lies a natural family $\mathcal{X} \to \mathcal{H}$, $\mathcal{X} \subset \mathbb{P}^N \times \mathcal{H}$, and the fibres over $\mathcal{H}^0 \supset \mathcal{H}_0$ have at most RDP's as singularities.

A simple but powerful observation is that, in order to analyse deformation equivalence, one may restrict oneself to the case where $dim(T) = 1$: since two points in a complex space $T \subset \mathbb{C}^n$ belong to the same irreducible component of T if and only if they belong to an irreducible curve $T' \subset T$.

One may further reduce to the case where T is smooth simply by taking the normalization $T^0 \to T_{red} \to T$ of the reduction T_{red} of T, and taking the pull-back of the family to T^0.

This procedure is particularly appropriate in order to study the closure of subsets of the pseudomoduli space. But in order to show openness of certain subsets, the optimal strategy is to consider the small deformations of the canonical models (this is like Columbus' egg: the small deformations of the minimal models are sometimes too complicated to handle, as shown by Burns and Wahl [B-W74] already for surfaces in $\mathbb{P}^3$).

The basic tool is the generalization due to Grauert of Kuranishi's theorem ([Gra74], see also [Sern06], cor. 1.1.11 page 18, prop. 2.4.8 , page 70)

Theorem 3.28 (Grauert's Kuranishi type theorem for complex spaces) *Let* X *be a compact complex space: then*

(1) there is a semiuniversal deformation $\pi : (\mathcal{X}, X_0) \to (T, t_0)$ *of* X, *i.e., a deformation such that every other small deformation* $\pi' : (\mathcal{X}', X'_0) \to (T', t'_0)$ *is the pull-back of* π *for an appropriate morphism* $f : (T', t'_0) \to (T, t_0)$ *whose*

differential at t_0' is uniquely determined. (II) (T, t_0) is unique up to isomorphism, and is a germ of analytic subspace of the vector space $\operatorname{Ext}^1(\Omega^1_X, \mathcal{O}_X)$, inverse image of the origin under a local holomorphic map (called obstruction map and denoted by ob) ob : $\operatorname{Ext}^1(\Omega^1_X, \mathcal{O}_X) \to T^2(X)$ whose differential vanishes at the origin (the point corresponding to the point t_0).

The obstruction space $T^2(X)$ equals $\operatorname{Ext}^2(\Omega^1_X, \mathcal{O}_X)$ if X is a local complete intersection.

The theorem of Kuranishi [Kur62, Kur65] dealt with the case of compact complex manifolds, and in this case $\operatorname{Ext}^j(\Omega^1_X, \mathcal{O}_X) \cong H^j(X, \Theta_X)$, where $\Theta_X := Hom(\Omega^1_X, \mathcal{O}_X)$ is the sheaf of holomorphic vector fields. In this case the quadratic term in the Taylor development of *ob*, given by the cup product $H^1(X, \Theta_X) \times H^1(X, \Theta_X) \to H^2(X, \Theta_X)$, is easier to calculate.

3.4 Isolated Singularities, Simultaneous Resolution

The main reason in the last subsection to consider deformations of compact complex spaces was the aim to have a finite dimensional base T for the semiuniversal deformation (this would not have been the case in general).

Things work in a quite parallel way if one considers germs of isolated singularities of complex spaces (X, x_0). The definitions are quite similar, and there is an embedding $\mathcal{X} \to \mathbb{C}^n \times T$ such that π is induced by the second projection. There is again a completely similar general theorem by Grauert ([Gra72] and again see [Sern06], cor. 1.1.11 page 18, prop. 2.4.8 , page 70)

Theorem 3.29 (Grauert's theorem for deformations of isolated singularities) *Let (X, x_0) be a germ of an isolated singularity of a complex space: then*

(I) There is a semiuniversal deformation $\pi : (\mathcal{X}, X_0, x_0) \to (\mathbb{C}^n, 0) \times (T, t_0)$ of X, i.e., a deformation such that every other small deformation π' : $(\mathcal{X}', X_0', x_0') \to (\mathbb{C}^n, 0) \times (T', t_0')$ is the pull-back of π for an appropriate morphism $f : (T', t_0') \to (T, t_0)$ whose differential at t_0' is uniquely determined.

(II) (T, t_0) is unique up to isomorphism, and is a germ of analytic subspace of the vector space $\operatorname{Ext}^1(\Omega^1_X, \mathcal{O}_X)$, inverse image of the origin under a local holomorphic map (called obstruction map and denoted by ob) ob : $\operatorname{Ext}^1(\Omega^1_X, \mathcal{O}_X) \to T^2(X)$ whose differential vanishes at the origin (the point corresponding to the point t_0).

The obstruction space $T^2(X)$ equals $\operatorname{Ext}^2(\Omega^1_X, \mathcal{O}_X)$ if X is a local complete intersection.

One derives easily from the above a previous result of G. Tjurina concerning the deformations of isolated hypersurface singularities.

For, assume that $(X, 0) \subset (\mathbb{C}^{n+1}, 0)$ is the zero set of a holomorphic function f, $X = \{z | f(z) = 0\}$ and therefore, if $f_j = \frac{\partial f}{\partial z_j}$, the origin is the only point in the locus $\mathcal{S} = \{z | f_j(z) = 0 \ \forall j\}$.

We have then the exact sequence

$$0 \to \mathcal{O}_X \overset{(f_j)}{\to} \mathcal{O}_X^{n+1} \to \Omega_X^1 \to 0$$

which yields $\mathrm{Ext}^j(\Omega_X^1, \mathcal{O}_X) = 0$ for $j \geq 2$, and

$$\mathrm{Ext}^1(\Omega_X^1, \mathcal{O}_X) \cong \mathcal{O}_{\mathbb{C}^{n+1},0}/(f, f_1, \ldots f_{n+1}) := T^1.$$

In this case the basis of the semiuniversal deformation is just the vector space T^1, called the *Tjurina Algebra*, and one obtains the following

Corollary 3.30 (Tjurina's deformation) *Given $(X, 0) \subset (\mathbb{C}^{n+1}, 0)$ an isolated hypersurface singularity $X = \{z | f(z) = 0\}$, let $g_1, \ldots g_\tau$ be a basis of the Tjurina Algebra $T^1 = \mathcal{O}_{\mathbb{C}^{n+1},0}/(f, f_1, \ldots f_{n+1})$ as a complex vector space.*
Then $\mathcal{X} \subset \mathbb{C}^{n+1} \times \mathbb{C}^\tau$, $\mathcal{X} := \{z | F(z, t) := f(z) + \sum_j t_j g_j(z) = 0\}$ is the semiuniversal deformation of $(X, 0)$.

A similar result holds more generally (with the same proof) when X is a complete intersection of r hypersurfaces $X = \{z | \phi_1(z) = \cdots = \phi_r(z) = 0\}$, and then one has a semiuniversal deformation of the form $\mathcal{X} \subset \mathbb{C}^{n+1} \times \mathbb{C}^\tau$, $\mathcal{X} := \{z | F_i(z, t) := \phi_i(z) + \sum_j t_j G_{i,j}(z) = 0, i = 1, \ldots r\}$.
In both cases the singularity admits a so-called *smoothing*, given by the Milnor fibre (cf. [Mil68])

Definition 3.31 *Given a hypersurface singularity $(X, 0)$, $X = \{z | f(z) = 0\}$, the Milnor fibre $\mathfrak{M}_{\delta,\epsilon}$ is the intersection of the hypersurface $\{z | f(z) = \epsilon\}$ with the ball $\overline{B(0, \delta)}$ with centre the origin and radius $\delta << 1$, when $|\epsilon| << \delta$.*
$\mathfrak{M} := \mathfrak{M}_{\delta,\epsilon}$ is a manifold with boundary whose diffeomorphism type is independent of ϵ, δ when $|\epsilon| << \delta << 1$.
More generally, for a complete intersection, the Milnor fibre is the intersection of the ball $\overline{B(0, \delta)}$ with centre the origin and radius $\delta << 1$ with a smooth level set $X_\epsilon := \{z | \phi_1(z) = \epsilon_1, \ldots \phi_r(z) = \epsilon_r\}$.

Remark 3.32 *Milnor defined the Milnor fibre $\mathfrak{M}$ in a different way, as the intersection of the sphere $S(0, \delta)$ with centre the origin and radius $\delta << 1$ with the set $\{z | f(z) = \eta | f(z) |\}$, for $|\eta| = 1$.*
In this way the complement $S(0, \delta) \backslash X$ is fibred over S^1 with fibres diffeomorphic to the interiors of the Milnor fibres; using Morse theory Milnor showed that $\mathfrak{M}$ has the homotopy type of a bouquet of μ spheres of dimension n, where μ, called the Milnor number, is defined as the dimension of the Milnor algebra $M^1 = \mathcal{O}_{\mathbb{C}^{n+1},0}/(f_1, \ldots f_{n+1})$ as a complex vector space.

The Milnor algebra and the Tjurina algebra coincide in the case of a weighted homogeneous singularity (this means that there are weights $m_0, \ldots m_n$ such that f contains only monomials $z_0^{i_0} \ldots z_n^{i_n}$ of weighted degree $\sum_j i_j m_j = d$), by Euler's rule $\sum_j m_j z_j f_j = df$.

This is the case, for instance, for the Rational Double Points, the singularities which occur on the canonical models of surfaces of general type. Moreover, for these, the Milnor number μ is easily seen to coincide with the index i in the label for the singularity (i.e., $i = n$ for an A_n-singularity), which in turn corresponds to the number of vertices of the corresponding Dynkin diagram.

Therefore, by the description we gave of the minimal resolution of singularities of a RDP, we see that this is also homotopy equivalent to a bouquet of μ spheres of dimension 2. This is in fact no accident, it is just a manifestation of the fact that there is a so-called simultaneous resolution of singularities (cf. [Tju70, Briesk68-b, Briesk71])

Theorem 3.33 (Simultaneous resolution according to Brieskorn and Tjurina) *Let $T := \mathbb{C}^\mu$ be the basis of the semiuniversal deformation of a Rational Double Point $(X, 0)$. Then there exists a finite ramified Galois cover $T' \to T$ such that the pull-back $\mathcal{X}' := \mathcal{X} \times_T T'$ admits a simultaneous resolution of singularities $p : \mathcal{S}' \to \mathcal{X}'$ (i.e., p is bimeromorphic and all the fibres of the composition $\mathcal{S}' \to \mathcal{X}' \to T'$ are smooth and equal, for t'_0, to the minimal resolution of singularities of $(X, 0)$.*

We shall give Tjurina's proof for the case of A_n-singularities.
Proof. Assume that we have the A_n-singularity

$$\{(x, y, z) \in \mathbb{C}^3 | xy = z^{n+1}\}.$$

Then the semiuniversal deformation is given by

$$\mathcal{X} := \{((x, y, z), (a_2, \ldots a_{n+1})) \in \mathbb{C}^3 \times \mathbb{C}^n | xy = z^{n+1} + a_2 z^{n-1} + \ldots a_{n+1}\},$$

the family corresponding to the natural deformations of the simple cyclic covering.

We take a ramified Galois covering with group $\mathcal{S}_{n+1}$ corresponding to the splitting polynomial of the deformed degree $n + 1$ polynomial

$$\mathcal{X}' := \{((x, y, z), (\alpha_1, \ldots \alpha_{n+1})) \in \mathbb{C}^3 \times \mathbb{C}^{n+1} | \sum_j \alpha_j = 0, \ xy = \prod_j (z - \alpha_j)\}.$$

One resolves the new family $\mathcal{X}'$ by defining $\phi_i : \mathcal{X}' \dashrightarrow \mathbb{P}^1$ as

$$\phi_i := (x, \prod_{j=1}^{i}(z - \alpha_j))$$

and then taking the closure of the graph of $\Phi := (\phi_1, \ldots \phi_n) : \mathcal{X}' \dashrightarrow (\mathbb{P}^1)^n$. $\square$

We shall consider now in detail the case of a node, i.e., an A_1 singularity. This singularity and its simultaneous resolution was considered also in the course by Seidel, and will occur once more when dealing with Lefschetz pencils (but then in lower dimension).

Example 3.34 *Consider a node, i.e., an A_1 singularity.*

Here, we write $f = z^2 - x^2 - y^2$, and the total space of the semiuniversal deformation $\mathcal{X} = \{(x, y, z, t)|f - t = 0\} = \{(x, y, z, t)|z^2 - x^2 - y^2 = t\}$ is smooth. The base change $t = w^2$ produces a quadratic nondegenerate singularity at the origin for $\mathcal{X}' = \{(x, y, z, w)|z^2 - x^2 - y^2 = w^2\} = \{(x, y, z, w)|z^2 - x^2 = y^2 + w^2\}$.

The closure of the graph of $\psi := \frac{z-x}{w+iy} = \frac{w-iy}{z+x}$ yields a so-called small resolution, replacing the origin by a curve isomorphic to $\mathbb{P}^1$.

In the Arbeitstagung of 1958 Michael Atiyah made the observation that this procedure is nonunique, since one may also use the closure of the rational map $\tilde{\psi} := \frac{z-x}{w-iy} = \frac{w+iy}{z+x}$ to obtain another small resolution. An alternative way to compare the two resolutions is to blow up the origin, getting the big resolution (with exceptional set $\mathbb{P}^1 \times \mathbb{P}^1$) and view each of the two small resolutions as the contraction of one of the two rulings of $\mathbb{P}^1 \times \mathbb{P}^1$.

Atiyah showed in this way (see also [BPV84]) that the moduli space for K3 surfaces is non Hausdorff.

Remark 3.35 *The first proof of Theorem 3.33 was given by G. Tjurina. It had been observed that the Galois group G of the covering $T' \to T$ in the above theorem is the Weyl group corresponding to the Dynkin diagram of the singularity, defined as follows. If $\mathcal{G}$ is the simple algebraic group corresponding to the Dynkin diagram (see [Hum75]), and H is a Cartan subgroup, N_H its normalizer, then the Weyl group is the factor group $W := N_H/H$. For example, A_n corresponds to the group $SL(n+1, \mathbb{C})$, its Cartan subgroup is the subgroup of diagonal matrices, which is normalized by the symmetric group $\mathcal{S}_{n+1}$, and N_H is here a semidirect product of H with $\mathcal{S}_{n+1}$.*

As we already mentioned, E. Brieskorn [Briesk71] found a direct explanation of this interesting phenomenon, according to a conjecture of Grothendieck. He proved that an element $x \in \mathcal{G}$ is unipotent and subregular iff the morphism $\Psi : \mathcal{G} \to H/W$, sending x to the conjugacy class of its semisimple part x_s, factors around x as the composition of a submersion with the semiuniversal deformation of the corresponding RDP singularity.

With the aid of Theorem 3.33 we can now prove that deformation equivalence for minimal surfaces of general type is the same as restricted deformation equivalence for their canonical models (i.e., one allows only deformations whose fibres have at most canonical singularities).

Idea of the Proof of Theorem 3.27.

It suffices to observe that

(0) if we have a family $p: \mathcal{S} \to \Delta$ where $\Delta \subset \mathbb{C}$ is the unit disk, and the fibres are smooth surfaces, if the central fibre is minimal of general type, then so are all the others.

(1) If we have a family $p : \mathcal{S} \to \Delta$, where $\Delta \subset \mathbb{C}$ is the unit disk, and the fibres are smooth minimal surfaces of general type, then their canonical models form a flat family $\pi : \mathcal{X} \to \Delta$.

(2) If we have a flat family $\pi : \mathcal{X} \to \Delta$ whose fibres X_t have at most Rational Double Points and K_{X_t} is ample, then for each $t \in \Delta$ there is a ramified covering $f : (\Delta, 0) \to (\Delta, t)$ such that the pull back $f^* \mathcal{X}$ admits a simultaneous resolution.

(0) is a consequence of Kodaira's theorem on the stability of -1-curves by deformation (see [Kod63-b]) and of the two following facts:

(i) that a minimal surface S with $K_S^2 > 0$ is either of general type, or isomorphic to $\mathbb{P}^2$ or to a Segre-Hirzebruch surface $\mathbb{F}_n$ ($n \neq 1$, $\mathbb{F}_0 \cong \mathbb{P}^1 \times \mathbb{P}^1$)

(ii) that $\mathbb{P}^2$ is rigid (every deformation of $\mathbb{P}^2$ is a product), while $\mathbb{F}_n$ deforms only to $\mathbb{F}_m$, with $n \equiv m \pmod 2$.

(2) is essentially the above quoted theorem, (1) is a consequence of Bombieri's theorem, since $p_*(\mathcal{O}_{\mathcal{X}}(5K_{\mathcal{X}})$ is generated by global sections and a trivialization of this sheaf provides a morphism $\phi : \mathcal{X} \to \Delta \times \mathbb{P}^N$ which induces the 5-canonical embedding on each fibre. □

We end this section by describing the results of Riemenschneider [Riem74] on the semiuniversal deformation of the quotient singularity $\frac{1}{4}(1,1)$ described in Example 3.3, and a generalization thereof.

More generally, Riemenschneider considers the singularity Y_{k+1}, a quotient singularity of the RDP (Rational Double Point) A_{2k+1} $\{uv - z^{2k+2} = 0\}$ by the involution multiplying (u, v, z) by -1. Indeed, this is a quotient singularity of type $\frac{1}{4k+4}(1, 2k+1)$, and the A_{2k+1} singularity is the quotient by the subgroup $2\mathbb{Z}/(4k+4)\mathbb{Z}$.

We use here the more general concept of Milnor fibre of a smoothing which the reader can find in Definition 4.5.

Theorem 3.36 (Riemenschneider) *The basis of the semiuniversal deformation of the singularity Y_{k+1}, quotient of the RDP A_{2k+1} by multiplication by -1, consists of two smooth components T_1, T_2 intersecting transversally. Both components yield smoothings, but only the smoothing over T_1 admits a simultaneous resolution. The Milnor fibre over T_1 has Milnor number $\mu = k+1$, the Milnor fibre over T_2 has Milnor number $\mu = k$.*

For the sake of simplicity, we shall explicitly describe the two families in the case $k = 0$ of the quotient singularity $\frac{1}{4}(1,1)$ described in Example 3.3. We use for this the two determinantal presentations of the singularity.

(1) View the singularity as $\mathbb{C}[y_0, \dots, y_4]/J$, where J is the ideal generated by the 2×2 minors of the matrix $\begin{pmatrix} y_0 & y_1 & y_2 & y_3 \\ y_5 & y_6 & y_7 & y_4 \end{pmatrix}$ and by the three functions $f_i := y_i - y_{4+i}$, for $i = 1, 2, 3$ (geometrically, this amounts to viewing the rational normal curve of degree 4 as a linear section of the Segre four-fold $\mathbb{P}^1 \times \mathbb{P}^3$). We get the family T_1, with base $\mathbb{C}^3$, by changing the level sets of the three functions f_i , $f_i(y) = t_i$, for $t = (t_1, t_2, t_3) \in \mathbb{C}^3$.

(2) View the singularity as $\mathbb{C}[y_0, \dots, y_4]/I$, where I is the ideal generated by the 2×2 minors of the matrix $\begin{pmatrix} y_0 & y_1 & y_2 \\ y_1 & y_5 & y_3 \\ y_2 & y_3 & y_4 \end{pmatrix}$ and by the function $f := y_5 - y_2$.

In this second realization the cone over a rational normal curve of degree 4 (in $\mathbb{P}^4$) is viewed as a linear section of the cone over the Veronese surface.

We get the family T_2, with base $\mathbb{C}$, by changing the level set of the function f, $y_5 - y_2 = t$, for $t \in \mathbb{C}$.

We see in the latter case that the Milnor fibre is just the complement to a smooth conic in the complex projective plane $\mathbb{P}^2$, therefore its Milnor number (equal by definition to the second Betti number) is equal to 0. Indeed the Milnor fibre is homotopically equivalent to the real projective plane, but this is better seen in another way which allows a great generalization.

In fact, as we already observed, the singularities Y_k are a special case ($n = 2, d = k + 1, a = 1$) of the following

$$\text{Cyclic quotient singularities } \frac{1}{dn^2}(1, dna - 1) = A_{dn-1}/\mu_n.$$

These are quotients of $\mathbb{C}^2$ by a cyclic group of order dn^2 acting with the indicated characters $(1, dna - 1)$, but can also be viewed as quotients of the Rational Double Point A_{dn-1} of equation $uv - z^{dn} = 0$ by the action of the group μ_n of n-roots of unity acting in the following way:

$$\xi \in \mu_n \text{ acts by} : (u, v, z) \to (\xi u, \xi^{-1}v, \xi^a z).$$

This quotient action gives rise to a quotient family $\mathcal{X} \to \mathbb{C}^d$, where $\mathcal{X} = \mathcal{Y}/\mu_n$, $\mathcal{Y}$ is the hypersurface in $\mathbb{C}^3 \times \mathbb{C}^d$ of equation

$$(***) \quad uv - z^{dn} = \Sigma_{k=0}^{d-1} t_k z^{kn}$$

and the action of μ_n is extended trivially on the factor $\mathbb{C}^d$.

We see in this way that the Milnor fibre is the quotient of the Milnor fibre of the Rational Double Point A_{dn-1} by a cyclic group of order n acting freely. In particular, in the case $n = 2, d = 1, a = 1$, it is homotopically equivalent to the quotient of S^2 by the antipodal map, and we get $\mathbb{P}_{\mathbb{R}}^2$.

Another important observation is that $\mathcal{Y}$, being a hypersurface, is Gorenstein (this means that the canonical sheaf $\omega_{\mathcal{Y}}$ is invertible). Hence, such a quotient $\mathcal{X} = \mathcal{Y}/\mu_n$ by an action which is unramified in codimension 1, is (by definition) $\mathbb{Q}$-Gorenstein.

Remark 3.37 *These smoothings were considered by Kollár and Shepherd Barron ([K-SB88], 3.7-3.8-3.9, cf. also [Man90]), who pointed out their relevance in the theory of compactifications of moduli spaces of surfaces, and showed that, conversely, any $\mathbb{Q}$-Gorenstein smoothing of a quotient singularity is induced by the above family (which has a smooth base, $\mathbb{C}^d$).*

Returning to the cyclic quotient singularity $\frac{1}{4}(1, 1)$, the first description that we gave of the $\mathbb{Q}$-Gorenstein smoothing (which does obviously not admit a simultaneous resolution since its Milnor number is 0) makes clear that an alternative way is to view the singularity (cf. Example 3.3) as a bidouble cover of the plane branched on three lines passing through the origin, and then this smoothing (T_2) is simply obtained by deforming these three lines till they meet in three distinct points.

4 Lecture 3: Deformation and Diffeomorphism, Canonical Symplectic Structure for Surfaces of General Type

Summarizing some of the facts we saw up to now, given a birational equivalence class of surfaces of general type, this class contains a unique (complete) smooth minimal surface S, called the *minimal model*, such that K_S is nef $(K_S \cdot C \geq 0$ for every effective curve C); and a unique surface X with at most Rational Double Points as singularities, and such that the invertible sheaf ω_X is ample, called the *canonical model*.

S is the minimal resolution of the singularities of X, and every pluricanonical map of S factors through the projection $\pi : S \to X$.

The basic numerical invariants of the birational class are $\chi := \chi(\mathcal{O}_S) = \chi(\mathcal{O}_X) = 1 - q + p_g$ $(p_g = h^0(\mathcal{O}_S(K_S)) = h^0(\omega_X))$ and $K_S^2 = K_X^2$ (here K_X is a Cartier divisor such that $\omega_X \cong \mathcal{O}_X(K_X)$).

The totality of the canonical models of surfaces with fixed numerical invariants $\chi = x, K^2 = y$ are parametrized (not uniquely, because of the action of the projective group) by a quasi projective scheme $\mathcal{H}_0(x, y)$, which we called the *pseudo moduli space*.

The connected components of the pseudo moduli spaces $\mathcal{H}_0(x, y)$ are the deformation types of the surfaces of general type, and a basic question is whether one can find some invariant to distinguish these. While it is quite easy to find invariants for the irreducible components of the pseudo moduli space, just by using the geometry of the fibre surface over the generic point, it is less easy to produce effective invariants for the connected components. Up to now the most effective invariant to distinguish connected components has been the *divisibility index* r of the canonical class (r is the divisibility of $c_1(K_S)$ in $H^2(S, \mathbb{Z})$) (cf. [Cat86])

Moreover, as we shall try to illustrate more amply in the next lecture, there is another fundamental difference between the curve and the surface case. Given a curve, the genus g determines the topological type, the differentiable type, and the deformation type, and the moduli space $\mathfrak{M}_g$ is irreducible.

In the case of surfaces, the pseudo moduli space $\mathcal{H}_0(x, y)$ is defined over $\mathbb{Z}$, whence the absolute Galois group $Gal(\bar{\mathbb{Q}}, \mathbb{Q})$ operates on it. In fact, it operates by possibly changing the topology of the surfaces considered, in particular the fundamental group may change !

Therefore the algebro-geometric study of moduli spaces cannot be reduced only to the study of isomorphism classes of complex structures on a fixed differentiable manifold.

We shall now recall how the deformation type determines the differentiable type, and later we shall show that each surface of general type S has a symplectic structure (S, ω), unique up to symplectomorphism, such that the cohomology class of ω is the canonical class $c_1(K_S)$.

4.1 Deformation Implies Diffeomorphism

Even if well known, let us recall the theorem of Ehresmann [Ehr43]

Theorem 4.1 (Ehresmann) *Let $\pi : \mathcal{X} \to T$ be a proper submersion of differentiable manifolds with T connected: then π is a differentiable fibre bundle, in particular all the fibre manifolds X_t are diffeomorphic to each other.*

The idea of the proof is to endow $\mathcal{X}$ with a Riemannian metric, so that a local vector field ξ on the base T has a unique differentiable lifting which is orthogonal to the fibres. Then, in the case where T has dimension 1, one integrates the lifted vector field. The general case is proven by induction on $dim_{\mathbb{R}} T$.

The same argument allows a variant with boundary of Ehresmann's theorem

Lemma 4.2 *Let $\pi : \mathcal{M} \to T$ be a proper submersion where $\mathcal{M}$ is a differentiable manifold with boundary, such that also the restriction of π to $\partial \mathcal{M}$ is a submersion. Assume that T is a ball in $\mathbb{R}^n$, and assume that we are given a fixed trivialization ψ of a closed family $\mathcal{N} \to T$ of submanifolds with boundary. Then we can find a trivialization of $\pi : \mathcal{M} \to T$ which induces the given trivialization ψ.*

Proof. It suffices to take on $\mathcal{M}$ a Riemannian metric where the sections $\psi(p, T)$, for $p \in \mathcal{N}$, are orthogonal to the fibres of π. Then we use the customary proof of Ehresmann's theorem, integrating liftings orthogonal to the fibres of standard vector fields on T. $\square$

Ehresmann's theorem implies then the following

Proposition 4.3 *Let X, X' be two compact complex manifolds which are deformation equivalent. Then they are diffeomorphic by a diffeomorphism $\phi : X' \to X$ preserving the canonical class (i.e., such that $\phi^* c_1(K_X) = c_1(K_{X'})$).*

Proof. The result follows by induction once it is established for X, X' fibres of a family $\pi : \mathcal{X} \to \Delta$ over a 1-dimensional disk. Ehresmann's theorem provides a differentiable trivialization $\mathcal{X} \cong X \times \Delta$. Notice that, since the normal bundle to a fibre is trivial, the canonical divisor of a fibre K_{X_t} is the restriction of the canonical divisor $K_{\mathcal{X}}$ to X_t. It follows that the trivialization provides a diffeomorphism ϕ which preserves the canonical class. $\square$

Remark 4.4 *Indeed, by the results of Seiberg Witten theory, an arbitrary diffeomorphism between differentiable 4-manifolds carries $c_1(K_X)$ either to $c_1(K_{X'})$ or to $-c_1(K_{X'})$ (cf. [Wit94] or [Mor96]). Thus deformation equivalence imposes only ϵ more than diffeomorphism only.*

4.2 Symplectic Approximations of Projective Varieties with Isolated Singularities

The variant 4.2 of Ehresmann's theorem will now be first applied to the Milnor fibres of smoothings of isolated singularities.

Let (X, x_0) be the germ of an isolated singularity of a complex space, which is pure dimensional of dimension $n = dim_{\mathbb{C}} X$, assume $x_0 = 0 \in X \subset \mathbb{C}^{n+m}$, and consider as above the ball $\overline{B(x_0, \delta)}$ with centre the origin and radius δ. Then, for all $0 < \delta << 1$, the intersection $\mathcal{K}_0 := X \cap S(x_0, \delta)$, called the *link* of the singularity, is a smooth manifold of real dimension $2n - 1$.

Consider the semiuniversal deformation $\pi : (\mathcal{X}, X_0, x_0) \rightarrow (\mathbb{C}^{n+m}, 0) \times (T, t_0)$ of X and the family of singularity links $\mathcal{K} := \mathcal{X} \cap (S(x_0, \delta) \times (T, t_0))$. By a uniform continuity argument it follows that $\mathcal{K} \rightarrow T$ is a trivial bundle if we restrict T suitably around the origin t_0 (it is a differentiably trivial fibre bundle in the sense of stratified spaces, cf. [Math70]).

We can now introduce the concept of Milnor fibres of (X, x_0).

Definition 4.5 *Let (T, t_0) be the basis of the semiuniversal deformation of a germ of isolated singularity (X, x_0), and let $T = T_1 \cup \cdots \cup T_r$ be the decomposition of T into irreducible components. T_j is said to be a smoothing component if there is a $t \in T_j$ such that the corresponding fibre X_t is smooth. If T_j is a smoothing component, then the corresponding Milnor fibre is the intersection of the ball $\overline{B(x_0, \delta)}$ with the fibre X_t, for $t \in T_j$, $|t| < \eta << \delta << 1$.*

Whereas the singularity links form a trivial bundle, the Milnor fibres form only a differentiable bundle of manifolds with boundary over the open set $T_j^0 := \{t \in T_j, |t - t_0| < \eta | X_t \text{ is smooth}\}$.

Since however T_j is irreducible, T_j^0 is connected, and the Milnor fibre is unique up to smooth isotopy, in particular up to diffeomorphism.

We shall now apply again Lemma 4.2 in order to perform some surgeries to projective varieties with isolated singularities.

Theorem 4.6 *Let $X_0 \subset \mathbb{P}^N$ be a projective variety with isolated singularities admitting a smoothing component.*

Assume that for each singular point $x_h \in X$, we choose a smoothing component $T_{j(h)}$ in the basis of the semiuniversal deformation of the germ (X, x_h). Then (obtaining different results for each such choice) X can be approximated by symplectic submanifolds W_t of $\mathbb{P}^N$, which are diffeomorphic to the glueing of the 'exterior' of X_0 (the complement to the union $B = \cup_h B_h$ of suitable (Milnor) balls around the singular points) with the Milnor fibres $\mathcal{M}_h$, glued along the singularity links $\mathcal{K}_{h,0}$.

A pictorial view of the proof is contained in Fig. 3.
Proof.
First of all, for each singular point $x_h \in X$, we choose a holomorphic path $\Delta \rightarrow T_{j(h)}$ mapping 0 to the distinguished point corresponding to the germ (X, x_h), and with image of $\Delta \backslash 0$ inside the smoothing locus $T_{j(h)}^0 \cap \{t | |t| < \eta\}$.

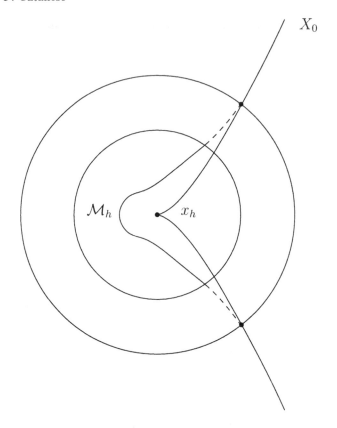

Fig. 3. Glueing the 'exterior' of X_0 (to the Milnor Ball around x_h) with a smaller Milnor fibre $\mathcal{M}_h$

We apply then Lemma 4.2 once more in order to thicken the trivialization of the singularity links to a closed tubular neighbourhood in the family $\mathcal{X}$.

Now, in order to simplify our notation, and without loss of generality, assume that X_0 has only one singular point x_0, and let $B := B(x_0, \delta)$ be a Milnor ball around the singularity. Moreover, for $t \neq 0, t \in \Delta \cap B(0, \eta)$ we consider the Milnor fibre $\mathcal{M}_{\delta,\eta}(t)$, whereas we have the two Milnor links

$$\mathcal{K}_0 := X_0 \cap S(x_0, \delta) \text{ and } \mathcal{K}_t := \mathcal{X}_t \cap S(x_0, \delta - \epsilon)$$

We can consider the Milnor collars $\mathcal{C}_0(\epsilon) := X_0 \cap (\overline{B(x_0, \delta)} \backslash B(x_0, \delta - \epsilon))$, and $\mathcal{C}_t(\epsilon) := \mathcal{X}_t \cap (\overline{B(x_0, \delta)} \backslash B(x_0, \delta - \epsilon))$.

The Milnor collars fill up a complex submanifold of dimension $dim X_0 + 1 := n + 1$ of $\mathbb{C}^{n+m} \times \Delta$.

We glue now $X \backslash B(x_0, \delta - \epsilon))$ and the Milnor fibre $\mathcal{M}_{\delta,\eta}(t)$ by identifying the Milnor collars $\mathcal{C}_0(\epsilon)$ and $\mathcal{C}_t(\epsilon)$.

We obtain in this way an abstract differentiable manifold W which is independent of t, but we want now to give an embedding $W \to W_t \subset \mathbb{C}^{n+m}$ such that $X \backslash B(x_0, \delta))$ maps through the identity, and the complement of the collar inside the Milnor fibre maps to $\mathcal{M}_{\delta,\eta}(t)$ via the restriction of the identity.

As for the collar $\mathcal{C}_0(\delta)$, its outer boundary will be mapped to $\mathcal{K}_0$, while its inner boundary will be mapped to $\mathcal{K}_t$ (i.e., we join the two different singularity links by a differentiable embedding of the abstract Milnor collar).

For $\eta << \delta$ the tangent spaces to the image of the abstract Milnor collar can be made very close to the tangent spaces of the Milnor collars $\mathcal{M}_{\delta,\epsilon}(t)$, and we can conclude the proof via Lemma 2.6.   □

The following well known theorem of Moser guarantees that, once the choice of a smoothing component is made for each $x_h \in Sing(X)$, then the approximating symplectic submanifold W_t is unique up to symplectomorphism.

Theorem 4.7 (Moser) *Let $\pi : \mathcal{X} \to T$ be a proper submersion of differentiable manifolds with T connected, and assume that we have a differentiable 2-form ω on $\mathcal{X}$ with the property that*

(*) $\forall t \in T$ $\omega_t := \omega|_{X_t}$ yields a symplectic structure on X_t whose class in $H^2(X_t, \mathbb{R})$ is locally constant on T (e.g., if it lies on $H^2(X_t, \mathbb{Z})$).

Then the symplectic manifolds (X_t, ω_t) are all symplectomorphic.

The unicity of the symplectic manifold W_t will play a crucial role in the next subsection.

4.3 Canonical Symplectic Structure for Varieties with Ample Canonical Class and Canonical Symplectic Structure for Surfaces of General Type

Theorem 4.8 *A minimal surface of general type S has a canonical symplectic structure, unique up to symplectomorphism, and stable by deformation, such that the class of the symplectic form is the class of the canonical sheaf $\Omega_S^2 = \mathcal{O}_S(K_S)$. The same result holds for any projective smooth variety with ample canonical bundle.*

Proof.

Let V be a smooth projective variety of dimension n whose canonical divisor K_V is ample.

Then there is a positive integer m (depending only on n) such that mK_V is very ample (any $m \geq 5$ does by Bombieri's theorem in the case of surfaces, for higher dimension we can use Matsusaka's big theorem, cf. [Siu93] for an effective version).

Therefore the mth pluricanonical map $\phi_m := \phi_{|mK_V|}$ is an embedding of V in a projective space $\mathbb{P}^{P_m - 1}$, where $P_m := dim H^0(\mathcal{O}_V(mK_V))$.

We define then ω_m as follows: $\omega_m := \frac{1}{m}\phi_m^*(FS)$ (where FS is the Fubini-Study form $\frac{i}{2\pi}\partial\bar{\partial}log|z|^2$), hence ω_m yields a symplectic form as desired.

One needs to show that the symplectomorphism class of (V, ω_m) is independent of m. To this purpose, suppose that the integer r has also the property that ϕ_r yields an embedding of V: the same holds also for rm, hence it suffices to show that (V, ω_m) and (V, ω_{mr}) are symplectomorphic.

To this purpose we use first the well known and easy fact that the pull back of the Fubini-Study form under the rth Veronese embedding v_r equals the rth multiple of the Fubini-Study form. Second, since $v_r \circ \phi_m$ is a linear projection of ϕ_{rm}, by Moser's Theorem follows the desired symplectomorphism. Moser's theorem implies also that if we have a deformation $\pi : \mathcal{V} \to T$ where T is connected and all the fibres have ample canonical divisor, then all the manifolds V_t, endowed with their canonical symplectic structure, are symplectomorphic.

Assume now that S is a minimal surface of general type and that K_S is not ample: then for any $m \geq 5$ (by Bombieri's cited theorem) ϕ_m yields an embedding of the canonical model X of S, which is obtained by contracting the finite number of smooth rational curves with selfintersection number $= -2$ to a finite number of Rational Double Point singularities. For these, the base of the semiuniversal deformation is smooth and yields a smoothing of the singularity.

By the quoted Theorem 3.33 on simultaneous resolution, it follows that

(1) S is diffeomorphic to any smoothing S' of X (but it can happen that X does not admit any global smoothing, as shown by many examples which one can find for instance in [Cat89]).

(2) S is diffeomorphic to the manifold obtained glueing the exterior $X \backslash B$ (B being the union of Milnor balls of radius δ around the singular points of X) together with the respective Milnor fibres, i.e., S is diffeomorphic to each of the symplectic submanifolds W of projective space which approximate the embedded canonical model X according to Theorem 4.6.

We already remarked that W is unique up to symplectomorphism, and this fact ensures that we have a unique canonical symplectic structure on S (up to symplectomorphism).

Clearly moreover, if X admits a global smoothing, we can then take S' sufficiently close to X as our approximation W. Then S' is a surface with ample canonical bundle, and, as we have seen, the symplectic structure induced by (a submultiple of) the Fubini Study form is the canonical symplectic structure.

The stability by deformation is again a consequence of Moser's theorem. $\square$

4.4 Degenerations Preserving the Canonical Symplectic Structure

Assume once more that we consider the minimal surfaces S of general type with fixed invariants $\chi = x$ and $K^2 = y$, and their 5-canonical models Σ_5, which are surfaces with Rational Double Points and of degree $25K^2$ in a fixed projective space $\mathbb{P}^N$, where $N = \chi + 10K^2 - 1$.

The choice of S and of a projective basis for $\mathbb{P}H^0(5K_S)$ yields, as we saw, a point in the 5-pseudo moduli space of surfaces of general type with given invariants $\chi = x$ and $K^2 = y$, i. e., the locally closed set $\mathcal{H}_0(x, y)$ of the corresponding Hilbert scheme $\mathcal{H}$, which is the closed subset

$$\mathcal{H}_0(x, y) := \{\Sigma \in \mathcal{H}^0 | \omega_\Sigma^{\otimes 5} \cong \mathcal{O}_\Sigma(1)\}$$

of the open set

$$\mathcal{H}^0(x, y) := \{\Sigma | \Sigma \text{ is reduced with only R.D.P.'s as singularities}\}.$$

In fact, even if this pseudo moduli space is conceptually clear, it is computationally more complex than just an appropriate open subset of $\mathcal{H}^0(x, y)$, which we denote by $\mathcal{H}^{00}(x, y)$ and parametrizes triples

$$(S, L, \mathcal{B})$$

where

(i) S is a minimal surface of general type with fixed invariants $\chi = x$ and $K^2 = y$

(ii) $L \in Pic^0(S)$ is a topologically trivial holomorphic line bundle

(iii) $\mathcal{B}$ is a a projective basis for $\mathbb{P}H^0(5K_S + L)$.

To explain how to define $\mathcal{H}^{00}(x, y)$, let $\mathcal{H}^n(x, y) \subset \mathcal{H}^0(x, y)$ be the open set of surfaces Σ with $K_\Sigma^2 = y$. Let H be the hyperplane divisor, and observe that by the Riemann Roch theorem $P_\Sigma(m) = \chi(\mathcal{O}_\Sigma) + 1/2 \, mH \cdot (mH - K_\Sigma)$, while by definition $P_\Sigma(m) = x + 1/2(5m - 1)5my$. Hence, $H^2 = 25y$, $H \cdot K_\Sigma = 5y$, $\chi(\mathcal{O}_\Sigma) = x$, and by the Index theorem $K_\Sigma^2 \leq y$, equality holding if and only if $H \sim 5K_\Sigma$.

Since the group of linear equivalence classes of divisors which are numerically equivalent to zero is parametrized by $Pic^0(\Sigma) \times Tors(H^2(\Sigma, \mathbb{Z}))$, we get that the union of the connected components of $\mathcal{H}^n(x, y)$ containing $\mathcal{H}_0(x, y)$ yields an open set $\mathcal{H}^{00}(x, y)$ as described above.

Since $Pic^0(S)$ is a complex torus of dimension $q = h^1(\mathcal{O}_S)$, it follows that indeed there is a natural bijection, induced by inclusion, between irreducible (resp. connected) components of $\mathcal{H}_0(x, y)$ and of $\mathcal{H}^{00}(x, y)$. Moreover, $\mathcal{H}_0(x, y)$ and $\mathcal{H}^{00}(x, y)$ coincide when $q = 0$.

As we shall see, there are surfaces of general type which are diffeomorphic, or even canonically symplectomorphic, but which are not deformation equivalent.

Even if $\mathcal{H}^{00}(x, y)$ is highly disconnected, and not pure dimensional, one knows by a general result by Hartshorne [Hart66], that the Hilbert scheme $\mathcal{H}$ is connected, and one may therefore ask

(A) is $\overline{\mathcal{H}^{00}(x, y)}$ connected?

(B) which kind of singular surfaces does one have to consider in order to connect different components of $\mathcal{H}^{00}(x, y)$?

The latter question is particular significant, since first of all any projective variety admits a flat deformation to a scheme supported on the projective cone over its hyperplane section (iterating this procedure, one reduces to the so-called *stick figures*, which in this case would be supported on a finite union of planes. Second, because when going across badly singular surfaces, then the topology can change drastically (compare Example 5.12, page 329 of [K-SB88]).

We refer to [K-SB88] and to [Vieh95] for a theory of compactified moduli spaces of surfaces of general type. We would only like to mention that the theory describes certain classes of singular surfaces which are allowed, hence a certain open set in the Hilbert scheme $\mathcal{H}$.

One important question is, however, which degenerations of smooth surfaces do not change the canonical symplectomorphism class. In other words, which surgeries do not affect the canonical symplectic structure.

A positive result is the following theorem, which is used in order to show that the Manetti surfaces are canonically symplectomorphic (cf. [Cat06])

Theorem 4.9 *Let* $\mathcal{X} \subset \mathbb{P}^N \times \Delta$ *and* $\mathcal{X}' \subset \mathbb{P}^N \times \Delta'$ *be two flat families of normal surfaces over the disc of radius 2 in* $\mathbb{C}$.

Denote by $\pi : \mathcal{X} \to \Delta$ *and by* $\pi' : \mathcal{X}' \to \Delta$ *the respective projections and make the following assumptions on the respective fibres of* π, π':

(1) the central fibres X_0 *and* X_0' *are surfaces with cyclic quotient singularities and the two flat families yield* $\mathbb{Q}$-*Gorenstein smoothings of them.*

(2) the other fibres X_t, X_t', *for* $t, t' \neq 0$ *are smooth.*

Assume moreover that

(3) the central fibres X_0 *and* X_0' *are projectively equivalent to respective fibres* $(X_0 \cong Y_0$ *and* $X_0' \cong Y_1)$ *of an equisingular projective family* $\mathcal{Y} \subset \mathbb{P}^N \times \Delta$ *of surfaces.*

Set $X := X_1$, $X' := X_1'$: *then*

(a) X *and* X' *are diffeomorphic*

(b) if FS *denotes the symplectic form inherited from the Fubini-Study Kähler metric on* $\mathbb{P}^N$, *then the symplectic manifolds* (X, FS) *and* (X', FS) *are symplectomorphic.*

The proof of the above is based on quite similar ideas to those of the proof of Theorem 4.6.

Remark 4.10 *Theorem 4.9 holds more generally for varieties of higher dimension with isolated singularities under the assumption that, for each singular point* x_0 *of* X_0, *letting* $y_0(t)$ *be the corresponding singularity of* Y_t

(i) $(X_0, x_0) \cong (Y_t, y_0(t))$

(ii) the two smoothings $\mathcal{X}, \mathcal{X}'$, *correspond to paths in the same irreducible component of* $Def(X_0, x_0)$.

5 Lecture 4: Irrational Pencils, Orbifold Fundamental Groups, and Surfaces Isogenous to a Product

In the previous lecture we considered the possible deformations and mild degenerations of surfaces of general type. In this lecture we want to consider a very explicit class of surfaces (and higher dimensional varieties), those which admit an unramified covering which is a product of curves (and are said to be *isogenous to a product*). For these one can reduce the description of the moduli space to the description of certain moduli spaces of curves with automorphisms.

Some of these varieties are rigid, i.e., they admit no nontrivial deformations; in any case these surfaces S have the weak rigidity property that any surface homeomorphic to them is deformation equivalent either to S or to the conjugate surface $\bar{S}$.

Moreover, it is quite interesting to see which is the action of complex conjugation on the moduli space: it turns out that it interchanges often two distinct connected components. In other words,there are surfaces such that the complex conjugate surface is not deformation equivalent to the surface itself (this phenomenon has been observed by several authors independently, cf. [F-M94] Theorem 7.16 and Corollary 7.17 on p. 208, completed in [Fried05] for elliptic surfaces, cf. [KK02,Cat03,BCG05] for the case of surfaces of general type). However, in this case we obtained surfaces which are diffeomorphic to each other, but only through a diffeomorphism not preserving the canonical class.

Other reasons to include these examples are not only their simplicity and beauty, but also the fact that these surfaces lend themself quite naturally to reveal the action of the Galois group $Gal(\overline{\mathbb{Q}}, \mathbb{Q})$ on moduli spaces.

In the next section we shall recall some basic results on fibred surfaces which are used to treat the class of surfaces isogenous to a product.

5.1 Theorem of Castelnuovo–De Franchis, Irrational Pencils and the Orbifold Fundamental Group

We recall some classical and some new results (see [Cat91] and [Cat03b] for more references)

Theorem 5.1 (Castelnuovo–de Franchis) *Let X be a compact Kähler manifold and $U \subset H^0(X, \Omega^1_X)$ be an isotropic subspace (for the wedge product) of dimension ≥ 2. Then there exists a fibration $f : X \to B$, where B is a curve, such that $U \subset f^*(H^0(B, \Omega^1_B))$ (in particular, the genus $g(B)$ of B is at least 2).*

Idea of proof

Let ω_1, ω_2 be two $\mathbb{C}$-linearly independent 1-forms $\in H^0(X, \Omega^1_X)$ such that $\omega_1 \wedge \omega_2 \equiv 0$. Then their ratio defines a nonconstant meromorphic function F with $\omega_1 = F\omega_2$.

After resolving the indeterminacy of the meromorphic map $F : X \dashrightarrow \mathbb{P}^1$ we get a morphism $\tilde{F} : \tilde{X} \to \mathbb{P}^1$ which does not need to have connected fibres, so we let $f : \tilde{X} \to B$ be its Stein factorization.

Since holomorphic forms are closed, $0 = d\omega_1 = dF \wedge \omega_2$ and the forms ω_j restrict to zero on the fibres of f. A small ramification calculation shows then that the two forms ω_j are pull back of holomorphic one forms on B, whence B has genus at least two. Since every map of $\mathbb{P}^1 \to B$ is constant, we see that f is indeed holomorphic on X itself. $\square$

Definition 5.2 *Such a fibration f as above is called an irrational pencil.*

Using Hodge theory and the Künneth formula, the Castelnuovo–de Franchis theorem implies (see [Cat91]) the following

Theorem 5.3 *(Isotropic subspace theorem). (1) Let X be a compact Kähler manifold and $U \subset H^1(X, \mathbb{C})$ be an isotropic subspace of dimension ≥ 2. Then there exists an irrational pencil $f : X \to B$, such that $U \subset f^*(H^1(B, \mathbb{C}))$.*
(2) There is a 1-1 correspondence between irrational pencils $f : X \to B$, $g(B) = b \geq 2$, and subspaces $V = U \oplus \bar{U}$, where U is maximal isotropic of dimension b.

Proof.
(1) Using the fact that $H^1(X, \mathbb{C}) = H^0(X, \Omega_X^1) \oplus \overline{H^0(X, \Omega_X^1)}$ we write a basis of U as $(\phi_1 = \omega_1 + \overline{\eta_1}, \dots, \phi_b = \omega_b + \overline{\eta_b})$.
Since again Hodge theory gives us the direct sum

$$H^2(X, \mathbb{C}) = H^0(X, \Omega_X^2) \oplus H^1(X, \Omega_X^1) \oplus \overline{H^0(X, \Omega_X^2)}$$

the isotropicity condition $\phi_i \wedge \phi_j = 0 \in H^2(X, \mathbb{C})$ reads:

$$\omega_i \wedge \omega_j \equiv 0, \eta_i \wedge \eta_j \equiv 0, \omega_i \wedge \overline{\eta_j} + \overline{\eta_i} \wedge \omega_j \equiv 0, \ \forall i, j.$$

The first two identities show that we are done if one can apply the theorem of Castelnuovo–de Franchis to the ω_j's, respectively to the η_j's , obtaining two irrational pencils $f : X \to B, f' : X \to B'$. In fact, if the image of $f \times f' : X \to B \times B'$ is a curve, then the main assertion is proven. Else, $f \times f'$ is surjective and the pull back f^* is injective. But then $\omega_i \wedge \overline{\eta_j} + \overline{\eta_i} \wedge \omega_j \equiv 0$ contradicts the Künneth formula.
Hence, there is only one case left to consider, namely that, say, all the ω_j's are $\mathbb{C}$-linearly dependent. Then we may assume $\omega_j \equiv 0$, $\forall j \geq 2$ and the above equation yields $\omega_1 \wedge \overline{\eta_j} = 0$, $\forall j \geq 2$. But then $\omega_1 \wedge \eta_j \equiv 0$, since if ξ is the Kähler form, $|\omega_1 \wedge \eta_j|^2 = \int_X \omega_1 \wedge \overline{\eta_j} \wedge \overline{\omega_1 \wedge \overline{\eta_j}} \wedge \xi^{n-2} = 0$.
(2) Follows easily from (1) as follows.
The correspondence is given by $f \mapsto V := f^*(H^1(B, \mathbb{C}))$.
In fact, since $f : X \to B$ is a continuous map which induces a surjection of fundamental groups, then the algebra homomorphism f^* is injective when restricted to $H^1(B, \mathbb{C})$ (this statement follows also without the Kähler hypothesis) and $f^*(H^1(B, \mathbb{C})) \subset H^1(X, \mathbb{C})$ contains many isotropic subspaces U of dimension b with $U \oplus \bar{U} = f^*(H^1(B, \mathbb{C}))$. If such subspace U is not maximal isotropic, then it is contained in U', which determines an irrational pencil f' to a curve B' of genus $> b$, and f factors through f' in view of the fact that every curve of positive genus is embedded in its Jacobian. But this contradicts the fact that f has connected fibres. $\square$

To give an idea of the power of the above result, let us show how the following result due to Gromov ([Grom89], see also [Cat94] for details) follows as a simple consequence

Corollary 5.4 *Let X be a compact Kähler manifold and assume we have a surjective morphism $\pi_1(X) \to \Gamma$, where Γ has a presentation with n generators, m relations, and with $n - m \geq 2$. Then there is an irrational pencil $f : X \to B$, such that $2g(B) \geq n - m$ and $H^1(\Gamma, \mathbb{C}) \subset f^*(H^1(B, \mathbb{C})$.*

Proof. By the argument we gave in (2) above, $H^1(\Gamma, \mathbb{C})$ injects into $H^1(X, \mathbb{C})$ and we claim that each vector v in $H^1(\Gamma, \mathbb{C})$ is contained in a nontrivial isotropic subspace. This follows because the classifying space $Y := K(\Gamma, 1)$ is obtained by attaching n 1-cells, m 2-cells, and then only cells of higher dimension. Hence $h^2(\Gamma, \mathbb{Q}) = h^2(Y, \mathbb{Q}) \leq m$, and $w \to w \wedge v$ has a kernel of dimension ≥ 2 on $H^1(\Gamma, \mathbb{C})$. The surjection $\pi_1(X) \to \Gamma$ induces a continuous map $F : X \to Y$, and each vector in the pull back of $H^1(\Gamma, \mathbb{C})$ is contained in a nontrivial maximal isotropic subspace, thus, by (2) above , in a subspace $V := f^*(H^1(B, \mathbb{C}))$ for a suitable irrational pencil f. Now, the corresponding subspaces V are defined over $\mathbb{Q}$ and $H^1(\Gamma, \mathbb{C})$ is contained in their union. Hence, by Baire's theorem, $H^1(\Gamma, \mathbb{C})$ is contained in one of them. $\square$

In particular, Gromov's theorem applies to a surjection $\pi_1(X) \to \Pi_g$, where $g \geq 2$, and Π_g is the fundamental group of a compact complex curve of genus g. But in general the genus b of the target curve B will not be equal to g, and we would like to detect b directly from the fundamental group $\pi_1(X)$. For this reason (and for others) we need to recall a concept introduced by Deligne and Mostow ([D-M93], see also [Cat00]) in order to extend to higher dimensions some standard arguments about Fuchsian groups.

Definition 5.5 *Let Y be a normal complex space and let D be a closed analytic set. Let $D_1, \ldots, D_r$ be the divisorial (codimension 1) irreducible components of D, and attach to each D_j a positive integer $m_j > 1$.*
    Then the orbifold fundamental group $\pi_1^{orb}(Y \backslash D, (m_1, \ldots m_r))$ *is defined as the quotient of $\pi_1(Y \backslash (D_1 \cup \cdots \cup D_r))$ by the subgroup normally generated by the $\{\gamma_1^{m_1}, \ldots, \gamma_r^{m_r}\}$, where γ_i is a simple geometric loop around the divisor D_i (this means, γ_i is the conjugate via a simple path δ of a local loop γ which, in a local coordinate chart where Y is smooth and $D_i = \{(z)|z_1 = 0\}$, is given by $\gamma(\theta) := (exp(2\pi i\theta), 0, \ldots 0)$, $\forall \theta \in [0, 1]$.*
    We observe in fact that another choice for γ_i gives a conjugate element, so the group is well defined.

Example 5.6 *Let $Y = \mathbb{C}$, $D = \{0\}$: then $\pi_1^{orb}(\mathbb{C} \backslash \{0\}, m) \cong \mathbb{Z}/m$ and its subgroups correspond to the subgroups $H \subset \mathbb{Z}$ such that $H \supset m\mathbb{Z}$, i.e., $H = d\mathbb{Z}$, where d divides m.*

The above example fully illustrates the meaning of the orbifold fundamental group, once we use once more the well known theorem of Grauert and Remmert [GR58]

Remark 5.7 *There is a bijection between*

$$\text{Monodromies } \mu : \pi_1^{orb}(Y\backslash D, (m_1, \ldots m_r)) \to \mathcal{S}(M)$$

and normal locally finite coverings $f : X \to Y$, *with general fibre* $\cong M$, *and such that for each component* R_i *of* $f^{-1}(D_i)$ *the ramification index divides* m_i.

We have moreover (see [Cat00]) the following

Proposition 5.8 *Let* X *be a complex manifold, and* G *a group of holomorphic automorphisms of* X, *acting properly discontinuously. Let* D *be the branch locus of* $\pi : X \to Y := X/G$, *and for each divisorial component* D_i *of* D *let* m_i *be the branching index. Then we have an exact sequence*

$$1 \to \pi_1(X) \to \pi_1^{orb}(Y\backslash D, (m_1, \ldots m_r)) \to G \to 1.$$

Remark 5.9 *(I) In order to extend the above result to the case where* X *is only normal (then* $Y := X/G$ *is again normal), it suffices to define the orbifold fundamental group of a normal variety* X *as*

$$\pi_1^{orb}(X) := \pi_1(X\backslash Sing(X)).$$

(II) Taking the monodromy action of $\pi_1^{orb}(Y\backslash D, (m_1, \ldots m_r))$ *acting on itself by translations, we see that there exists a universal orbifold covering space* $\overline{(Y\backslash D, (m_1, \ldots m_r))}$ *with a properly discontinuous action of* $\pi_1^{orb}(Y\backslash D, (m_1, \ldots m_r))$ *having* Y *as quotient, and the prescribed ramification.*

(III) Obviously the universal orbifold covering space $\overline{(Y\backslash D, (m_1, \ldots m_r))}$ *is (connected and) simply connected.*

Example 5.10 *(a) Let* Y *be a compact complex curve of genus* g, $D = \{p_1, \ldots p_r\}$: *then* $\Gamma := \pi_1^{orb}(Y\backslash\{p_1, \ldots p_r\}, (m_1, \ldots m_r))$ *has a presentation*

$$\Gamma :=< \gamma_1, \ldots, \gamma_r, \alpha_1, \beta_1, \ldots \alpha_g, \beta_g | \gamma_1 \cdot \ldots \cdot \gamma_r \cdot \prod_{i=1}^{g} [\alpha_i, \beta_i] = 1, \gamma_j^{m_j} = 1 >$$

(b) Γ *acts on a simply connected complex curve* Σ, *with* $\Sigma/\Gamma \cong Y$. *By the uniformization theorem* $\Sigma \cong \mathbb{P}^1$ *iff* Σ *is compact, i.e., iff* Γ *is finite (then* $Y \cong \mathbb{P}^1$). *If instead* Γ *is infinite, then there is a finite index subgroup* Γ' *acting freely on* Σ. *Then correspondingly we obtain* $C' := \Sigma/\Gamma' \to Y$ *a finite covering with prescribed ramification* m_i *at each point* p_i.

Example 5.11 *Triangle groups We let* $Y = \mathbb{P}^1$, $r = 3$, *without loss of generality* $D = \{\infty, 0, 1\}$. *Then the orbifold fundamental group in this case reduces to the previously defined triangle group* $T(m_1, m_2, m_3)$ *which has a presentation*

$$T(m_1, m_2, m_3) :=< \gamma_1, \gamma_2, \gamma_3 | \gamma_1 \cdot \gamma_2 \cdot \gamma_3 = 1, \gamma_1^{m_1} = 1, \gamma_2^{m_2} = 1, \gamma_3^{m_3} = 1 > .$$

The triangle group is said to be of elliptic type *iff* $\Sigma \cong \mathbb{P}^1$, *of* parabolic
type *iff* $\Sigma \cong \mathbb{C}$, *of* hyperbolic type *iff* $\Sigma \cong \mathbb{H} := \{\tau | Im(\tau) > 0\}$.

*It is classical (and we have already seen the first alternative as a conse-
quence of Hurwitz' formula in lecture 2) that the three alternatives occur*

- *Elliptic* $\Leftrightarrow \sum_i \frac{1}{m_i} > 1 \Leftrightarrow$ *(2,2,n) or (2,3,n) (n = 3,4,5)*
- *Parabolic* $\Leftrightarrow \sum_i \frac{1}{m_i} = 1 \Leftrightarrow$ *(3,3,3) or (2,3,6) or (2,4,4)*
- *Hyperbolic* $\Leftrightarrow \sum_i \frac{1}{m_i} < 1$

We restrict here to the condition $1 < m_i < \infty$, *else for instance there is also
the parabolic case* $(2, 2, \infty)$, *where the uniformizing function is* $\cos: \mathbb{C} \to \mathbb{P}^1_{\mathbb{C}}$.

The group $T(m_1, m_2, m_3)$, *which was described for the elliptic case in lecture 2, is in the parabolic case a semidirect product of the period lattice* Λ *of
an elliptic curve by its group* μ_n *of linear automorphisms*

- *(3,3,3)* : $\Lambda = \mathbb{Z} \oplus \zeta_3\mathbb{Z}$, ζ_3 *a generator of* μ_3
- *(2,3,6)* : $\Lambda = \mathbb{Z} \oplus \zeta_3\mathbb{Z}$, $-\zeta_3$ *a generator of* μ_6
- *(2,4,4)* : $\Lambda = \mathbb{Z} \oplus i\mathbb{Z}$, i *a generator of* μ_4.

*There is a good reason to call the above 'triangle groups'. Look in fact at
the ramified covering* $f : \Sigma \to \mathbb{P}^1$, *branched in* $\{\infty, 0, 1\}$. *Complex conjugation
on* $\mathbb{P}^1$ *lifts to the covering, as we shall see later in more detail. Consider then
a connected component* Δ *of* $f^{-1}(\mathbb{H})$. *We claim that it is a triangle (in the
corresponding geometry: elliptic, resp. Euclidean, respective hyperbolic) with
angles* $\pi/m_1, \pi/m_2, \pi/m_3$.

*In fact, take a lift of complex conjugation which is the identity on one of
the three sides of* Δ: *then it follows that this side is contained in the fixed
locus of an antiholomorphic automorphism of* Σ, *and the assertion follows
then easily.*

In terms of this triangle (which is unique up to automorphisms of Σ
*in the elliptic and hyperbolic case) it turns out that the three generators of
$T(m_1, m_2, m_3)$ are just rotations around the vertices of the triangle, while the
triangle group $T(m_1, m_2, m_3)$ sits as a subgroup of index 2 inside the group
generated by the reflections on the sides of the triangle.*

Let us leave for the moment aside the above concepts, which will be of
the utmost importance in the forthcoming sections, and let us return to the
irrational pencils.

Definition 5.12 *Let X be a compact Kähler manifold and assume we have
a pencil $f : X \to B$. Assume that $t_1, \dots t_r$ are the points of B whose fibres
$F_i := f^{-1}(t_i)$ are the multiple fibres of f. Denote by m_i the multiplicity of F_i,
i.e., the G.C.D. of the multiplicities of the irreducible components of F_i. Then
the orbifold fundamental group of the fibration $\pi_1(f) := \pi_1(b, m_1, \dots m_r)$ is
defined as the quotient of $\pi_1(B\backslash\{t_1, \dots t_r\})$ by the subgroup normally generated
by the $\gamma_i^{m_i}$'s, where γ_i is a geometric loop around t_i.*

The orbifold fundamental group is said to be of hyperbolic type *if the corresponding universal orbifold (ramified) covering of B is the upper half plane.*

The orbifold fundamental group of a fibration is a natural object in view of the following result (see [CKO03, Cat03b])

Proposition 5.13 *Given a fibration $f : X \to B$ of a compact Kähler manifold onto a compact complex curve B, we have the orbifold fundamental group exact sequence $\pi_1(F) \to \pi_1(X) \to \pi_1(b, m_1, \ldots m_r) \to 0$, where F is a smooth fibre of f.*

The previous exact sequence leads to following result, which is a small generalization of Theorem 4.3. of [Cat03b] and a variant of several other results concerning fibrations onto curves (see [Cat00, Cat03b]), valid more generally for quasi-projective varieties (in this case the starting point is the closedness of logarithmic forms, proven by Deligne in [Del70], which is used in order to obtain extensions of the theorem of Castelnuovo and De Franchis to the non complete case, see [Bau97, Ara97]).

Theorem 5.14 *Let X be a compact Kähler manifold and let $(b, m_1, \ldots m_r)$ be a hyperbolic type. Then there is a bijection between pencils $f : X \to B$ of type $(b, m_1, \ldots m_r)$ and epimorphisms $\pi_1(X) \to \pi_1(b, m_1, \ldots m_r)$ with finitely generated kernel.*

Proof. One direction follows right away from proposition 5.13, so assume that we are given such an epimorphism. Since $\pi_1(b, m_1, \ldots m_r)$ is of hyperbolic type, it contains a normal subgroup H of finite index which is isomorphic to a fundamental group Π_g of a compact curve of genus $g \geq 2$.

Let H' be the pull back of H in $\pi_1(X)$ under the given surjection, and let $X' \to X$ the corresponding Galois cover, with Galois group $G \cong \pi_1(b, m_1, \ldots m_r)/H$.

By the isotropic subspace theorem, there is an irrational pencil $f' : X' \to C$, where the genus of C is at least g, corresponding to the surjection $\psi : \pi_1(X') = H' \to H \cong \Pi_g$. The group G acts on X' leaving the associated cohomology subspace $(f'^*(H^1(C, \mathbb{C}))$ invariant, whence G acts on C preserving the fibration, and we get a fibration $f : X \to B := C/G$.

By Theorem 4.3 of [Cat03b], since the kernel of ψ is finitely generated, it follows that $\psi = f'_* : \pi_1(X') \to \Pi_g = \pi_1(C)$. G operates freely on X' and effectively on C: indeed G acts nontrivially on Π_g by conjugation, since a hyperbolic group has trivial centre. Thus we get an action of $\pi_1(b, m_1, \ldots m_r)$ on the upper half plane $\mathbb{H}$ whose quotient equals $C/G := B$, which has genus b.

We use now again a result from Theorem 4.3 of [Cat03b], namely, that f' has no multiple fibres. Since the projection $C \to B$ is branched in r points with ramification indices equal to $(m_1, \ldots m_r)$, it follows immediately that the orbifold fundamental group of f is isomorphic to $\pi_1(b, m_1, \ldots m_r)$. $\square$

Remark 5.15 *The crucial property of Fuchsian groups which is used in [Cat03b] is the so called NINF property, i.e., that every normal nontrivial subgroup of infinite index is not finitely generated. From this property follows that, given a fibration $f : X' \to C$, the kernel of $f_* : \pi_1(X') \to \pi_1(C)$ is finitely generated (in the hyperbolic case) if and only if there are no multiple fibres.*

5.2 Varieties Isogenous to a Product

Definition 5.16 *A complex algebraic variety X of dimension n is said to be isogenous to a higher product if and only if there is a finite étale cover $C_1 \times \dots C_n \to X$, where $C_1, \dots, C_n$ are compact Riemann surfaces of respective genera $g_i := g(C_i) \geq 2$.*

In fact, X is isogenous to a higher product if and only if there is a finite étale Galois cover of X isomorphic to a product of curves of genera at least two, ie., $X \cong (C_1 \times \dots C_n)/G$, where G is a finite group acting freely on $C_1 \times \dots C_n$.

Moreover, one can prove that there exists a unique minimal such Galois realization $X \cong (C_1 \times \dots C_n)/G$ (see [Cat00]).

In proving this plays a key role a slightly more general fact:

Remark 5.17 *The universal covering of a product of curves $C_1 \times \dots C_n$ of hyperbolic type as above is the polydisk $\mathbb{H}^n$.*

The group of automorphisms of $\mathbb{H}^n$ is a semidirect product of the normal subgroup $Aut(\mathbb{H})^n$ by the symmetric group $\mathcal{S}_n$ (cf. [Ves84] VIII, 1 pages 236–238). This result is a consequence of three basic facts:

(i) Using the subgroup $Aut(\mathbb{H})^n$ we may reduce to consider only automorphisms which leave the origin invariant

(ii) We use the Hurwitz trick to show that the tangent representation is faithful: if $g(z) = z + F_m(z) + \dots$ is the Taylor development at the origin and with mth order term $F_m(z) \neq 0$, then for the rth iterate of g we get $z \to z + rF_m(z) + \dots$, contradicting the Cauchy inequality for the rth iterate when $r \gg 0$

(iii) Using the circular invariance of the domain $(z \to \lambda z, |\lambda| = 1)$, one sees that the automorphisms which leave the origin invariant are linear, since, if $g(0) = 0$, then $g(z)$ and $\lambda^{-1} g(\lambda z)$ have the same derivative at the origin, whence by ii) they are equal

A fortiori, the group of automorphisms of such a product, $Aut(C_1 \times \dots C_n)$ has as normal subgroup $Aut(C_1) \times \dots Aut(C_n)$, and with quotient group a subgroup of $\mathcal{S}_n$.

The above remark leads to the following

Definition 5.18 *A variety isogenous to a product is said to be* unmixed *if in its minimal realization $G \subset Aut(C_1) \times \dots Aut(C_n)$. If $n = 2$, the condition of minimality is equivalent to requiring that $G \to Aut(C_i)$ is injective for $i = 1, 2$.*

The characterization of varieties X isogenous to a (higher) product becomes simpler in the surface case. Hence, assume in the following $X = S$ to be a surface: then

Theorem 5.19 *(see [Cat00]). (a) A projective smooth surface is isogenous to a higher product if and only if the following two conditions are satisfied:*
(1) there is an exact sequence

$$1 \to \Pi_{g_1} \times \Pi_{g_2} \to \pi = \pi_1(S) \to G \to 1,$$

where G is a finite group and where Π_{g_i} denotes the fundamental group of a compact curve of genus $g_i \geq 2$;
(2) $e(S)(= c_2(S)) = \frac{4}{|G|}(g_1 - 1)(g_2 - 1)$.
(b) Any surface X with the same topological Euler number and the same fundamental group as S is diffeomorphic to S. The corresponding subset of the moduli space, $\mathfrak{M}_S^{top} = \mathfrak{M}_S^{diff}$, corresponding to surfaces orientedly homeomorphic, resp. orientedly diffeomorphic to S, is either irreducible and connected or it contains two connected components which are exchanged by complex conjugation.

In particular, if X is orientedly diffeomorphic to S, then X is deformation equivalent to S or to $\bar{S}$.

Sketch of the Proof.
The necessity of conditions (1) and (2) of (a) is clear, since there is an étale Galois cover of S which is a product, and then $e(S) \cdot |G| = e(C_1 \times C_2) = e(C_1) \cdot e(C_2) = 4(g_1 - 1)(g_2 - 1)$.

Conversely, take the étale Galois cover S' of S with group G corresponding to the exact sequence (1). We need to show that S' is isomorphic to a product.

By Theorem 5.14 the two projections of the direct product $\Pi_{g_1} \times \Pi_{g_2}$ yield two holomorphic maps to curves of respective genera g_1, g_2, hence we get a holomorphic map $F : S' \to C_1 \times C_2$, such that $f_j := p_j \circ F : S' \to C_j$ is a fibration. Let h_2 be the genus of the fibres of f_1: then since Π_{g_2} is a quotient of the fundamental group of the fibre, it follows right away that $h_2 \geq g_2$.

We use then the classical (cf. [BPV84], Proposition 11.4, page 97).

Theorem of Zeuthen-Segre *Let $f : S \to B$ be a fibration of an algebraic surface onto a curve of genus b, with fibres of genus g: then*

$$e(S) \geq 4(g - 1)(b - 1),$$

equality holding iff all the fibres are smooth, or , if $g = 1$, all the fibres are multiple of smooth curves.

Hence $e(S) \geq 4(g_1 - 1)(h_2 - 1) \geq (g_1 - 1)(g_2 - 1) = e(S)$, equality holds, $h_2 = g_2$, all the fibres are smooth and F is then an isomorphism.

Part (b): we consider first the unmixed case. This means that the group G does not mix the two factors, whence the individual subgroups Π_{g_i} are normal in $\pi_1(S)$, and moding out by the second of them one gets the exact sequence

$$1 \to \Pi_{g_1} \to \pi_1(S)/\Pi_{g_2} \to G \to 1,$$

which is easily seen to be the orbifold exact sequence for the quotient map $C_1 \rightarrow C_1/G$. This immediately shows that the differentiable structure of the action of G on the product $C_1 \times C_2$ is determined, hence also the differentiable structure of the quotient S is determined by the exact sequence (1) in 5.19.

We have now to choose complex structures on the respective manifolds C_i, which make the action of G holomorphic. Note that the choice of a complex structure implies the choice of an orientation, and that once we have fixed the isomorphism of the fundamental group of C_i with Π_{g_i} and we have chosen an orientation (one of the two generators of $H^2(\Pi_{g_i}, \mathbb{Z})$) we have a marked Riemann surface. Then the theory of Teichmüller spaces shows that the space of complex structures on a marked Riemann surface of genus $g \geq 2$ is a complex manifold $\mathcal{T}_g$ of dimension $3(g-1)$ diffeomorphic to a ball. The finite group G, whose differentiable action is specified, acts on $\mathcal{T}_g$, and the fixed point set equals the set of complex structures for which the action is holomorphic. The result follows then from Proposition 4.13 of [Cat00], which is a slight generalization of one of the solutions [Tro96] of the Nielsen realization problem.

Proposition 5.20 (Connectivity of Nielsen realization) *Given a differentiable action of a finite group G on a fixed oriented and marked Riemann surface of genus g, the fixed locus $Fix(G)$ of G on $\mathcal{T}_g$ is non empty, connected and indeed diffeomorphic to an euclidean space.*

Let us first explain why the above proposition implies part (b) of the theorem (in the unmixed case). Because the moduli space of such surfaces is then the image of a surjective holomorphic map from the union of two connected complex manifolds. We get two such manifolds because of the choice of orientations on both factors which together must give the fixed orientation on our algebraic surface. Now, if we change the choice of orientations, the only admissible choice is the one of reversing orientations on both factors, which is exactly the result of complex conjugation.

Idea of proof Let us now comment on the underlying idea for the above proposition: as already said, Teichmüller space $\mathcal{T}_g$ is diffeomorphic to an Euclidean space of dimension $6g-6$, and admits a Riemannian metric, the Weil-Petersson metric, concerning which Wolpert and Tromba proved the existence of a C^2-function f on $\mathcal{T}_g$ which is proper, G-invariant, non negative ($f \geq 0$), and finally such that f is strictly convex for the given metric (i.e., strictly convex along the W-P geodesics).

Recall that, G being a finite group, its action can be linearized at the fixed points, in particular $Fix(G)$ is a smooth submanifold.

The idea is to use Morse theory for the function f which is strictly convex, and proper, thus it always has a minimum when restricted to a submanifold of $\mathcal{T}_g$

(1) There is a unique critical point x_o for f on $\mathcal{T}_g$, which is an absolute minimum on $\mathcal{T}_g$ (thus $\mathcal{T}_g$ is diffeomorphic to an euclidean space).

(2) If we are given a connected component M of Fix(G), then a critical point y_o for the restriction of f to M is also a critical point for f on T_g: in fact f is G invariant, thus df vanishes on the normal space to M at y_o.

(3) Thus every connected component M of Fix(G) contains x_o, and, Fix(G) being smooth, it is connected. Fix(G) is nonempty since x_0, being the unique minimum, belongs to Fix(G).

(4) Since f is strictly convex, and proper on Fix(G), then by Morse theory Fix(G) is diffeomorphic to an euclidean space.

□

In the mixed case there is a subgroup G^o of index 2 consisting of transformations which do not mix the two factors, and a corresponding subgroup π^o of $\pi = \pi_1(S)$ of index 2, corresponding to an étale double cover S' yielding a surface of unmixed type. By the first part of the proof, it will suffice to show that, once we have found a lifting isomorphism of π^o with a subgroup Γ^o of $Aut(\mathbb{H}) \times Aut(\mathbb{H})$, then the lifting isomorphism of π with a subgroup Γ of $Aut(\mathbb{H} \times \mathbb{H})$ is uniquely determined.

The transformations of Γ^o are of the form $(x, y) \rightarrow (\gamma_1(x), \gamma_2(y))$. Pick any transformation in $\Gamma \backslash \Gamma^o$: it will be a transformation of the form $(a(y), b(x))$. Since it normalizes Γ^o, for each $\delta \in \Gamma^o$ there is $\gamma \in \Gamma^o$ such that

$$a\gamma_2 = \delta_1 a, \quad b\gamma_1 = \delta_2 b.$$

We claim that a, b are uniquely determined. For instance, if a' would also satisfy $a'\gamma_2 = \delta_1 a'$, we would obtain

$$a'a^{-1} = \delta_1(a'a^{-1})\delta_1^{-1}.$$

This would hold in particular for every $\delta_1 \in \Pi_{g_1}$, but since only the identity centralizes such a Fuchsian group, we conclude that $a' = a$.

Remark 5.21 *A completely similar result holds in higher dimension, but the Zeuthen-Segre theorem allows an easier formulation in dimension two.*

One can moreover weaken the hypothesis on the fundamental group, see Theorem B of [Cat00].

5.3 Complex Conjugation and Real Structures

The interest of Theorem 5.19 lies in its constructive aspect.

Theorem 5.19 shows that in order to construct a whole connected component of the moduli space of surfaces of general type, given by surfaces isogenous to a product, it suffices, in the unmixed type, to provide the following data:

(i)   A finite group G

(ii) Two orbifold fundamental groups $A_1 := \pi_1(b_1, m_1, \ldots m_r)$, $A_2 := \pi_1(b_2, n_1, \ldots n_h)$

(iii) Respective surjections $\rho_1 : A_1 \to G$, $\rho_2 : A_2 \to G$ such that
(iv) If we denote by Σ_i the image under ρ_i of the conjugates of the powers of the generators of A_i of finite order, then

$$\Sigma_1 \cap \Sigma_2 = \{1_G\}$$

(v) Each surjection ρ_i is order preserving, in the sense for instance that a generator of $A_1 := \pi_1(b_1, m_1, \ldots m_r)$ of finite order m_i has as image an element of the same order m_i.

In fact, if we take a curve C'_1 of genus b_1, and r points on it, to ρ_1 corresponds a Galois covering $C_i \to C'_i$ with group G, and the elements of G which have a fixed point on C_i are exactly the elements of Σ_i. Therefore we have a diagonal action of G on $C_1 \times C_2$ (i.e., such that $g(x, y) = (\rho_1(g)(x), \rho_2(g)(y))$, and condition iv) is exactly the condition that G acts freely on $C_1 \times C_2$.

There is some arbitrariness in the above choice, namely, in the choice of the isomorphism of the respective orbifold fundamental groups with A_1, A_2, and moreover one can compose each ρ_i simultaneously with the same automorphism of G (i.e., changing G up to isomorphism). Condition (v) is technical, but important in order to calculate the genus of the respective curves C_i.

In order to pass to the complex conjugate surface (this is an important issue in Theorem 5.19), it is clear that we take the conjugate curve of each C'_i, and the conjugate points of the branch points, but we have to be more careful in looking at what happens with the homomorphisms ρ_i.

For this reason, it is worthwhile to recall some basic facts about complex conjugate structures and real structures.

Definition 5.22 *Let X be an almost complex manifold, i.e., the pair of a differentiable manifold M and an almost complex structure J: then the complex conjugate almost complex manifold $\bar{X}$ is given by the pair $(M, -J)$. Assume now that X is a complex manifold, i.e., that the almost complex structure is integrable. Then the same occurs for $-J$, because, if $\chi : U \to \mathbb{C}^n$ is a local chart for X, then $\bar{\chi} : U \to \mathbb{C}^n$ is a local chart for $\bar{X}$.*

In the case where X is a projective variety $X \subset \mathbb{P}^N$, then we easily see that $\bar{X}$ equals $\sigma(X)$, where $\sigma : \mathbb{P}^N \to \mathbb{P}^N$ is given by complex conjugation, and the homogeneous ideal of $\bar{X} = \sigma(X)$ is the complex conjugate of the homogeneous ideal I_X of X, namely:

$$I_{\bar{X}} = \{P \in \mathbb{C}[z_0, \ldots z_N] | \overline{P(\bar{z})} \in I_X\}.$$

Definition 5.23 *Given complex manifolds X, Y let $\phi : X \to \bar{Y}$ be a holomorphic map. Then the same map of differentiable manifolds defines an antiholomorphic map $\bar{\phi} : X \to Y$ (also, equivalently, an antiholomorphic map $\phi^{**} : \bar{X} \to \bar{Y}$).*

A map $\phi : X \to Y$ is said to be dianalytic if it is either holomorphic or antiholomorphic. ϕ determines also a dianalytic map $\phi^{**} : \bar{X} \to \bar{Y}$ which is holomorphic iff ϕ is holomorphic.

The reason to distinguish between the maps ϕ, $\bar{\phi}$ and $\bar{\phi}^{**}$ in the above definition lies in the fact that maps between manifolds are expressed locally as maps in local coordinates, so in these terms $\bar{\phi}(x)$ is indeed the antiholomorphic function $\overline{\phi(x)}$, while $\bar{\phi}^{**}(x) = \phi(\bar{x})$.

With this setup notation, we can further proceed to define the concept of a real structure on a complex manifold.

Definition 5.24 *Let X be a complex manifold.*

(1) The Klein Group of X, denoted by $Kl(X)$ or by $Dian(X)$, is the group of dianalytic automorphisms of X.

(2) A real structure on X is an antiholomorphic automorphism $\sigma : X \to X$ such that $\sigma^2 = Id_X$.

Remark 5.25 *We have a sequence*

$$0 \to Bihol(X) := Aut(X) \to Dian(X) := Kl(X) \to \mathbb{Z}/2 \to 0$$

which is exact if and only if X is biholomorphic to $\bar{X}$, and splits if and only if X admits a real structure.

Example 5.26 *Consider the anharmonic elliptic curve corresponding to the Gaussian integers: $X := \mathbb{C}/(\mathbb{Z} \oplus i\mathbb{Z})$.*

Obviously X is real, since $z \to \bar{z}$ is an antiholomorphic involution.

But there are infinitely many other real structures, since if we take an antiholomorphism σ we can write $\sigma(z) = i^r \bar{z} + \mu$, $\mu = a + ib$, with $a, b \in \mathbb{R}/\mathbb{Z}$ and the condition $\sigma(\sigma(z)) \equiv z (mod \ \mathbb{Z} \oplus i\mathbb{Z})$ is equivalent to

$$i^r \bar{\mu} + \mu = n + im, \ n, m \in \mathbb{Z} \Leftrightarrow a + ib + i^r a - i^{r+1} b = n + im$$

and has the following solutions:

- $r = 0, a \in \{0, 1/2\}, b$ arbitrary
- $r = 1, a = -b$ arbitrary
- $r = 2, a$ arbitrary, $b \in \{0, 1/2\}$
- $r = 3, a = b$ arbitrary.

In the above example the group of biholomorphisms is infinite, and we have an infinite number of real structures, but many of these are isomorphic, as the number of isomorphism classes of real structures is equal to the number of conjugacy classes (for $Aut(X)$) of such splitting involutions.

For instance, in the genus 0 case, there are only two conjugacy classes of real structures on $\mathbb{P}^1_{\mathbb{C}}$:

$$\sigma(z) = \bar{z}, \ \sigma(z) = -\frac{1}{\bar{z}}.$$

They are obviously distinguished by the fact that in the first case the set of real points $X(\mathbb{R}) = Fix(\sigma)$ equals $\mathbb{P}^1_{\mathbb{R}}$, while in the second case we have an empty set. The sign is important, because the real structure $\sigma(z) = \frac{1}{\bar{z}}$, which has $\{z||z| = 1\}$ as set of real points, is conjugated to the first. Geometrically, in the first case we have the circle of radius 1, $\{(x, y, z) \in \mathbb{P}^2_{\mathbb{C}} | x^2 + y^2 + z^2 = 1\}$, in the second the imaginary circle of radius -1, $\{(x, y, z) \in \mathbb{P}^2_{\mathbb{C}} | x^2 + y^2 + z^2 = -1\}$.

It is clear from the above discussion that there can be curves C which are isomorphic to their conjugate, yet do not need to be real: this fact was discovered by C. Earle, and shows that the set of real curves is only a semialgebraic set of the complex moduli space, because it does not coincide with the set $\mathfrak{M}_g(\mathbb{R})$ of real points of $\mathfrak{M}_g$.

We want now to give some further easy example of this situation.

We observe preliminarily that C is isomorphic to $\bar{C}$ if and only in there is a finite group G of automorphisms such that C/G has a real structure which lifts to an antiholomorphism of C (in fact, if $C \cong \bar{C}$ it suffices to take $Aut(C) = G$ if $g(C) \geq 2$).

We shall denote this situation by saying that the covering $C \to C/G$ is real.

Definition 5.27 *We shall say that the covering* $C \to C/G$ *is an* n-angle *covering if* $C/G \cong \mathbb{P}^1$ *and the branch points set consists of* n *points.*

We shall say that C *is an* n-angle *curve if* $C \to C/Aut(C)$ *is an* n-angle *covering.*

Remark 5.28 *(a) Triangle coverings furnish an example of a moduli space* (C, G), *of the type discussed above, which consists of a single point.*

(b) If $C \to C/G$ *is an* n-angle *covering with* n *odd, then the induced real structure on* $C/G \cong \mathbb{P}^1$ *has a non empty set of real points (the branch locus* B *is indeed invariant), thus we may assume it to be the standard complex conjugation* $z \mapsto \bar{z}$.

Example 5.29 *We construct here examples of families of real quadrangle covers* $C \to C/G$ *such that* $(C/G)(\mathbb{R}) = \emptyset$, *and such that, for a general curve in the family,* $G = Aut(C)$, *and the curve* C *is not real. The induced real structure on* $C/G \cong \mathbb{P}^1$ *is then* $\sigma(z) = -\frac{1}{\bar{z}}$, *and the quotient* $(C/G)/\sigma \cong \mathbb{P}^2_{\mathbb{R}}$.

We choose then as branch set $B \subset \mathbb{P}^1_{\mathbb{C}}$ *the set* $\{\infty, 0, w, -\frac{1}{\bar{w}}\}$, *and denote by* $0, u$ *the corresponding image points in* $\mathbb{P}^2_{\mathbb{R}}$.

Observe now that

$$\pi_1(\mathbb{P}^2_{\mathbb{R}} \setminus \{0, u\}) = \langle a, b, x | ab = x^2 \rangle \cong \langle a, x \rangle$$

and the étale double covering $\mathbb{P}^1_{\mathbb{C}} \to \mathbb{P}^2_{\mathbb{R}}$ *corresponds to the quotient obtained by setting* $a = b = 1$, *thus* $\pi_1(\mathbb{P}^1_{\mathbb{C}} \setminus B)$ *is the free group of rank 3*

$$\pi_1(\mathbb{P}^1_{\mathbb{C}} \setminus B) = \langle a, x^2, x^{-1}ax \rangle \cong \langle a, b, a' := x^{-1}ax, b' := x^{-1}bx | ab = a'b' \rangle.$$

We let G' *be the group* $(\mathbb{Z}/2n) \oplus (\mathbb{Z}/m)$, *and let* C *be the Galois cover of* $\mathbb{P}^2_{\mathbb{R}}$ *branched in* $\{0, u\}$ *corresponding to the epimorphisms such that* $x \mapsto (1, 0), a \mapsto (0, 1)$. *It follows that* C *is a 4-angle covering with group* $G \cong (2\mathbb{Z}/2n\mathbb{Z}) \oplus (\mathbb{Z}/m\mathbb{Z})$. *It is straightforward to verify the following*

Claim: G' contains no antiholomorphism of order 2, if n is even.

Thus it follows that C is not real, provided that $G = Aut(C)$. To simplify things, let $n = 4, m = 2$. By Hurwitz' formula C has genus 3, and $8 = |G| = 4(g-1)$. Assume that $Aut(C) \neq G$. If G has index 2, then we get an involution on $\mathbb{P}^1$ preserving the branch set B. But the cross-ratio of the four points equals exactly $-\frac{1}{|w|^2}$, and this is not anharmonic for w general (i.e., $\neq 2, -1, 1/2$). If instead $C \rightarrow C/Aut(C)$ is a triangle curve, then we get only a finite number of curves, and again a finite set of values of w, which we can exclude.

Now, since $|Aut(C)| > 8(g-1)$, if $C \rightarrow C/Aut(C)$ is not a triangle curve, then the only possibility, by the Hurwitz' formula, is that we have a quadrangle cover with branching indices $(2, 2, 2, 3)$. But this is absurd, since a ramification point of order 4 for $C \rightarrow C/G$ must have a higher order of ramification for the map $C \rightarrow C/Aut(C)$.

There is however one important special case when a curve isomorphic to its conjugate must be real, we have namely the following

Proposition 5.30 *Let $C \rightarrow C/G$ be a triangle cover which is real and has distinct branching indices $(m_1 < m_2 < m_3)$: then C is real (i.e., C has a real structure).*

Proof. Let σ be the real structure on $C/G \cong \mathbb{P}^1$. The three branch points of the covering must be left fixed by σ, since the branching indices are distinct (observe that $\mu(\sigma_* \gamma_i)$ is conjugate to $\mu(\gamma_i)$, whence it has the same order). Thus, without loss of generality we may assume that the three branch points are real, and indeed equal to $\{0, 1, \infty\}$, while $\sigma(z) = \bar{z}$.

Choose 2 as base point, and a basis of the fundamental group as in Fig. 4:

$$\pi_1(\mathbb{P}^1 \setminus \{0, 1, \infty\}, 2) = \langle \alpha, \beta, \gamma | \alpha\beta\gamma = 1 \rangle, \ \sigma_* \alpha = \alpha^{-1}, \sigma_* \gamma = \gamma^{-1}.$$

Now, σ lifts if and only if the monodromy μ of the G-covering is equivalent to the one of $\mu \circ \sigma_*$ by an inner automorphism $Int(\phi)$ of the symmetric group which yields a group automorphism $\psi \colon G \rightarrow G$. Set $a := \mu(\alpha), b := \mu(\beta)$. Then these two elements generate G, and since $\psi(a) = a^{-1}, \psi(b) = b^{-1}$ it follows that ψ has order 2, as well as the corresponding covering transformation. We have shown the existence of the desired real structure. $\square$

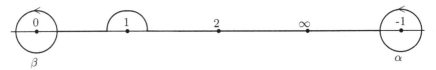

Fig. 4. The loops α and β

We shall now give a simple example of a nonreal triangle cover, based on the following

Lemma 5.31 *Let G be the symmetric group $\mathfrak{S}_n$ in $n \geq 7$ letters, let $a :=$ $(5, 4, 1)(2, 6)$, $c := (1, 2, 3)(4, 5, 6, \ldots, n)$.*
Assume that n is not divisible by 3: then
(1) There is no automorphism ψ of G carrying $a \to a^{-1}$, $c \to c^{-1}$
(2) $\mathfrak{S}_n = <a, c>$
(3) The corresponding triangle cover is not real

Proof. (1) Since $n \neq 6$, every automorphism of G is an inner one. If there is a permutation g conjugating a to a^{-1}, c to c^{-1}, g would leave each of the sets $\{1, 2, 3\}$, $\{4, 5, \ldots, n\}$, $\{1, 4, 5\}$, $\{2, 6\}$ invariant. By looking at their intersections we conclude that g leaves the elements $1, 2, 3, 6$ fixed. But then $gcg^{-1} \neq c^{-1}$.

(2) Observe that a^3 is a transposition: hence, it suffices to show that the group generated by a and c is 2-transitive. Transitivity being obvious, let us consider the stabilizer of 3. Since n is not divisible by 3, the stabilizer of 3 contains the cycle $(4, 5, 6, \ldots, n)$; since it contains the transposition $(2, 6)$ as well as $(5, 4, 1)$, this stabilizer is transitive on $\{1, 2, 4, \ldots, n\}$.

(3) We have $ord(a) = 6$, $ord(c) = 3(n - 3)$, $ord(b) = ord(ca) = ord((1, 6, 3)(4, 2, 7, \ldots n)) = LCM(3, (n - 4))$. Thus the orders are distinct and the nonexistence of such a ψ implies that the triangle cover is not real. □

We can now go back to Theorem 5.19, where the surfaces homeomorphic to a given surface isogenous to a product were forming one or two connected components in the moduli space. The case of products of curves is an easy example where we get one irreducible component, which is self conjugate. We show now the existence of countably many cases where there are two distinct connected components.

Theorem 5.32 *Let $S = (C_1 \times C_2)/G$ be a surface isogenous to a product of unmixed type, with $g_1 \neq g_2$. Then S is deformation equivalent to $\bar{S}$ if and only if (C_j, G) is deformation equivalent to $(\overline{C_j}, G)$ for $j = 1, 2$. In particular, if (C_1, G) is rigid, i.e., $C_1 \to C_1/G$ is a triangle cover, S is deformation equivalent to $\bar{S}$ only if (C_1, G) is isomorphic to $(\overline{C_1}, G)$. There are infinitely many connected components of the moduli space of surfaces of general type for which S is not deformation equivalent to $\bar{S}$.*

Proof. $\bar{S} = \overline{(C_1 \times C_2)/G} = (\overline{(C_1} \times \overline{C_2})/G$ and since $g_1 \neq g_2$ the normal subgroups Π_{g_j} of the fundamental group $\pi_1(\bar{S})$ are uniquely determined. Hence $\bar{S}$ belongs to the same irreducible connected component containing S (according to the key Proposition) if and only if $(\overline{C_j}, G)$ belongs to the same irreducible connected component containing (C_j, G).

We consider now cases where $C_1 \rightarrow C_1/G$ is a triangle cover, but not isomorphic to $(\overline{C_1}, G)$: then clearly S is not deformation equivalent to $\bar{S}$.

We let , for $n \geq 7, n \neq 0 (mod 3)$, $C_1 \rightarrow C_1/G$ be the nonreal triangle cover provided by Lemma 5.31. Let g_1 be the genus of C_1, observe that $2g_1 - 2 \leq (5/6)\, n!$ and consider an arbitrary integer $g \geq 2$ and a surjection $\Pi_g \rightarrow S_n$ (this always exists since Π_g surjects onto a free group with g generators).

The corresponding étale covering of a curve C of genus g is a curve C_2 with genus $g_2 > g_1$ since $2g_2 - 2 \geq (2g - 2)\, n! \geq 2\, n!$. The surfaces $S = C_1 \times C_2$ are our desired examples, the action of $G = S_n$ on the product is free since the action on the second factor is free.   □

Kharlamov and Kulikov gave [KK02, KK02-b] rigid examples of surfaces S which are not isomorphic to their complex conjugate, for instance they considered a $(\mathbb{Z}/5)^2$ covering of the plane branched on the nine lines in the plane $\mathbb{P}^2$ dual to the nine flexes of a cubic, the Fermat cubic for example. These examples have étale coverings which were constructed by Hirzebruch ([Hirz83], see also [BHH87]) in order to produce simple examples of surfaces on the Bogomolov Miyaoka Yau line $K^2 = 3c_2$, which, by results of Yau and Miyaoka [Yau77, Miya83] have the unit ball in $\mathbb{C}^2$ as universal covering, whence they are strongly rigid according to a theorem of Mostow [Most73]: this means that any surface homotopically equivalent to them is either biholomorphic or antibiholomorphic to them.

Kharlamov and Kulikov prove that the Klein group of such a surface S consists only of the above group $(\mathbb{Z}/5)^2$ of biholomorphic transformations, for an appropriate choice of the $(\mathbb{Z}/5)^2$ covering, such that to pairs of conjugate lines correspond pairs of elements of the group which cannot be obtained from each other by the action of a single automorphism of the group $(\mathbb{Z}/5)^2$.

In the next section we shall show how to obtain rigid examples with surfaces isogenous to a product.

5.4 Beauville Surfaces

Definition 5.33 *A surface S isogenous to a higher product is called a Beauville surface if and only if S is rigid.*

This definition is motivated by the fact that Beauville constructed such a surface in [Bea78] , as a quotient $F \times F$ of two Fermat curves of degree 5 (and genus 6). Rigidity was observed in [Cat00].

Example 5.34 ('The' Beauville surfaces) *Let F be the plane Fermat 5-ic $\{x^5 + y^5 + z^5 = 0\}$. The group $(\mathbb{Z}/5)^2$ has a projective action obtained by multiplying the coordinates by fifth roots of unity. The set of stabilizers is given by the multiples of $a := e_1, b := e_2, c := e_1 + e_2$, where $e_1(x, y, z) = (\epsilon x, y, z)$, $e_2(x, y, z) = (x, \epsilon y, z), \epsilon := exp(2\pi i/5)$. In other words, F is a triangle cover of $\mathbb{P}^1$ with group $(\mathbb{Z}/5)^2$ and generators $e_1, e_2, -(e_1+e_2)$. The set σ of stabilizers* *is the union of three lines in the vector space* $(\mathbb{Z}/5)^2$, *corresponding to three points in* $\mathbb{P}^1_{\mathbb{Z}/5}$. *Hence, there is an automorphism* ψ *of* $(\mathbb{Z}/5)^2$ *such that* $\psi(\Sigma) \cap \Sigma = \{0\}$. *Beauville lets then* $(\mathbb{Z}/5)^2$ *act on* $F \times F$ *by the action* $g(P, Q) := (gP, \psi(g)Q)$, *which is free and yields a surface* S *with* $K_S^2 = 8$, $p_g = q = 0$. *It is easy to see that such a surfaces is not only real, but defined over* $\mathbb{Q}$. *It was pointed out in [BaCa04] that there are exactly two isomorphism classes of such Beauville surfaces.*

Let us now construct some Beauville surfaces which are not isomorphic to their complex conjugate.

To do so, we observe that the datum of an unmixed Beauville surface amounts to a purely group theoretical datum, of two systems of generators $\{a, c\}$ and $\{a', c'\}$ for a finite group G such that, defining b through the equation $abc = 1$, and the stabilizer set $\Sigma(a, c)$ as

$$\cup_{i \in \mathbb{N}, g \in G} \{ga^i g^{-1}, gb^i g^{-1}, gc^i g^{-1}\}$$

the following condition must be satisfied, assuring that the diagonal action on the product of the two corresponding triangle curves is free

$$\Sigma(a, c) \cap \Sigma(a', c') = \{1_G\}.$$

Example 5.35 *Consider the symmetric group* $\mathfrak{S}_n$ *for* $n \equiv 2 (mod\ 3)$, *define elements* $a, c \in \mathfrak{S}_n$ *as in Lemma 5.31, and define further* $a' := \sigma^{-1}$, $c' := \tau\sigma^2$, *where* $\tau := (1, 2)$ *and* $\sigma := (1, 2, \dots, n)$. *It is obvious that* $\mathfrak{S}_n = < a', c' >$. *Assuming* $n \geq 8$ *and* $n \equiv 2(3)$, *it is easy to verify that* $\Sigma(a, c) \cap \Sigma(a', c') = \{1\}$, *since one observes that elements which are conjugate in* $\mathfrak{S}_n$ *have the same type of cycle decomposition. The types in* $\Sigma(a, c)$ *are derived from* (6), $(3n - 9)$, $(3n - 12)$, *(as for instance* (3), (2), $(n - 4)$ *and* $(n - 3))$ *since we assume that* 3 *does neither divide* n *nor* $n - 1$, *whereas the types in* $\Sigma(a', c')$ *are derived from* (n), $(n - 1)$, *or* $(\frac{n-1}{2}, \frac{n+1}{2})$.

One sees therefore (since $g_1 \neq g_2$) *that the pairs* $(a, c), (a', c')$ *determine Beauville surfaces which are not isomorphic to their complex conjugates.*

Our knowledge of Beauville surfaces is still rather unsatisfactory, for instance the following question is not yet completely answered.

Question 5.36 *Which groups* G *can occur?*

It is easy to see (cf. [BCG05]) that if the group G is abelian, then it can only be $(\mathbb{Z}/n)^2$, where $G.C.D.(n, 6) = 1$.

Together with I. Bauer and F. Grunewald, we proved in [BCG05] (see also [BCG06]) the following results:

Theorem 5.37 *(1) The following groups admit unmixed Beauville structures*
(a) $\mathfrak{A}_n$ *for large* n
(b) $\mathfrak{S}_n$ *for* $n \in \mathbf{N}$ *with* $n \geq 7$
(c) $\mathbf{SL}(2, \mathbb{F}_p)$, $\mathbf{PSL}(2, \mathbb{F}_p)$ *for* $p \neq 2, 3, 5$

After checking that all finite simple nonabelian groups of order ≤ 50000, with the exception of $\mathfrak{A}_5$, admit unmixed Beauville structures, we were led to the following

Conjecture 5.38 *All finite simple nonabelian groups except $\mathfrak{A}_5$ admit an unmixed Beauville structure.*

Beauville surfaces were extensively studied in [BCG05] (cf. also [BCG06]) with special regard to the effect of complex conjugation on them.

Theorem 5.39 *There are Beauville surfaces S not biholomorphic to $\bar{S}$ with group*

(1) The symmetric group $\mathfrak{S}_n$ for any $n \geq 7$
(2) The alternating group $\mathfrak{A}_n$ for $n \geq 16$ and $n \equiv 0$ mod 4, $n \equiv 1$ mod 3, $n \not\equiv 3, 4$ mod 7

We got also examples of isolated real points in the moduli space which do not correspond to real surfaces:

Theorem 5.40 *Let $p > 5$ be a prime with $p \equiv 1$ mod 4, $p \not\equiv 2, 4$ mod 5, $p \not\equiv 5$ mod 13 and $p \not\equiv 4$ mod 11. Set $n := 3p + 1$. Then there is a Beauville surface S with group $\mathfrak{A}_n$ which is biholomorphic to its conjugate $\bar{S}$, but is not real.*

Beauville surfaces of the mixed type also exist, but their construction turns out to be quite more complicated (see [BCG05]). Indeed (cf. [BCG06-b]) the group of smallest order has order 512.

6 Lecture 5: Lefschetz Pencils, Braid and Mapping Class Groups, and Diffeomorphism of ABC-Surfaces

6.1 Surgeries

The most common surgery is the connected sum, which we now describe.

Let M be a manifold of real dimension m, thus for each point $p \in M$ there is an open set U_p containing p and a homeomorphism (local coordinate chart) $\psi_p : U_p \to V_p \subset \mathbb{R}^m$ onto an open set V_p of $\mathbb{R}^m$ such that (on its domain of definition)

$\psi_{p'} \circ \psi_p^{-1}$ is a:

- Homeomorphism (onto its image) if M is a *topological manifold*
- Diffeomorphism (onto its image) if M is a *differentiable manifold*
- Biholomorpism (onto its image) if M is a *complex manifold* (in this last case $m = 2n$, $\mathbb{R}^m = \mathbb{C}^n$).

Definition 6.1 *The operation of connected sum $M_1 \sharp M_2$ can be done for two differentiable or topological manifolds of the same dimension.*

Choose respective points $p_i \in M_i$ and local charts

$$\psi_{p_i} : U_{p_i} \xrightarrow{\cong} B(0, \epsilon_i) := \{x \in \mathbb{R}^m \,||x| < \epsilon_i\}.$$

Fix positive real numbers $r_i < R_i < \epsilon_i$ such that

$$(\ast\ast) \; R_2/r_2 = R_1/r_1$$

and set $M_i^* := M_i \backslash \psi_{p_i}^{-1}(\overline{B(0, r_i)})$: then M_1^* and M_2^* are glued together through the diffeomorphism $\psi : N_1 := B(0, R_1) \backslash \overline{B(0, r_1)} \to N_2 := B(0, R_2) \backslash \overline{B(0, r_2)}$ such that $\psi(x_1) = \frac{R_2 r_1}{|x_1|} \tau(x_1)$ where either $\tau(x) = x$, or $\tau(x)$ is an orientation reversing linear isometry (in the case where the manifolds M_i are oriented, we might prefer, in order to furnish the connected sum $M_1 \sharp M_2$ of a compatible orientation, to have that ψ be orientation preserving).

In other words the connected sum $M_1 \sharp M_2$ is the quotient space of the disjoint union $(M_1^*) \cup^o (M_2^*)$ through the equivalence relation which identifies $y \in \psi_{p_1}^{-1}(N_1))$ to $w \in \psi_{p_2}^{-1}(N_2))$ iff

$$w = \psi_{p_2}^{-1} \circ \psi \circ \psi_{p_1}(y).$$

We have the following

Theorem *The result of the operation of connected sum is independent of the choices made.*

An elementary and detailed proof in the differentiable case (the one in which we are more interested) can be found in [B-J90], pages 101–110.

Example 6.2 *The most intuitive example (see Fig. 5) is the one of two compact orientable Riemann surfaces M_1, M_2 of respective genera g_1, g_2: $M_1 \sharp M_2$ has then genus $g_1 + g_2$. In this case, however, if M_1, M_2 are endowed of a complex structure, we can even define a connected sum as complex manifolds, setting $\psi(z_1) = e^{2\pi i \theta} \frac{R_2 r_1}{z_1}$.*

Here, however, the complex structure is heavily dependent on the parameters p_1, p_2, $e^{2\pi i\theta}$, and $R_2 r_1 = R_1 r_2$.

In fact, if we set $t := R_2 r_1 e^{2\pi i\theta} \in \mathbb{C}$, we see that $z_1 z_2 = t$, and if $t \to 0$ then it is not difficult to see that the limit of $M_1 \sharp M_2$ is the singular curve obtained from M_1, M_2 by glueing the points p_1, p_2 to obtain the node $z_1 z_2 = 0$.

This interpretation shows that we get in this way all the curves near the boundary of the moduli space $\mathfrak{M}_g$. It is not clear to us in this moment how big a subset of the moduli space one gets through iterated connected sum operations. One should however point out that many of the conjectures made about the stable cohomology ring $H^*(\mathfrak{M}_g, \mathbb{Z})$ were suggested by the possibility of interpreting the connected sum as a sort of H-space structure on the union of all the moduli spaces $\mathfrak{M}_g$ (cf. [Mum83]).

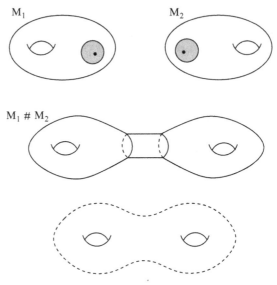

Fig. 5. The connected sum

Remark 6.3 *(1) One cannot perform a connected sum operation for complex manifolds of dimension > 1. The major point is that there is no biholomorphism bringing the inside boundary of the ring domain N_1 to the outside boundary of N_2. The reason for this goes under the name of holomorphic convexity: if $n \geq 2$ every holomorphic function on N_1 has, by Hartogs' theorem, a holomorphic continuation to the ball $B(0, R_1)$. While, for each point p in the outer boundary, there is a holomorphic function f on N_1 such that $lim_{z \to p}|f(z)| = \infty$.*

(2) The operation of connected sum makes the diffeomorphism classes of manifolds of the same dimension m a semigroup: associativity holds, and as neutral element we have the sphere $S^m := \{x \in \mathbb{R}^{m+1}||x| = 1\}$.

(3) A manifold M is said to be irreducible *if $M \cong M_1 \sharp M_2$ implies that either M_1 or M_2 is homotopically equivalent to a sphere S^m.*

A further example is the more general concept of

Definition 6.4 (SURGERY) *For $i = 1, 2$, let $N_i \subset M_i$ be a differentiable submanifold.*

Then there exists (if $M_i = \mathbb{R}^N$ this is an easy consequence of the implicit function theorem) an open set $U_i \supset N_i$ which is diffeomorphic to the normal bundle ν_{N_i} of the embedding $N_i \to M_i$, and through a diffeomorphism which carries N_i onto the zero section of ν_{N_i}.

Suppose now that we have diffeomorphisms $\phi : N_1 \to N_2$, and $\psi : (\nu_{N_1} - N_1) \to (\nu_{N_2} - N_2)$, the latter compatible with the projections $p_i : \nu_{N_i} \to N_i$ (i.e., $p_2 \circ \psi = \phi \circ p_1$), and with the property of being orientation reversing on the *fibres. We can then define as before a manifold $M_1 \natural_\psi M_2$, the quotient of the disjoint union $(M_1 - N_1) \cup^o (M_2 - N_2)$ by the equivalence relation identifying $(U_1 - N_1)$ with $(U_2 - N_2)$ through the diffeomorphism induced by ψ.*

Remark 6.5 *This time the result of the operation depends upon the choice of ϕ and ψ.*

The two surgeries described above combine together in the special situation of the fibre sum.

Definition 6.6 (FIBRE SUM) *For $i = 1, 2$, let $f_i \colon M_i \to B_i$ be a proper surjective differentiable map between differentiable manifolds, let $p_i \in B_i$ be a noncritical value, and let $N_i \subset M_i$ be the corresponding smooth fibre $N_i := f_i^{-1}(p_i)$.*

Then there exists a natural trivialization (up to a constant matrix) of the normal bundle ν_{N_i} of the embedding $N_i \to M_i$, and if we assume as before that we have a diffeomorphism $\phi : N_1 \to N_2$ we can perform a surgery $M := M_1 \natural_\phi M_2$, and the new manifold M admits a proper surjective differentiable map onto the connected sum $B := B_1 \sharp B_2$.

The possibility of variations on the same theme is large: for instance, given $f_i \colon M_i \to B_i$ $(i = 1, 2)$ proper surjective differentiable maps between differentiable manifolds with boundary, assume that $\partial M_i \to \partial B_i$ is a fibre bundle, and there are compatible diffeomorphisms $\phi : \partial B_1 \to \partial B_2$ and $\psi : \partial M_1 \to \partial M_2$: then we can again define the fibre sum $M := M_1 \natural_\psi M_2$ which admits a proper surjective differentiable map onto $B := B_1 \natural_\phi B_2$.

In the case where $(B_2, \partial B_2)$ is an euclidean ball with a standard sphere as boundary, and $M_2 = F \times B_2$, the question about unicity (up to diffeomorphism) of the surgery procedure is provided by a homotopy class. Assume in fact that we have two attaching diffeomorphisms $\psi, \psi' : \partial M_1 \to F \times \partial B_2$. Then from them we construct $\Psi := \psi' \circ \psi^{-1} : F \times \partial B_2 \to F \times \partial B_2$, and we notice that $\Psi(x, t) = (\Psi_1(x, t), \Psi_2(t))$, where $\Psi_2(t) = \phi' \circ \phi^{-1}$. We can then construct a classifying map $\chi \colon \partial B_2 \cong S^{n-1} \to Diff(F)$ such that

$$\Psi_1(x, t) = \chi(\Psi_2(t))(x).$$

We get in this way a free homotopy class $[\chi]$, on which the diffeomorphism class of the surgery depends. If this homotopy class is a priori trivial, then the result is independent of the choices made: this is the case for instance if F is a compact complex curve of genus $g \geq 1$.

In order to understand better the unicity of these surgery operations, and of their compositions, we therefore see the necessity of a good understanding of isotopies of diffeomorphisms. To this topic is devoted the next subsection.

6.2 Braid and Mapping Class Groups

E. Artin introduced the definition of the *braid group* (cf. [Art26, Art65]), thus allowing a remarkable extension of Riemann's concept of monodromy of algebraic functions. Braids are a powerful tool, even if not so easy to handle, and especially appropriate for the study of the differential topology of algebraic varieties, in particular of algebraic surfaces.

Remark 6.7 *We observe that the subsets* $\{w_1, \ldots, w_n\} \subset \mathbb{C}$ *of* n *distinct points in* $\mathbb{C}$ *are in one to one correspondence with monic polynomials* $P(z) \in \mathbb{C}[z]$ *of degree* n *with non vanishing discriminant* $\delta(P)$.

Definition 6.8 *Let* $\mathbb{C}[z]_n^1$ *be the affine space of monic polynomials of degree* n. *Then the group*
$$\mathcal{B}_n := \pi_1(\mathbb{C}[z]_n^1 \setminus \{P|\delta(P) = 0\}),$$
i.e., the fundamental group of the space of monic polynomials of degree n *having* n *distinct roots, is called* Artin's braid group.

Usually, one takes as base point the polynomial $P(z) = (\prod_{i=1}^n (z - i)) \in \mathbb{C}[z]_n^1$ (or the set $\{1, \ldots, n\}$).

To a closed (continuous) path $\alpha : [0, 1] \to (\mathbb{C}[z]_n^1 \setminus \{P|\delta(P) = 0\})$ one can associate the subset $B_\alpha := \{(z, t) \in \mathbb{C} \times \mathbb{R} \mid \alpha_t(z) := \alpha(t)(z) = 0\}$ of $\mathbb{R}^3$, which gives a visually suggestive representation of the associated braid.

It is however customary to view a braid as moving from up to down, that is, to associate to α the set $B'_\alpha := \{(z, t)|(z, -t) \in B_\alpha\}$.

Figure 6 below shows two realizations of the same braid.

Remark 6.9 *There is a lifting of* α *to* $\mathbb{C}^n$, *the space of ordered* n-*tuples of roots of monic polynomials of degree* n, *hence there are (continuous) functions* $w_i(t)$ *such that* $w_i(0) = i$ *and* $\alpha_t(z) = \prod_{i=1}^n (z - w_i(t))$.

It follows that to each braid is naturally associated a permutation $\tau \in \mathfrak{S}_n$ *given by* $\tau(i) := w_i(1)$.

Even if it is not a priori evident, a very powerful generalization of Artin's braid group was given by M. Dehn (cf. [Dehn38], we refer also to the book [Bir74]).

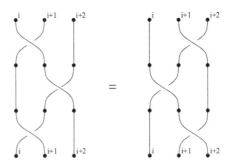

Fig. 6. Relation $aba = bab$ on braids

Definition 6.10 *Let M be a differentiable manifold, then the* mapping class group *(or Dehn group) of M is the group*

$$Map(\mathrm{M}) := \pi_0(Diff(\mathrm{M})) = (Diff(\mathrm{M})/Diff^0(\mathrm{M})),$$

where $Diff^0(\mathrm{M})$, the connected component of the identity, is the subgroup of diffeomorphisms of M isotopic to the identity (i.e., they are connected to the identity by a path in $Diff(\mathrm{M})$).

Remark 6.11 *If M is oriented then we often tacitly take $Diff^+(\mathrm{M})$, the group of orientation preserving diffeomorphisms of M instead of $Diff(\mathrm{M})$, in the definition of the mapping class group. But it is more accurate to distinguish in this case between $Map^+(\mathrm{M})$ and $Map(\mathrm{M})$.*

If M is a compact complex curve of genus g, then its mapping class group is denoted by Map_g. The representation of $M = C_g$ as the $K(\pi, 1)$ space $\mathbb{H}/\Pi_g$, i.e., as a quotient of the (contractible) upper halfplane $\mathbb{H}$ by the free action of a Fuchsian group isomorphic to $\Pi_g \cong \pi_1(C_g)$, immediately yields the isomorphism $Map_g \cong Out(\Pi_g) = Aut(\Pi_g)/Int(\Pi_g)$.

In this way the orbifold exact sequences considered in the previous lecture

$$1 \to \Pi_{g_1} \to \pi_1^{orb} \to G \to 1$$

determine the topological action of G since the homomorphism $G \to Map_g$ is obtain by considering, for $g \in G$, the automorphisms obtained via conjugation by a lift $\tilde{g} \in \pi_1^{orb}$ of g.

The relation between Artin's and Dehn's definition is the following:

Theorem 6.12 *The braid group $\mathcal{B}_n$ is isomorphic to the group*

$$\pi_0(Map^\infty(\mathbb{C}\backslash\{1,\ldots n\})),$$

where $Map^\infty(\mathbb{C}\backslash\{1,\ldots n\})$ is the group of diffeomorphisms which are the identity outside the disk with centre 0 and radius $2n$.

In this way Artin's standard generators σ_i of $\mathcal{B}_n$ $(i = 1,\ldots n-1)$ can be represented by the so-called half-twists.

Definition 6.13 *The* half-twist σ_j *is the diffeomorphism of $\mathbb{C}\backslash\{1,\ldots n\}$ isotopic to the homeomorphism given by:*

– Rotation of $180°$ on the disk with centre $j + \frac{1}{2}$ and radius $\frac{1}{2}$

– On a circle with the same centre and radius $\frac{2+t}{4}$ the map σ_j is the identity if $t \geq 1$ and rotation of $180(1-t)$ degrees, if $t \leq 1$

Now, it is obvious from Theorem 6.12 that $\mathcal{B}_n$ acts on the free group $\pi_1(\mathbb{C}\backslash\{1,\ldots n\})$, which has a geometric basis (we take as base point the complex number $p := -2ni$) $\gamma_1, \ldots \gamma_n$ as illustrated in Fig. 7.

This action is called the *Hurwitz action of the braid group* and has the following algebraic description

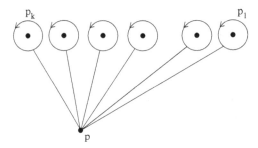

Fig. 7. A geometric basis of $\pi_1(\mathbb{C} - \{1, \ldots n\})$

- $\sigma_i(\gamma_i) = \gamma_{i+1}$
- $\sigma_i(\gamma_i \gamma_{i+1}) = \gamma_i \gamma_{i+1}$, whence $\sigma_i(\gamma_{i+1}) = \gamma_{i+1}^{-1} \gamma_i \gamma_{i+1}$
- $\sigma_i(\gamma_j) = \gamma_j$ for $j \neq i, i+1$

Observe that the product $\gamma_1 \gamma_2 \ldots \gamma_n$ is left invariant under this action.

Definition 6.14 *Let us consider a group G and its cartesian product G^n. The map associating to each $(g_1, g_2, \ldots, g_n)$ the product $g := g_1 g_2 \ldots, g_n \in G$ gives a partition of G^n, whose subsets are called* factorizations *of an element $g \in G$.*

$\mathcal{B}_n$ acts on G^n leaving invariant the partition, and its orbits are called the Hurwitz equivalence classes of factorizations.

We shall use the following notation for a factorization: $g_1 \circ g_2 \circ \cdots \circ g_n$, which should be carefully distinguished from the product $g_1 g_2 \ldots g_n$, which yields an element of G.

Remark 6.15 *A broader equivalence relation for the set of factorizations is obtained considering the equivalence relation generated by Hurwitz equivalence and by* simultaneous conjugation. *The latter, using the following notation $a_b := b^{-1} a b$, corresponds to the action of G on G^n which carries $g_1 \circ g_2 \circ \cdots \circ g_n$ to $(g_1)_b \circ (g_2)_b \circ \cdots \circ (g_n)_b$.*

Observe that the latter action carries a factorization of g to a factorization of the conjugate g_b of g, hence we get equivalence classes of factorizations for conjugacy classes of elements of G.

The above equivalence relation plays an important role in several questions concerning plane curves and algebraic surfaces, as we shall soon see.

Let us proceed for the meantime considering another interesting relation between the braid groups and the Mapping class groups.

This relation is based on the topological model provided by the hyperelliptic curve C_g of equation

$$w^2 = \prod_{i=1}^{2g+2} (z - i)$$

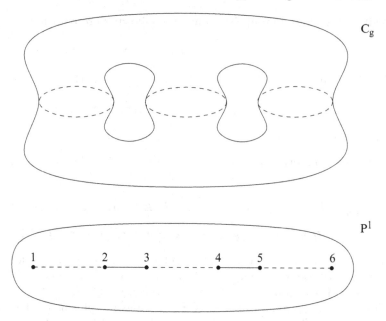

Fig. 8. Hyperelliptic curve of genus 2

(see Fig. 8 describing a hyperelliptic curve of genus $g = 2$).

Observe that, if Y is the double unramified covering of $(\mathbb{P}^1 - \{1, \ldots 2g+2\})$, inverse image of $(\mathbb{P}^1 - \{1, \ldots 2g+2\})$ in C_g, C_g is the natural compactification of Y obtained by adding to Y the *ends* of Y (i.e., in such a compactification one adds to Y the following $lim_{K \subset\subset Y} \pi_0(Y - K)$).

This description makes it clear that every homeomorphism of $(\mathbb{P}^1 - \{1, \ldots 2g + 2\})$ which leaves invariant the subgroup associated to the covering Y admits a lifting to a homeomorphism of Y, whence also to a homeomorphism of its natural compactification C_g.

Such a lifting is not unique, since we can always compose with the nontrivial automorphism of the covering.

We obtain in this way a central extension

$$1 \to \mathbb{Z}/2 =< H > \to \mathcal{M}ap_g^h \to \mathcal{M}ap_{0,2g+2} \to 1$$

where

- H is the hyperelliptic involution $w \to -w$ (the nontrivial automorphism of the covering)
- $\mathcal{M}ap_{0,2g+2}$ is the Dehn group of $(\mathbb{P}^1 - \{1, \ldots 2g + 2\})$
- $\mathcal{M}ap_g^h$ is called the hyperelliptic subgroup of the mapping class group $\mathcal{M}ap_g$, which consists of all the possible liftings.
  If $g \geq 3$, it is a proper subgroup of $\mathcal{M}ap_g$.

While Artin's braid group $\mathcal{B}_{2g+2}$ has the following presentation:

$$\langle \sigma_1, \ldots \sigma_{2g+1} | \sigma_i \sigma_j = \sigma_j \sigma_i \ for |i - j| \geq 2, \ \sigma_i \sigma_{i+1} \sigma_i = \sigma_{i+1} \sigma_i \sigma_{i+1} \rangle,$$

Dehn's group of $(\mathbb{P}^1 - \{1, \ldots 2g + 2\})$ $\mathcal{M}ap_{0,2g+2}$ has the presentation:

$$\langle \sigma_1, \ldots \sigma_{2g+1} | \sigma_1 \cdots \sigma_{2g+1} \sigma_{2g+1} \cdots \sigma_1 = 1, (\sigma_1 \ldots \sigma_{2g+1})^{2g+2} = 1,$$

$$\sigma_i \sigma_j = \sigma_j \sigma_i \ for |i - j| \geq 2, \ \sigma_i \sigma_{i+1} \sigma_i = \sigma_{i+1} \sigma_i \sigma_{i+1} \rangle,$$

finally the hyperelliptic mapping class group $\mathcal{M}ap_g^h$ has the presentation:

$$\langle \xi_1, \ldots \xi_{2g+1}, H | \xi_1 \cdots \xi_{2g+1} \xi_{2g+1} \cdots \xi_1 = H, H^2 = 1, (\xi_1 \ldots \xi_{2g+1})^{2g+2} = 1,$$

$$H\xi_i = \xi_i H \ \forall i, \xi_i \xi_j = \xi_j \xi_i \ for |i - j| \geq 2, \ \xi_i \xi_{i+1} \xi_i = \xi_{i+1} \xi_i \xi_{i+1} \rangle.$$

We want to illustrate the geometry underlying these important formulae. Observe that σ_j yields a homeomorphism of the disk U with centre $j + 1/2$ and radius $3/4$, which permutes the two points $j, j + 1$.

Therefore there are two liftings of σ_j to homeomorphisms of the inverse image V of U in C_g: one defines then ξ_j as the one of the two liftings which acts as the identity on the boundary ∂V, which is a union of two loops (see Fig. 9).

ξ_j is called the *Dehn twist* and corresponds geometrically to the diffeomorphism of a truncated cylinder which is the identity on the boundary, a rotation by 180° on the equator, and on each parallel at height t is a rotation by $t\,360°$ (where $t \in [0, 1]$).

One can define in the same way a Dehn twist for each loop in C_g (i.e., a subvariety diffeomorphic to S^1):

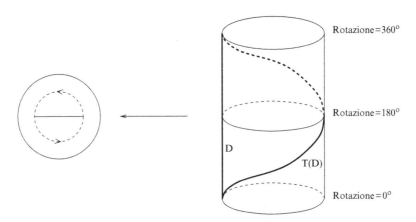

Fig. 9. At the *left*, a half twist; at the *right*: its lift, the Dehn-Twist-T, and its action on the segment D

Definition 6.16 *Let C be an oriented Riemann surface. Then a positive Dehn twist T_α with respect to a simple closed curve α on C is an isotopy class of a diffeomorphism h of C which is equal to the identity outside a neighbourhood of α orientedly homeomorphic to an annulus in the plane, while inside the annulus h rotates the inner boundary of the annulus by 360° to the right and damps the rotation down to the identity at the outer boundary.*

Dehn's fundamental result [Dehn38] was the following

Theorem 6.17 *The mapping class group $\mathcal{Map}_g$ is generated by Dehn twists.*

Explicit presentations of $\mathcal{Map}_g$ have later been given by Hatcher and Thurston [HT80], and an improvement of the method lead to the simplest available presentation, due to Wajnryb ([Waj83], see also [Waj99]).

We shall see in the next subsection how the Dehn twists are related to the theory of Lefschetz fibrations.

6.3 Lefschetz Pencils and Lefschetz Fibrations

The method introduced by Lefschetz for the study of the topology of algebraic varieties is the topological analogue of the method of hyperplane sections and projections of the classical italian algebraic geometers.

An excellent exposition of the theory of Lefschetz pencils is the article by Andreotti and Frankel [A-F69], that we try to briefly summarize here.

Let $X \subset \mathbb{P}^N$ be projective variety, which for simplicity we assume to be smooth, and let $L \cong \mathbb{P}^{N-2} \subset \mathbb{P}^N$ be a general linear subspace of codimension 2. L is the base locus of a pencil of hyperplanes $H_t, t \in \mathbb{P}^1$, and the indeterminacy locus of a rational map $\phi : \mathbb{P}^N \backslash L \to \mathbb{P}^1$.

The intersection $Z : X \cap L$ is smooth, and the blow up of X with centre Z yields a smooth variety X' with a morphism $f \colon X' \to \mathbb{P}^1$ whose fibres are isomorphic to the hyperplane sections $Y_t := X \cap H_t$, while the exceptional divisor is isomorphic to the product $Z \times \mathbb{P}^1$ and on it the morphism f corresponds to the second projection.

Definition 6.18 *The dual variety $W^\vee \subset \mathbb{P}^{N\vee}$ of a projective variety W is defined as the closure of the set of hyperplanes which contain the tangent space TW_p at a smooth point $p \in W$. A pencil of hyperplanes $H_t, t \in \mathbb{P}^1$, is said to be a Lefschetz pencil if the line L' dual to the subspace L*

(1) Does not intersect $W^\vee$ if $W^\vee$ is not a hypersurface
(2) Intersects $W^\vee$ transversally in $\mu := \deg(W^\vee)$ points otherwise

An important theorem is the
Biduality theorem: $(W^\vee)^\vee = W$.

It follows from the above theorem and the previous definition that if $W^\vee$ is not a hypersurface, f is a differentiable fibre bundle, while in case (2) all the fibres are smooth, except μ fibres which correspond to tangent hyperplanes

H_{t_j}. And for these Y_{t_j} has only one singular point p_j, which has an ordinary quadratic singularity as a hypersurface in X (i.e., there are local holomorphic coordinates $(z_1, \ldots z_n)$ for X such that locally at p_h

$$Y_{t_h} = \{z | \sum_j z_j^2 = 0\}).$$

Writing $z_j = u_j + iv_j$, the equation $\sum_j z_j^2 = \rho$ for $\rho \in \mathbb{R}$ reads out as $\sum_j u_j v_j = 0$, $\sum_j (u_j^2 - v_j^2) = \rho$. In vector notation, and assuming $\rho \in \mathbb{R}_{\geq 0}$, we may rewrite as

$$\langle u, v \rangle = 0, |u|^2 = \rho + |v|^2.$$

Definition 6.19 *The* vanishing cycle *is the sphere $\Sigma_{t_h + \rho}$ of $Y_{t_h + \rho}$ given, for $\rho \in \mathbb{R}_{>0}$, by $\{u + iv | |u|^2 = \rho, v = 0\}$.*

The normal bundle of the vanishing cycle Σ_t in Y_t is easily seen, in view of the above equations, to be isomorphic to the tangent bundle to the sphere S^{n-1}, whence we can identify a tubular neighbourhood of Σ_t in Y_t to the unit ball in the tangent bundle of the sphere S^{n-1}. We follow now the definition given in [Kas80] of the corresponding Dehn twist.

Definition 6.20 *Identify the sphere $\Sigma = S^{n-1} = \{u | |u| = 1\}$ to the zero section of its unit tangent bundle $Y = \{(u, v) | \langle u, v \rangle = 0, |u| = 1, |v| \leq 1\}$.*
Then the Dehn twist $T := T_\Sigma$ is the diffeomorphism of Y such that, if we let $\gamma_{u,v}(t)$ be the geodesic on S^{n-1} with initial point u, initial velocity v, then

$$T(u, v) := -(\gamma_{u,v}(\pi|v|), \frac{d}{dt}\gamma_{u,v}(\pi|v|)).$$

We have then: (1) T is the antipodal map on Σ
(2) T is the identity on the boundary $\partial Y = \{(u, v) | \langle u, v \rangle = 0, |u| = 1 = |v|\}$.

One has the

Picard-Lefschetz Theorem *The Dehn twist T is the local monodromy of the family Y_t (given by the level sets of the function $\sum_j z_j^2$).*
Moreover, by the classical Ehresmann theorem, one sees that a singular fibre Y_{t_j} is obtained from a smooth fibre by substituting a neighbourhood of the vanishing cycle Σ with the contractible intersection of the complex quadratic cone $\sum_j z_j^2 = 0$ with a ball around p_j. Hence

Theorem 6.21 (Generalized Zeuthen Segre formula) *The number μ of singular fibres in a Lefschetz pencil, i.e., the degree of the dual variety $X^\vee$, is expressed as a sum of topological Euler numbers*

$$e(X) + e(Z) = 2e(Y) + (-1)^n \mu,$$

where Y is a smooth hyperplane section, and $Z = L \cap X$ is the base locus of the pencil.

Proof. (idea) Replacing Z by $Z \times \mathbb{P}^1$ we see that we replace $e(Z)$ by $2e(Z)$, hence the left hand side expresses the Euler number of the blow up X'.

This number can however be computed from the mapping f: since the Euler number is multiplicative for fibre bundles, we would have that this number were $2e(Y)$ if there were no singular fibre. Since however for each singular fibre we replace something homotopically equivalent to the sphere S^{n-1} by a contractible set, we have to subtract $(-1)^{n-1}$ for each singular fibre. □

Lefschetz pencils were classically used to describe the homotopy and homology groups of algebraic varieties.

The main point is that the finite part of X', i.e., $X' - Y_\infty$, has the socalled 'Lefschetz spine' as homotopy retract.

In order to explain what this means, assume, without loss of generality from the differentiable viewpoint, that the fibres Y_0 and Y_∞ are smooth fibres, and that the singular fibres occur for some roots of unity t_j, which we can order in counterclockwise order.

Definition 6.22 *Notation being as before, define the relative vanishing cycle Δ_j as the union, over t in the segment $[0, t_j]$, of the vanishing cycles $\Sigma_{t,j}$: these are defined, for t far away from t_j, using a trivialization of the fibre bundle obtained restricting $f: X' \to \mathbb{P}^1$ to the half open segment $[0, t_j)$.*

The Lefschetz spine of the Lefschetz pencil is the union of the fibre Y_0 with the μ relative vanishing cycles Δ_j.

Theorem 6.23 (Lefschetz' theorems I and II) *(1) The Lefschetz spine is a deformation retract of $X' - Y_\infty$.*

(2) The affine part $X - Y_\infty$ has the homotopy type of a cell complex of dimension n.

(3) The inclusion $\iota: Y_0 \to X$ induces homology homomorphisms $H_i(\iota): H_i(Y_0, \mathbb{Z}) \to H_i(X, \mathbb{Z})$ which are

(3i) Bijective for $i < n - 1$

(3ii) Surjective if $i = n - 1$; moreover

(4) The kernel of $H_{n-1}(\iota)$ is generated by the vanishing cycles, i.e., by the images of $H_{n-1}(\Sigma_{0,j}, \mathbb{Z})$.

Comment on the proof:

(1) Follows by using the Ehresmann's theorem outside of the singularities, and by retracting locally a neighbourhood of the singularities partly on a smooth fibre Y_t, with $t \in (0, t_j)$, and partly on the union of the vanishing cycles. Then one goes back all the way to Y_0.

For (2) we simply observe that $X - Y_\infty$ has $(Y_0 \backslash Z) \cup (\cup_j \Delta_j)$ as deformation retract. Hence, it is homotopically equivalent to a cell complex obtained by attaching μ n-cells to $Y_0 \backslash Z$, and (2) follows then by induction on n.

(3) and (4) are more delicate and require some diagram chasing, which can be found in [A-F69], and which we do not reproduce here. □

In the 1970s Moishezon and Kas realized (see e.g. [Moi77] and [Kas80]), after the work of Smale on the smoothing of handle attachments, that Lefschetz fibrations could be used to investigate the differential topology of algebraic varieties, and especially of algebraic surfaces.

For instance, they give a theoretical method, which we shall now explain, for the extremely difficult problem to decide whether two algebraic surfaces which are not deformation equivalent are in fact diffeomorphic [Kas80].

Definition 6.24 *Let M be a compact differentiable (or even symplectic) manifold of real even dimension $2n$*

A Lefschetz fibration is a differentiable map $f : M \to \mathbb{P}^1_{\mathbb{C}}$ which

(a) is of maximal rank except for a finite number of critical points $p_1, \ldots p_m$ which have distinct critical values $b_1, \ldots b_m \in \mathbb{P}^1_{\mathbb{C}}$.

(b) has the property that around p_i there are complex coordinates $(z_1, \ldots z_n) \in \mathbb{C}^n$ such that locally $f = \sum_j z_j^2 + const.$ (in the symplectic case, we require the given coordinates to be Darboux coordinates, i.e., such that the symplectic form ω of M corresponds to the natural constant coefficients symplectic structure on $\mathbb{C}^n$).

Remark 6.25 *(1) A similar definition can be given if M is a manifold with boundary, replacing $\mathbb{P}^1_{\mathbb{C}}$ by a disc $D \subset \mathbb{C}$.*

(2) An important theorem of Donaldson [Don99] asserts that for symplectic manifolds there exists (as for the case of projective manifolds) a Lefschetz pencil, i.e., a Lefschetz fibration $f : M' \to \mathbb{P}^1_{\mathbb{C}}$ on a symplectic blow up M' of M (see [MS98] for the definition of symplectic blow-up).

(3) A Lefschetz fibration with smooth fibre $F_0 = f^{-1}(b_0)$ and with critical values $b_1, \ldots b_m \in \mathbb{P}^1_{\mathbb{C}}$, once a geometric basis $\gamma_1, \gamma_2, \ldots, \gamma_m$ of $\pi_1(\mathbb{P}^1_{\mathbb{C}} \backslash \{b_1, \ldots, b_m\}, b_0)$ is chosen, determines a factorization of the identity in the mapping class group $\mathcal{M}ap(F_0)$

$$\tau_1 \circ \tau_2 \circ \cdots \circ \tau_m = Id$$

as a product of Dehn twists.

(4) Assume further that $b_0, b_1, \ldots b_m \in \mathbb{C} = \mathbb{P}^1 \backslash \{\infty\}$: then the Lefschetz fibration determines also a homotopy class of an arc λ between $\tau_1 \tau_2 \ldots \tau_m$ and the identity in $Diff^0(F_0)$. This class is trivial when $F_0 = C_g$, a compact Riemann surface of genus $g \geq 1$.

(5) More precisely, the Lefschetz fibration f determines isotopy classes of embeddings $\phi_j : S^{n-1} \to F_0$ and of bundle isomorphisms ψ_j between the tangent bundle of S^{n-1} and the normal bundle of the embedding ϕ_j; τ_j corresponds then to the Dehn twist for the embedding ϕ_j.

We are now ready to state the theorem of Kas (cf. [Kas80]).

Theorem 6.26 *Two Lefschetz fibrations (M, f), (M', f') are equivalent (i.e., there are two diffeomorphisms $u : M \to M', v : \mathbb{P}^1 \to \mathbb{P}^1$ such that $f' \circ u = v \circ f$) if and only if*

(1) The corresponding invariants

$$(\phi_1, \ldots \phi_m), (\psi_1, \ldots \psi_m); (\phi'_1, \ldots \phi'_m), (\psi'_1, \ldots \psi'_m)$$

correspond to each other via a diffeomorphism of F_0 and a diffeomorphism v of $\mathbb{P}^1$. This implies in particular

(1') The two corresponding factorizations of the identity in the mapping class group are equivalent (under the equivalence relation generated by Hurwitz equivalence and by simultaneous conjugation).

(2) The respective homotopy classes λ, λ' correspond to each other under the above equivalence.

Conversely, given $(\phi_1, \ldots \phi_m)(\psi_1, \ldots \psi_m)$ such that the corresponding Dehn twists $\tau_1, \tau_2, \ldots \tau_m$ yield a factorization of the identity, and given a homotopy class λ of a path connecting $\tau_1 \tau_2 \ldots \tau_m$ to the identity in $Diff(F_0)$, there exists an associated Lefschetz fibration.

If the fibre F_0 is a Riemann surface of genus $g \geq 2$ then the Lefschetz fibration is uniquely determined by the equivalence class of a factorization of the identity

$$\tau_1 \circ \tau_2 \circ \cdots \circ \tau_m = Id$$

as a product of Dehn twists.

Remark 6.27 *(1) A similar result holds for Lefschetz fibrations over the disc and we get a factorization*

$$\tau_1 \circ \tau_2 \circ \cdots \circ \tau_m = \phi$$

of the monodromy ϕ of the fibration over the boundary of the disc D.

(2) A Lefschetz fibration with fibre C_g admits a symplectic structure if each Dehn twist in the factorization is positively oriented (see Sect. 2 of [A-B-K-P-00]).

Assume that we are given two Lefschetz fibrations over $\mathbb{P}^1_{\mathbb{C}}$: then we can consider the fibre sum of these two fibrations, which depends as we saw on a diffeomorphism chosen between two respective smooth fibers (cf. [G-S99] for more details).

This operation translates (in view of the above quoted theorem of Kas) into the following definition of 'conjugated composition' of factorization:

Definition 6.28 *Let $\tau_1 \circ \tau_2 \circ \cdots \circ \tau_m = \phi$ and $\tau'_1 \circ \tau'_2 \circ \cdots \circ \tau'_r = \phi'$ be two factorizations: then their composition conjugated by ψ is the factorization*

$$\tau_1 \circ \tau_2 \circ \ldots \tau_m \circ (\tau'_1)_\psi \circ (\tau'_2)_\psi \circ \cdots \circ (\tau'_r)_\psi = \phi \circ (\phi')_\psi.$$

Remark 6.29 *(1) If ψ and ϕ' commute, we obtain a factorization of $\phi\phi'$.*

(2) A particular case is the one where $\phi = \phi' = id$ and it corresponds to Lefschetz fibrations over $\mathbb{P}^1$.

No matter how beautiful the above results are, for a general X projective or M symplectic, one has Lefschetz pencils, and not Lefschetz fibrations, and a natural question is to which extent the surgery corresponding to the blowup does indeed simplify the differentiable structure of the manifold. In the next subsection we shall consider results by Moishezon somehow related to this question.

6.4 Simply Connected Algebraic Surfaces: Topology Versus Differential Topology

In the case of compact topological manifolds of real dimension 4 the methods of Morse theory and of simplification of cobordisms turned out to encounter overwhelming difficulties, and only in 1982 M. Freedman [Free82], using new ideas in order to show the (topological) triviality of certain handles introduced by Casson, was able to obtain a complete classification of the simply connected compact topological 4-manifolds.

Let M be such a manifold, fix an orientation of M, and let

$$q_M \colon H_2(M, \mathbb{Z}) \times H_2(M, \mathbb{Z}) \to \mathbb{Z}$$

be the intersection form, which is unimodular by Poincaré duality.

Theorem 6.30 (Freedman's theorem) *Let M be an oriented compact simply connected topological manifold: then M is determined by its intersection form and by the Kirby-Siebenmann invariant $\alpha(M) \in \mathbb{Z}/2$, which vanishes if and only if $M \times [0, 1]$ admits a differentiable structure.*

The basic invariants of q_m are its signature $\sigma(M) := b^+(M) - b^-(M)$, and its parity ($q_m$ is said to be *even* iff $q_m(x, x) \equiv 0 \ (mod 2) \ \forall x \in H_2(M, \mathbb{Z})$).

A basic result by Serre [Ser64] says that if q_M is indefinite then it is determined by its rank, signature and parity.

The corollary of Freedman's theorem for complex surfaces is the following

Theorem 6.31 *Let S be a compact simply connected complex surface, and let r be the divisibility index of the canonical class $c_1(K_X) \in H^2(X, \mathbb{Z})$.*
S is said to be EVEN if q_S is EVEN, and this holds iff $r \equiv 0 (mod 2)$, else S is said to be ODD. Then

- *(EVEN) If S is EVEN, then S is topologically a connected sum of copies of $\mathbb{P}^1_{\mathbb{C}} \times \mathbb{P}^1_{\mathbb{C}}$ and of a K3 surface if the signature of the intersection form is negative, and of copies of $\mathbb{P}^1_{\mathbb{C}} \times \mathbb{P}^1_{\mathbb{C}}$ and of a K3 surface with opposed orientation in the case where the signature is positive.*
- *(ODD) S is ODD: then S is topologically a connected sum of copies of $\mathbb{P}^2_{\mathbb{C}}$ and of $\mathbb{P}^{2}_{\mathbb{C}}{}^{opp}$.*

Proof. S has a differentiable structure, whence $\alpha(S) = 0$, and the corollary follows from Serre's result if the intersection form is indefinite.

We shall now show that if the intersection form is definite, then $S \cong \mathbb{P}^2_\mathbb{C}$. Observe that $q = 0$, since S is simply connected, and therefore $b^+(S) = 2p_g + 1$, in particular the intersection form is positive, $b_2 = 2p_g + 1$, hence $e(S) = 2\chi(S) + 1$, and $K^2_S = 10\chi(S) - 1$ by Noether's formula.

By the Yau Miyaoka inequality $q = 0$ implies $K^2_S \leq 9\chi(S)$, whence $\chi(S) \leq 1$ and $p_g = 0$.

Therefore $\chi(S) = 1$, and $K^2_S = 9$. Applying again Yau's theorem [Yau77] we see that $S = \mathbb{P}^2_\mathbb{C}$. In fact, if S were of general type its universal cover would be the unit ball in $\mathbb{C}^2$, contradicting simple connectivity. □

Remark 6.32 $\mathbb{P}^2_\mathbb{C}{}^{opp}$ *is the manifold* $\mathbb{P}^2_\mathbb{C}$ *with opposed orientation.*

A K3 surface is (as we already mentioned) a surface S orientedly diffeomorphic to a nonsingular surface X of degree 4 in $\mathbb{P}^3_\mathbb{C}$, for instance

$$X = \{(x_0, x_1, x_2, x_3) \in \mathbb{P}^3_\mathbb{C} | x_0^4 + x_1^4 + x_2^4 + x_3^4 = 0\}.$$

(by a theorem of Kodaira, cf. [Kod63], S is also deformation equivalent to such a surface X).

Not only $\mathbb{P}^2_\mathbb{C}$ is the only algebraic surface with a definite intersection form, but Donaldson showed that a result of a similar flavour holds for differentiable manifolds, i.e., if we have a positive definite intersection form, then we have topologically a connected sum of copies of $\mathbb{P}^2_\mathbb{C}$.

There are several restrictions for the intersection forms of differentiable manifolds, the oldest one being Rokhlin's theorem stating that the intersection form in the even case is divisible by 16. Donaldson gave other restrictions for the intersection forms of differentiable 4-manifolds (see [D-K90]), but the socalled 11/8 conjecture is still unproven: it states that if the intersection form is even, then we have topologically a connected sum as in the case (EVEN) of Theorem 6.31.

More important is the fact that Donaldson's theory has made clear in the 1980s [Don83, Don86, Don90, Don92] how drastically homeomorphism and diffeomorphism differ in dimension 4, and especially for algebraic surfaces.

Later on, the Seiberg-Witten theory showed with simpler methods the following result (cf. [Wit94] o [Mor96]):

Theorem 6.33 *Any diffeomorphism between minimal surfaces (a fortiori, an even surface is necessarily minimal) S, S' carries $c_1(K_S)$ either to $c_1(K_{S'})$ or to $-c_1(K_{S'})$*

Corollary 6.34 *The divisibility index r of the canonical class $c_1(K_S) \in H^2(S, \mathbb{Z})$ is a differentiable invariant of S.*

Since only the parity $r(mod 2)$ of the canonical class is a topological invariant it is then not difficult to construct examples of simply connected algebraic surfaces which are homeomorphic but not diffeomorphic (see [Cat86]).

Let us illustrate these examples, obtained as simple bidouble covers of $\mathbb{P}^1 \times \mathbb{P}^1$.

These surfaces are contained in the geometric vector bundle whose sheaf of holomorphic sections is $\mathcal{O}_{\mathbb{P}^1 \times \mathbb{P}^1}(a, b) \oplus \mathcal{O}_{\mathbb{P}^1 \times \mathbb{P}^1}(c, d)$ and are described there by the following two equations:

$$z^2 = f(x, y),$$

$$w^2 = g(x, y),$$

where f and g are bihomogeneous polynomials of respective bidegrees $(2a, 2b)$, $(2c, 2d)$ (f is a section of $\mathcal{O}_{\mathbb{P}^1 \times \mathbb{P}^1}(2a, 2b)$, g is a section of $\mathcal{O}_{\mathbb{P}^1 \times \mathbb{P}^1}(2c, 2d)$).

These Galois covers of $\mathbb{P}^1 \times \mathbb{P}^1$, with Galois group $(\mathbb{Z}/2\mathbb{Z})^2$, are smooth if and only if the two curves $C := \{f = 0\}$ and $D := \{g = 0\}$ in $\mathbb{P}^1 \times \mathbb{P}^1$ are smooth and intersect transversally.

The holomorphic invariants can be easily calculated, since, if $p \colon X \to \mathbb{P} := \mathbb{P}^1 \times \mathbb{P}^1$ is the finite Galois cover, then

$$p_* \mathcal{O}_X \cong \mathcal{O}_{\mathbb{P}^1 \times \mathbb{P}^1} \oplus z\mathcal{O}_{\mathbb{P}^1 \times \mathbb{P}^1}(-a, -b) \oplus w\mathcal{O}_{\mathbb{P}^1 \times \mathbb{P}^1}(-c, -d) \oplus zw\mathcal{O}_{\mathbb{P}^1 \times \mathbb{P}^1}(-a-c, -b-d).$$

Hence $h^1(\mathcal{O}_X) = 0$, whereas $h^2(\mathcal{O}_X) = (a - 1)(b - 1) + (c - 1)(d - 1) + (a + c - 1)(b + d - 1)$. Assume that X is smooth: then the ramification formula yields

$$\mathcal{O}_X(K_X) = \mathcal{O}_X(p^* K_{\mathbb{P}^1 \times \mathbb{P}^1} + R) = p^*(\mathcal{O}_{\mathbb{P}^1 \times \mathbb{P}^1}(a + c - 2, b + d - 2))$$

since $R = div(z) + div(w)$. In particular, $K_X^2 = 8(a + c - 2)(b + d - 2)$ and the holomorphic invariants of such coverings depend only upon the numbers $(a + b - 2)(c + d - 2)$ and $ab + cd$.

Theorem 6.35 *Let S, S' be smooth bidouble covers of $\mathbb{P}^1 \times \mathbb{P}^1$ of respective types $(a, b)(c, d), (a', b')(c', d')$.*

Then S is of general type for $a + c \geq 3, b + d \geq 3$, and is simply connected. Moreover, the divisibility $r(S)$ of the canonical class K_S is equal to G.C.D.$((a + c - 2), (b + d - 2))$.

S and S' are (orientedly) homeomorphic if and only if $r(S) \equiv r(S')(mod 2)$ and

$$(a + b - 2)(c + d - 2) = (a' + b' - 2)(c' + d' - 2) \text{ and } ab + cd = a'b' + c'd'.$$

S and S' are not diffeomorphic if $r(S) \neq r(S')$, and for each integer h, we can find such surfaces $S_1, \ldots S_h$ which are pairwise homeomorphic but not diffeomorphic.

Idea of the proof. Set for simplicity $u := (a + c - 2), v := (b + d - 2)$ so that $\mathcal{O}_S(K_S) = p^*(\mathcal{O}_{\mathbb{P}^1 \times \mathbb{P}^1}(u, v))$ is ample whenever $u, v \geq 1$.

The property that S is simply connected (cf. [Cat84] for details) follows once one establishes that the fundamental group $\pi_1((\mathbb{P}^1 \times \mathbb{P}^1) \setminus (C \cup D))$ is abelian. To establish that the group is abelian, since it is generated by a product of simple geometric loops winding once around a smooth point of $C \cup D$, it suffices to show that these loops are central. But this follows from considering a Lefschetz pencil having C (respectively, D) as a fibre (in fact, an S^1 bundle over a punctured Riemann surface is trivial).

Since this group is abelian, it is generated by two elements γ_C, γ_D which are simple geometric loops winding once around C, resp. D. The fundamental group $\pi_1(S \backslash R)$ is then generated by $2\gamma_C$ and $2\gamma_D$, but these two elements lie in the kernel of the surjection $\pi_1(S \backslash R) \rightarrow \pi_1(S)$ and we conclude the triviality of this latter group.

The argument for the divisibility of K_S is more delicate, and we refer to [Cat86] for the proof of the key lemma asserting that $p^*(H^2(\mathbb{P}^1 \times \mathbb{P}^1, \mathbb{Z})) = H^2(S, \mathbb{Z})^G$ where G is the Galois group $G = (\mathbb{Z}/2)^2$ (the proof uses arguments of group cohomology and is valid in greater generality). Thus, the divisibility of K_S equals the one of $c_1(\mathcal{O}_{\mathbb{P}^1 \times \mathbb{P}^1}(u, v))$, i.e., $G.C.D.(u, v)$.

Now, resorting to Freedman's theorem, it suffices to observe that rank and signature of the intersection form are given by $e(S) - 2, \sigma(S)$, and these, as we saw in the first lecture, equal $12\chi(S) - K_S^2, K_S^2 - 8\chi(S)$. In this case $K_S^2 = 8uv$, $\chi(S) = uv + (ab + cd)$.

There remain to find h such surfaces, and for this purpose, we use Bombieri's argument (appendix to [Cat84]): namely, let $u_i' v_i' = 6^n$ be h distinct factorizations and, for a positive number T, set $u_i := T u_i', v_i := T v_i'$. It is clear that $G.C.D.(u_i, v_i) = T(G.C.D.(u_i', v_i'))$ and these G.C.D.'s are distinct since the given factorizations are distinct (as unordered factorizations), and they are even integers if each u_i', v_i' is even.

It suffices to show that there are integers w_i, z_i such that, setting $a_i := (u_i + w_i)/2 + 1, c_i := (u_i - w_i)/2 + 1, b_i := (v_i - z_i)/2 + 1, d_i := (v_i + z_i)/2 + 1$, then $a_i b_i + c_i d_i = constant$ and the required inequalities $a_i, b_i, c_i, d_i \geq 3$ are verified.

This can be done by the box principle.

It is important to contrast the existence of homeomorphic but not diffeomorphic algebraic surfaces to an important theorem established at the beginning of the study of 4-manifolds by C.T.C. Wall [Wall62]:

Theorem 6.36 (C.T.C. Wall) *Given two simply connected differentiable 4-manifolds M, M' with isomorphic intersection forms, then there exists an integer k such that the iterated connected sums $M \sharp k(\mathbb{P}^1 \times \mathbb{P}^1)$ and $M' \sharp k(\mathbb{P}^1 \times \mathbb{P}^1)$ are diffeomorphic.*

Remark 6.37 *(1) If we take $\mathbb{P}_{\mathbb{C}}^{2 \, opp}$, i.e., $\mathbb{P}^2$ with opposite orientation, then the selfintersection of a line equals -1, just as for the exceptional curve of a blow up. It is easy to see that blowing up a point of a smooth complex surface S is the same differentiable operation as taking the connected sum $S \sharp \mathbb{P}_{\mathbb{C}}^{2 \, opp}$.*

(2) Recall that the blowup of the plane $\mathbb{P}^2$ in two points is isomorphic to the quadric $\mathbb{P}^1 \times \mathbb{P}^1$ blown up in a point. Whence, for the connected sum

calculus, $M\sharp(\mathbb{P}^1 \times \mathbb{P}^1)\ \sharp\mathbb{P}^{2\ opp}_{\mathbb{C}} \cong M\sharp(\mathbb{P}^2)\ \sharp 2\mathbb{P}^{2\ opp}_{\mathbb{C}}$. *From Wall's theorem follows then (consider Wall's theorem for M^{opp}) that for any simply connected 4-manifold M there are integers k, p, q such that $M\sharp(k+1)(\mathbb{P}^2)\ \sharp(k)\mathbb{P}^{2\ opp}_{\mathbb{C}} \cong p(\mathbb{P}^2)\ \sharp(q)\mathbb{P}^{2\ opp}_{\mathbb{C}}$.*

The moral of Wall's theorem was that homeomorphism of simply connected 4-manifolds implies *stable* diffeomorphism (i.e., after iterated connected sum with some basic manifolds as $(\mathbb{P}^1 \times \mathbb{P}^1)$ or, with both $\mathbb{P}^2, \mathbb{P}^{2\ opp}_{\mathbb{C}}$).

The natural question was then how many such connected sums were indeed needed, and if there were needed at all. As we saw, the Donaldson and Seiberg Witten invariants show that some connected sum is indeed needed.

Boris Moishezon, in collaboration with Mandelbaum, studied the question in detail [Moi77, M-M76, M-M80] for many concrete examples of algebraic surfaces, and gave the following

Definition 6.38 *A differentiable simply connected 4-manifold M is completely decomposable if there are integers p, q with $M \cong p(\mathbb{P}^2)\ \sharp(q)\mathbb{P}^{2\ opp}_{\mathbb{C}}$, and almost completely decomposable if $M\sharp(\mathbb{P}^2)$ is completely decomposable (note that the operation yields a manifold with odd intersection form, and if M is an algebraic surface $\neq \mathbb{P}^2$, then we get an indefinite intersection form.*

Moishezon and Mandelbaum [M-M76] proved almost complete decomposability for smooth hypersurfaces in $\mathbb{P}^3$, and Moishezon proved [Moi77] almost complete decomposability for simply connected elliptic surfaces. Observe that rational surfaces are obviously completely decomposable, and therefore one is only left with simply connected surfaces of general type, for which as far as I know the question of almost complete decomposability is still unresolved.

Donaldson's work clarified the importance of the connected sum with $\mathbb{P}^2$, showing the following results (cf. [D-K90] pages 26–27).

Theorem 6.39 (Donaldson) *If M_1, M_2 are simply connected differentiable 4-manifolds with $b^+(M_i) > 0$, then the Donaldson polynomial invariants $q_k \in S^d(H^2(M, \mathbb{Z})$ are all zero for $M = M_1\sharp M_2$. If instead M is an algebraic surfaces, then the Donaldson polynomials q_k are $\neq 0$ for large k. In particular, an algebraic surface cannot be diffeomorphic to a connected sum $M_1\sharp M_2$ with M_1, M_2 as above (i.e., with $b^+(M_i) > 0$).*

6.5 ABC Surfaces

This subsection is devoted to the diffeomorphism type of certain series of families of bidouble covers, depending on three integer parameters (a,b,c) (cf. [Cat02, CW04]).

Let us make some elementary remark, which will be useful in order to understand concretely the last part of the forthcoming definition.

Consider the projective line $\mathbb{P}^1$ with homogeneous coordinates (x_0, x_1) and with nonhomogeneous coordinate $x := x_1/x_0$. Then the homogeneous polynomials of degree m $F(x_0, x_1)$ are exactly the space of holomorphic sections of

$\mathcal{O}_{\mathbb{P}^1}(m)$: in fact to such an F corresponds the pair of holomorphic functions $f_0(x) := \frac{F(x_0,x_1)}{x_0^m}$ on $U_0 := \mathbb{P}^1 \backslash \{\infty\}$, and $f_1(1/x) := \frac{F(x_0,x_1)}{x_1^m}$ on $U_1 := \mathbb{P}^1 \backslash \{0\}$. They satisfy the cocycle condition $f_0(x)x^m = f_1(1/x)$.

We assumed here m to be a positive integer, because $\mathcal{O}_{\mathbb{P}^1}(-m)$ has no holomorphic sections, if $m > 0$. On the other hand, sheaf theory (the exponential sequence and the partition of unity argument) teaches us that the cocycle x^{-m} for $\mathcal{O}_{\mathbb{P}^1}(-m)$ is cohomologous, if we use differentiable functions, to $\bar{x}^m$ (indeed $x^{-m} = \frac{\bar{x}^m}{|x|^{2m}}$, a formula which hints at the homotopy $\frac{\bar{x}^m}{|x|^{2mt}}$ of the two cocycles).

This shows in particular that the polynomials $F(\bar{x}_0, \bar{x}_1)$ which are homogeneous of degree m are differentiable sections of $\mathcal{O}_{\mathbb{P}^1}(-m)$.

Since sometimes we shall need to multiply together sections of $\mathcal{O}_{\mathbb{P}^1}(-m)$ with sections of $\mathcal{O}_{\mathbb{P}^1}(m)$, and get a global function, we need the cocycles to be the inverses of each other. This is not a big problem, since on a circle of radius R we have $\bar{x}x = R^2$. Hence to a polynomial $F(\bar{x}_0, \bar{x}_1)$ we associate the two functions

$$f_0(\bar{x}) := \frac{F(\bar{x}_0, \bar{x}_1)}{\bar{x}_0^m} \text{ on } \{x||x| \le R\}$$

$$f_1(1/\bar{x}) := R^{2m} \frac{F(\bar{x}_0, \bar{x}_1)}{\bar{x}_1^m} \text{ on } \{x||x| \ge R\}$$

and this trick allows to carry out local computations comfortably.

Let us go now to the main definition:

Definition 6.40 *An (a, b, c) surface is the minimal resolution of singularities of a simple bidouble cover S of $(\mathbb{P}^1 \times \mathbb{P}^1)$ of type $((2a, 2b), (2c, 2b))$ having at most Rational Double Points as singularities.*

An $(a, b, c)^{nd}$ surface is defined more generally as (the minimal resolution of singularities of) a natural deformation of an (a, b, c) surface with R.D.P.'s : i.e., the canonical model of an $(a, b, c)^{nd}$ surface is embedded in the total space of the direct sum of two line bundles L_1, L_2 (whose corresponding sheaves of sections are $\mathcal{O}_{\mathbb{P}^1 \times \mathbb{P}^1}(a, b), \mathcal{O}_{\mathbb{P}^1 \times \mathbb{P}^1}(c, b)$), and defined there by a pair of equations

$$(***) \quad z_{a,b}^2 = f_{2a,2b}(x, y) + w_{c,b}\phi_{2a-c,b}(x, y)$$

$$w_{c,b}^2 = g_{2c,2b}(x, y) + z_{a,b}\psi_{2c-a,b}(x, y)$$

where f, g, ϕ, ψ, are bihomogeneous polynomials , belonging to respective vector spaces of sections of line bundles: $f \in H^0(\mathbb{P}^1 \times \mathbb{P}^1, \mathcal{O}_{\mathbb{P}^1 \times \mathbb{P}^1}(2a, 2b))$, $\phi \in H^0(\mathbb{P}^1 \times \mathbb{P}^1, \mathcal{O}_{\mathbb{P}^1 \times \mathbb{P}^1}(2a - c, b))$ and $g \in H^0(\mathbb{P}^1 \times \mathbb{P}^1, \mathcal{O}_{\mathbb{P}^1 \times \mathbb{P}^1}(2c, 2d))$, $\psi \in H^0(\mathbb{P}^1 \times \mathbb{P}^1, \mathcal{O}_{\mathbb{P}^1 \times \mathbb{P}^1}(2c - a, b))$.

A perturbation of an (a, b, c) surface is an oriented smooth 4-manifold defined by equations as $(***)$, but where the sections ϕ, ψ are differentiable, and we have a dianalytic perturbation if ϕ, ψ are polynomials in the variables $x_i, y_j, \overline{x}_i, \overline{y}_j$, according to the respective positivity or negativity of the entries of the bidegree.

Remark 6.41 *By the previous formulae,*
(1)(a, b, c) surfaces have the same invariants $\chi(S) = 2(a + c - 2)(b - 1) + b(a + c), K_S^2 = 16(a + c - 2)(b - 1)$
(2) the divisibility of their canonical class is G.C.D.$((a + c - 2), 2(b - 1))$
(3) Moreover, we saw that (a, b, c) surfaces are simply connected, thus
(4) Once we fix b and the sum $(a + c) = s$, the corresponding (a, b, c) surfaces are all homeomorphic

As a matter of fact, once we fix b and the sum $(a + c)$, the surfaces in the respective families are homeomorphic by a homeomorphism carrying the canonical class to the canonical class. This fact is a consequence of the following proposition, which we learnt from [Man96]

Proposition 6.42 *Let S, S' be simply connected minimal surfaces of general type such that $\chi(S) = \chi(S') \geq 2$, $K_S^2 = K_{S'}^2$, and moreover such that the divisibility indices of K_S and $K_{S'}$ are the same.*
Then there exists a homeomorphism F between S and S', unique up to isotopy, carrying $K_{S'}$ to K_S.

Proof. By Freedman's theorem ([Free82], cf. especially [F-Q90], page 162) for each isometry $h : H_2(S, \mathbb{Z}) \to H_2(S', \mathbb{Z})$ there exists a homeomorphism F between S and S', unique up to isotopy, such that $F_* = h$. In fact, S and S' are smooth 4-manifolds, whence the Kirby-Siebenmann invariant vanishes.

Our hypotheses that $\chi(S) = \chi(S')$, $K_S^2 = K_{S'}^2$ and that $K_S, K_{S'}$ have the same divisibility imply that the two lattices $H_2(S, \mathbb{Z})$, $H_2(S', \mathbb{Z})$ have the same rank, signature and parity, whence they are isometric since S, S' are algebraic surfaces. Finally, by Wall's theorem [Wall62] (cf. also [Man96], page 93) such isometry h exists since the vectors corresponding to the respective canonical classes have the same divisibility and by Wu's theorem they are characteristic: in fact Wall's condition $b_2 - |\sigma| \geq 4$ (σ being the signature of the intersection form) is equivalent to $\chi \geq 2$. $\square$

We come now to the main result of this section (see [CW04] for details)

Theorem 6.43 *Let S be an (a, b, c)-surface and S' be an $(a + 1, b, c - 1)$-surface. Moreover, assume that $a, b, c - 1 \geq 2$. Then S and S' are diffeomorphic.*

Idea of the Proof.
Before we dwell into the proof, let us explain the geometric argument which led me to conjecture the above theorem in 1997.

Assume that the polynomials f, g define curves C, D which are union of vertical and horizontal lines. Fix for simplicity affine coordinates in $\mathbb{P}^1$. Then we may assume, without loss of generality, that the curve C is constituted by the horizontal lines $y = 1, \ldots y = 2b$, and by the vertical lines $x = 2, \ldots x = 2a + 1$, while the curve D is formed by the horizontal lines $y = -1, \ldots y = -2b$, and by the vertical lines $x = 0$, $x = 1/4$, $x = 2a + 2, \ldots x = 2a + 2c - 1$. The corresponding surface X has double points as singularities, and its minimal resolution is a deformation of a smooth (a, b, c)-surface (by the cited results of Brieskorn and Tjurina).

Likewise, we let X' be the singular surface corresponding to the curve C' constituted by the horizontal lines $y = 1, \ldots y = 2b$, and by the vertical lines $x = 0$, $x = 1/4$, $x = 2, \ldots x = 2a + 1$, and to the curve D' formed by the horizontal lines $y = -1, \ldots y = -2b$, and by the $(2c - 2)$ vertical lines $x = 2a + 2, \ldots x = 2a + 2c - 1$.

We can split X as the union $X_0 \cup X_\infty$, where $X_0 := \{(x, y, z, w)| \, |x| \leq 1\}$, $X_\infty := \{(x, y, z, w)| \, |x| \geq 1\}$, and similarly $X' = X'_0 \cup X'_\infty$.

By our construction, we see immediately that $X'_\infty = X_\infty$, while there is a natural diffeomorphism Φ of $X_0 \cong X'_0$.

It suffices in fact to set $\Phi(x, y, z, w) = (x, -y, w, z)$.

The conclusion is that both S and S' are obtained glueing the same two 4-manifolds with boundary S_0, S_∞ glueing the boundary $\partial X_0 = \partial X_\infty$ once through the identity, and another time through the diffeomorphism Φ. It will follow that the two 4-manifolds are diffeomorphic if the diffeomorphism $\Phi|_{\partial S_0}$ admits an extension to a diffeomorphism of S_0.

(1) The relation with Lefschetz fibrations comes from the form of Φ, since Φ does not affect the variable x, but it is essentially given by a diffeomorphism Ψ of the fibre over $x = 1$,

$$\Psi(y, z, w) = (-y, w, z).$$

Now, the projection of an (a, b, c) surface onto $\mathbb{P}^1$ via the coordinate x is not a Lefschetz fibration, even if f, g are general, since each time one of the two curves C, D has a vertical tangent, we shall have two nodes on the corresponding fibre. But a smooth general natural deformation

$$z^2 = f(x, y) + w\phi(x, y) \tag{1}$$
$$w^2 = g(x, y) + z\psi(x, y),$$

would do the game if $\phi \neq 0$ (i.e., $2a - c > 0$) and $\psi \neq 0$ (i.e., $2c - a > 0$).

Otherwise, it is enough to take a perturbation as in the previous definition (a dianalytic one suffices), and we can realize both surfaces S and S' as symplectic Lefschetz fibrations (cf. also [Don99, G-S99]).

(2) The above argument about S, S' being the glueing of the same two manifolds with boundary S_0, S_∞ translates directly into the property that the corresponding Lefschetz fibrations over $\mathbb{P}^1$ are fibre sums of the same pair of Lefschetz fibrations over the respective complex discs $\{x| \, |x| \leq 1\}, \{x| \, |x| \geq 1\}$.

(3) Once the first fibre sum is presented as composition of two factorizations and the second as twisted by the 'rotation' Ψ, (i.e., as we saw, the same composition of factorizations, where the second is conjugated by Ψ), in order to prove that the two fibre sums are equivalent, it suffices to apply a very

simple lemma, which can be found in [Aur02], and that we reproduce here because of its beauty

Lemma 6.44 (Auroux) *Let τ be a Dehn twist and let F be a factorization of a central element $\phi \in \mathcal{M}ap_g$, $\tau_1 \circ \tau_2 \circ \cdots \circ \tau_m = \phi$.*

If there is a factorization F' such that F is Hurwitz equivalent to $\tau \circ F'$, then $(F)_\tau$ is Hurwitz equivalent to F.

In particular, if F is a factorization of the identity, $\Psi = \Pi_h \tau'_h$, and $\forall h \; \exists F'_h$ such that $F \cong \tau'_h \circ F'_h$, then the fibre sum with the Lefschetz pencil associated with F yields the same Lefschetz pencil as the fibre sum twisted by Ψ.

Proof.
  If $\cong$ denotes Hurwitz equivalence, then

$$(F)_\tau \cong \tau \circ (F')_\tau \cong F' \circ \tau \cong (\tau)_{(F')^{-1}} \circ F' = \tau \circ F' \cong F.$$

$\square$

Corollary 6.45 *Notation as above, assume that $F\colon \tau_1 \circ \tau_2 \circ \cdots \circ \tau_m = \phi$ is a factorization of the Identity and that Ψ is a product of some Dehn twists τ_i appearing in F. Then the fibre sum with the Lefschetz pencil associated with F yields the same result as the same fibre sum twisted by Ψ.*

Proof. We need only to verify that for each h, there is F'_h such that $F \cong \tau_h \circ F'_h$.

But this is immediately obtained by applying $h - 1$ Hurwitz moves, the first one between τ_{h-1} and τ_h, and proceeding further to the left till we obtain τ_h as first factor. $\square$

(4) It suffices now to show that the diffeomorphism Ψ is in the subgroup of the mapping class group generated by the Dehn twists which appear in the first factorization.

Figure 10 below shows the fibre C of the fibration in the case $2b = 6$: it is a bidouble cover of $\mathbb{P}^1$, which we can assume to be given by the equations $z^2 = F(y)$, $w^2 = F(-y)$, where the roots of F are the integers $1, \ldots, 2b$.

Moreover, one sees that the monodromy of the fibration at the boundary of the disc is trivial, and we saw that the map Ψ is the diffeomorphism of order 2 given by $y \mapsto -y$, $z \mapsto w$, $w \mapsto z$, which in our figure is given as a rotation of $180°$ around an axis inclined in direction north-west.

The figure shows a dihedral symmetry, where the automorphism of order 4 is given by $y \mapsto -y$, $z \mapsto -w$, $w \mapsto z$.

(5) A first part of the proof, which we skip here, consists in identifying the Dehn twists which appear in the first factorization.

It turns out that, among the Dehn twists which appear in the first factorization, there are those which correspond to the inverse images of the segments between two consecutive integers (cf. Fig. 10). These circles can be organized on the curve C in six chains (not disjoint) and finally one reduces oneself to the computational heart of the proof: showing that the isotopy class of Ψ is

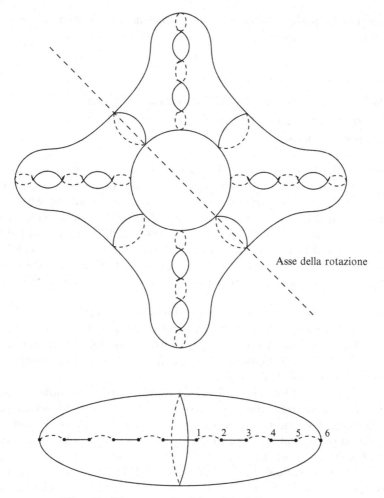

Fig. 10. The curve C with a dihedral symmetry

the same as the product Ψ' of the six Coxeter elements associated to such chains.

We recall here that, given a chain of curves $\alpha_1, \ldots \alpha_n$ on a Riemann surface, the *Coxeter element associated to the chain* is the product

$$\Delta := (T_{\alpha_1})(T_{\alpha_2} T_{\alpha_1}) \ldots (T_{\alpha_n} T_{\alpha_{n-1}} \ldots T_{\alpha_1})$$

of the Dehn twists associated to the curves of the chain.

In order to finally prove that Ψ' (the product of such Coxeter elements) and Ψ are isotopic, one observes that if one removes the above cited chains of circles from the curve C, one obtains 4 connected components which are diffeomorphic to circles. By a result of Epstein it is then sufficient to verify that Ψ and Ψ' send each such curve to a pair of isotopic curves: this last step needs a list of lengthy (though easy) verifications, for which it is necessary to have explicit drawings.

For details we refer to the original paper [CW04].

7 Epilogue: Deformation, Diffeomorphism and Symplectomorphism Type of Surfaces of General Type

As we repeatedly said, one of the fundamental problems in the theory of complex algebraic surfaces is to understand the moduli spaces of surfaces of general type, and in particular their connected components, which, as we saw in the third lecture, parametrize the deformation equivalence classes of minimal surfaces of general type, or equivalently of their canonical models.

We remarked that deformation equivalence of two minimal models S, S' implies their canonical symplectomorphism and a fortiori an oriented diffeomorphism preserving the canonical class (a fortiori, a homeomorphism with such a property).

In the late eighties Friedman and Morgan (cf. [F-M94]) made the bold conjecture that two algebraic surfaces are diffeomorphic if and only if they are deformation equivalent. We will abbreviate this conjecture by the acronym *def = diff*. Indeed, I should point out that I had made the opposite conjecture in the early eighties (cf. [Katata83]).

Later in this section we shall briefly describe the first counterexamples, due to M. Manetti (cf. [Man01]): these have the small disadvantage of providing nonsimplyconnected surfaces, but the great advantage of yielding non deformation equivalent surfaces which are canonically symplectomorphic (see [Cat02, Cat06] for more details).

We already described in Lecture 4 some easy counterexamples to this conjecture (cf. [Cat03, KK02, BCG05]), given by pairs of complex conjugate surfaces, which are not deformation equivalent to their complex conjugate surface.

We might say that, although describing some interesting phenomena, the counterexamples contained in the cited papers by Catanese, Kharlamov–Kulikov, Bauer–Catanese–Grunewald are 'cheap', since the diffeomorphism carries the canonical class to its opposite. I was recently informed [Fried05] by R. Friedman that also he and Morgan were aware of such 'complex conjugate' counterexamples, but for the case of some elliptic surfaces having an infinite fundamental group.

After the examples by Manetti it was however still possible to weaken the conjecture def = diff in the following way.

Question 7.1 *Is the speculation* def = diff *true if one requires the diffeomorphism* $\phi : S \to S'$ *to send the first Chern class* $c_1(K_S) \in H^2(S, \mathbb{Z})$ *in* $c_1(K_{S'})$ *and moreover one requires the surfaces to be simply connected?*

But even this weaker question turned out to have a negative answer, as it was shown in our joint work with Wajnryb [CW04].

Theorem 7.2 ([CW04]) *For each natural number h there are simply connected surfaces $S_1, \ldots, S_h$ which are pairwise diffeomorphic, but not deformation equivalent.*

The following remark shows that the statement of the theorem implies a negative answer to the above question.

Remark 7.3 *If two surfaces are deformation equivalent, then there exists a diffeomorphism sending the canonical class $c_1(K_S) \in H^2(S, \mathbb{Z})$ to the canonical class $c_1(K_{S'})$. On the other hand, by the cited result of Seiberg-Witten theory we know that a diffeomorphism sends the canonical class of a minimal surface S to $\pm c_1(K_{S'})$. Therefore, if one gives at least three surfaces, which are pairwise diffeomorphic, one finds at least two surfaces with the property that there exists a diffeomorphism between them sending the canonical class of one to the canonical class of the other.*

7.1 Deformations in the Large of ABC Surfaces

The above surfaces $S_1, \ldots, S_h$ in Theorem 7.2 belong to the class of the so-called (a, b, c)-surfaces, whose diffeomorphism type was shown in the previous Lecture to depend only upon the integers $(a + c)$ and b.

The above Theorem 7.2 is thus implied by the following result:

Theorem 7.4 *Let S, S' be simple bidouble covers of $\mathbf{P}^1 \times \mathbf{P}^1$ of respective types $((2a, 2b), (2c, 2b)$, and $(2a + 2k, 2b), (2c - 2k, 2b)$, and assume*

- *(I) a, b, c, k are strictly positive even integers with $a, b, c - k \geq 4$*
- *(II) $a \geq 2c + 1$*
- *(III) $b \geq c + 2$ and either*
- *(IV1) $b \geq 2a + 2k - 1$ or (IV2) $a \geq b + 2$*

Then S and S' are not deformation equivalent.

The theorem uses techniques which have been developed in a series of papers by the author and by Manetti [Cat84, Cat87, Cat86, Man94, Man97]. They use essentially the local deformation theory a' la Kuranishi for the canonical models, normal degenerations of smooth surfaces and a study of quotient singularities of rational double points and of their smoothings (this method was used in [Cat87] in order to study the closure in the moduli space of a class of bidouble covers of $\mathbb{P}^1 \times \mathbb{P}^1$ satisfying other types of inequalities).

Although the proof can be found in [Cat02, CW04], and in the Lecture Notes by Manetti in this volume, I believe it worthwhile to sketch the main ideas and arguments of the proof.

Main arguments of the Proof.

These are the three main steps of the proof:

Step I: determination of a subset $\mathfrak{N}_{a,b,c}$ of the moduli space

Step II: proof that $\mathfrak{N}_{a,b,c}$ is an open set

Step III: proof that $\mathfrak{N}_{a,b,c}$ is a closed set

Let us first of all explain the relevance of hypothesis (2) for step III. If we consider the natural deformations of (a, b, c) surfaces, which are parametrized by a quadruple of polynomials (f, g, ϕ, ψ) and given by the two equations

$$z^2 = f(x, y) + w\phi(x, y),$$

$$w^2 = g(x, y) + z\psi(x, y),$$

we observe that f and g are polynomials of respective bidegrees $(2a, 2b)$, $(2c, 2b)$, while ϕ and ψ have respective bidegrees $(2a - c, b)$, $(2c - a, b)$. Hence $a \geq 2c+1$, implies that $\psi \equiv 0$, therefore every small deformation preserves the structure of an iterated double cover. This means that the quotient Y of our canonical model X by the involution $z \mapsto -z$ admits an involution $w \mapsto -w$, whose quotient is indeed $\mathbb{P}^1 \times \mathbb{P}^1$.

This fact will play a special role in the study of limits of such $(a, b, c)^{nd}$ surfaces, showing that this iterated double cover structure passes in a suitable way to the limit, hence $\mathfrak{N}_{a,b,c}$ is a closed subset of the moduli space.

Step I.

The family $(\mathfrak{N}_{a,b,c})$ consists of all the (minimal resolutions of the) natural deformations of simple bidouble covers of the Segre-Hirzebruch surfaces $\mathbb{F}_{2h}$ which have only Rational Double Points as singularities and are of type $((2a, 2b),(2c,2b))$.

In order to explain what this means, let us recall, as in [Cat82] pages 105–111, that a basis of the Picard group of $\mathbb{F}_{2h}$ is provided, for $h \geq 1$, by the fibre F of the projection to $\mathbb{P}^1$, and by $F' := \sigma_\infty + hF$, where σ_∞ is the unique section with negative self-intersection $= -2h$. Observe that $F^2 = F'^2 = 0, FF' = 1$, and that F is nef, while $F' \cdot \sigma_\infty = -h$.

We set $\sigma_0 := \sigma_\infty + 2hF$, so that $\sigma_\infty \sigma_0 = 0$, and we observe (cf. Lemma 2.7 of [Cat82]) that $|m\sigma_0 + nF|$ has no base point if and only if $m, n \geq 0$. Moreover, $|m\sigma_0 + nF|$ contains σ_∞ with multiplicity ≥ 2 if $n < -2h$.

At this moment, the above remarks and the inequalities (II), (III), (IV) can be used to imply that all natural deformations have the structure of an iterated double covering, since their canonical models are defined by the following two equations:

$$z^2 = f(x, y) + w\phi(x, y),$$

$$w^2 = g(x, y).$$

Step II.

A key point here is to look only at the deformation theory of the canonical models.

To prove that the family of canonical models $(\mathfrak{N}_{a,b,c})$ yields an open set in the moduli space it suffices to show that, for each surface X, the Kodaira Spencer map is surjective.

In fact, one can see as in in [Cat82] that the family $(\mathfrak{N}_{a,b,c})$ is parametrized by a smooth variety which surjects onto $H^1(\Theta_{\mathbb{F}})$.

Observe that the tangent space to the Deformations of X is provided by $\mathrm{Ext}^1_{\mathcal{O}_X}(\Omega^1_X, \mathcal{O}_X)$.

Denoting by $\pi : X \to \mathbb{F} := \mathbb{F}_{2h}$ the projection map and differentiating equations (7) we get an exact sequence for Ω^1_X

$$0 \to \pi^*(\Omega^1_{\mathbb{F}}) \to \Omega^1_X \to \mathcal{O}_{R_z}(-R_z) \oplus \mathcal{O}_{R_w}(-R_w) \to 0$$

as in (1.7) of [Man94], where $R_z = div(z), R_w = div(w)$.

Applying the derived exact sequence for $\mathrm{Hom}_{\mathcal{O}_X}(\ldots, \mathcal{O}_X)$ we obtain the same exact sequence as Theorem (2.7) of [Cat82], and (1.9) of [Man94], namely:

$$(**) \ 0 \to H^0(\Theta_X) \to H^0(\pi^*\Theta_{\mathbb{F}}) \to H^0(\mathcal{O}_{R_z}(2R_z)) \oplus H^0(\mathcal{O}_{R_w}(2R_w)) \to$$

$$\to \mathrm{Ext}^1_{\mathcal{O}_X}(\Omega^1_X, \mathcal{O}_X) \to H^1(\pi^*\Theta_{\mathbb{F}}).$$

There is now some technical argument, quite similar to the one given in [Cat82], and where our inequalities are used in order to show that $H^1(\pi^*\Theta_{\mathbb{F}}) = H^1(\Theta_{\mathbb{F}} \otimes \pi_*(\mathcal{O}_X))$ equals $H^1(\Theta_{\mathbb{F}})$: we refer to [CW04] for details.

Summarizing the proof of step II, we observe that the smooth parameter space of our family surjects onto $H^1(\Theta_{\mathbb{F}})$, and its kernel, provided by the natural deformations with fixed base $\mathbb{F}_{2h}$, surjects onto $H^0(\mathcal{O}_{R_z}(2R_z)) \oplus H^0(\mathcal{O}_{R_w}(2R_w))$. Thus the Kodaira Spencer is onto and we get an open set in the moduli space.

Step III.

We want now to show that our family $\mathfrak{N}_{a,b,c}$ yields a closed set in the moduli space.

It is clear at this moment that we obtained an irreducible component of the moduli space. Let us consider the surface over the generic point of the base space of our family: then it has $\mathbb{Z}/2$ in the automorphism group (sending $z \to -z$, as already mentioned).

As shown in [Cat82], this automorphism acts then biregularly on the canonical model X_0 of each surface corresponding to a point in the closure of our open set. This holds in fact more generally for the action of any finite group G: the representation of G on $H^0(S, \mathcal{O}(5K_S))$ depends on discrete data, whence it is fixed in a family, and then the set of fixed points in the pseudomoduli space $\{X | g(X) = X \ \forall g \in G\}$ is a closed set.

We use now the methods of [Cat87, Man97], and more specifically we can apply Theorem 4.1 of [Man97] to conclude with

Claim III .1 If X_0 is a canonical model which is a limit of canonical models X_t of surfaces S_t in our family, then the quotient Y_0 of X_0 by the subgroup $\mathbb{Z}/2 \subset Aut(X_0)$ mentioned above is a surface with Rational Double Points.

Claim III .2 The family of such quotients Y_t has a $\mathbb{Z}/2$-action over the generic point, and dividing by it we get (cf. [Man97, Theorem 4.10]) as quotient Z_0 a Hirzebruch surface. Thus our surface X_0 is also an iterated double cover of some $\mathbb{F}_{2h}$, hence it belongs to the family we constructed.

Argument for claim III.1 Since smooth canonical models are dense, we may assume that X_0 is a limit of a 1-parameter family X_t of smooth canonical models; for the same reason we may assume that the quotient Y_0 is the limit of smooth surfaces $Y_t = X_t/(\mathbb{Z}/2)$ (of general type if $c, b \geq 3$).

Whence,

(1) Y_0 has singularities which are quotient of Rational Double Points by $(\mathbb{Z}/2)$.

(2) Y_t is a smoothing of Y_0, and since we assume the integers c, b to be even, the canonical divisor of Y_t is 2-divisible.

Now, using Theorem 3.6, the involutions acting on RDP's can be classified (cf. [Cat87] for this and the following), and it turns out that the quotient singularities are again RDP's, with two possible exceptions:

Type (c): the singularity of Y_0 is a quotient singularity of type $\frac{1}{4k+2}(1, 2k)$, and X_0 is the A_{2k} singularity, quotient by the subgroup $2\mathbb{Z}/(4k+2)\mathbb{Z}$.

Type (e): the singularity of Y_0 is a quotient singularity of type $\frac{1}{4k+4}(1, 2k+1)$, and X_0 is the A_{2k+1} singularity, quotient by the subgroup $2\mathbb{Z}/(4k+4)\mathbb{Z}$.

The versal families of deformations of the above singularities have been described by Riemenschneider in [Riem74], who showed:

(C) In the case of type (c), the base space is smooth, and it yields a smoothing admitting a simultaneous resolution.

(E) In the case of type (e), the base space consists of two smooth components intersecting transversally, $T_1 \cup T_2$. T_1 yields a smoothing admitting a simultaneous resolution (we denote this case by 'case (E1)').

Hypothesis (2), of 2-divisibility of the canonical divisor of Y_t, is used in two ways. The first consequence is that the intersection form on $H^2(Y_t, \mathbb{Z})$ is even; since however the Milnor fibre of the smoothing is contained in Y_t, it follows that no 2-cycle in the Milnor fibre can have odd selfintersection number. This then excludes case (C), and also case (E1) for $k \geq 1$.

In case (E2) we have a socalled $\mathbb{Z}$-Gorenstein smoothing, namely, the T_2 family is the quotient of the hypersurface

$$(* * *) \quad uv - z^{2n} = \Sigma_{h=0}^1 t_h z^{hn}$$

by the involution sending $(u, v, z) \mapsto (-u, -v, -z)$.

The result is that the Milnor fibre has a double étale cover which is the Milnor fibre of A_{n-1} ($n = k + 1$), in particular its fundamental group equals $\mathbb{Z}/2$. The universal cover corresponds to the cohomology class of the canonical divisor. This however contradicts condition (2), and case (E2) is excluded too.

For case (E1) $k = 0$ we argue similarly: the involution acts trivially on the parameter t, and in the central fibre it has an isolated fixed point. Because of simultaneous resolution, the total space $\cup_t X_t$ may be taken to be smooth, and then the set of fixed points for the involution is a curve mapping isomorphically on the parameter space $\{t\}$. Then the Milnor fibre should have a double cover ramified exactly in one point, but this is absurd since by van Kampen's theorem the point complement is simply connected.

Argument for claim III.2

Here, $Z_t := (Y_t/\mathbb{Z}/2) \cong \mathbb{P}^1 \times \mathbb{P}^1 = \mathbb{F}_0$ and again the canonical divisor is 2-divisible. Whence, the same argument as before applies, showing that Z_0 has necessarily Rational Double Points as singularities. But again, since the Milnor fibre embeds in $\mathbb{P}^1 \times \mathbb{P}^1 = \mathbb{F}_0$, the intersection form must have negativity at most 1, and be even. This leaves only the possibility of an A_1 singularity. This case can be again excluded by the same argument given for the case (E1) $k = 0$ above.

Proof that Theorem 7.4 implies Theorem 7.2.

It suffices to show what we took up to now for granted: the irreducible component $\mathfrak{N}_{a,b,c}$ uniquely determines the numbers a, b, c up to the obvious permutations: $a \leftrightarrow c$, and , if $a = c$, the possibilities of exchanging a with b.

It was shown more generally in [Cat84] Theorem 3.8 that the natural deformations of bidouble covers of type $(2a, 2b)(2c, 2d)$ yield an irreducible component of the moduli space, and that these are distinct modulo the obvious permutations (exchange type $(2a, 2b)(2c, 2d)$ with type $(2c, 2d)(2a, 2b)$ and with type $(2b, 2a)(2d, 2c)$). This follows from geometrical properties of the canonical map at the generic point.

However, the easiest way to see that the irreducible component $\mathfrak{N}_{a,b,c}$ determines the numbers a, b, c, under the given inequalities (II0, III), (IV) is to observe that the dimension of $\mathfrak{N}_{a,b,c}$ equals $M := (b + 1)(4a + c + 3) + 2b(a + c + 1) - 8$. Recall in fact that $K^2/16 = (a + c - 2)(b - 1)$, and $(8\chi - K^2)/8 = b(a + c)$: setting $\alpha = a + c, \beta = 2b$, we get that α, β are then the roots of a quadratic equation, so they are determined up to exchange, and uniquely if we restrict our numbers either to the inequality $a \geq 2b$ or to the inequality $b \geq a$.

Finally $M = (\frac{\beta}{2} + 1)(\alpha + 3) + \beta(\alpha + 1) - 8 + 3a(\frac{\beta}{2} + 1)$ then determines a, whence the ordered triple (a, b, c). $\square$

Remark 7.5 *If, as in [Cat02], we assume*

(IV2) $a \geq b + 2$,

then the connected component $\mathfrak{N}_{a,b,c}$ of the moduli space contains only iterated double covers of $\mathbb{P}^1 \times \mathbb{P}^1$.

7.2 Manetti Surfaces

Manetti in [Man01] considers surfaces which are desingularization of certain $(\mathbb{Z}/2)^r$ covers X of rational surfaces Y which are blowup of the quadric $Q := \mathbb{P}^1 \times \mathbb{P}^1$ at n points $P_1, \ldots P_n$.

His construction is made rather complicated, not only by the desire to construct an arbitrarily high number of surfaces which are pairwise diffeomorphic but not deformation equivalent, but also by the crucial target to obtain that every small deformation is again such a Galois $(\mathbb{Z}/2)^r$ cover. This requirement makes the construction not very explicit (Lemma 3.6 ibidem).

Let us briefly recall the structure of normal finite $(\mathbb{Z}/2)^r$ covers with smooth base Y (compare [Par91, Man01], and also [BC06] for a description in terms of the monodromy homomorphism).

We denote by $G = (\mathbb{Z}/2)^r$ the Galois group, and by σ an element of G. We denote by $G^\vee := Hom(G, \mathbb{C}^*)$ the dual group of characters, $G^\vee \cong (\mathbb{Z}/2)^r$, and by χ an element of $G^\vee$. As for any flat finite abelian covering $f : X \to Y$ we have

$$f_*\mathcal{O}_X = \bigoplus_{\chi \in G^\vee} \mathcal{O}_Y(-L_\chi) = \mathcal{O}_Y \oplus (\bigoplus_{\chi \in G^\vee \backslash \{0\}} \mathcal{O}_Y(-L_\chi)).$$

To each element of the Galois group $\sigma \in G$ one associates a divisor D_σ, such that $2D_\sigma$ is the direct image divisor $f_*(R_\sigma)$, R_σ being the divisorial part of the set of fixed points for σ.

Let x_σ be a section such that $div(x_\sigma) = D_\sigma$: then the algebra structure on $f_*\mathcal{O}_X$ is given by the following symmetric bilinear multiplication maps:

$$\mathcal{O}_Y(-L_\chi) \otimes \mathcal{O}_Y(-L_\eta) \to \mathcal{O}_Y(-L_{\chi+\eta})$$

associated to the section

$$x_{\chi,\eta} \in H^0(Y, \mathcal{O}_Y(L_\eta + L_\chi - L_{\chi+\eta})), \quad x_{\chi,\eta} := \prod_{\chi(\sigma)=\eta(\sigma)=1} x_\sigma.$$

Associativity follows since, given characters χ, η, θ, $\{\sigma | (\chi+\eta)(\sigma) = \theta(\sigma) = 1\}$ is the disjoint union of $\{\sigma | \chi(\sigma) = \theta(\sigma) = 1, \eta(\sigma) = 0\}$ and of $\{\sigma | \eta(\sigma) = \theta(\sigma) = 1, \chi(\sigma) = 0\}$, so that

$$\mathcal{O}_Y(-L_\chi) \otimes \mathcal{O}_Y(-L_\eta) \otimes \mathcal{O}_Y(-L_\theta) \to \mathcal{O}_Y(-L_{\chi+\eta+\theta})$$

is given by the section $\prod_{\sigma \in \Sigma} x_\sigma$, where

$$\Sigma := \{\sigma | \chi(\sigma) = \eta(\sigma) = 1, \text{or } \chi(\sigma) = \theta(\sigma) = 1, \text{or } \eta(\sigma) = \theta(\sigma) = 1\}.$$

In particular, the covering $f : X \to Y$ is embedded in the vector bundle $\mathbb{V}$ which is the direct sum of the line bundles whose sheaves of sections are the $\mathcal{O}_Y(-L_\chi)$, and is there defined by equations

$$z_\chi z_\eta = z_{\chi+\eta} \prod_{\chi(\sigma)=\eta(\sigma)=1} x_\sigma.$$

Noteworthy is the special case $\chi = \eta$, where $\chi + \eta$ is the trivial character 1, and $z_1 = 1$.

In particular, let $\chi_1, \ldots \chi_r$ be a basis of $G^\vee \cong (\mathbb{Z}/2)^r$, and set $z_i := z_{\chi_i}$. We get then the r equations

$$(\sharp) \quad z_i^2 = \prod_{\chi_i(\sigma)=1} x_\sigma.$$

These equations determine the field extension, hence one gets X as the normalization of the Galois cover given by $(\sharp)$.

We can summarize the above discussion in the following

Proposition 7.6 *A normal finite $G \cong (\mathbb{Z}/2)^r$ covering of smooth variety Y is completely determined by the datum of*
(1) reduced effective divisors D_σ, $\forall \sigma \in G$, which have no common components
(2) divisor classes $L_1, \ldots L_r$, for $\chi_1, \ldots \chi_r$ a basis of $G^\vee$, such that we have the following linear equivalence
(3)

$$2L_i \equiv \sum_{\chi_i(\sigma)=1} D_\sigma.$$

Conversely, given the datum of (1) and (2), if (3) holds, we obtain a normal scheme X with a finite $G \cong (\mathbb{Z}/2)^r$ covering $f \colon X \to Y$.

Idea of the proof
It suffices to determine the divisors L_χ for the other elements of $G^\vee$. But since any χ is a sum of basis elements, it suffices to exploit the fact that the linear equivalences

$$L_{\chi+\eta} \equiv L_\eta + L_\chi - \sum_{\chi(\sigma)=\eta(\sigma)=1} D_\sigma$$

must hold, and apply induction. Since the covering is well defined as the normalization of the Galois cover given by $(\sharp)$, each L_χ is well defined. Then the above formulae determine explicitly the ring structure of $f_*\mathcal{O}_X$, hence X. $\square$

A natural question is of course when the scheme X is a variety, i.e., X being normal, when X is connected, or equivalently irreducible. The obvious answer is that X is irreducible if and only if the monodromy homomorphism

$$\mu \colon H_1(Y \backslash (\cup_\sigma D_\sigma), \mathbb{Z}) \to G$$

is surjective.

Remark 7.7 *As a matter of fact, we know, from the cited theorem of Grauert and Remmert, that μ determines the covering. It is therefore worthwhile to see how μ determines the datum of (1) and (2).*
Write for this purpose the branch locus $D := \sum_\sigma D_\sigma$ as a sum of irreducible components D_i. To each D_i corresponds a simple geometric loop γ_i around

D_i, and we set $\sigma_i := \mu(\gamma_i)$. Then we have that $D_\sigma := \sum_{\sigma_i=\sigma} D_i$. For each character χ, yielding a double covering associated to the composition $\chi \circ \mu$, we must find a divisor class L_χ such that $2L_\chi \equiv \sum_{\chi(\sigma)=1} D_\sigma$.

Consider the exact sequence

$$H^{2n-2}(Y, \mathbb{Z}) \to H^{2n-2}(D, \mathbb{Z}) = \oplus_i \mathbb{Z}[D_i] \to H_1(Y \backslash D, \mathbb{Z}) \to H_1(Y, \mathbb{Z}) \to 0$$

and the similar one with $\mathbb{Z}$ replaced by $\mathbb{Z}/2$. Denote by Δ the subgroup image of $\oplus_i \mathbb{Z}/2[D_i]$. The restriction of μ to Δ is completely determined by the knowledge of the σ_i's, and we have

$$0 \to \Delta \to H_1(Y \backslash D, \mathbb{Z}/2) \to H_1(Y, \mathbb{Z}/2) \to 0.$$

Dualizing, we get

$$0 \to H^1(Y, \mathbb{Z}/2) \to H^1(Y \backslash D, \mathbb{Z}/2) \to Hom(\Delta, \mathbb{Z}/2) \to 0.$$

The datum of μ, extending $\mu|_\Delta$ is then seen to correspond to an affine space over the vector space $H^1(Y, \mathbb{Z}/2)$: and since $H^1(Y, \mathbb{Z}/2)$ classifies divisor classes of 2-torsion on Y, we infer that the different choices of L_χ such that $2L_\chi \equiv \sum_{\chi(\sigma)=1} D_\sigma$ correspond bijectively to all the possible choices for μ.

Corollary 7.8 *Same notation as in proposition 7.6. Then the scheme X is irreducible if $\{\sigma | D_\sigma > 0\}$ generates G.*

Proof. We have seen that if $D_\sigma \geq D_i \neq 0$, then $\mu(\gamma_i) = \sigma$, whence we infer that μ is surjective. $\square$

An important role plays again here the concept of *natural deformations*. This concept was introduced for bidouble covers in [Cat84], Definition 2.8, and extended to the case of abelian covers in [Par91], Definition 5.1. However, the two definitions do not coincide, because Pardini takes a much larger parameter space. We propose therefore to call Pardini's case the case of *extended natural deformations*.

Definition 7.9 *Let $f : X \to Y$ be a finite $G \cong (\mathbb{Z}/2)^r$ covering with Y smooth and X normal, so that X is embedded in the vector bundle $\mathbb{V}$ defined above and is defined by equations*

$$z_\chi z_\eta = z_{\chi+\eta} \prod_{\chi(\sigma)=\eta(\sigma)=1} x_\sigma.$$

Let $\psi_{\sigma,\chi}$ be a section $\psi_{\sigma,\chi} \in H^0(Y, \mathcal{O}_Y(D_\sigma - L_\chi))$, given $\forall \sigma \in G, \chi \in G^\vee$. To such a collection we associate an extended natural deformation, *namely, the subscheme of $\mathbb{V}$ defined by equations*

$$z_\chi z_\eta = z_{\chi+\eta} \prod_{\chi(\sigma)=\eta(\sigma)=1} \left(\sum_\theta \psi_{\sigma,\theta} \cdot z_\theta \right).$$

We have instead a (restricted) natural deformation *if we restrict ourselves to the* θ*'s such that* $\theta(\sigma) = 0$, *and we consider only an equation of the form*

$$z_\chi z_\eta = z_{\chi+\eta} \prod_{\chi(\sigma)=\eta(\sigma)=1} \left(\sum_{\theta(\sigma)=0} \psi_{\sigma,\theta} \cdot z_\theta \right).$$

The deformation results which we explained in the last lecture for simple bidouble covers work out also for $G \cong (\mathbb{Z}/2)^r$ which are locally simple, i.e., enjoy the property that for each point $y \in Y$ the σ's such that $y \in D_\sigma$ are a linear independent set. This is a good notion since (compare [Cat84], Proposition 1.1) if also X is smooth the covering is indeed locally simple.

One has the following result (see [Man01], Sect. 3)

Proposition 7.10 *Let* $f : X \to Y$ *be a locally simple* $G \cong (\mathbb{Z}/2)^r$ *covering with* Y *smooth and* X *normal. Then we have the exact sequence*

$$\oplus_{\chi(\sigma)=0}(H^0(\mathcal{O}_{D_\sigma}(D_\sigma - L_\chi))) \to \mathrm{Ext}^1_{\mathcal{O}_X}(\Omega^1_X, \mathcal{O}_X) \to \mathrm{Ext}^1_{\mathcal{O}_X}(f^*\Omega^1_Y, \mathcal{O}_X).$$

In particular, every small deformation of X *is a natural deformation if*
  (i) $H^1(\mathcal{O}_Y(-L_\chi)) = 0$
  (ii) $\mathrm{Ext}^1_{\mathcal{O}_X}(f^*\Omega^1_Y, \mathcal{O}_X) = 0$
If moreover
  (iii) $H^0(\mathcal{O}_Y(D_\sigma - L_\chi)) = 0 \ \forall \sigma \in G, \chi \in G^\vee$,
every small deformation of X *is again a* $G \cong (\mathbb{Z}/2)^r$ *covering.*

Comment on the proof.
  In the above proposition condition (i) ensures that $H^0(\mathcal{O}_Y(D_\sigma - L_\chi)) \to H^0(\mathcal{O}_{D_\sigma}(D_\sigma - L_\chi))$ is surjective.
  Condition (ii) and the above diagram imply then that the natural deformations are parametrized by a smooth manifold and have surjective Kodaira Spencer map, whence they induce all the infinitesimal deformations. $\square$
  In Manetti's application one needs an extension of the above result. In fact (ii) does not hold, since the manifold Y is not rigid (one can move the points $P_1, \dots P_n$ which are blown up in the quadric Q). But the moral is the same, in the sense that one can show that all the small deformations of X are G-coverings of a small deformation of Y.
  Before we proceed to the description of the Manetti surfaces, we consider some simpler surfaces, which however clearly illustrate one of the features of Manetti's construction.

Definition 7.11 *A singular bidouble Manetti surface of type* (a, b) *and triple of order* n *is a singular bidouble cover of* $Q := \mathbb{P}^1 \times \mathbb{P}^1$ *branched on three smooth curves* C_1, C_2, C_3 *belonging to the linear system of sections of the sheaf* $\mathcal{O}_Q(a, b)$ *and which intersect in* n *points* $p_1, \dots p_n$, *with distinct tangents.*
  A smooth bidouble Manetti surface of type (a, b) *and triple of order* n *is the minimal resolution of singularities* S *of such a surface* X *as above.*

Remark 7.12 *(1) With such a branch locus, a Galois group of type $G = (\mathbb{Z}/2)^r$ can be only $G = (\mathbb{Z}/2)^3$ or $G = (\mathbb{Z}/2)^2$ (we can exclude the uninteresting case $G = (\mathbb{Z}/2)$). The case $r = 3$ can only occur if the class of the three curves (a, b) is divisible by two since, as we said, the homology group of the complement $Q\backslash(\cup_i C_i)$ is the cokernel of the map $H^2(Q, \mathbb{Z}) \to \oplus_1^3(\mathbb{Z}C_i)$. The case $r = 3$ is however uninteresting, since in this case the elements $\phi(\gamma_i)$ are a basis, thus over each point p_i we have a nodal singularity of the covering surface, which obviously makes us remain in the same moduli space as the one where the three curves have no intersection points whatsoever.*

(2) Assume that $r = 2$, and consider the case where the monodromy μ is such that the $\mu(\gamma_i)$'s are the three nontrivial elements of the group $G = (\mathbb{Z}/2)^2$.

Let $p = p_i$ be a point where the three smooth curves C_1, C_2, C_3 intersect with distinct tangents: then over the point p there is a singularity (X, x) of the type considered in Example 3.3, namely, a quotient singularity which is analytically the cone over a rational curve of degree 4.

If we blow up the point p, and get an exceptional divisor E, the loop γ around the exceptional divisor E is homologous to the sum of the three loops $\gamma_1, \gamma_2, \gamma_3$ around the respective three curves C_1, C_2, C_3. Hence it must hold $\mu(\gamma) = \sum_i \mu(\gamma_i) = 0$, and the pull back of the covering does not have E in the branch locus. The inverse image A of E is a $(\mathbb{Z}/2)^2$ covering of E branched in three points, and we conclude that A is a smooth rational curve of self-intersection -4.

One sees (compare [Cat99]) that

Proposition 7.13 *Let X be a singular bidouble Manetti surface of type (a, b) and triple of order n: then if S is the minimal resolution of the singularities $x_1, \ldots x_n$ of X, then S has the following invariants:*
$$K_S^2 = 18ab - 24(a + b) + 32 - n$$
$$\chi(S) = 4 + 3(ab - a - b).$$
Moreover S is simply connected if (a, b) is not divisible by 2.

Idea of the proof For $n = 0$ these are the standard formulae since $2K_S = f^*(3a - 4, 3b - 4)$, and $\chi(\mathcal{O}_Q(-a, -b)) = 1 + 1/2(a(b - 2) + b(a - 2))$.

For $n > 0$, each singular point x_n lowers K_S^2 by 1, but leaves $\chi(S)$ invariant. In fact again we have $2K_X = f^*(3a - 4, 3b - 4)$, but $2K_S = 2K_X - \sum_i A_i$. For $\chi(S)$, one observes that x_i is a rational singularity, whence $\chi(\mathcal{O}_X) = \chi(\mathcal{O}_S)$.

It was proven in [Cat84] that S is simply connected for $n = 0$ when (a, b) is not divisible by 2 (in the contrary case the fundamental group equals $\mathbb{Z}/2$.) Let us then assume that $n \geq 1$.

Consider now a 1-parameter family $C_{3,t}$, $t \in T$, such that for $t \neq 0$ $C_{3,t}$ intersects C_1, C_2 transversally, while $C_{3,0} = C_3$. We get a corresponding family X_t of bidouble covers such that X_t is smooth for $t \neq 0$ and, as we just saw, simply connected. Then S is obtained from $X_t, t \neq 0$ replacing the Milnor fibres by tubular neighbourhoods of the exceptional divisors A_i, $i = 1, \ldots n$. Since A_i is smooth rational, these neighbourhoods are simply connected, and the result follows then easily by the first van Kampen theorem, which implies that $\pi_1(S)$ is a quotient of $\pi_1(X_t), t \neq 0$.   □

The important fact is that the above smooth bidouble Manetti surfaces of type (a, b) and triple of order n are parametrized, for $b = la, l \geq 2, n = la(2a - c), 0 < 2c < a$, by a disconnected parameter space ([Man01], Corollary 2.12: observe that we treat here only the case of $k = 3$ curves).

We cannot discuss here the method of proof, which relies on the socalled Brill Noether theory of special divisors: we only mention that Manetti considers the two components arising form the respective cases where $\mathcal{O}_{C_1}(p_1 + \ldots p_n) \cong \mathcal{O}_{C_1}(a - c, b)$, $\mathcal{O}_{C_1}(p_1 + \ldots p_n) \cong \mathcal{O}_{C_1}(a, b - lc)$, and shows that the closures of these loci yield two distinct connected components.

Unfortunately, one sees easily that smooth bidouble Manetti surfaces admit natural deformations which are not Galois coverings of the blowup Y of Q in the points $p_1, \ldots p_n$, hence Manetti is forced to take more complicated $G \cong (\mathbb{Z}/2)^r$ coverings (compare Sect. 6 of [Man01], especially page 68, but compare also the crucial Lemma 3.6).

The Galois group is chosen as $G = (\mathbb{Z}/2)^r$, where $r := 2 + n + 5$ (once more we make the simplifying choice $k = 1$ in 6.1 and foll. of [Man01]).

Definition 7.14 *(1) Let $G_1 := (\mathbb{Z}/2)^2$, $G_2 := (\mathbb{Z}/2)^n$, $G' := G_1 \oplus G_2 \oplus (\mathbb{Z}/2)^4$, $G := G' \oplus (\mathbb{Z}/2)$.*
(2) Let $D : G' \rightarrow Pic(Y)$ be the mapping sending

- *The three nonzero elements of G_1 to the classes of the proper transforms of the curves C_i, i.e., of $\pi^*(C_j) - \sum_i A_i$*
- *The canonical basis of G_2 to the classes of the exceptional divisors A_i*
- *The first two elements of the canonical basis of $(\mathbb{Z}/2)^4$ to the pull back of the class of $\mathcal{O}_Q(1, 0)$, the last two to the pull back of the class of $\mathcal{O}_Q(0, 1)$*
- *The other elements of G' to the zero class.*

With the above setting one has (Lemma 3.6 of [Man01])

Proposition 7.15 *There is an extension of the map $D : G' \rightarrow Pic(Y)$ to $D : G \rightarrow Pic(Y)$, and a map $L : G^\vee \rightarrow Pic(Y)$, $\chi \mapsto L_\chi$ such that*
(i) The cover conditions $2L_\chi \equiv \sum_{\chi(\sigma)=1} D_\sigma$ are satisfied
(ii) $-D_\sigma + L_\chi$ is an ample divisor
(iii) D_σ is an ample divisor for $\sigma \in G \backslash G'$

Definition 7.16 *Let now S be a G-covering of Y associated to the choice of some effective divisors D_σ in the given classes. S is said to be a Manetti surface.*

For simplicity we assume now that these divisors D_σ are smooth and intersect transversally, so that S is smooth.

Condition (iii) guarantees that S is connected, while condition (ii) and an extension of the argument of Proposition 7.10 shows that all the small deformations are G-coverings of such a rational surface Y, blowup of Q at n points.

We are going now only to very briefly sketch the rest of the arguments:

Step A It is possible to choose one of the D_σ's to be so positive that the group of automorphisms of a generic such surface S is just the group G.

Step B Using the natural action of G on any such surface, and using again arguments similar to the ones described in Step III of the last lecture, one sees that we get a closed set of the moduli space.

Step C The families of surfaces thus described fibre over the corresponding families of smooth bidouble Manetti surfaces: since for the latter one has more than one connected component, the same holds for the Manetti surfaces.

In the next section we shall show that the Manetti surfaces corresponding to a fixed choice of the extension D are canonically symplectomorphic.

In particular, they are a strong counterexample to the Def=Diff question.

7.3 Deformation and Canonical Symplectomorphism

We start discussing a simpler case:

Theorem 7.17 *Let S and S' be the respective minimal resolutions of the singularities of two singular bidouble Manetti surfaces X, X' of type (a, b), both triple of the same order n: then S and S' are diffeomorphic, and indeed symplectomorphic for their canonical symplectic structure.*

Proof. In order to set up our notation, we denote by C_1, C_2, C_3 the three smooth branch curves for $p : X \to Q$, and denote by $p_1, .., p_n$ the points where these three curves intersect (with distinct tangents): similarly the covering $p' : X' \to Q$ determines C_1', C_2', C_3' and $p_1', .., p_n'$. Let Y be the blow up of the quadric Q at the n points $p_1, .., p_n$, so that S is a smooth bidouble cover of Y, similarly S' of Y'.

Without loss of generality we may assume that C_1, C_2 intersect transversally in $2ab$ points, and similarly C_1', C_2'.

We want to apply Theorem 4.9 to S, S' (i.e., the X, X' of Theorem 4.9 are our S, S'). Let $\hat{C}_3$ be a general curve in the pencil spanned by C_1, C_2, and consider the pencil $C(t) = tC_3 + (1 - t)\hat{C}_3$. For each value of t, $C_1, C_2, C(t)$ meet in $p_1, .., p_n$, while for $t = 0$ they meet in $2ab$ points, again with distinct tangents by our generality assumption. We omit the other finitely many t's for which the intersection points are more than n, or the tangents are not distinct. After blowing up $p_1, .., p_n$ and taking the corresponding bidouble covers, we obtain a family S_t with $S_1 = S$, and such that S_0 has exactly $2ab - n := h$ singular points, quadruple of the type considered in Example 3.3.

Similarly, we have a family S_t', and we must find an equisingular family $Z_u, u \in U$, containing S_0 and S_0'.

Let $\mathbb{P}$ be the linear system $\mathbb{P}(H^0(Q, \mathcal{O}_Q(a, b)))$, and consider a general curve in the Grassmannian $Gr(1, \mathbb{P})$, giving a one dimensional family $C_1[w], C_2[w]$,

$w \in W$, of pairs of points of $\mathbb{P}$ such that $C_1[w]$ and $C_2[w]$ intersect transversally in $2ab$ points of Q.

Now, the covering of W given by

$$\{(w, p_1(w), \ldots p_n(w)) | p_1(w), \ldots p_n(w) \in C_1[w] \cap C_2[w], \ p_i(w) \neq p_j(w) \text{for } i \neq j\}$$

is irreducible. This is a consequence of the General Position Theorem (see [ACGH85], page 112) stating that if C is a smooth projective curve, then for each integer n the subset $C_{dep}^n \subset C^n$,

$$C_{dep}^n := \{(p_1, \ldots p_n) | p_i \neq p_j \text{for } i \neq j, \ p_1, \ldots p_n \text{ are linearly dependent}\}$$

is smooth and irreducible.

We obtain then a one dimensional family with irreducible basis U of rational surfaces $Y(u)$, obtained blowing up Q in the n points $p_1(w(u)), \ldots p_n(w(u))$, and a corresponding family Z_u of singular bidouble covers of $Y(u)$, each with $2ab - n$ singularities of the same type described above.

We have then the situation of Theorem 4.9, whence it follows that S, S', endowed with their canonical symplectic structures, are symplectomorphic. □

The same argument , mutatis mutandis, shows (compare [Cat02, Cat06])

Theorem 7.18 *Manetti surfaces of the same type (same integers $a, b, n, r = 2n + 7$, same divisor classes $[D_\sigma]$) are canonically symplectomorphic.*

Manetti indeed gave the following counterexample to the Def= Diff question:

Theorem 7.19 (Manetti) *For each integer $h > 0$ there exists a surface of general type S with first Betti number $b_1(S) = 0$, such that the subset of the moduli space corresponding to surfaces which are orientedly diffeomorphic to S contains at least h connected components.*

Remark 7.20 *Manetti proved the diffeomorphism of the surfaces which are here called Manetti surfaces using some results of Bonahon ([Bon83]) on the diffeotopies of lens spaces.*

We have given a more direct proof also because of the application to canonical symplectomorphism.

Corollary 7.21 *For each integer $h > 0$ there exist surfaces of general type $S_1, \ldots S_h$ with first Betti number $b_1(S_j) = 0$, socalled Manetti surfaces, which are canonically symplectomorphic, but which belong to h distinct connected components of the moduli space of surfaces diffeomorphic to S_1.*

In spite of the fact that we begin to have quite a variety of examples and counterexamples, there are quite a few interesting open questions, the first one concerns the existence of simply connected surfaces which are canonically symplectomorphic, but not deformation equivalent:

Question 7.22 *Are the diffeomorphic (a, b, c)-surfaces of Theorem 6.43, endowed with their canonical symplectic structure, indeed symplectomorphic?*

Remark 7.23 *A possible way of showing that the answer to the question above is yes (and therefore exhibiting symplectomorphic simply connected surfaces which are not deformation equivalent) goes through the analysis of the braid monodromy of the branch curve of the 'perturbed' quadruple covering of $\mathbb{P}^1 \times \mathbb{P}^1$ (the composition of the perturbed covering with the first projection $\mathbb{P}^1 \times \mathbb{P}^1 \to \mathbb{P}^1$ yields the Lefschetz fibration). One would like to see whether the involution ι on $\mathbb{P}^1$, $\iota(y) = -y$ can be written as the product of braids which show up in the factorization.*

This approach turns out to be more difficult than the corresponding analysis which has been made in the mapping class group, because the braid monodromy contains very many 'tangency' factors which do not come from local contributions to the regeneration of the branch curve from the union of the curves $f = 0, g = 0$ counted twice.

Question 7.24 *Are there (minimal) surfaces of general type which are orientedly diffeomorphic through a diffeomorphism carrying the canonical class to the canonical class, but, endowed with their canonical symplectic structure, are not canonically symplectomorphic?*

Are there such examples in the simply connected case?

The difficult question is then: how to show that diffeomorphic surfaces (diffeomorphic through a diffeomorphism carrying the canonical class to the canonical class) are not symplectomorphic?

We shall briefly comment on this in the next section, referring the reader to the other Lecture Notes in this volume (for instance, the one by Auroux and Smith) for more details.

7.4 Braid Monodromy and Chisini' Problem

Let $B \subset \mathbb{P}^2_{\mathbb{C}}$ be a plane algebraic curve of degree d, and let P be a general point not on B. Then the pencil of lines L_t passing through P determines a one parameter family of d-uples of points of $\mathbb{C} \cong L_t \backslash \{P\}$, namely, $L_t \cap B$.

Blowing up the point P we get the projection $\mathbb{F}_1 \to \mathbb{P}^1$, whence the braid at infinity is a full rotation, corresponding to the generator of the (infinite cyclic) centre of the braid group $\mathcal{B}_d$,

$$(\Delta_d^2) := (\sigma_{d-1}\sigma_{d-2}\ldots\sigma_1)^d.$$

Therefore one gets a factorization of Δ_d^2 in the braid group $\mathcal{B}_d$, and the equivalence class of the factorization does neither depend on the point P (if P is chosen to be general), nor does it depend on B, if B varies in an equisingular family of curves.

Chisini was mainly interested in the case of *cuspidal* curves (compare [Chi44, Chi55]), mainly because these are the branch curves of a generic projection $f : S \to \mathbb{P}^2_{\mathbb{C}}$, for any smooth projective surface $S \subset \mathbb{P}^r$.

More precisely, a *generic projection* $f : S \to \mathbb{P}^2_{\mathbb{C}}$ has the following properties:

- It is a finite morphism whose branch curve B has only nodes and cusps as singularities, and moreover
- The local monodromy around a smooth point of the branch curve is a transposition

Maps with those properties are called *generic coverings*: for these the local monodromies are only $\mathbb{Z}/2 = \mathfrak{S}_2$ (at the smooth points of the branch curve B), $\mathfrak{S}_3$ at the cusps, and $\mathbb{Z}/2 \times \mathbb{Z}/2$ at the nodes.

In such a case we have a *cuspidal* factorization, i.e., all factors are powers of a half twist, with respective exponents $1, 2, 3$.

Chisini posed the following daring

Conjecture 7.25 (Chisini's conjecture) *Given two generic coverings* $f : S \to \mathbb{P}^2_{\mathbb{C}}$, $f' : S' \to \mathbb{P}^2_{\mathbb{C}}$, *both of degree at least 5, assume that they have the same branch curve B. Is it then true that f and f' are equivalent?*

Observe that the condition on the degree is necessary, since a counterexample for $d \leq 4$ is furnished by the dual curve B of a smooth plane cubic (as already known to Chisini, cf. [Chi44]). Chisini in fact observed that there are two generic coverings, of respective degrees 3 and 4, and with the given branch curve. Combinatorially, we have a triple of transpositions corresponding in one case to the sides of a triangle ($d = 3$, and the monodromy permutes the vertices of the triangle), and in the other case to the three medians of the triangle ($d = 4$, and the monodromy permutes the vertices of the triangle plus the barycentre).

While establishing a very weak form of the conjecture [Cat86]. I remarked that the dual curve B of a smooth plane cubic is also the branch curve for three nonequivalent generic covers of the plane from the Veronese surface (they are distinct since they determine three distinct divisors of 2-torsion on the cubic).

The conjecture seems now to have been almost proven (i.e., it is not yet proven in the strongest possible form), after that it was first proven by Kulikov (cf. [Kul99]) under a rather complicated assumption, and that shortly later Nemirovskii [Nem01] noticed (just by using the Miyaoka-Yau inequality) that Kulikov's complicated assumption was implied by the simpler assumption $d \geq 12$.

Kulikov proved now [Kul06] the following

Theorem 7.26 (Kulikov) *Two generic projections with the same cuspidal branch curve B are isomorphic unless if the projection $p : S \to \mathbb{P}^2$ of one of them is just a linear projection of the Veronese surface.*

Chisini's conjecture concerns a fundamental property of the fundamental group of the complement $\mathbb{P}^2 \backslash B$, namely to admit only one conjugacy class of surjections onto a symmetric group $\mathcal{S}_n$, satisfying the properties of a generic covering.

In turn, the fundamental group $\pi_1(\mathbb{P}^2 \backslash B)$ is completely determined by the braid monodromy of B, i.e., the above equivalence class (modulo Hurwitz equivalence and simultaneous conjugation) of the above factorization of Δ_d^2. So, a classical question was: which are the braid monodromies of cuspidal curves?

Chisini found some necessary conditions, and proposed some argument in order to show the sufficiency of these conditions, which can be reformulated as

Chisini's problem: (cf. [Chi55]).

Given a cuspidal factorization, which is regenerable to the factorization of a smooth plane curve, is there a cuspidal curve which induces the given factorization?

Regenerable means that there is a factorization (in the equivalence class) such that, after replacing each factor σ^i $(i = 2, 3)$ by the i corresponding factors (e.g. , σ^3 is replaced by $\sigma \circ \sigma \circ \sigma$) one obtains the factorization belonging to a non singular plane curve.

A negative answer to the problem of Chisini was given by B. Moishezon in [Moi94].

Remark 7.27 *(1) Moishezon proves that there exist infinitely many non equivalent cuspidal factorizations observing that $\pi_1(\mathbb{P}_{\mathbb{C}}^2 \backslash B)$ is an invariant defined in terms of the factorization alone and constructing infinitely many non isomorphic such groups. On the other hand, the family of cuspidal curves of a fixed degree is an algebraic set, hence it has a finite number of connected components. These two statements together give a negative answer to the above cited problem of Chisini.*

The examples of Moishezon have been recently reinterpreted in [ADK03], with a simpler treatment, in terms of symplectic surgeries.

Now, as conjectured by Moishezon, a cuspidal factorization together with a generic monodromy with values in $\mathfrak{S}_n$ induces a covering $M \to \mathbb{P}_{\mathbb{C}}^2$, where the fourmanifold M has a unique symplectic structure (up to symplectomorphism) with class equal to the pull back of the Fubini Study form on $\mathbb{P}^2$ (see for instance [A-K00]).

What is more interesting (and much more difficult) is however the converse.

Extending Donaldson's techniques (for proving the existence of symplectic Lefschetz fibrations) Auroux and Katzarkov [A-K00] proved that each symplectic 4-manifold is in a natural way 'asymptotically' realized by such a generic covering.

They show that, given a symplectic fourmanifold (M, ω) with $[\omega] \in H^2(M, \mathbb{Z})$, there exists a multiple m of a line bundle L with $c_1(L) = [\omega]$ and three general sections s_0, s_1, s_2 of $L^{\otimes m}$, which are ϵ-holomorphic with many of their derivatives (that a section s is ϵ-holomorphic means, very roughly speaking, that once one has chosen a compatible almost complex structure, $|\bar{\partial}s| < \epsilon |\partial s|$) yielding a finite covering of the plane $\mathbb{P}^2$ which is generic and with branch curve a symplectic subvariety whose singularities are only nodes and cusps.

The only price they have to pay is to allow also negative nodes, i.e., nodes which in local holomorphic coordinates are defined by the equation

$$(y - \bar{x})(y + \bar{x}) = 0.$$

The corresponding factorization in the braid group contains then only factors which are conjugates of σ_1^j, with $j = -2, 1, 2, 3$.

Moreover, the factorization is not unique, because it may happen that two consecutive nodes, one positive and one negative, may disappear, and the corresponding two factors disappear from the factorization. In particular, $\pi_1(\mathbb{P}_{\mathbb{C}}^2 \backslash B)$ is no longer an invariant and the authors propose to use an appropriate quotient of $\pi_1(\mathbb{P}_{\mathbb{C}}^2 \backslash B)$ in order to produce invariants of symplectic structures.

It seems however that, in the computations done up to now, even the groups $\pi_1(\mathbb{P}_{\mathbb{C}}^2 \backslash B)$ allow only to detect homology invariants of the projected fourmanifold [ADKY04].

Let us now return to the world of surfaces of general type.

Suppose we have a surface S of general type and a pluricanonical embedding $\psi_m \colon X \to \mathbb{P}^N$ of the canonical model X of S. Then a generic linear projection of the pluricanonical image to $\mathbb{P}_{\mathbb{C}}^2$ yields, if $S \cong X$, a generic covering $S \to \mathbb{P}_{\mathbb{C}}^2$ (else the singularities of X create further singularities for the branch curve B and other local coverings).

By the positive solution of Chisini's conjecture, the branch curve B determines the surface S uniquely (up to isomorphism). We get moreover the equivalence class of the braid monodromy factorization, and this does not change if S varies in a connected family of surfaces with K_S ample (i.e., the surfaces equal their canonical models).

Motivated by this observation of Moishezon, Moishezon and Teicher in a series of technically difficult papers (see e.g. [MT92]) tried to calculate fundamental groups of complements $\pi_1(\mathbb{P}_{\mathbb{C}}^2 \backslash B)$, with the intention of distinguishing connected components of the moduli spaces of surfaces of general type.

Indeed, it is clear that these groups are invariants of the connected components of the open set of the moduli space corresponding to surfaces with ample canonical divisor K_S. Whether one could even distinguish connected components of moduli spaces would in my opinion deserve a detailed argument, in view of the fact that several irreducible components consist of surfaces whose canonical divisor is not ample (see for instance [Cat89] for several series of examples).

But it may be that the information carried by $\pi_1(\mathbb{P}_{\mathbb{C}}^2 \backslash B)$ be too scanty, so one could look at further combinatorial invariants, rather than the class of the braid monodromy factorization for B.

In fact a generic linear projection of the pluricanonical image to $\mathbb{P}^3_{\mathbb{C}}$ gives a surface Σ with a double curve Γ'. Now, projecting further to $\mathbb{P}^2_C$ we do not only get the branch curve B, but also a plane curve Γ, image of Γ'.

Even if Chisini's conjecture tells us that from the holomorphic point of view B determines the surface S and therefore the curve Γ, it does not follow that the fundamental group $\pi_1(\mathbb{P}^2_{\mathbb{C}} \setminus B)$ determines the group $\pi_1(\mathbb{P}^2_{\mathbb{C}} \setminus (B \cup \Gamma))$.

It would be interesting to calculate this second fundamental group, even in special cases.

Moreover, generalizing a proposal done by Moishezon in [Moi83], one can observe that the monodromy of the restriction of the covering $\Sigma \to \mathbb{P}^2$ to $\mathbb{P}^2_{\mathbb{C}} \setminus (B \cup \Gamma))$ is more refined, since it takes values in a braid group $\mathcal{B}_n$, rather than in a symmetric group $\mathcal{S}_n$.

One could proceed similarly also for the generic projections of symplectic fourmanifolds.

But in the symplectic case one does not have the advantage of knowing a priori an explicit number $m \leq 5$ such that ψ_m is a pluricanonical embedding for the general surface S in the moduli space.

Acknowledgments

I would like to thank the I.H.P. for its hospitality in november 2005, and the I.H.E.S. for its hospitality in march 2006, which allowed me to write down some parts of these Lecture Notes.

I am grateful to Fabio Tonoli for his invaluable help with the figures, and to Ingrid Bauer, Michael Lönne, Paolo Oliverio, Roberto Pignatelli and Sönke Rollenske for comments on a preliminary version.

The research of the author was supported by the SCHWERPUNKT "Globale Methoden in der komplexen Geometrie", a visit to IHES by contract nr RITA-CT-2004-505493.

References

[A-B-K-P-00] J. Amoros, F. Bogomolov, L. Katzarkov, T.Pantev *Symplectic Lefschetz fibrations with arbitrary fundamental group*, With an appendix by Ivan Smith, **J. Differential Geom. 54** no. 3, (2000), 489–545.

[A-F69] A. Andreotti, T. Frankel, *The second Lefschetz theorem on hyperplane sections,* in 'Global Analysis (Papers in Honor of K. Kodaira)' **Univ. Tokyo Press, Tokyo** (1969), 1–20.

[Andr73] A. Andreotti, *Nine lectures on complex analysis,* in 'Complex analysis (Centro Internaz. Mat. Estivo C.I.M.E., I Ciclo, Bressanone, 1973)', **Edizioni Cremonese, Roma**(1974), 1–175.

[Ara97] D. Arapura, *Geometry of cohomology support loci for local systems I,* **J. Algebraic Geom. 6** no. 3 (1997), 563–597.

[ACGH85] E. Arbarello, M. Cornalba, P.A. Griffiths, J. Harris, *Geometry of algebraic curves Volume I,* **Grundlehren der mathematischen Wissenschaften 267, Springer-Verlag, New York.** (1985), XVI + 386 p.

[Art26] E. Artin, *Theorie der Zöpfe,* **Hamburg Univ. Math. Seminar Abhandlungen 4–5** (1926), 47–72.

[Art65] E. Artin, *The collected papers of Emil Artin,* edited by Serge Lang and John T. Tate, **Addison–Wesley Publishing Co., Inc.**, Reading, Mass. London (1965).

[ArtM66] M. Artin, *On isolated rational singularities of surfaces,* **Amer. J. Math. 88** (1966), 129–136.

[A-K00] D. Auroux, L. Katzarkov , *Branched coverings of* $\mathbb{CP}^2$ *and invariants of symplectic 4-manifolds* **Inv. Math. 142** (2000), 631–673.

[Aur02] D. Auroux, *Fiber sums of genus 2 Lefschetz fibrations.* **Turkish J. Math. 27, no. 1** (2003), 1–10 .

[ADK03] D. Auroux, S. Donaldson, L. Katzarkov, *Luttinger surgery along Lagrangian tori and non-isotopy for singular symplectic plane curves.* **Math. Ann. 326 no. 1** (2003), 185–203.

[ADKY04] D. Auroux, S. Donaldson, L. Katzarkov and M. Yotov, *Fundamental groups of complements of plane curves and symplectic invariants* **Topology 43, no. 6** (2004), 1285–1318.

[BPV84] W. Barth, C. Peters, A. Van de Ven, *Compact complex surfaces.* **Ergebnisse der Mathematik und ihrer Grenzgebiete (3). Springer-Verlag**, Berlin,(1984).

[BHH87] G. Barthel, F. Hirzebruch, T. Höfer, *Geradenkonfigurationen und Algebraische Flächen.* **Aspekte der Math. D4, Vieweg** (1987).

[Bau97] I. Bauer, *Irrational pencils on non-compact algebraic manifolds,* **Internat. J. Math. 8 , no. 4** (1997), 441–450.

[BaCa04] I. Bauer, F. Catanese, *Some new surfaces with* $p_g = q = 0$, The Fano Conference, **Univ. Torino, Torino**, (2004), 123–142.

[BCG05] I. Bauer, F. Catanese, F. Grunewald, *Beauville surfaces without real structures.* In: Geometric methods in algebra and number theory, **Progr. Math., 235**, Birkhäuser (2005), 1–42.

[BCG06] I. Bauer, F. Catanese, F. Grunewald, *Chebycheff and Belyi polynomials, dessins d'enfants, Beauville surfaces and group theory.* **Mediterranean J. Math. 3, no.2**, (2006) 119–143.

[BCG06-b]   I. Bauer, F. Catanese, F. Grunewald, *The classification of sur-*
faces with $p_g = q = 0$ isogenous to a product of curves,
math.AG/0610267.

[BC06]   I. Bauer, F. Catanese, *A volume maximizing canonical surface in*
3-space, math.AG/0608020, to appear in **Comm. Math. Helvetici.**

[Bea78]   A. Beauville, *Surfaces algébriques complexes,* **Asterisque 54,** Soc.
Math. France (1978).

[Bir74]   J.S. Birman, *Braids, links, and mapping class groups,* **Annals of**
Mathematics Studies, No. 82, Princeton University Press,
Princeton, N.J.; University of Tokyo Press, Tokyo (1974).

[Bom73]   E. Bombieri, *Canonical models of surfaces of general type,* **Publ.**
Math. I.H.E.S., 42 (1973), 173–219.

[B-H75]   E. Bombieri, D. Husemoller, *Classification and embeddings of sur-*
faces, in 'Algebraic geometry, Humboldt State Univ., Arcata, Calif.,
1974', **Proc. Sympos. Pure Math., Vol. 29, Amer. Math.**
Soc., Providence, R.I. (1975), 329–420.

[Bon83]   F. Bonahon, *Difféotopies des espaces lenticulaires,* **Topology**
22(1983), 305–314.

[Briesk68-a]   E. Brieskorn, *Rationale Singularitäten komplexer Flächen,* **Invent.**
Math. 4 (1967/1968), 336–358.

[Briesk68-b]   E. Brieskorn, *Die Auflösung der rationalen Singularitäten holomor-*
pher Abbildungen, **Math. Ann. 178** (1968), 255–270.

[Briesk71]   E. Brieskorn, *Singular elements of semi-simple algebraic groups,*
'Actes du Congrés International des Mathématiciens' (Nice, 1970),
Tome 2, **Gauthier-Villars,** Paris (1971), 279–284.

[B-J90]   T. Bröcker, K. Jänich *Einführung in die Differentialtopologie* , **Hei-**
delberger Taschenbücher Band 143, Springer Verlag (1990).

[B-W74]   D. Burns, J. Wahl, *Local contributions to global deformations of sur-*
faces, **Invent. Math. 26**(1974), 67–88.

[Car57]   H. Cartan, *Quotient d'un espace analytique par un groupe*
d'automorphismes, in 'A symposium in honor of S. Lefschetz, Algebraic geometry and topology' **Princeton University Press,**
Princeton, N. J. (1957), 90–102.

[Cat82]   F. Catanese, *Moduli of surfaces of general type.* in 'Algebraic geometry – open problems' (Ravello, 1982), **Lecture Notes in Math.,**
997 Springer, Berlin-New York (1983), 90–112.

[Cat84]   F. Catanese, *On the Moduli Spaces of Surfaces of General Type,*
J. Differential Geom. 19 (1984), 483–515.

[Cat86]   F. Catanese, *On a problem of Chisini,* **Duke Math. J. 53**(1986),
33–42.

[Cat87]   F. Catanese, *Automorphisms of Rational Double Points and Mod-*
uli Spaces of Surfaces of General Type, **Comp. Math. 61** (1987),
81–102.

[Cat86]   F. Catanese, *Connected Components of Moduli Spaces,* **J. Differen-**
tial Geom. 24 (1986), 395–399.

[Cat88]   F. Catanese, *Moduli of algebraic surfaces.* Theory of moduli (Montecatini Terme, 1985), **Lecture Notes in Math. 1337,** Springer,
Berlin (1988), 1–83.

[Cat89]   F. Catanese, *Everywhere non reduced Moduli Spaces* , **Inv. Math.**
98, 293–310 (1989).

[Cat91]   F. Catanese, *Moduli and classification of irregular Kähler manifolds (and algebraic varieties) with Albanese general type fibrations.* **Invent. math. 104**, (1991), 263–289.

[Cat94]   F. Catanese, *Fundamental groups with few relations,* in 'Higher dimensional complex varieties. Proceedings of the international conference, Trento, Italy, June 15–24, 1994' **Walter de Gruyter, Berlin** (1996), 163–165.

[C-F-96]   F. Catanese, M. Franciosi, *Divisors of small genus on algebraic surfaces and projective embeddings,* in 'Proceedings of the Hirzebruch 65 Conference on Algebraic Geometry (Ramat Gan, 1993)', **Israel Math. Conf. Proc., 9,** BarIlan Univ., Ramat Gan, (1996), 109–140.

[CFHR99]   F. Catanese, M. Franciosi, K. Hulek, M. Reid, *Embeddings of curves and surfaces,* **Nagoya Math. J. 154** (1999), 185–220.

[Cat99]   F. Catanese, *Singular bidouble covers and the construction of interesting algebraic surfaces.* **AMS Cont. Math. bf 241** (1999), 97–120.

[Cat00]   F. Catanese, *Fibred surfaces, varieties isogenous to a product and related moduli spaces.,* **Amer. J. Math. 122, no.1** (2000), 1–44.

[Cat03]   F. Catanese, *Moduli Spaces of Surfaces and Real Structures,* **Ann. Math. (2) 158, no.2** (2003), 577–592.

[Cat03b]   F. Catanese, *Fibred Kähler and quasi projective groups,* Advances in Geometry, suppl., Special Issue dedicated to A. Barlotti's 80-th birthday (2003), **Adv. Geom. Suppl.** (2003) S13–S27.

[CKO03]   F. Catanese, J. Keum, K. Oguiso, *Some remarks on the universal cover of an open K3 surface.* **Math. Ann. 325, No. 2,** (2003), 279–286.

[Cat02]   F. Catanese, *Symplectic structures of algebraic surfaces and deformation,* 14 pages, math.AG/0207254.

[CW04]   F. Catanese, B. Wajnryb, *Diffeomorphism of simply connected algebraic surfaces.* **math.AG/0405299, J. Differential Geom. 76 no. 2** (2007), 177–213.

[Cat06]   F. Catanese, *Canonical symplectic structures and deformations of algebraic surfaces,* 11 pages, math.AG/0608110 to appear in Communications in Contemporary Mathematics.

[Chi44]   O. Chisini, *Sulla identità birazionale delle funzioni algebriche di due variabili dotate di una medesima curva di diramazione,* **Ist. Lombardo Sci. Lett. Cl. Sci. Mat. Nat. Rend. (3) 8(77)** (1944), 339–356.

[Chi55]   O. Chisini, *Il teorema di esistenza delle trecce algebriche. I-III,* **Atti Accad. Naz. Lincei. Rend. Cl. Sci. Fis. Mat. Nat. (8), 17–18**(1954), 143–149; (1955), 307–311; (1955), 8–13.

[Corn76]   M. Cornalba, *Complex tori and Jacobians.* Complex analysis and its applications (Lectures, Internat. Sem., Trieste, 1975), Vol. II, **Internat. Atomic Energy Agency,** Vienna, (1976), 39–100.

[Dehn38]   M. Dehn, *Die Gruppe der Abbildungsklassen. (Das arithmetische Feld auf Flächen.)* **Acta Math. 69** (1938), 135–206.

[Del70]   P. Deligne, *Équations différentielles á points singuliers réguliers,* **Lecture Notes in Mathematics, Vol. 163. Springer-Verlag, Berlin-New York**(1970), iii+133 pp.

[D-M93]   P. Deligne, G.D. Mostow, *Commensurabilities among lattices in* PU$(1, n)$, **Annals of Mathematics Studies, 132, Princeton University Press, Princeton, NJ** (1993), viii+183 pp.

[Dolg82]   I. Dolgachev, *Weighted projective varieties,* in 'Group actions and vector fields (Vancouver, B.C., 1981)', **Lecture Notes in Math., 956, Springer**, Berlin, (1982), 34–71.

[Don83]   S.K. Donaldson, *An Application of Gauge Theory to Four-Dimensional Topology,* **J. Differential Geom. 18** (1983), 279–315.

[Don86]   S.K. Donaldson, *Connections, cohomology and the intersection forms of 4-manifolds.* **J. Differential Geom. 24** (1986), 275–341.

[Don90]   S.K. Donaldson, *Polynomial invariants for smooth four-manifolds.* **Topology 29, 3** (1990), 257–315.

[Don92]   S.K. Donaldson, *Gauge theory and four-manifold topology.* [CA] Joseph, A. (ed.) et al., First European congress of mathematics (ECM), Paris, France, July 6–10, 1992. Volume I: Invited lectures (Part 1). **Basel: Birkhäuser, Prog. Math. 119** (1994) 121–151.

[Don96]   S.K. Donaldson, *The Seiberg-Witten Equations and 4-manifold topology.* **Bull. Am. Math. Soc., (N S) 33, 1** (1996) 45–70.

[Don96-2]   S.K. Donaldson, *Symplectic submanifolds and almost-complex geometry.* **J. Differential Geom. 44, 4** (1996), 666–705.

[Don99]   S.K. Donaldson, *Lefschetz pencils on symplectic manifolds,* **J. Differential Geom. 53, no. 2**(1999), 205–236.

[D-K90]   S.K. Donaldson, P.B. Kronheimer *The geometry of four-manifolds.* **Oxford Mathematical Monographs. Oxford: Clarendon Press,** (1990) ix, 440 p.

[Durf79]   A.H. Durfee, *Fifteen characterizations of rational double points and simple critical points,* **Enseign. Math. (2) 25, no. 1–2** (1979), 131–163.

[Ehr43]   C. Ehresmann, *Sur les espaces fibrés différentiables.* **C.R. Acad. Sci. Paris 224** (1947), 1611–1612.

[Eke88]   T. Ekedahl, *Canonical models of surfaces of general type in positive characteristic,* **Inst. Hautes Études Sci. Publ. Math. No. 67** (1988), 97–144.

[Enr49]   F. Enriques, *Le Superficie Algebriche.* **Zanichelli**, Bologna (1949).

[Free82]   M. Freedman, *The topology of four-dimensional manifolds.,* **J. Differential Geom. 17, n. 3** (1982), 357–454.

[F-Q90]   M. Freedman and F.Quinn, *Topology of 4-manifolds.,* **Princeton Math. Series 39** (1990).

[F-M-M87]   R. Friedman, B.G. Moishezon, J.W.Morgan, *On the C^∞ invariance of the canonical classes of certain algebraic surfaces.,* **Bull. Amer. Math. Soc. 17** (1987), 283–286.

[F-M88]   R. Friedman and J.W.Morgan, *Algebraic surfaces and four-manifolds: some conjectures and speculations.,* **Bull. Amer. Math. Soc. 18** (1988) 1–19.

[F-M94]   R. Friedman; J. W. Morgan, *Smooth four-manifolds and complex surfaces,* **Ergebnisse der Mathematik und ihrer Grenzgebiete (3), 27 Springer-Verlag**, Berlin, (1994), x+520 pp.

[F-M97]   R. Friedman and J.W.Morgan, *Algebraic surfaces and Seiberg-Witten invariants.* **J. Algebr. Geom. 6, No.3** (1997) 445–479.

[Fried05]   R. Friedman. *Letter to the author*, August 9 2005.

[Gie77]   D. Gieseker, *Global moduli for surfaces of general type*, **Invent. Math. 43, no. 3**(1977), 233–282.

[Gompf95]   R.E. Gompf, *A new construction of symplectic manifolds* **Ann. of Math. 142** 3 (1995), 527–595.

[Gompf01]   R.E. Gompf, *The topology of symplectic manifolds*. **Turkish J. Math. 25** no. 1, (2001), 43–59.

[G-S99]   R.E. Gompf, A. Stipsicz, *4-manifolds and Kirby calculus*. **Graduate Studies in Mathematics, 20. American Mathematical Society, Providence, RI**, (1999) xvi+558 pp.

[GR58]   H. Grauert, R. Remmert, *Komplexe Räume*, **Math. Ann. 136** (1958), 245–318.

[Gra72]   H. Grauert, *Über die Deformation isolierter Singularitäten analytischer Mengen*, **Invent. Math. 15** (1972), 171–198.

[Gra74]   H. Grauert, *Der Satz von Kuranishi für kompakte komplexe Räume*, **Invent. Math. 25**(1974), 107–142.

[Grom89]   M. Gromov, *Sur le groupe fondamental d'une variété kählérienne*, **C. R. Acad. Sci., Paris, Sér. I 308, No. 3**(1989), 67–70.

[Groth60]   A. Grothendieck, *Techniques de construction et théoremes d'existence en géométrie algébrique. IV, Les schemas de Hilbert*, Sem. Bourbaki, Vol. 13, (1960–1961), 1–28.

[Hart77]   R. Hartshorne, *Algebraic geometry*, **Springer GTM 52** (1977).

[Hart66]   R. Hartshorne, *Connectedness of Hilbert scheme*, **Publ. Math., Inst. Hautes Étud. Sci. 29**(1966), 5–48.

[HT80]   A. Hatcher, W.Thurston, *A presentation for the mapping class group of a closed orientable surface*, **Topology 19, no. 3** (1980), 221–237.

[Hir64]   H. Hironaka, *Resolution of singularities of an algebraic variety over a field of characteristic zero. I, II.* **Ann. of Math. (2) 79**(1964), 109–203; ibid. 205–326.

[Hirz83]   F. Hirzebruch, *Arrangements of lines and algebraic surfaces*,in 'Arithmetic and geometry, vol. II', **Progr. math 36, Birkhäuser** (1983), 113–140.

[Hor75]   E. Horikawa, *On deformations of quintic surfaces*, **Invent. Math. 31** (1975), 43–85.

[Hum75]   J. E. Humphreys, *Linear algebraic groups*, **Graduate Texts in Mathematics 21, Springer-Verlag** New York - Heidelberg - Berlin, (1975), XV, 247 p.

[Huy05]   D. Huybrechts, *Complex geometry. An introduction. Universitext. Springer-Verlag*, Berlin, (2005), xii+309 pp.

[J-Y85]   J. Jost, S.T. Yau *A strong rigidity theorem for a certain class of compact complex analytic surfaces*, **Math. Ann. 271** (1985), 143–152.

[Kas80]   A. Kas, *On the handlebody decomposition associated to a Lefschetz fibration*. **Pacific J. Math. 89** no. 1, (1980), 89–104.

[Katata83]   *Open problems: Classification of algebraic and analytic manifolds.* Classification of algebraic and analytic manifolds, Proc. Symp. Katata/Jap. 1982. Edited by Kenji Ueno. **Progress in Mathematics, 39. Birkhäuser, Boston, Mass.** (1983), 591–630.

[KK02]   V. M. Kharlamov, V.S. Kulikov, *On real structures of real surfaces.*
            Izv. Ross. Akad. Nauk Ser. Mat. **66**, no. 1, 133–152 (2002); transla-
            tion in Izv. Math. **66**, no. 1, 133–150 (2002).

[KK02-b]   V. Kharlamov, V. Kulikov, *Deformation inequivalent complex conju-
            gated complex structures and applications,* **Turkish J. Math. 26,
            no. 1**(2002), 1–25.

[Klein1884] F. Klein, *Lectures on the icosahedron and the solution of equations
            of the fifth degree,* second and revised edition, **Dover Publications,
            Inc.**, New York, N.Y. (1956), xvi+289 pp.

[K-S58]   K. Kodaira,; D. C. Spencer, *On deformations of complex analytic
            structures. I, II,* **Ann. of Math. (2) 67** (1958), 328–466.

[Kod60]   K. Kodaira, *On compact complex analytic surfaces, I,* **Ann. of
            Math. 71** (1960), 111–152.

[Kod63]   K. Kodaira, *On compact analytic surfaces, III,* **Ann. Math. (2) 78**
            (1963), 1–40.

[Kod63-b]   K. Kodaira, *On stability of compact submanifolds of complex mani-
            folds,* **Am. J. Math. 85**(1963), 79–94.

[Kod64]   K. Kodaira, *On the structure of compact complex analytic surfaces.
            I.,* **Amer. J. Math. 86** (1964), 751–798.

[Kod66]   K. Kodaira, *On the structure of compact complex analytic surfaces.
            II.* **Amer. J. Math. 88** (1966), 682–721.

[K-SB88]   J. Kollár and N.I. Shepherd Barron, *Threefolds and deformations of
            surface singularities.* **Inv. Math. 91** (1988) 299–338.

[Kul99]   V.S. Kulikov, *On Chisini's conjecture,* **Izv. Ross. Akad. Nauk
            Ser. Mat. 63, no. 6**(1999), 83–116; translation in **Izv. Math. 63,
            no. 6** (1999), 1139–1170.

[Kul06]   V.S. Kulikov, *On Chisini's conjecture. II* **math. AG/0610356v1.**

[Kur62]   M. Kuranishi, *On the locally complete families of complex analytic
            structures,* **Ann. Math. (2) 75** (1962), 536–577.

[Kur65]   M. Kuranishi, *New proof for the existence of locally complete families
            of complex structures,* Proc. Conf. Complex Analysis, Minneapolis
            1964 (1965), 142–154.

[L-W86]   E. Looijenga and J. Wahl, *Quadratic functions and smoothing surface
            singularities* **Topology 25** (1986), 261–291.

[M-M76]   R. Mandelbaum, B. Moishezon, *On the topological structure of non-
            singular algebraic surfaces in CP^3,* **Topology 15**(1976), 23–40.

[M-M80]   R. Mandelbaum, B. Moishezon, *On the topology of simply-connected
            algebraic surfaces,* **Trans. Am. Math. Soc. 260** (1980), 195–222.

[Man90]   M. Manetti, *$\mathbb{Q}$-Gorenstein smoothings of quotient singularities,*
            Preprint Scuola Normale Pisa (1990).

[Man94]   M. Manetti, *On some Components of the Moduli Space of Surfaces
            of General Type,* **Comp. Math. 92** (1994) 285–297.

[Man96]   M. Manetti, *Degenerations of Algebraic Surfaces and Applications
            to Moduli Problems,* **Tesi di Perfezionamento Scuola Normale
            Pisa** (1996) 1–142.

[Man97]   M. Manetti, *Iterated Double Covers and Connected Components of
            Moduli Spaces,* **Topology 36, 3** (1997) 745–764.

[Man01]   M. Manetti, *On the Moduli Space of diffeomorphic algebraic surfaces,*
            Inv. Math. 143 (2001), 29–76.

[Math70]   J. Mather, *Notes on topological stability*, **Harvard University Math. Notes** (1970).

[MS98]   D. McDuff, D. Salamon, *Introduction to symplectic topology.* **Oxford Mathematical Monographs, The Clarendon Press, Oxford University Press, New York** (1998).

[Miya83]   Y. Miyaoka, *Algebraic surfaces with positive indices,* in 'Classification of algebraic and analytic manifolds', Proc. Symp. Katata/Jap. 1982, **Prog. Math. 39** (1983), 281–301.

[Mil68]   J. W. Milnor, *Singular points of complex hypersurfaces,* **Annals of Mathematics Studies 61, Princeton University Press and the University of Tokyo Press** (1968), 122 p.

[Moi77]   B. Moishezon, *Complex surfaces and connected sums of complex projective planes,* **Lecture Notes in Math., 603. Springer**, Berlin (1977).

[Moi81]   B. Moishezon, *Stable branch curves and braid monodromies,* Algebraic geometry, Proc. Conf., Chicago Circle 1980, Lect. Notes Math. 862, 107–192 (1981).

[Moi83]   B. Moishezon, *Algebraic surfaces and the arithmetic of braids I,* in 'Arithmetic and geometry, Pap. dedic. I. R. Shafarevich, Vol. II: Geometry' **Prog. Math. 36, Birkhäuser** (1983), 199–269.

[MT92]   B. Moishezon, M. Teicher, *Finite fundamental groups, free over $\mathbb{Z}/c\mathbb{Z}$, for Galois covers of $\mathbb{CP}^2$.* **Math. Ann. 293, no. 4**(1992), 749–766.

[Moi94]   B. Moishezon, *The arithmetics of braids and a statement of Chisini,* in 'Geometric Topology, Haifa 1992', **Contemp. Math. 164, Amer. Math. Soc., Providence, RI** (1994), 151–175.

[Mor96]   J. W. Morgan, *The Seiberg-Witten equations and applications to the topology of smooth four-manifolds.* **Mathematical Notes 44. Princeton Univ. Press** vi (1996).

[Mori88]   S. Mori, *Flip theorem and the existence of minimal models for 3-folds,* **J. Amer. Math. Soc. 1** , no. 1 (1988), 117–253.

[Mos65]   J. Moser, *On the volume elements on a manifold.* **Trans. Amer. Math. Soc. 120** (1965), 286–294.

[Most73]   G.D. Mostow, *Strong rigidity of locally symmetric spaces,* **Annals of Mathematics Studies, no. 78, Princeton University Press, Princeton, N.J.; University of Tokyo Press, Tokyo** (1973).

[Mum62]   D. Mumford, *The canonical ring of an algebraic surface,* **Ann. of Math. (2) 76** (1962) 612–615, appendix to *The theorem of Riemann-Roch for high multiples of an effective divisor on an algebraic surface,* by O. Zariski, ibid. 560–611.

[Mum76]   D. Mumford, *Algebraic geometry. I. Complex projective varieties.* Reprint of the 1976 edition. **Classics in Mathematics. Springer-Verlag**, Berlin (1995), x+186 pp.

[Mum83]   D. Mumford, *Towards an enumerative geometry of the moduli space of curves,* in 'Arithmetic and geometry, Pap. dedic. I. R. Shafarevich, Vol. II: Geometry', **Prog. Math. 36, Birkhäuser** (1983), 271–328.

[Nem01]   S.Y. Nemirovskii, *On Kulikov's theorem on the Chisini conjecture.* **Izv. Ross. Akad. Nauk Ser. Mat. 65, no. 1** (2001), 77–80; translation in **Izv. Math. 65** , no. 1(2001), 71–74.

[Par91]   R. Pardini, *Abelian covers of algebraic varieties,* **J. Reine Angew. Mathematik 417** (1991), 191–213.

[Ram72-4]   C.P. Ramanujam, *Remarks on the Kodaira vanishing theorem*, **J. Indian Math. Soc. 36** (1972), 41–51; supplement **J. Indian Math. Soc. 38** (1974), 121–124.

[Reid80]   M. Reid, *Canonical 3-folds,* in 'Journées de géometrie algébrique, Angers/France 1979', **Sijthoff and Nordhoff** (1980), 273–310.

[Reid87]   M. Reid, *Young person's guide to canonical singularities,* in 'Algebraic geometry, Proc. Summer Res. Inst., Brunswick/Maine 1985', part 1, **Proc. Symp. Pure Math. 46** (1987), 345–414.

[Riem74]   O. Riemenschneider, *Deformationen von Quotientensingularitäten (nach zyklischen Gruppen),* **Math. Ann. 209** (1974), 211–248.

[Sern06]   Sernesi, Edoardo *Deformations of algebraic schemes.* Grundlehren der Mathematischen Wissenschaften **Fundamental Principles of Mathematical Sciences, 334.** Springer-Verlag, Berlin, 2006. xii+339 pp.

[FAC55]   J.P. Serre, *Faisceaux algébriques cohérents,* **Ann. of Math. (2) 61** (1955), 197–278.

[Gaga55-6]   J.P. Serre, *Géométrie algébrique et géométrie analytique.* **Ann. Inst. Fourier, Grenoble 6** (1955–1956), 1–42.

[Ser59]   J.P. Serre, *Groupes algébriques et corps de classes.* **Publications de l'institut de mathématique de l'université de Nancago, VII. Hermann**, Paris (1959), 202 pp.

[Ser64]   J.P. Serre, *Exemples de variétés projectives conjuguées non homéomorphes.* **C. R. Acad. Sci. Paris 258**, (1964) 4194–4196.

[Shaf74]   I.R. Shafarevich, I. R. *Basic algebraic geometry.* Translated from the Russian by K. A. Hirsch. Revised printing of **Grundlehren der mathematischen Wissenschaften, Vol. 213, 1974. Springer Study Edition. Springer-Verlag**, Berlin-New York, (1977), xv+439 pp.

[Sieg73]   C.L. Siegel, *Topics in complex function theory. Vol. III. Abelian functions and modular functions of several variables.* Translated from the German by E. Gottschling and M. Tretkoff. With a preface by Wilhelm Magnus. Reprint of the 1973 original. **Wiley Classics Library. A Wiley-Interscience Publication. John Wiley and Sons, Inc.**, New York (1989), x+244 pp.

[Siu93]   Y.T. Siu, *An effective Matsusaka big theorem*, **Ann. Inst. Fourier 43** (1993), 1387–1405.

[STY02]   I. Smith, R. Thomas, S.T. Yau, *Symplectic conifold transitions,* **J. Diff. Geom. 62** (2002), 209–242.

[ST03]   I. Smith, R. Thomas, *Symplectic surgeries from singularities,* **Turkish J. Math. 27** (2003), 231–250.

[Thur76]   W.P. Thurston, *Some simple examples of symplectic manifolds,* **Proc. Amer. Math. Soc. 55, no. 2** (1976), 467–468.

[Tju70]   G.N. Tjurina, *Resolution of singularities of plane (= flat) deformations of double rational points* , **Funk. Anal. i Prilozen 4** (1970), 77–83.

[Tro96]   A. Tromba, *Dirichlet's energy on Teichmüller's moduli space and the Nielsen realization problem,* **Math. Z. 222** (1996), 451–464.

[Ves84]   E. Vesentini, *Capitoli scelti della teoria delle funzioni olomorfe,* **Unione Matematica Italiana** (1984).

[Vieh95]   E. Viehweg, *Quasi-projective moduli for polarized manifolds,* **Ergebnisse der Mathematik und ihrer Grenzgebiete, 3. Folge, 30,** **Springer-Verlag, Berlin** (1995), viii + 320 pp.

[Vois02]   C. Voisin, *Théorie de Hodge et géométrie algébrique complexe,* **Cours Spécialisés , 10, Société Mathématique de France,** Paris, (2002), viii+595 pp.

[Waj83]   B. Wajnryb, *A simple presentation for the mapping class group of an orientable surface,* **Israel J. Math. 45, no. 2–3** (1983), 157–174.

[Waj99]   B. Wajnryb, *An elementary approach to the mapping class group of a surface,* **Geom. Topol. 3**(1999), 405–466.

[Wall62]   C.T.C. Wall, *On the orthogonal groups of unimodular quadratic forms,* **Math. Ann. 147** (1962), 328–338.

[Wit94]   E. Witten, *Monopoles and Four-Manifolds,* **Math. Res. Lett. 1** (1994), 809–822.

[Yau77]   S.T. Yau, *Calabi's conjecture and some new results in algebraic geometry,* **Proc. Natl. Acad. Sci. USA 74** (1977), 1798–1799 .

[Yau78]   Yau, S. T.: S.T. Yau, *On the Ricci curvature of a compact Kähler manifold and the complex Monge-Ampère equation. I.* **Comm. Pure Appl. Math. 31, no. 3** (1978), 339–411.

Smoothings of Singularities and Deformation Types of Surfaces

Marco Manetti

Dipartimento di Matematica "G. Castelnuovo", Università di Roma
"La Sapienza", Piazzale Aldo Moro 5, I-00185 Roma, Italy
manetti@mat.uniroma1.it, http://www.mat.uniroma1.it/people/manetti/

1 Introduction

The aim of these lectures is to illustrate some major development in the understanding of the connectedness properties of the moduli space of surfaces of general type. We are especially interested in the results obtained after the last C.I.M.E. course on this topic, course made by F. Catanese in 1985 [24]; the reading of the C.I.M.E. notes [24] provides also an ideal starting point for this chapter.

In order to introduce and to give motivation, we begin by recalling the most important notion and some results.

We shall say that two smooth surfaces S_1, S_2 are direct deformation each other if there exists a proper flat family of smooth surfaces $f\colon X \to C$, where C is an irreducible smooth curve and there exists two fibres of f respectively isomorphic to S_1, S_2. Direct deformation is a relation in the set of isomorphism classes of smooth surfaces; the equivalence relation generated is called *deformation equivalence* or *deformation in the large* and will be denoted by $\overset{def}{\sim}$.

There are several properties of smooth surfaces which are invariant under deformation equivalence, for example.

(A) By Ehresmann fibration theorem [41, 72] two deformation equivalent surfaces have the same differential structure. In particular all the topological and differential invariants of the underlying oriented four-manifold are invariants under $\overset{def}{\sim}$.

We recall that for a complex projective surface S the numbers q, p_g, K_S^2 and $\chi(\mathcal{O}_S)$ are invariants of the oriented underlying topological four-manifold. More precisely

$$b_1 = 2q, \qquad b_+ = 2p_g + 1 \qquad \text{(Hodge index theorem)},$$

$$2\chi(\mathcal{O}_S) = 1 - b_1(S) + b_+, \quad K_S^2 = 12\chi(\mathcal{O}_S) - e(S) = b_+ - b_- + 8\chi(\mathcal{O}_S)$$

where $e(S)$ is the topological Euler–Poincaré characteristic, b_1 is the first Betti number and b_+, b_- are respectively the number of positive and negative eigenvalues of the intersection form on $H_2(S, \mathbb{R})$.

(B) If $S_1 \overset{def}{\sim} S_2$ then S_1 and S_2 have the same Kodaira dimension. In fact holds the following stronger result proved by Shigeru Iitaka as a (non trivial) consequence of Enriques–Kodaira classification of surfaces:

Theorem 1.1 (Iitaka [67]). *The positive plurigenera $P_n(S) = h^0(S, nK_S)$, for $n \geq 0$, of a smooth compact complex surface S, are invariant under arbitrary holomorphic deformations.*

The deformations of Segre-Hirzebruch surfaces $\mathbb{F}_q$ (see [24, 72]) provide examples where the negative plurigenera are not preserved.

The point **B** is also a consequence of point **A** and the fact that the positive plurigenera and the Kodaira dimension are differential invariants of smooth algebraic surfaces (this is the well known Van de Ven conjecture, proved in 1994–1995 by several people, see, e.g. [55, 107]).

(C) If $S_1 \overset{def}{\sim} S_2$ and S_1 is minimal of general type then also S_2 is minimal of general type.

This follows from Iitaka theorem because a surface of general type is minimal if and only if $P_2 = \chi + K^2$ [12, VII.5.5]. Note that, without the assumption that the fibres of f are of general type, the above statement is false: consider for instance the Segre-Hirzebruch surface $\mathbb{F}_3$ which deforms to the blow up of $\mathbb{P}^2$ at a point.

(D) If S is a minimal surface of general type then $K_S^2 > 0$, in particular the canonical class $k_S = c_1(K_S) = -c_1(S)$ does not belong to the torsion subgroup of $H^2(S, \mathbb{Z})$ and then it is well defined its divisibility

$$r(S) = \max\{r \in \mathbb{N} \mid r^{-1}c_1(S) \in H^2(S, \mathbb{Z})/\text{Torsion}\}.$$

If $f\colon X \to C$ is a proper flat family of smooth surfaces we can find a covering $C = \cup U_i$ of open contractible subsets such that every restriction $X_i \to U_i$ is topologically trivial. In particular for every $t \in U_i$ the map $H^2(X_i, \mathbb{Z}) \to H^2(f^{-1}(t), \mathbb{Z})$ is an isomorphism and, by adjunction formula, preserves the canonical classes. In particular r is a deformation invariant. It is still true, although non trivial to prove that, if $f\colon S_1 \to S_2$ is an orientation preserving diffeomorphism of compact complex surfaces, then $f^*c_1(S_2) = \pm c_1(S_1)$ [55]. In particular the divisibility r is also a differential invariant (cf. also [137]).

Remark 1.1 *The parity of the divisibility $r(S)$ is a topological invariant of a simply connected compact complex surface S. If fact, the group $H^2(S, \mathbb{Z})$ is torsion free and the intersection product $q\colon H^2(S, \mathbb{Z}) \times H^2(S, \mathbb{Z}) \to \mathbb{Z}$ is symmetric and unimodular. Moreover the (mod 2) reduction of the canonical class $k_S \in H^2(S, \mathbb{Z})$ is exactly the Wu class [101] and then $q(k_S, x) \equiv q(x, x)$*

(mod 2) *for every x. This implies that, if $k_S \neq 0$, then the unimodular form q is either even or odd according to the parity of $r(S)$.*
Every orientation preserving homeomorphism between two simply connected surfaces S_1, S_2 induces an isometry between $H^2(S_1, \mathbb{Z})$ and $H^2(S_2, \mathbb{Z})$ and therefore $r(S_1) \equiv r(S_2)$ (mod 2).

It is well known that two compact Riemann surfaces are deformation equivalent if and only if they have the same genus. The seek for a similar statement for compact complex surfaces is expressed as the **def = P**? problem, where P is a set of deformation invariant properties of surfaces. The def = P? problem asks essentially how many deformation types of surface exist with given invariants P.
For examples, one can consider the following three (nested) problems:

1. **def=top?** problem: how many deformation types of orientedly homeomorphic complex surfaces exist?
2. **def=top+r?** problem: how many deformation types of orientedly homeomorphic complex surfaces with the same divisibility of the canonical class exist?
3. **def=diff?** problem: how many deformation types of orientedly diffeomorphic complex surfaces exist?

In these notes we restrict our attention to minimal surfaces of general type. According to Bombieri's theorem about pluricanonical maps [15], there exists at most a finite number of deformation types of surfaces with given Chern numbers c_1^2, c_2. Since c_1^2, c_2 are topological invariants, this gives a partial answer to the def=top? problem.

For every simply connected compact oriented topological four-manifolds X the group $H^2(X, \mathbb{Z})$ is free of finite rank and the intersection product $q \colon H^2(X, \mathbb{Z}) \times H^2(X, \mathbb{Z}) \to \mathbb{Z}$ is a symmetric unimodular bilinear form. We have the following criterion for homeomorphism.

Theorem 1.2 (Freedman [50, 1.5+addendum]). *Let X_1 and X_2 be two simply connected compact oriented differentiable four-manifolds and let $f^* \colon H^2(X_2, \mathbb{Z}) \to H^2(X_1, \mathbb{Z})$ be an isometry with respect the intersection forms. Then f^* is induced by an orientation preserving homeomorphism $f \colon X_1 \to X_2$.*

For every symmetric bilinear form $q \colon \mathbb{Z}^n \times \mathbb{Z}^n \to \mathbb{Z}$ its rank and its signature are defined respectively as the rank and the signature of the extended form $q_{\mathbb{R}} \colon \mathbb{R}^n \times \mathbb{R}^n \to \mathbb{R}$. We shall say that the parity of q is even if $q(x, x) \in 2\mathbb{Z}$ for every $x \in \mathbb{Z}^n$ and odd otherwise.

Proposition 1.2 (Eichler's theorem [123, p.92], [135]) *Two unimodular indefinite symmetric bilinear forms defined over the integers are isometric if and only if they have the same rank, signature and parity.*

For definite forms this is not true but in the geometric case this does not give any problem, in fact we have:

Theorem 1.3 (Donaldson [36, 1.3.1]). *If the intersection form of a simply connected oriented compact smooth four-manifold X is definite positive then the intersection form q is represented by the identity matrix in some basis of $H^2(X, \mathbb{Z})$.*

Theorem 1.4 (Kodaira–Yau [12]). *The projective plane is the only simply connected compact complex surface with definite intersection product.*

By Noether's formula and index theorem, the invariants K_S^2 and $\chi(\mathcal{O}_S)$ determine the rank and the signature of q_S. Putting together all these facts we have:

Corollary 1.3 *Two simply connected minimal surfaces of general type are orientedly homeomorphic if and only if they have the same K^2, χ and the same parity of the divisibility of the canonical class r.*

In [22] F. Catanese was able to find k distinct homeomorphic minimal surfaces of general type $S_1, \ldots, S_k$ with different $r(S_i)$, for arbitrarily large values of k. Since the divisibility $r(S)$ of the canonical class is invariant under deformation, we have the following result.

Theorem 1.5 (Catanese [22], Salvetti [118, 119]). *The number of deformation types of simply connected homeomorphic minimal surfaces of general type can be arbitrarily large.*

Given an unimodular quadratic form q of rank b and signature σ over an integral lattice Λ, a primitive vector $v \in \Lambda$ is called of characteristic type if $q(v, x) \equiv q(x, x) \pmod{2}$ for every $x \in \Lambda$; otherwise v is called of ordinary type. If the quadratic form is even than every primitive vector is of ordinary type.

A theorem of C.T.C. Wall [135] says that if $b - |\sigma| \geq 4$, then the group of isometric automorphisms of Λ acts transitively on the set of primitive vectors of fixed norm and type. If $\Lambda = H^2(S, \mathbb{Z})$ is the second cohomology group of a simply connected compact complex surface S, then the condition $b - |\sigma| \geq 4$ is equivalent to $\chi(\mathcal{O}_S) > 1$; moreover the primitive root of k_S is characteristic if and only if $r(S)$ is an odd integer.

In conclusion there exists a homeomorphism $f \colon S \to S'$ between simply connected algebraic surfaces with $\chi > 1$ matching up the canonical classes if and only if S, S' have the same invariants K^2, χ, r.

Theorem 1.6 (Manetti [93]). *Let $\delta(K^2, \chi, r)$ denote the number of deformation types of simply connected minimal surfaces of general type with invariants K^2, χ, r. Then there is no polynomial in K^2, χ, r which bounds (from above) the number $\delta(K^2, \chi, r)$.*

The main ideas of [93] is the use of a particular class of simply connected algebraic surfaces called *simple iterated double covers of* $\mathbb{P}^1 \times \mathbb{P}^1$.

Take an integer $n \geq 2$ and a sequence of n pairs of positive integer $(a_1, b_1), \ldots, (a_n, b_n)$. Then consider a Galois cover $X \to \mathbb{P}^1 \times \mathbb{P}^1$ of degree 2^n obtained by taking the square roots of n generic bihomogeneous polynomials $p_1, \ldots, p_n$ of respective bidegrees $(2a_1, 2b_1), \ldots, (2a_n, 2b_n)$ (see Definition 5.8 for a precise definition). We call the surface X a simple iterated double cover associate to the numerical sequence $(a_1, b_1), \ldots, (a_n, b_n)$.

For $n = 2$ such surfaces where previously introduced by Catanese in [21]; in the same paper it is proved, under some mild condition on (a_1, b_1) and (a_2, b_2) (see Theorem 3.8 in [21]), that these surfaces are simply connected and that surfaces corresponding to different (up to permutations) numerical sequences belongs to different irreducible components of the moduli space. Moreover Catanese made the suggestion [26] to consider such surfaces as possible counterexamples to the **def=top+r?** and the **def=diff?** problems. The answer to the **def=top+r?** follows from the results of [23, 91]: more precisely, two simple bidouble covers of $\mathbb{P}^1 \times \mathbb{P}^1$ are not deformation equivalent whenever they are associated to different numerical sequences satisfying the conditions $a_1 > \max(2a_2, b_1 + 1)$, $b_2 > \max(2b_1, a_2 + 1)$ and $b_1, a_2 \geq 4$.

For $n > 2$ we have some similar results; one of the most important is the following theorem.

Theorem 1.7 (Manetti [93, Thm. C]). *Consider simple iterated double covers of* $\mathbb{P}^1 \times \mathbb{P}^1$ *associated to sequences* $(a_1, b_1), \ldots, (a_n, b_n)$ *satisfying the numerical inequalities*

1. $a_i, b_i \geq 3$ *for every* $i = 1, \ldots, n$
2. $\min(2a_i - a_j, 2b_i - b_j) < 0$ *for every* $j < i$
3. $a_n \geq b_n + 2$ *and* $b_{n-1} \geq a_{n-1} + 2$
4. a_i, b_i *are even for* $i = 2, \ldots, n$
5. $2a_i - a_{i+1} \geq 2$ *and* $2b_i - b_{i+1} \geq 2$ *for every* $i < n$

Then different sequences of pairs give deformation nonequivalent surfaces.

The conditions of Theorem 1.7 allow to set $a_1 = a_2 = \cdots = a_n = a$ and a simple calculations [93, Remark 6.2] shows that the invariants K^2, χ and r depends only by n, a and $\sum_i b_i$. The proof of Theorem 1.6 is a consequence of this remark and of a easy combinatorial argument [93, p. 763].

The **def=diff?** problem was first solved in [96]

Theorem 1.8 ([96]). *The number of deformation types of diffeomorphic minimal surfaces of general type can be arbitrarily large.*

The surfaces considered in the proof of Theorem 1.8 are abelian covers of certain rational surfaces. They are regular but not simply connected.

More recent counterexamples to the **def=diff?** problem are given in [29, 31, 69].

In particular, the paper [31] contains the proof that, for simple bidouble covers of $\mathbb{P}^1 \times \mathbb{P}^1$ of type $(a, b), (a, c)$, the differential type depends only on the two numbers a and $b + c$. Following the terminology introduced in [30, Definition 2.4], we shall call these surfaces (a, b, c)-surfaces and we refer to Catanese's course in this volume for more details.

The goal of this notes is to explain the principal ideas and the main techniques used in the proof of Theorems 1.7 and 1.8. In order to avoid a too technical paper, which may obscure the underlying geometric ideas, we do not work in the full generality but we restrict to some significant examples. For example, we prove Theorem 1.7 only in the particular case of (a, b, c)-surfaces.

2 Deformation Equivalence of Surfaces

2.1 Rational Double Points

Let (X, p) be a normal two-dimensional singularity: a well known theorem (for a historical sketch [84]), generalized to higher dimension by Hironaka, says that there exists a resolution $\delta \colon (S, E) \to (X, p)$ where S is a smooth complex surface, δ is a proper holomorphic map, $E = \delta^{-1}(p)$ is a reduced curve, called exceptional divisor, and $\delta \colon S - E \to X - \{p\}$ is biholomorphic. (for proofs see also [12, 80]).

Since (X, p) is normal, we have $\delta_* \mathcal{O}_S = \mathcal{O}_X$, E is connected and, if $E = \cup E_i$ is the irreducible decomposition, then by Grauert–Mumford theorem [103] the intersection matrix $E_i \cdot E_j$ is negative definite. By Levi's extension theorem [12, I.8.7] every meromorphic function on $S - E$ extends to a meromorphic function on S.

If $\delta' \colon (S', E') \to (X, p)$ is another resolution, then we can find a third one $\nu \colon (W, F) \to (X, p)$ and morphisms $\tau \colon (W, F) \to (S, E)$ and $\tau' \colon (W, F) \to (S', E')$ such that $\nu = \delta \tau = \delta' \tau'$. From Castelnuovo criterion of decomposition of bimeromorphic maps, both τ and τ' are composition of blow-ups of points of the exceptional divisors. If $C_1, C_2 \subset E$ are (-1)-curves then $(C_1 + C_2)^2 < 0$, $C_1 \cdot C_2 \leq 0$ and then either $C_1 = C_2$ or $C_1 \cap C_2 = \emptyset$.

A resolution $(S, E) \to (X, p)$ is called *minimal* if E does not contain (-1)-curves. The above argument shows that every normal two-dimensional singularity has a minimal resolution which is unique up to isomorphism.

A resolution $(S, E) \to (X, p)$ is called *global normal crossing* if E satisfies the following conditions:

GNC1 All the irreducible components of E are smooth and intersect transversally

GNC2 Not more than two components pass through any given point

GNC3 Any two different components intersect at most once

According to the desingularization theorem of curves in surfaces, global normal crossing resolutions always exist, although they are not minimal in general. Anyhow there exists a resolution (S, E), called the *minimal canonical resolution*, which is unique up to isomorphism and minimal in the class of resolutions satisfying the above conditions GNC1 and GNC2.

We now introduce some invariants of a normal two-dimensional singularity (X, p). Let us denote by $\delta \colon (S, E) \to (X, p)$ its minimal canonical resolution and by $E = \cup_{i=1}^n E_i$ the irreducible decomposition of the exceptional divisor.

The *genus* of the singularity is defined as $g(X, p) = h^0(R^1 \delta_* \mathcal{O}_S)$. If X is a Stein representative of the singularity, then by Leray spectral sequence it follows that $g(X, p) = h^1(\mathcal{O}_S)$. Since the irregularity of a smooth surface is invariant under blow-up, the definition of the genus is independent from the choice of the resolution.

Definition 2.1 *A rational surface singularity is a normal surface singularity of genus 0.*

The *Dynkin diagram* D_X is the weighted dual graph of its minimal canonical resolution. More precisely, D_X is the graph whose vertices corresponds to irreducible components E_i and the number of edges connecting E_i to E_j is equal to the intersection number $E_i \cdot E_j$. Every vertex E_i is decorated with its selfintersection E_i^2 and its geometric genus $g(E_i)$.
The Dynkin diagram depends only on the topological type of the pair (S, E) and it is an invariant of the analytic germ (X, p). If the singularity is determined up to isomorphism by D_X then it is called a *taut singularity*.

It is an easy exercise in linear algebra to prove (cf. [6, Prop.2]) that there exists a unique minimal effective divisor $Z = \sum a_i E_i$ such that $Z \neq 0$ and $Z \cdot E_i \leq 0$ for every $i = 1, \dots, n$. This divisor is called the *fundamental cycle*.

If S is a smooth complex, possibly non compact, surface we denote by $K_S \in \mathrm{Pic}(S)$ the canonical line bundle and by $k_S \in H^2(S, \mathbb{Z})$ its first Chern class.
If D is a divisor is S with compact support and $\mathcal{L} \in \mathrm{Pic}(S)$, then the intersection product $D \cdot \mathcal{L}$ is well defined and depends only on the cohomology classes $[D] \in H_2(S, \mathbb{Z}) = H_c^2(S, \mathbb{Z}), c_1(\mathcal{L}) \in H^2(S, \mathbb{Z})$.
The arithmetic genus of D is by definition

$$p_a(D) = 1 + \frac{1}{2} D \cdot (D + K_S).$$

When D is an irreducible curve this definition is the same of the usual arithmetic genus $h^1(\mathcal{O}_D)$, while for general effective divisor we have $p_a(D) = 1 - \chi(\mathcal{O}_D)$ [12, II.11].

Theorem 2.1 (M. Artin). *A normal surface singularity is rational if and only if the arithmetic genus of its fundamental cycle is 0.*

For a proof we refer to the original paper [6] or to the books [12, Ch. III], [10].

Corollary 2.2 *The minimal canonical resolution of a rational singularity is minimal and global normal crossing; the irreducible components of the exceptional divisor are smooth rational curves and the Dynkin diagram is a weighted tree.*

Proof. Let $\delta\colon (S,E) \to (X,p)$ be the minimal canonical resolution. We have $p_a(A+B) = p_a(A) + p_a(B) + A \cdot B - 1$ for every pair of divisors A, B supported in E, while if $H < Z$ is an effective divisor there exists a component E_i such that $H \cdot E_i > 0$ and then $H + E_i \leq Z$, $p_a(H + E_i) \geq p_a(H)$; by induction we get $p_a(H) \leq 0$ for every effective divisor $H \leq Z$. In particular, every irreducible component has arithmetic genus 0 and then it is smooth rational, while for every pair of irreducible components E_1, E_2 we have $E_1 \cdot E_2 = p_a(E_1 + E_2) - p_a(E_1) - p_a(E_2) + 1 \leq 1$.

To see that the E does not contain (-1)-curves we assume the contrary, say $E_1^2 = -1$. By assumption E_1 intersects at least 3 irreducible components, say E_2, E_3 and E_4.

Since $Z \cdot E_1 \leq 0$, the multiplicity of E_1 in Z is at least 2 but $p_a(2E_1 + E_2 + E_3 + E_4) = 1$.

Similarly the arithmetic genus of a loop of smooth rational curves is 1 and then the Dynkin diagram is a tree. $\square$

Note that we can recognize if a singularity is rational from its Dynkin diagram D_X: in fact, if all the components of E are smooth rational then the fundamental cycle and its genus depend only on D_X.

Theorem 2.2 (M. Artin). *Let $(S,E) \xrightarrow{\delta} (X,p)$ be the minimal resolution of a rational singularity with fundamental cycle Z. Then:*

1. $\delta^*(\mathfrak{m}^k) = \mathcal{O}_S(-kZ)$ *for every* $k > 0$, *where* $\mathfrak{m} \subset \mathcal{O}_X$ *is the ideal sheaf of the point* p.
2. *The multiplicity of X at p is equal to* $-Z^2$.
3. *The embedding dimension of X at p is equal to* $-Z^2 + 1$.

Proof. See, e.g. [6]. $\square$

Definition 2.3 *A rational singularity with fundamental cycle Z is called a rational n-point if* $-Z^2 = K_S \cdot Z + 2 = n$.

Therefore a rational 1-point is smooth and a rational double point (RDP from now on) is defined in $\mathbb{C}^3$ by a function of multiplicity 2.

A rational singularity is a RDP if and only if every component of the exceptional divisor has selfintersection -2. In fact, if E is the exceptional curve of the minimal resolution of a RDP then, by minimality, $E_i \cdot K_S \geq 0$ for every component E_i. By definition of RDP $K_S \cdot Z = -2 - Z^2 + p_a(Z) = 0$ and then $K_S \cdot E_i = 0$, $E_i^2 = -2$. Conversely, if $E_i^2 = -2$ for every i, then $K_S \cdot E_i = 0$, $K_S \cdot Z = 0$ and $Z^2 = -2 - K_S \cdot Z = -2$.

Lemma 2.4 *A twodimensional rational singularity is Gorenstein if and only if it is a rational double point.*

Proof. Every rational double point is a hypersurface singularity and then it is Gorenstein. Conversely if $(S, E) \to (X, 0)$ is the minimal resolution of a Gorenstein rational singularity then the canonical bundle of $S - E$ is trivial and K_S is linearly equivalent to $A - B$ for some pair of effective divisors A, B supported in E and without common components.

By minimality $0 \leq K_S \cdot A = A^2 - A \cdot B \leq A^2 \leq 0$ and therefore $A^2 = 0$ which implies $A = 0$. If $B \neq 0$ then $B \cdot E_i = -K_S \cdot E_i \leq 0$ and, by definition of fundamental cycle, $B = Z + H$ for some effective divisor H. We have $-2 = Z^2 + K_S \cdot Z = Z^2 - B \cdot Z = -H \cdot Z$ and then $Z \cdot H = 2$ which is a contradiction.

In conclusion K_S is trivial and every component of E is a rational curve with selfintersection -2. $\square$

The next table contains the complete classification (made first by Du Val [40]) of rational double points.

Name	Canonical equation	Dynkin diagram and fundamental cycle
A_n	$z^2 = x^2 + y^{n+1}$	
D_n	$z^2 = xy^2 + x^{n-1}$	
E_6	$z^2 = x^3 + y^4$	
E_7	$z^2 = x^3 + xy^3$	
E_8	$z^2 = x^3 + y^5$	

2.2 Quotient Singularities

Let G be a finite group of automorphisms of a complex analytic space X. In [19] Cartan proved that the orbit space has a natural structure of analytic space. The main ingredient of his proof was the following beautiful result nowadays known as "Cartan's Lemma".

Lemma 2.5 (Cartan) *Let (X, x) be an analytic singularity with Zariski tangent space T and let G be a finite group of automorphisms of (X, x).*
Then there exists a G-equivariant embedding $(X, x) \to (T, 0)$, in particular the induced representation $G \to GL(T)$ is faithful.

Proof (Sketch of proof). (cf. [19, Lemma 2, p. 98]) Let $\mathfrak{m}$ be the maximal ideal of the local ring $\mathcal{O}_{X,x}$. By definition T is the dual vector space of $\mathfrak{m}/\mathfrak{m}^2$ and there exists a G-equivariant exact sequence

$$0 \longrightarrow \mathfrak{m}^2 \longrightarrow \mathfrak{m} \xrightarrow{\ \pi\ } \frac{\mathfrak{m}}{\mathfrak{m}^2} \longrightarrow 0.$$

Since G is reductive there exists a G-equivariant linear map $s\colon \mathfrak{m}/\mathfrak{m}^2 \to \mathfrak{m}$ such that $\pi s = Id$. The embedding $(X, x) \to (T, 0)$ induced by s is G-equivariant. $\square$

As an application of Cartan's lemma we prove two results that we shall use in the next lectures.

Lemma 2.6 *Let G be a finite group acting on a normal complex variety X and let $D \subset X$ be a prime divisor. Then the subgroup*

$$H = \{g \in G \mid g(x) = x \ \text{ for every } \ x \in D\}$$

is cyclic.

Proof. Let $x \in X$ be a generic point of D, then both D and X are smooth at x and H acts on the singularity (X, x). By Cartan's lemma the action of H the tangent space $T_{x,X}$ is faithful and trivial on the hyperplane $T_{x,D}$. Then H is a finite subgroup of linear automorphisms of the one-dimensional vector space $T_{x,X}/T_{x,D}$. $\square$

Lemma 2.7 *Let $f\colon (\mathbb{C}^n, 0) \to (\mathbb{C}, 0)$ be a germ of holomorphic function of multiplicity ≥ 3; denote by $(X, 0) \subset (\mathbb{C} \times \mathbb{C}^n, 0)$ the hypersurface singularity of equation $x_0^2 = f(x_1, \ldots, x_n)$ and by $\tau\colon (X, 0) \to (X, 0)$ the involution $\tau(x_0) = -x_0$.*
Assume that G is a finite group of automorphisms of $(X, 0)$ and $\tau \in G$. Then τ belongs to the center of G and therefore the quotient group $G/\langle\tau\rangle$ acts on $(\mathbb{C}^n, 0)$ preserving the hypersurface singularity of equation $f = 0$.

Proof. The finite group G acts on the Zariski tangent space $T_{x,X} = \mathbb{C}^{n+1}$ and the hyperplane $H = \{x_0 = 0\}$, being the tangent cone of $(X, 0)$, is G-invariant. If $T_{x,X} = V \oplus H$ is a G-decomposition then τ is the identity on H, $-$identity on V and therefore belongs to the center of G.

The last part of the lemma follows from the fact that $(\mathbb{C}^n, 0)$ is the quotient $(X, 0)/\langle \tau \rangle$ and $f = 0$ is the equation of the branching locus of the projection $(X, 0) \to (\mathbb{C}^n, 0)$. $\square$

Definition 2.8 *A quotient singularity is a singularity isomorphic to $(\mathbb{C}^n, 0)/G$ for some finite group $G \subset \mathrm{Aut}(\mathbb{C}^n, 0)$.*

According to Cartan lemma we can assume without loss of generality $G \subset GL(n, \mathbb{C})$.

By a theorem of Chevalley [32] the quotient $\mathbb{C}^n/G$ is smooth if and only if G is generated by complex reflections, where by definition $g \in GL(n, \mathbb{C})$ is a complex reflection if g has finite order and the rank of $g - Id$ is 1. Therefore, if $H \subset G$ is the subgroup generated by all the complex reflections then, replacing G with G/H, we can also assume that G does not contain complex reflections [17, 114].

For $n = 1$ every finite subgroup $G \subset GL(1)$ is cyclic and $\mathbb{C}/G = \mathbb{C}$. On the other hand all the elements of G are complex reflections.

For $n = 2$ a finite subgroup $G \subset GL(2, \mathbb{C})$ does not contain complex reflections if and only if acts freely on $\mathbb{C}^2 - \{0\}$.

From now on we consider only quotient two-dimensional singularities. In the paper [17], Brieskorn proved that *every quotient singularity is rational.* He also proved in the same paper that *every quotient singularity is taut* and that there exists rational singularities that are not taut.

The paper [17] also contains the complete classification of the Dynkin diagrams of quotient two-dimensional singularities, while the complete classification of the Dynkin diagrams of taut two-dimensional singularities has been done by Laufer [79].

Example 2.9 (Rational double points) *If $G \subset SL(2, \mathbb{C})$ then $X = \mathbb{C}^2/G$ is Gorenstein. In fact $\omega = dz_1 \wedge dz_2$ is a G-invariant nowhere vanishing holomorphic two-form in $\mathbb{C}^2 - \{0\}$ and since G acts freely on $\mathbb{C}^2 - \{0\}$ $\omega \in H^0(X - \{0\}, K_X)$. Thus K_X is the trivial line bundle on $X - \{0\}$ and X is Gorenstein.*

Since every quotient singularity is rational and every rational Gorenstein singularity is a RDP, we have that every X as above is a rational double point. Conversely every RDP is isomorphic to the quotient of $\mathbb{C}^2$ by a finite subgroup of $SL(2, \mathbb{C})$ (see, e.g. [87]).

Example 2.10 (Cyclic singularities) *By a cyclic singularity of type $\frac{1}{n}(a, b)$ we mean the quotient of $\mathbb{C}^2$ by the action of the cyclic group generated by the diagonal automorphism with eigenvalues $\exp(2\pi i a/n), \exp(2\pi i b/n)$.*

Since the quotient of $\mathbb{C}^2$ by a complex reflection is again smooth it is not restrictive to assume g.c.d.$(a, n) =$ g.c.d.$(b, n) = 1$ and then, possibly changing generator, we finally get that every cyclic singularity is isomorphic to a cyclic singularity of type $\dfrac{1}{n}(1, q)$ with $0 < q < n$, g.c.d.$(q, n) = 1$.

Example 2.11 (Singularities of class T) *It is possible to define singularities of class T in several ways, the most elementary is:*

Definition 2.12 *A cyclic quotient singularity is said to be of class T if it is of type $\dfrac{1}{dn^2}(1, dna - 1)$ with g.c.d.$(n, a) = 1$.*

Let $(S, E) \to (X, p)$ be the minimal canonical resolution of a normal surface singularity and let $E_1, \ldots, E_n$ be the irreducible components of the exceptional divisor. The β-invariant is defined as the rational number $\beta(X, p) = n + F^2$, where $F = \sum a_i E_i$, $a_i \in \mathbb{Q}$, is the solution of the linear system

$$F \cdot E_i = 2g(E_i) - 2 - E_i^2, \qquad i = 1, \ldots, n.$$

It is not difficult to prove (see [90, Thm. 17], [92, IV.3.1] and [132, 2.8.2]) that the β-invariant of a cyclic singularity is an integer if and only if the singularity is of class T.

Example 2.13 *The cyclic singularity X of type $\dfrac{1}{4}(1, 1)$ is by definition the* Spec *of the subalgebra $A \subset \mathbb{C}[z_0, z_1]$ generated by homogeneous polynomials of degree $4k$ and then it is isomorphic to the cone over the rational normal curve of degree 4 in $\mathbb{P}^4$.*
A possible set of generators of the $\mathbb{C}$-algebra A is

$$\begin{cases} x = z_0^4 + z_1^4 \\ y = 2z_0^2 z_1^2 \\ w_1 = z_0^4 - z_1^4 \\ w_2 = \sqrt{2}(z_0^3 z_1 - z_0 z_1^3) \\ w_3 = \sqrt{2}(z_0^3 z_1 + z_0 z_1^3) \end{cases}$$

Setting $f_1 = y$, $f_2 = x + y$ and $f_3 = x - y$, the singularity X is embedded in $\mathbb{C}^5$, with affine coordinates x, z, w_1, w_2, w_3, by the six equations:

$$\begin{cases} w_i^2 = f_j f_k \\ w_j w_k = w_i f_i \end{cases} \qquad \{i, j, k\} = \{1, 2, 3\}. \tag{1}$$

Conversely, if $f_1, f_2, f_3 \colon (\mathbb{C}^2, 0) \to (\mathbb{C}, 0)$ are germs of holomorphic functions of multiplicity 1 with pairwise distinct tangent lines at 0, then up to a holomorphic change of variable (see, e.g. [12, p. 63]) we may assume $f_1 = y$, $f_2 = x + y$ and $f_3 = x - y$ and therefore the six equations (1) define the cyclic singularity $(1, 1)/4$.

2.3 RDP-Deformation Equivalence

The definition of $\overset{def}{\sim}$ generalizes in a natural way to the class of normal projective surfaces with at most rational double points; we shall call these surfaces *RDP-surfaces*.

Two RDP-surfaces S_1, S_2 are deformation each other in the large if there exists a proper flat family of RDP-surfaces $f \colon X \to C$, where C is an irreducible smooth curve and there exists two fibres of f respectively isomorphic to S_1, S_2. The equivalence relation generated is called *RDP-deformation equivalence*.

Remark 2.14 *If $X \to Y$ is a flat family of RDP-surfaces over a connected quasiprojective variety Y, then any two fibres of it are RDP-deformation equivalent; in fact it is not restrictive to assume Y irreducible, taking the restriction of the family to a suitable subvariety of Y we can assume that Y is a reduced connected curve and then by transitivity we may reduce to the case where Y is a reduced irreducible curve. Taking if necessary the base change over the normalization of Y we get a family of RDP-surfaces over a smooth curve.*

Definition 2.15 *Let $f \colon X \to T$ be a flat family of normal surfaces over a (possibly non reduced) complex space T. We shall say that f admits a simultaneous resolution if there exist a complex space Y and a proper map $Y \to X$ such that the composition $Y \to T$ is flat and $Y_t \to X_t$ is the minimal resolution of singularities for every $t \in T$.*

Notice that, if $T' \to T$ is a holomorphic map and $X \to T$ admits a simultaneous resolution, then also the pull-back family $X \times_T T' \to T'$ admits a simultaneous resolution.

The simultaneous resolutions do not exist in general and, if they exist, then they are not unique in general.

Theorem 2.3 (Brieskorn-Tyurina). *Let $X = \{X_t \mid t \in T\}$ be a flat family of normal surfaces over a (possibly non reduced) complex space T. Assume that every surface X_t has at most a finite number of singular points and that the restriction $\cup_t Sing(X_t) \to T$ is proper.*
If, for a point $t \in T$, the surface X_t has at most rational double points as singularities, then there exists an open neighbourhood $t \in U \subset T$ and a finite holomorphic surjective map $U' \to U$ such that X_s has at most rational double points for every $s \in U$ and the induced family $X \times_U U' \to U'$ admits a simultaneous resolution.

Proof. See [129]. □

Remark 2.16 *Looking at the proof of Theorem 2.3 and using Artin's approximation theorem [7], it is possible to prove that 2.3 holds also in the algebraic category, with U open in the etale topology (cf. also [8]).*

Corollary 2.17 *If two RDP-surfaces are RDP-deformation equivalent then their minimal resolution are deformation equivalent.*

Proof. Clear. □

The following example shows that the base change is unavoidable in the statement of Theorem 2.3

Example 2.18 (Atiyah [9]) *The affine variety $X \subset \mathbb{C}^4$ of equation $xy + z^2 = s$ can be considered as a flat family X_s of surfaces such that X_s is smooth for $s \neq 0$ and X_0 has an ordinary double point. This family does not admit simultaneous resolution: in fact if $\sigma : Y \to X$ is a simultaneous resolution then σ is ramified on codimension 2 and this is a contradiction because X is smooth.*
After the base change $s = t^2$, X becomes the affine variety $V \subset \mathbb{C}^4$ of equation $xy + z^2 = t^2$.
Let l_0, l_1 be homogeneous coordinates on $\mathbb{P}^1$ and consider $Y \subset \mathbb{C}^4 \times \mathbb{P}^1$ defined by the equations

$$l_0(z + t) = l_1 x \qquad l_0 y = l_1(t - z).$$

The projection on the first factor gives a surjective map $Y \to V$ and it is easily verified that for every t, $Y_t \to V_t$ is the minimal resolution of singularities. In particular in Y_0 there is a (-2)-curve which does not appear in the other fibres.
Changing the role of x and y in the above equation we get another simultaneous resolution $Z \to V$. It is an easy exercise to observe that the identity on V_0 lifts to an isomorphism $Y_0 \to Z_0$, while the identity on V does not lift to any isomorphism between Y and Z.

2.4 Relative Canonical Model

Given a smooth minimal surface of general type S, we denote by $X_n \subset \mathbb{P}(H^0(S, K_S^{\otimes n})^\vee)$ the image of its n-canonical map $\phi_n : S \to \mathbb{P}(H^0(S, K_S^{\otimes n})^\vee)$. According to [15], for every $n \geq 5$, the projective surface X_n has at most rational double points as singularities, $K_{X_n} = (\phi_n)_* K_S = \mathcal{O}_{X_n}(1)$ and ϕ_n is the minimal resolution of singularities. The isomorphism class of X_n is independent from n (provided that $n \geq 5$).

Definition 2.19 *In the notation above, the surface X_n, for $n \gg 0$, is called the canonical model of S.*

Let $f : \mathcal{X} \to T$ be a proper flat family of minimal surfaces of general type over a (possibly nonreduced) complex space T and denote $\omega_{\mathcal{X}/T} = \bigwedge^2 \Omega^1_{\mathcal{X}/T}$ where $\Omega^1_{\mathcal{X}/T}$ is the locally free sheaf of relative differentials. The line bundle $\omega_{\mathcal{X}/T}$ is flat over T and by adjunction formula its restriction to every fibre is the canonical bundle.
Since $H^1(S, nK_S) = 0$ for every minimal surface of general type S and every integer $n \geq 2$ [12, VII.5.5], we have that the direct image sheaf $f_* \omega_{\mathcal{X}/T}^{\otimes n}$ is locally free.

Assume that the sheaf $f_* \omega_{\mathcal{X}/T}^{\otimes n}$ is free for some $n \geq 5$, denote by $N+1$ its rank and let $s_0, \ldots, s_N \in H^0(\mathcal{X}, \omega_{\mathcal{X}/T}^{\otimes n})$ be a frame of it. This frame generates a base point free linear system and then a morphism $\phi \colon \mathcal{X} \to T \times \mathbb{P}^N$. By construction, the restriction of ϕ to every fibre of f is the n-canonical morphism and there exists a factorization $f \colon \mathcal{X} \xrightarrow{\phi} \mathcal{Y} = \phi(\mathcal{X}) \xrightarrow{g} T$.

Proposition 2.20 *In the notation above, the morphism g is flat and then $\mathcal{Y} \xrightarrow{g} T$ is a family of RDP-surfaces.*

Proof (Sketch of proof). By local flatness criterion [97] it is not restrictive to assume $T = \mathrm{Spec}(A)$, where A is a local Artinian $\mathbb{C}$-algebra with residue field $\mathbb{C}$.

Denote by $U_i = \{x \in \mathcal{X} \mid s_i(x) \neq 0\}$ and by $X \subset \mathcal{X}$, $Y \subset \mathcal{Y}$ the fibres over the closed point of T.

The functions $\dfrac{s_j}{s_i}$ generate the $\mathbb{C}$-algebra $H^0(U_i, \mathcal{O}_X)$. The surface Y has at most rational singularities and then, by Leray spectral sequence, $H^i(U_i, \mathcal{O}_X) = 0$ for every $i \geq 0$.

By considering the Cech cochain resolution of $\mathcal{O}_{\mathcal{X}}$ over a finite affine cover of U_i we get that $H^0(U_i, \mathcal{O}_{\mathcal{X}})$ is a free A-module generated by $\dfrac{s_j}{s_i}$ (cf. [134, 0.4]). In particular the coherent sheaf $\phi_* \mathcal{O}_{\mathcal{X}}$ is A-flat and the natural map $\mathcal{O}_{\mathcal{Y}} \xrightarrow{\alpha} \phi_* \mathcal{O}_{\mathcal{X}}$ is surjective.

If H_A is the kernel of α, then the A-flatness of $f_* \mathcal{O}_{\mathcal{X}}$ implies that $H_A \otimes_A \mathbb{C} = 0$; $H_A = 0$ by Nakayama's Lemma $H_A = 0$ and $\mathcal{O}_{\mathcal{Y}} = \phi_* \mathcal{O}_{\mathcal{X}}$ is flat over T.   □

In the general case we can find an open covering $T = \cup T_i$ such that the result of Proposition 2.20 applies to every restriction $\mathcal{X}_i \to T_i$. We then have proved the following

Theorem 2.4. *Two minimal surfaces of general type are deformation equivalent if and only if their canonical models are RDP-deformation equivalent.*

2.5 Automorphisms of Canonical Models

Let X be the canonical model of a minimal surface of general type S. We have $\mathrm{Aut}(X) = \mathrm{Aut}(S)$, where Aut denotes the group of biregular automorphisms. A classical results [2, 98] asserts that $\mathrm{Aut}(S)$ is a finite group for every smooth surface of general type.

For later use we need a relative version of the above result.

Theorem 2.5. *Let $\mathcal{X} \to S$ be a proper flat family of canonical models of surfaces of general type over a scheme of finite type S over $\mathbb{C}$.*

There exists a proper, unramified and quasifinite morphism of schemes $p \colon \mathrm{Aut}(\mathcal{X}/S) \to S$ and an isomorphism $u \colon p^* \mathcal{X} \to p^* \mathcal{X}$ over $\mathrm{Aut}(\mathcal{X}/S)$ with the following universal property:

For every morphism $f: T \to S$ and every isomorphism over T $v: f^*\mathcal{X} \to f^*\mathcal{X}$ there exists a unique morphism $g: T \to \mathrm{Aut}(\mathcal{X}/S)$ such that $f = pg$ and $v = g^*u$.

A proof can be found in [45]. If $T = s$ is a closed point of S and f is the inclusion we get that $p^{-1}(s) = \mathrm{Aut}(X_s)$. Since p is unramified we get in particular that *the cardinality of* $\mathrm{Aut}(X_s)$ *is an upper semicontinuos function on S.*

Corollary 2.21 *Let $f: \mathcal{X} \to C$ be a proper flat family of canonical models over a smooth reduced irreducible curve C. Assume that there exist a non-empty Zariski open subset $U \subset C$ and an automorphism u of $\mathcal{X}_{|U}$ such that $fu = f$. Then there exists a reduced irreducible curve B and a finite surjective morphism $q: B \to C$ such that q^*u extends to $q^*\mathcal{X}$.*

Proof. The automorphism u induces a morphism $g: U \to \mathrm{Aut}(\mathcal{X}/C)$. It is sufficient to take B as the irreducible component of $\mathrm{Aut}(\mathcal{X}/C)$ that contains $g(U)$. □

2.6 The Kodaira–Spencer Map

Definition 2.22 *Let $(B,0)$ be a germ of complex space, a deformation of a compact complex space M_0 over $(B,0)$ is the data of a complex space M and a pair of complex analytic maps $M_0 \xrightarrow{i} M \xrightarrow{f} B$ such that:*

1. $fi(M_0) = b_0$
2. *There exists an open neighbourhood $0 \in U \subset B$ such that $f: f^{-1}(U) \to U$ is proper and flat*
3. $i: M_0 \to f^{-1}(b_0)$ *is an isomorphism of complex spaces*

M is called the total space of the deformation and $(B,0)$ the base germ space.

Definition 2.23 *Two deformations of M_0 over the same base*

$$M_0 \xrightarrow{i} M \xrightarrow{f} (B,0), \qquad M_0 \xrightarrow{j} N \xrightarrow{g} (B,0),$$

are isomorphic if there exists an open neighbourhood $0 \in U \subset B$ and a commutative diagram of analytic maps

$$
\begin{array}{ccc}
M_0 & \xrightarrow{i} & f^{-1}(U) \\
{\scriptstyle j}\downarrow & \nearrow & \downarrow{\scriptstyle f} \\
g^{-1}(U) & \xrightarrow{g} & U
\end{array}
$$

with the diagonal arrow an analytic isomorphism.

For every germ of complex space $(B,0)$ we denote by $\text{Def}_{M_0}(B,0)$ the set of isomorphism classes of deformations of M_0 with base $(B,0)$. It is clear from the definition that if $0 \in U \subset B$ is open, then $\text{Def}_{M_0}(B,0) = \text{Def}_{M_0}(U,0)$.

There exists an action of the group $\text{Aut}(M_0)$ of holomorphic automorphisms of M_0 on the set $\text{Def}_{M_0}(B,0)$: if $g \in \text{Aut}(M_0)$ and $\xi : M_0 \xrightarrow{\;i\;} M \xrightarrow{\;f\;} (B,0)$ is a deformation we define

$$\xi^g : M_0 \xrightarrow{ig^{-1}} M \xrightarrow{\;f\;} (B,0).$$

It is clear that $\xi^g = \xi$ if and only if $g \colon f^{-1}(0) \to f^{-1}(0)$ can be extended to an isomorphism $\hat{g} \colon f^{-1}(V) \to f^{-1}(V)$, $0 \in V$ open neighbourhood, such that $f\hat{g} = f$.

If $\xi : M_0 \xrightarrow{\;i\;} M \xrightarrow{\;f\;} (B,0)$ is a deformation and $g \colon (C,0) \to (B,0)$ is a morphism of germs of complex spaces, then

$$g^*\xi : M_0 \xrightarrow{(i,0)} M \times_B C \xrightarrow{\;pr\;} (C,0)$$

is a deformation and its isomorphism class is well defined.

We denote by $(D,0) = \text{Spec}(\mathbb{C}[\epsilon]/(\epsilon^2))$ the double point. If $(B,0)$ is a germ of complex space, then there exists a natural bijection between the Zariski tangent space $T_{0,B}$ and the set of morphisms of germs $(D,0) \to (B,0)$.

If $\xi : M_0 \xrightarrow{\;i\;} M \xrightarrow{\;f\;} (B,0)$ is a deformation, the pull-back procedure gives a map $KS(\xi) \colon T_{0,B} \to \text{Def}_{M_0}(D,0)$. According to Schlessinger [120], the set $\text{Def}_{M_0}(D,0)$ has a canonical structure of vector space such that $KS(\xi)$ is a linear map for every ξ.

Definition 2.24 $KS(\xi)$ *is called the* Kodaira–Spencer map *of the deformation* ξ.

Definition 2.25 *A deformation* $\xi : M_0 \xrightarrow{\;i\;} M \xrightarrow{\;f\;} (B,0)$ *of a compact complex space* M_0 *is called* versal *if for every deformation* $\xi' : M_0 \xrightarrow{\;i\;} M' \xrightarrow{\;f'\;} (B',0)$ *there exists a morphism of germs* $g \colon (B',0) \to (B,0)$ *such that* ξ' *is isomorphic to* $g^*\xi$.
It is called universal *if in addiction* g *is unique.*
A deformation is called semiuniversal *if it is versal and its Kodaira–Spencer map is an isomorphism.*

It is obvious that universal $\Rightarrow$ semiuniversal $\Rightarrow$ versal.

Definition 2.26 *A deformation* $\xi : M_0 \xrightarrow{\;i\;} M \xrightarrow{\;f\;} (B,0)$ *of a compact complex space* M_0 *is called* minimal versal *if it is versal and for every versal deformation* $\xi' : M_0 \xrightarrow{\;i\;} M' \xrightarrow{\;f'\;} (B',0)$ *there exists a smooth morphism of germs* $g \colon (B',0) \to (B,0)$ *such that* ξ' *is isomorphic to* $g^*\xi$.

It is clear that any two minimal versal deformations of the same space are (non canonically) isomorphic.

The following theorem was proved by Kuranishi [78] for smooth manifolds and by Grauert [58] in full generality.

Theorem 2.6. *Every compact complex space X admits a semiuniversal and minimal versal deformation $X \xrightarrow{i} \mathcal{X} \xrightarrow{f} (\mathrm{Def}(X), 0)$.*

Proof. See, e.g. [38, 49, 58, 105, 108]. □

For later use we also need the following two lemmas

Lemma 2.27 *Assume that the group of analytic automorphisms $\mathrm{Aut}(X)$ of a compact complex space X is finite. Then its semiuniversal deformation is universal, the group $\mathrm{Aut}(X)$ acts on the germ $\mathrm{Def}(X)$, the quotient $\mathrm{Def}(X)/\mathrm{Aut}(X)$ is still a germ of complex space and the projection $\mathrm{Def}(X) \to \mathrm{Def}(X)/\mathrm{Aut}(X)$ is open.*

Proof. Since X is compact the assumption $\mathrm{Aut}(X)$ finite implies $H^0(X, \Theta_X) = 0$, where Θ denotes the tangent sheaf. According to Schlessinger [120] and Wavrik [136] the above vanishing implies that every semiuniversal deformation is universal. The last part of the statement follows from Cartan's theorem [19]. □

Lemma 2.28 *Let $\xi : \quad X \to \mathcal{X} \to (B, 0)$ be a deformation of a compact complex space X. If B is smooth and the Kodaira–Spencer map $KS(\xi) \colon T_{0,B} \to \mathrm{Def}_X(D, 0)$ is surjective then the ξ is a versal deformation and the induced morphism $h \colon (B, 0) \to (\mathrm{Def}(X), 0)$ is smooth.*

Proof. By definition the Kodaira–Spencer map of the semiuniversal deformation is bijective and then the differential of h is surjective. The smoothness of B implies the smoothness of h. □

If X is a reduced complex space there exists an $\mathrm{Aut}(X)$-equivariant isomorphism of vector spaces $\mathrm{Def}_X(D, 0) = \mathrm{Ext}^1_X(\Omega^1_X, \mathcal{O}_X)$ [38]. Thinking Ext^1 as the space of isomorphism classes of extensions of sheaves, the above isomorphism can be described in the following way:

Let $X \xhookrightarrow{i} Z \to \mathrm{Spec}(\mathbb{C}[\epsilon]/(\epsilon^2))$ be a first order deformation of X. The second exact sequence of differentials gives

$$\mathcal{O}_X \xrightarrow{\alpha} i^* \Omega^1_Z \longrightarrow \Omega^1_X \longrightarrow 0,$$

where α is the multiplication by $d\epsilon$. Since X is reduced $\mathcal{O}_X$ is without torsion and α is injective. Therefore we have associated to the deformation $X \xhookrightarrow{i} Z$ the element of $\mathrm{Ext}^1_X(\Omega^1_X, \mathcal{O}_X)$ corresponding to the extension

$$0 \longrightarrow \mathcal{O}_X \xrightarrow{\alpha} i^* \Omega^1_Z \longrightarrow \Omega^1_X \longrightarrow 0.$$

The action of $\mathrm{Aut}(X)$ on $\mathrm{Def}_X(D,0)$ has been already described, in particular the g-invariant deformations, $g \in \mathrm{Aut}(X)$ are exactly the deformations where g extends.

The natural action of $\mathrm{Aut}(X)$ on the vector space $\mathrm{Ext}^1_X(\Omega^1_X, \mathcal{O}_X)$ is given as follows: every $g \in \mathrm{Aut}(X)$ induces isomorphisms of sheaves

$$g\colon g^*\mathcal{O}_X \to \mathcal{O}_X, \qquad g\colon g^*\Omega^1_X \to \Omega^1_X$$

and transforms an extension

$$0 \longrightarrow \mathcal{O}_X \xrightarrow{\gamma} \mathcal{E} \xrightarrow{\delta} \Omega^1_X \longrightarrow 0$$

into

$$0 \longrightarrow \mathcal{O}_X \xrightarrow{\gamma g^{-1}} g^*\mathcal{E} \xrightarrow{g\delta} \Omega^1_X \longrightarrow 0.$$

3 Moduli Space for Canonical Surfaces

By a *canonical surface* we mean a normal projective surface X with at most rational double points as singularities such that the dualizing sheaf $\omega_X = \mathcal{O}_X(K_X)$ is an ample line bundle.

If $\nu\colon Y \to X$ is the minimal resolution, then Y is a minimal surface of general type and we have:

1. $K_X^2 = K_Y^2$, $\nu^*\omega_X = \omega_Y$ and $\nu_*\omega_Y = \omega_X$; in particular for every $i, n \in \mathbb{Z}$ we have $H^i(X, nK_X) = H^i(Y, nK_Y)$.
2. The complete linear system $|nK_X|$ is very ample for every $n \geq 5$.

If $n \geq 2$, then by Serre duality $h^2(Y, nK_Y) = h^0(Y, (1-n)K_Y) = 0$ and by Mumford–Ramanujam vanishing theorem $h^1(Y, nK_Y) = h^1(Y, (1-n)K_Y) = 0$ ([12, VII.5.5]). This implies, by Riemann–Roch formula, that for every $n \geq 2$

$$P_n(X) = P_n(Y) = \chi(\mathcal{O}_Y) + \frac{n(n-1)}{2}K_Y^2 = \chi(\mathcal{O}_X) + \frac{n(n-1)}{2}K_X^2.$$

$$H^1(X, nK_X) = H^2(Y, nK_Y) = 0.$$

We denote by $\mathcal{M}$ the set of isomorphism classes of canonical surfaces. Since every minimal surface of general type is the minimal resolution of a canonical surface we also have:

1. $\mathcal{M}$ is (isomorphic to) the set of isomorphism classes of minimal surfaces of general type.
2. $\mathcal{M}$ is (isomorphic to) the set of birational classes of surfaces of general type.

In this lecture we will see that $\mathcal{M}$ can be written as a countable disjoint union of complex quasiprojective schemes with the property that two minimal surfaces of general type are deformation equivalent if and only if they belongs to the same connected component of $\mathcal{M}$.

3.1 Gieseker's Theorem

There exists a good notion of family of canonical surfaces

Definition 3.1 *Let S be an algebraic scheme[1] over $\mathbb{C}$. A family of canonical surfaces over S is a proper flat morphism $f\colon \mathcal{X} \to S$ such that $f^{-1}(s)$ is a canonical surface for every closed point $s \in S$.*
Two families $f\colon \mathcal{X} \to S$, $g\colon \mathcal{Y} \to S$ are isomorphic if there exists an isomorphism $\mathcal{X} \to \mathcal{Y}$ commuting with f and g.

The above definition is clearly stable under base change and then it makes sense to consider the *moduli functor of canonical surfaces*. For every pair of integers x, y define

$$\mathcal{C}_{x,y}\colon \{\text{Algebraic Schemes}/\mathbb{C}\} \to \textbf{Sets}.$$

where $\mathcal{C}_{x,y}(S)$ is the set of isomorphism classes of families of canonical surfaces over S, whose fibres have numerical invariants $K^2 = x$ and $\chi = y$.
We are now ready to state the main theorem of the section.

Theorem 3.1 (Gieseker [56]). *For every pair of integers x, y, there exists a (possibly empty) complex quasiprojective scheme $\mathcal{M}_{x,y}$ which is a coarse moduli space for the functors $\mathcal{C}_{x,y}$.*

We recall, for the nonexpert reader, what coarse moduli space means.

- If $\text{Mor}(-, \mathcal{M}_{x,y})\colon \{\text{Algebraic Schemes}/\mathbb{C}\} \to \textbf{Sets}$ denotes the functors of morphisms of schemes, there exists a natural transformation $\Phi\colon \mathcal{C}_{x,y} \to \text{Mor}(-, \mathcal{M}_{x,y})$.
- $\Phi\colon \mathcal{C}_{x,y}(\text{Spec}(\mathbb{C})) \to \text{Mor}(\text{Spec}(\mathbb{C}), \mathcal{M}_{x,y})$ is a bijection.
- For every algebraic scheme $\mathcal{N}$ and every natural transformation $\Psi\colon \mathcal{C}_{x,y} \to \text{Mor}(-, \mathcal{N})$ there exists a unique morphism $\eta\colon \mathcal{M}_{x,y} \to \mathcal{N}$ such that $\Psi = \eta \circ \Phi$.

Remark 3.2

1. By definition $\mathcal{C}_{x,y}(\text{Spec}(\mathbb{C}))$ is the set of isomorphism classes of canonical surfaces with numerical invariants $K^2 = x$, $\chi = y$, while $\text{Mor}(\text{Spec}(\mathbb{C}), \mathcal{M}_{x,y})$ is the set of closed points of $\mathcal{M}_{x,y}$. In particular the set $\mathcal{M}$ is the disjoint union of all the set of closed points of $\mathcal{M}_{x,y}$.
2. For every proper flat family $f\colon \mathcal{X} \to S$ of canonical surfaces with invariants $K^2 = x$ and $\chi = y$, there exists a regular morphisms $\Phi(f)\colon S \to \mathcal{M}_{x,y}$. The naturality of Φ implies that for every closed point $s \in S$, the closed point of $\mathcal{M}_{x,y}$ corresponding to the isomorphism class of $f^{-1}(s)$ is exactly $\Phi(f)(s)$.

---

[1] An algebraic scheme is a separated scheme of finite type: if the reader prefer, he can replace algebraic schemes over $\mathbb{C}$ with (possibly reducible and nonreduced) complex algebraic varieties

Moreover the maps μ must be compatible with base change, this means that, if a flat family $f'\colon \mathcal{Y} \to T$ is induced from f by a morphism $\phi\colon T \to S$, then $\mu(f') = \mu(f) \circ \phi$.

The theorem of Gieseker is an almost "purely existence" theorem and it gives very few information about the structure of the moduli space. The only significant results that can be easily deduced from its construction are stated in the following theorem.

Theorem 3.2.   *1. Let X be a canonical surface and let $\mathrm{Def}(X)$ be the base space of the semiuniversal deformation of X. Then there exists an isomorphism of analytic singularities $(\mathcal{M}, [X]) \simeq \mathrm{Def}(X)/\mathrm{Aut}(X)$, where $\mathrm{Aut}(X)$ is the (finite) group of automorphisms of X.*
2. For every closed irreducible subset $Z \subset \mathcal{M}$ there exist a family of canonical model $f\colon \mathcal{X} \to S$ such that S is irreducible and $\Phi(f)(S) = Z$.
3. Two families $f\colon \mathcal{X} \to S$, $g\colon \mathcal{Y} \to S$ induce the same morphism $\Phi(f) = \Phi(g)\colon S \to \mathcal{M}$ if and only if there exists a Zariski open covering $S = \cup S_i$ and finite surjective morphisms $h_i\colon T_i \to S_i$ such that $\mathcal{X} \times_S T_i \to T_i$ is isomorphic to $\mathcal{Y} \times_S T_i \to T_i$ for every i.

From Theorem 3.2 follows in particular that two canonical surfaces are RDP-deformation equivalent if and only if they belong to the same connected component of $\mathcal{M}$.

An outline of the proof of Theorems 3.1 and 3.2 will be given in Sect. 3.3.

3.2 Constructing Connected Components: Some Strategies

Assume that it is given an irreducible quasiprojective variety Y, a Zariski open covering $Y = \cup U_i$ and, for every i, a flat family of canonical surfaces $f_i\colon \mathcal{X}_i \to U_i$.

Assume moreover that the induced morphisms $\Phi(f_i)\colon U_i \to \mathcal{M}$ coincide in the intersections $U_i \cap U_j$ and therefore give a regular morphism $\Phi\colon Y \to \mathcal{M}$.

It is clear that $\Phi(Y)$ is an irreducible constructible subset. It is a connected component if and only if $\Phi(Y)$ is open and closed in $\mathcal{M}$. The examples in literature have shown that the next two criteria are very useful.

Lemma 3.3 *In the notation above assume that Y is smooth and that for every i and $y \in U_i$ the Kodaira–Spencer map $T_{y,Y} \to \mathrm{Def}_{f_i^{-1}(y)}(D,0)$ is surjective. Then $\Phi(Y)$ is open.*

Proof. Since $\Phi(Y)$ is constructible it is sufficient to prove that it is open in the complex topology. If $y \in U_i$, then by assumption the morphism $(U_i, y) \to (\mathrm{Def}(f_i^{-1}(Y)), 0)$ is smooth and then

$$(U_i, y) \to (\mathcal{M}, \Phi(y)) = (\mathrm{Def}(f_i^{-1}(Y)), 0)/\mathrm{Aut}(f_i^{-1}(Y)), 0)$$

is open. $\square$

Lemma 3.4 *In the notation above, assume that there exists a nonempty Zariski open subset $N \subset Y$ with the following property:*
For every smooth pointed affine curve (C, p) and for every flat family of canonical surfaces $g \colon \mathcal{X} \to C$ such that its restriction

$$g \colon (\mathcal{X} - X_p) \to (C - \{p\})$$

is the pull-back of f_i for some i and some morphism $C - \{p\} \to U_i \cap N$, the isomorphism class of X_p belongs to $\Phi(Y)$.
Then $\Phi(Y)$ is closed in $\mathcal{M}$.

Proof. Since $\Phi(Y) \subset \overline{\Phi(N)}$, it is sufficient to prove that $\overline{\Phi(N)} \subset \Phi(Y)$. After a possible shrinking of N we can assume $N \subset U_i$ for some i and therefore that $\Phi \colon N \to \mathcal{M}$ is induced by a family $f \colon \mathcal{X} \to N$ of canonical models. Let $m \in \overline{\Phi(f)(N)}$ be a fixed element and let $g \colon \mathcal{Y} \to H$ be a family with H irreducible and $\overline{\Phi(f)(N)} = \Phi(g)(H)$. Since $\Phi(f)(N)$ is constructible, $\Phi(g)^{-1}(\Phi(f)(N))$ contains an open dense subset of H. Let $\overline{N}$ be a projective closure of N and $Z \subset H \times \overline{N}$ the closure of the fibred product of $\Phi(g)$ and $\Phi(f)$. The projection on the first factor $Z \to H$ is surjective and then there exists a smooth affine pointed curve (C, p) and a morphism $h \colon C \to Z$ such that $\Phi h(p) = m$ and $h(C - p) \subset H \times N$. Up to a possible base change, the pull back of the two families $\mathcal{X} \to N$ and $\mathcal{Y} \to H$ are isomorphic over $C - \{p\}$. $\square$

The Lemmas 3.3 and 3.4 allow to construct families $f_i \colon \mathcal{X}_i \to Y_i$ of canonical surfaces such that Y_i is irreducible and $Z_i := \overline{\Phi(f_i)(Y_i)}$ is a connected component of $\mathcal{M}$ for every i. The next problem to determine when $Z_i \neq Z_j$. A possible strategy consists of 4 steps.
The first, obvious step, is to look at the deformation invariants of the fibres of the families f_i. However, the examples will be much more interesting if the deformation invariants are the same.
The second step consists to determine the general properties of Z_i. For example, $Z_i \neq Z_j$ whenever $\dim Z_i \neq \dim Z_j$.
In the third step we look at the group G_i of biregular automorphisms of a generic surface of Z_i (this makes sense by Theorem 2.5). Different groups means different components.
The fourth step applies when the third fails, i.e. when $G_i = G_j$. In this case we look at the deformation invariants of the quotients $Y_i := X_i / \operatorname{Aut}(X_i)$, where X_i is a generic member of Z_i. If we are able to prove that the surfaces Y_i are canonical surfaces lying in different irreducible components of $\mathcal{M}$, then $Z_i \neq Z_j$.

3.3 Outline of Proof of Gieseker Theorem

Throughout this section $n \geq 5$ and $x, y > 0$ are fixed natural numbers. For simplicity we consider only canonical surfaces with numerical invariants $K^2 = x$, $\chi = y$ and then with n-plurigenus $P_n = y + n(n-1)x/2$.

Definition 3.5 *A n-canonically embedded surface is the data of a canonical surface X together a closed immersion $f: X \to \mathbb{P}^{P_n(X)-1}$ induced by the complete linear system $|\omega_X^n|$.*

Every canonical surface X admits a n-canonical embedding, and any two n-canonical embeddings of X differ by an automorphism of $\mathbb{P}^{P_n-1}$. In particular the set $\mathcal{M}_{x,y}$ of canonical surfaces is the quotient of the set $\mathcal{H}_{x,y}$ of n-canonically embedded surfaces under the natural action of the group $G = PGL(P_n, \mathbb{C})$.

The standard procedure of geometric invariant theory goes as follows:
Step 1. Give a structure of quasiprojective scheme on the set $\mathcal{H}_{x,y}$ such that the action of the algebraic group G is regular.
Step 2. Find a suitable n and a G-linearized polarization on $\mathcal{H}_{x,y}$ such that every point of $\mathcal{H}_{x,y}$ is G-stable (in the sense of Mumford's geometric invariant theory) and take $\mathcal{M}_{x,y} = \mathcal{H}_{x,y}/G$.

There exists a natural concept of family of n-canonically embedded surfaces: this is a consequence of the existence of the relative dualizing sheaf for a morphism.
More generally let $f: X \to Y$ be a flat family of normal surfaces and let $U \subset X$ be the (scheme theoretic) open subvariety of points where the map f is smooth. Then it is defined the relative dualizing sheaf $\omega_{X/Y}$ on X satisfying the following conditions [85], [132, 1.3]:

1. $\omega_{X/Y}$ is a coherent f-flat $\mathcal{O}_X$-module.
2. If $i: U \to X$ is the open immersion, then $\omega_{X/Y} = i_*(\bigwedge^2 \Omega_{U/Y}^1)$ where $\Omega_{U/Y}^1$ is the locally free sheaf of relative differentials.
3. The relative dualizing sheaf has the base change property. This means that for every morphism $Y' \to Y$ if $\pi: X' = X \times_Y Y' \to X$ denotes the projection, then $\omega_{X'/Y'} = \pi^*\omega_{X/Y}$.
4. If the fibres of f are Gorenstein, for example if f is a family of canonical surfaces, then the sheaf $\omega_{X/Y}$ is locally free.

Definition 3.6 *Let S be an algebraic scheme over $\mathbb{C}$ and denote by $p_1: S \times \mathbb{P}^{P_n-1} \to S$, $p_2: S \times \mathbb{P}^{P_n-1} \to \mathbb{P}^{P_n-1}$ the two projections.*
A family of n-canonically embedded surfaces is the data of a family $f: \mathcal{X} \to S$ of canonical surfaces together a closed embedding $\nu: \mathcal{X} \to S \times \mathbb{P}^{P_n-1}$ such that $f = p_1\nu$, $\nu^*p_2^*\mathcal{O}(1) = \omega_{\mathcal{X}/Y}^{\otimes n}$ and $\nu: f^{-1}(s) \to \mathbb{P}^{P_n-1}$ is a n-canonically embedded surface for every $s \in S$.

The above notion of family is stable under base change, therefore it makes sense the notion of universal family.

Proposition 3.7 (Tankeev [125]) *There exists an universal family $Z_n \subset \mathcal{H}_{x,y} \times \mathbb{P}^{P_n-1}$ of n-canonically embedded surfaces with $\mathcal{H}_{x,y}$ quasiprojective algebraic scheme.*

Proof (Idea of proof). Since the Hilbert polynomial $h(d)$ of an n-canonically embedded surface with invariants K^2 and χ is $h(d) = \chi + \frac{1}{2}dn(dn-1)K^2$, we start by considering the universal family $\mathcal{X} \to \text{Hilb}_{h(d)}(\mathbb{P}^{P_n-1})$ over the Hilbert scheme.

According to a theorem of Elkik [42, Thm. 4], the set $U_1 \subset \text{Hilb}_{h(d)}(\mathbb{P}^{P_n-1})$ of points u such that the fibre X_u is irreducible with at most rational singularities is open in the Hilbert scheme. The fibres of the restricted family $\mathcal{X} \to U_1$ are normal Cohen–Macaulay surfaces and then there exists the relative dualizing sheaf $\omega_{\mathcal{X}/U_1}$. The subset $U_2 \subset U_1$ of points u such that $\omega_{\mathcal{X}/U_1}$ is locally free of rank 1 at every point of X_u is an open subset. According to Lemma 2.4, a point $u \in \text{Hilb}_{h(d)}(\mathbb{P}^{P_n-1})$ belongs to U_2 if and only if X_u is irreducible and has at most rational double points as singularities.

Finally we take $\mathcal{H}_{x,y}$ as the closed subscheme of U_2 of points u such that the line bundle $nK_{X_u}(-1)$ is trivial.

Another possible way (used by Tankeev) to prove the openess of U_2 in the Hilbert scheme uses the explicit description of versal deformations of rational double points and Artin's approximation theorem [7].   □

The bulk of Gieseker paper [56] is devoted to prove the following proposition (written in the language of geometric invariant theory [57, 106, 122])

Proposition 3.8 (Gieseker [56]) *In the notation above, for n sufficiently large, there exists the geometric quotient $\mathcal{M}_{x,y} = \mathcal{H}_{x,y}/G$ and $\mathcal{M}_{x,y}$ is a quasiprojective scheme.*

For reader convenience we recall here the properties which characterize geometric quotients.

Let H be an algebraic scheme with a regular action of a linear algebraic group G. A geometric quotient is the data of an algebraic scheme M and a surjective G-invariant affine morphism $\phi\colon H \to M$ such that:

(1) Every fibre of ϕ contains exactly one G-orbit
(2) M is a categorical quotient, this means that for every G-invariant morphism $\psi\colon H \to N$ there exists an unique morphism $\eta\colon M \to N$ such that $\psi = \eta \circ \phi$
(3) For every open set $U \subset M$ there exists an isomorphism

$$\phi^*\colon \Gamma(U, \mathcal{O}_M) \to \Gamma(\phi^{-1}(U), \mathcal{O}_H)^G$$

(4) If $W \subset H$ is a closed G-invariant subset then $\phi(W)$ is closed. If $W_1, W_2 \subset H$ are disjoint G-invariant closed subsets then $\phi(W_1) \cap \phi(W_2) = \emptyset$

Let $\mathcal{M}_{x,y} = \mathcal{H}_{x,y}/G$ be the quotient as in Proposition 3.8 and let $f\colon \mathcal{X} \to S$ be a family of canonical surfaces, since on every fibre X_s of f the group $H^1(nK_{X_s})$ vanishes, by semicontinuity and base change the direct image sheaf $f_*\omega_{\mathcal{X}/S}^{\otimes n}$ is locally free and therefore there exists an open covering $S = \cup U_i$ and a structure of family of n-canonically embedded surfaces on every restriction $X \to U_i$.

Since $Z_n \to \mathcal{H}_{x,y}$ is universal, there exist maps $\mu_i \colon U_i \to \mathcal{H}_{x,y}$ inducing these families. Clearly their compositions with the projection map $\mathcal{H}_{x,y} \to \mathcal{M}_{x,y}$ can be glued and we obtain finally a map $\mu \colon S \to \mathcal{M}_{x,y}$. From this and from the general properties of geometric quotients it follows that $\mathcal{M}_{x,y} = \mathcal{H}_{x,y}/G$ is a coarse moduli space for canonical models of surfaces of general type with fixed invariants.

Since for every regular action the dimension of the orbits is a lower semi-continuos function, the fibres of the projection morphism $\phi \colon \mathcal{H}_{x,y} \to \mathcal{M}_{x,y}$ are irreducible of constant dimension. The properties of geometric quotients imply that a closed subset $V \subset \mathcal{M}_{x,y}$ is irreducible if and only if $\phi^{-1}(V)$ is irreducible.

If X_1, X_2 are canonical models of surfaces of general type belonging to the same irreducible component of $\mathcal{M}_{x,y}$ then there exists an irreducible component V of $\mathcal{H}_{x,y}$ such that S_1, S_2 are isomorphic to two fibres of the restriction to V of the universal family $Z_n \to \mathcal{H}_{x,y}$ and then they are deformation equivalent.

Let $[X] \in \mathcal{H}_{x,y}$ be the point corresponding to a n-canonically embedded surface $X \subset \mathbb{P}^{P_n-1}$; the universal family induces a holomorphic map between germ of complex spaces $h \colon (\mathcal{H}_{x,y}, [X]) \to (\mathrm{Def}(X), 0)$.

Lemma 3.9 *In the notation above, the map h is smooth and $h^{-1}(0)$ is the germ of the G-orbit of $[X]$.*

Proof (Sketch of proof). Let $X_A \to Spec(A)$ be an infinitesimal deformation of X, let $p \colon A \to B$ be a small extension of local Artinian $\mathbb{C}$-algebras and assume that $X_B = X_A \times_{Spec(A)} Spec(B)$ is n-canonically embedded. Since $H^1(\omega_X^{\otimes n}) = 0$, every section of $\omega_{X_B/Spec(B)}^{\otimes n}$ extends to a section of $\omega_{X_A/Spec(A)}^{\otimes n}$, extending the basis of $H^0(\omega_{X_B/Spec(B)}^{\otimes n})$ that gives the n-canonical embedding we obtain an embedding of X_A that extends the embedding of X_B.

Therefore we have proved that h is smooth and, since there exists a factorization $(\mathcal{H}_{x,y}, [X]) \xrightarrow{h} \mathrm{Def}(X) \to \mathcal{M}_{x,y}$, the germ $h^{-1}(0)$ is contained in the G-orbit.

Conversely it follows from the definition of the G-action on $\mathcal{H}_{x,y}$ that the restriction of the universal family to every G-orbit is a locally trivial family of n-canonically embedded surfaces and then $h^{-1}(0)$ contains the germ of the G-orbit. $\square$

The stabilizer $Stab([X]) \subset PGL(P_n, \mathbb{C})$ of $[X] \in \mathcal{H}_{x,y}$ is naturally isomorphic to the group of automorphisms of X and if $T \subset \mathcal{H}_{x,y}$ is the image of a section of h, then the induced action of $Stab([X])$ on T is compatible with the natural action of $\mathrm{Aut}(X)$ on the base space of the Kuranishi family $\mathrm{Def}(X)$.

Corollary 3.10 *Let X be the canonical model of a surface of general type, then the germ of $\mathcal{M}$ at $[X]$ is analytically isomorphic to the quotient $\mathrm{Def}(X)/\mathrm{Aut}(X)$.*

If S is a minimal surface of general type, the relative canonical model of its universal deformation induces a morphism of germs of complex spaces $\pi'\colon \mathrm{Def}(S) \to \mathrm{Def}(X)$, where X is the canonical model of S. The morphism π' is compatible with the actions of $\mathrm{Aut}(X) = \mathrm{Aut}(S)$ and then it is defined a natural map $\pi\colon \mathrm{Def}(S)/\mathrm{Aut}(S) \to \mathcal{M}$. Simultaneous resolution of RDP implies that π is finite and surjective but, according to Burns and Wahl results [18], in some cases (e.g. K_S not ample and $\mathrm{Aut}(S) = 0$) it is not an isomorphism. Note that if X is a n-canonically embedded surface with $q(X) = 0$ then locally at $[X]$, $\mathcal{H}_{x,y}$ is an open subscheme of the Hilbert scheme of $\mathbb{P}^{P_n-1}$. In fact in this case if $X_A \subset \mathbb{P}^{P_n-1} \times Spec(A)$ is an infinitesimal embedded deformation of X then there exists at most one extension on X_A of the line bundle $O_X(1) = \omega_X^{\otimes n}$ and then $O_{X_A}(1) = \omega_{X_A/Spec(A)}^{\otimes n}$.

4 Smoothings of Normal Surface Singularities

4.1 The Link of an Isolated Singularity

Let $(X,0)$ be an isolated complex singularity of dimension n and take a representative $0 \in X \subset U$, where U is a Stein manifold (e.g. an open polydisk in $\mathbb{C}^N$) and X is a closed analytic subset of U such that $X - 0$ is smooth. By a *norm* on X we mean a real analytic function $r\colon U \to [0, +\infty[$ such that $r^{-1}(0) = 0$. Such a function always exists, for instance $r = \sum |z_i|^2$ for a set of local holomorphic coordinates $z_1, \ldots, z_N$ of U. We denote for every $\epsilon > 0$

$$X_{r=\epsilon} = \{x \in X \mid r(x) = \epsilon\}, \quad X_{r\leq\epsilon} = \{x \in X \mid 0 < r(x) \leq \epsilon\}.$$

Lemma 4.1 *In the notation above there exist a real number $\epsilon > 0$ and a diffeomorphism $X_{r\leq\epsilon} \simeq X_{r=\epsilon} \times]0, \epsilon]$ such that the morphism r corresponds to the projection on the second factor.*
In particular, $X_{r=\epsilon}$ is a compact oriented manifold of real dimension $2n - 1$.

Proof. [87, 2.4]. □

Definition 4.2 *If r and ϵ satisfy the condition of Lemma 4.1, we shall say that the (real) analytic space with boundary $\{0\} \cup X_{r\leq\epsilon}$ is a good representative of the singularity $(X,0)$.*

The following proposition implies that the diffeomorphism class of the oriented manifold $X_{r=\epsilon}$ is independent from the choice of the norm and the representative X.

Proposition 4.3 *Let s, r be norms on X. Then there exists $\epsilon > 0$ such that:*

1. *$r\colon X_{r\leq\epsilon} \to]0, \epsilon]$ and $s\colon X_{s\leq\epsilon} \to]0, \epsilon]$ are proper submersions.*
2. *If $\delta < \epsilon$ and $X_{s\leq\delta} \subset X_{r<\epsilon}$ then there exist a smooth proper submersion $p\colon X_{r\leq\epsilon} - X_{s<\delta} \to [0,1]$ such that $p^{-1}(0) = X_{s=\delta}$ and $p^{-1}(1) = X_{r=\epsilon}$*

Proof. [87, 2.5]. □

Definition 4.4 *The real oriented manifold $X_{r=\epsilon}$, defined for $\epsilon << 1$ and up to orientation preserving diffeomorphism, it is called the* link *of the singularity $(X, 0)$.*

Example 4.5 (Lens spaces) *Let q, p be two relatively prime integers and let $S_R^3 = \{(x_1, x_2) \in \mathbb{C}^2 \mid |x_1|^2 + |x_2|^2 = R^2\}$ be the three-dimensional sphere of radius $R > 0$. If $p \geq 2$, then the lens space $L(p, q)$ is defined as the quotient S_R^3/μ_p, where the action is given by $\xi(x_1, x_2) = (\xi x_1, \xi^{-q} x_2)$. The action is free and orientation preserving, therefore $L(p, q)$ has a natural structure of oriented three-manifold.*

Note that $L(p, -q)$ is the link of the cyclic singularity X_0 of type $\frac{1}{p}(1, q)$; in fact it is sufficient to take as real analytic function $r \colon X_0 \to [0, +\infty[$ the factorization to the quotient of the usual norm of $\mathbb{C}^2$, which is clearly μ_p-invariant.

Let X be a good representative of a normal surface singularity $(X, 0)$. By definition the boundary ∂X is diffeomorphic to the link of the singularity and, since $(X, 0)$ is analytically irreducible, we have that ∂X is a compact connected oriented manifold of dimension 3.

For completeness of exposition we mention a remarkable result proved by Mumford [103]: *the link of a normal twodimensional singularity $(X, 0)$ is simply connected if and only if $(X, 0) = (\mathbb{C}^2, 0)$.* As a corollary Brieskorn proved [17] that *a normal surface singularity is quotient if and only if its link has finite fundamental group.*

Let S be a smooth compact connected oriented manifold of dimension 4 with boundary ∂S; denote by $S' = S \setminus \partial S$. The collar theorem [63] asserts that there is a neighbourhood of ∂S in S diffeomorphic to $\partial S \times [0, 1[$; in particular S, S' have the same homotopy type.

The duality theorems of Lefschetz and Poincaré give the isomorphisms (for every ring of coefficients A)

$$H_c^q(S', A) = H_{4-q}(S, A) = H^q(\overline{S}, \partial S, A).$$

Moreover the cup product induce a pairing

$$H^2(S, A) \times H^2(S, \partial S, A) \overset{\cup}{\longrightarrow} H^4(S, \partial S, A) = A$$

which is nondegenerate whenever $A = \mathbb{Q}, \mathbb{R}, \mathbb{C}$.

Taking $A = \mathbb{Z}$, the composition of the natural morphism $H^2(S, \partial S, \mathbb{Z}) \to H^2(S, \mathbb{Z})$ with the above pairing gives the intersection product

$$H^2(S, \partial S, \mathbb{Z}) \times H^2(S, \partial S, \mathbb{Z}) = H_2(S, \mathbb{Z}) \times H_2(S, \mathbb{Z}) \overset{q}{\longrightarrow} \mathbb{Z}.$$

Example 4.6 *Let ∂X be the link of a rational singularity with good representative X and let $S \to X$ be its minimal resolution with exceptional divisor $E = \cup E_i$, $i = 1, \ldots, r$: notice that $\partial X = \partial S$ and E is deformation retract of some neighbourhood [12, 103]. By Ehresmann theorem, applied to the proper submersion $r \colon \overline{S} \setminus E \to]0, \epsilon]$, we have that E is a deformation retract of S and then $H_1(S, \mathbb{Z}) = H^3(\overline{S}, \partial S, \mathbb{Z}) = 0$, $H_2(S, \mathbb{Z}) = \bigoplus \mathbb{Z}[E_i]$ and $H^2(S, \mathbb{Z}) = \bigoplus \mathbb{Z}\sigma_i$, where $\sigma_1, \ldots, \sigma_r$ is the dual basis of $E_1, \ldots, E_r$. The exact sequence of the pair $(S, \partial S)$ gives*

$$0 \longrightarrow H^1(\partial S, \mathbb{Z}) \longrightarrow H^2(\overline{S}, \partial S, \mathbb{Z}) \xrightarrow{\alpha} H^2(\overline{S}, Z) \longrightarrow H^2(\partial S, \mathbb{Z}) \longrightarrow 0$$

and the morphism α is represented by the intersection matrix $E_i \cdot E_j$. This implies that $H^1(\partial X, \mathbb{Z}) = H_2(\partial X, \mathbb{Z}) = 0$ and $H^2(\partial X, \mathbb{Z}) = H_1(\partial X, \mathbb{Z})$ is the cokernel of the morphism $\mathbb{Z}^r \to \mathbb{Z}^r$ induced by the intersection matrix of E.

Example 4.7 *The Example 4.6 shows that the link of the rational double point E_8 has the homology of the sphere S^3.*

4.2 The Milnor Fibre

A *deformation* of an analytic singularity $(V_0, 0)$ is a pair of morphisms of germs of complex spaces

$$(V_0, 0) \xrightarrow{i} (V, 0) \xrightarrow{f} (T, 0),$$

such that f is flat, $fi = 0$ and $i \colon (V_0, 0) \to (f^{-1}(0), 0)$ is an isomorphism. The deformation is called *trivial* if $(V, 0)$ is isomorphic to $(V_0 \times T, 0)$. The deformation is called a *smoothing* if T is reduced and there exists a representative $f \colon X \to T$ together a dense analytic open subset $U \subset T$ such that $f^{-1}(t)$ is smooth for every $t \in U$.

A one dimensional smoothing is a smoothing with base $(T, 0) = (\mathbb{C}, 0)$. If $(V_0, 0)$ is an isolated singularity, to give a one dimensional smoothing of $(V_0, 0)$ is the same of giving a disk $\Delta = \{t \in \mathbb{C} \mid |t| < \epsilon\}$, a complex space V and a flat map $f \colon V \to \Delta$, such that $(f^{-1}(0), 0) \simeq (V_0, 0)$, $V \setminus \{0\}$ is smooth and $f \colon V \setminus \{0\} \to \Delta$ has surjective differential at every point. In particular the fibre $V_t = f^{-1}(t)$ is nonsingular for every $t \in \Delta^* = \Delta \setminus \{0\}$.

From now on we shall simply say smoothing instead of one dimensional smoothing.

Assume $(V_0, 0)$ embedded in $(\mathbb{C}^N, 0)$, then there exists an embedding of $(V, 0)$ in $(\mathbb{C}^N \times \Delta, 0)$ such that the map f is induced by the projection on the second factor $\mathbb{C}^N \times \Delta \to \Delta$.

Let $\epsilon > 0$ be such that the sphere $S_r = \partial B_r$, $B_r = \{z \in \mathbb{C}^N \mid \|z\| < r\}$ intersects transversally V_0 for every $0 < r \leq \epsilon$. After a possible shrinking of

Δ we can assume that $f\colon V \cap S_\epsilon \times \Delta \to \Delta$ is proper and $S_\epsilon \times \Delta$ intersects transversally $f^{-1}(t)$ for every $t \in \Delta$. In this situation we set

$$X = V \cap (B_r \times \Delta) \qquad X_t = V_t \cap X \qquad \partial X_t = V_t \cap (S_r \times \Delta).$$

By Ehresmann's fibration theorem we have $\partial X = \cup_{t \in \Delta} \partial X_t \simeq \partial X_0 \times \Delta$ and the map $f\colon X \setminus X_0 \to \Delta^*$ is a locally trivial C^∞ fibre bundle with fibre F diffeomorphic to X_t for $t \neq 0$. We call F (resp. $\overline{F}$) the *Milnor fibre* (resp. *compact Milnor fibre*) of the smoothing f.

According to the basic theory of Milnor's fibration [87], the diffeomorphism class of F is independent of the embedding of V; in particular the topological invariants of F are invariants of the smoothing. By construction the boundary of F is diffeomorphic to the link of the singularity.

If n is the dimension of $(V_0, 0)$ then F is a Stein manifold of dimension n and then, by Andreotti-Frankel's theorem [99] it has the homotopy type of a n-dimensional CW complex. Considering homology and cohomology we have $H_i(F, \mathbb{Z}) = 0$ for $i > n$ and $H_n(F, \mathbb{Z})$ is a finitely generated free abelian group. Moreover, if $n \geq 2$, then the inclusion $\partial F \to \overline{F}$ induces a surjective morphism $\pi_1(\partial F) \to \pi_1(\overline{F})$.

Definition 4.8 *The integer* $\mu = \operatorname{rank} H_n(F, \mathbb{Z})$ *is called the* Milnor number *of the smoothing.*

From now on we consider only smoothings of normal surface singularities. Using the exact homotopy sequence of the fibration $f\colon X \setminus X_0 \to \Delta^*$ and Van Kampen theorem it is easy to see that the inclusion $F \subset X - \{0\}$ induces an isomorphism on π_1's (cf. [88, Lemma 5.1]).

If $(V, 0)$ is a rational double or triple point then, by Brieskorn-Tyurina's simultaneous resolution, the Milnor fibre F is diffeomorphic to a neighbourhood of the exceptional divisor in the minimal resolution. Therefore, if V_0 is of type A_r, D_r or E_r, then F has the homotopy type of a bouquet of r spheres and the intersection product

$$H^2(F, \partial F, \mathbb{R}) \times H^2(F, \partial F, \mathbb{R}) = H_2(F, \mathbb{R}) \times H_2(F, \mathbb{R}) \xrightarrow{q} \mathbb{R}$$

is negative definite. More generally, according to Steenbrink [124, Th. 2.24], for every smoothing of a rational surface singularity the intersection product q on the Milnor fibre is negative definite.

4.3 $\mathbb{Q}$-Gorenstein Singularities and Smoothings

Definition 4.9 *Let* $(Y, 0)$ *be a normal Cohen–Macaulay singularity with canonical divisor* K_Y. *We shall say that* $(Y, 0)$ *is* $\mathbb{Q}$-Gorenstein *of index* n *if there exists some nonzero integer* n' *such that the Weil divisor* $n' K_Y$ *is principal (= Cartier) and* n *is the smallest positive integer with this property.*

A deformation $(X, 0) \xrightarrow{\pi'} (\mathbb{C}, 0)$ *is called* $\mathbb{Q}$-Gorenstein *if the global space* $(X, 0)$ *is* $\mathbb{Q}$-Gorenstein.

According to Serre's criterion every normal surface singularity is Cohen–Macaulay and then, by [97, §23], the global space of a smoothing of a normal surface singularity is still normal and Cohen–Macaulay.

If a normal surface singularity $(X_0, 0)$ admits a $\mathbb{Q}$-Gorenstein smoothing then $(X_0, 0)$ is $\mathbb{Q}$-Gorenstein. The converse is generally false: for example, the quotient on $\mathbb{C}^3$ by the involution $(x, y, z) \mapsto (-x, -y, -z)$ is $\mathbb{Q}$-Gorenstein but does not has $\mathbb{Q}$-Gorenstein smothings, as follows from the next result.

Proposition 4.10 (Schlessinger) *Every deformation of an isolated quotient singularity of dimension ≥ 3 is trivial.*

Proof. [121]. □

Proposition 4.11 *Let $(X_0, 0)$ be a quotient two dimensional singularity admitting a $\mathbb{Q}$-Gorenstein smoothing of index n. Then $(X_0, 0)$ is*

1. *A smooth or a rational double points if $n = 1$*

2. *A cyclic singularity of type $\dfrac{1}{dn^2}(1, dna - 1)$ for $a, d > 0$ and a, n relatively prime, if $n \geq 2$*

Proof. This result is implicitly proved in the paper [88] and explicitly in [77, 3.10 and 3.11], [89].

It is easy to give examples of $\mathbb{Q}$-Gorenstein smoothing for each one of the above singularities, while the hardest part is to prove that no other singularity admits such a smoothing. This is done in [88] by studying the β-invariant, in [77] by using the theory of three-dimensional terminal singularity and in [89] by using the classification, made in [23] of finite groups of automorphisms of rational double points. □

Every smoothing of a rational double point is Gorenstein, i.e. $\mathbb{Q}$-Gorenstein of index 1. In order to describe the $\mathbb{Q}$-Gorenstein smoothing of cyclic singularities of class T we note that the singularity $\dfrac{1}{dn^2}(1, dna - 1)$, g.c.d.$(n, a) = 1$, is the quotient of the rational double point of type A_{dn-1}, defined in $\mathbb{C}^3$ by the equation $uv - y^{dn} = 0$, by the action of the group μ_n of nth roots of 1 given by

$$\mu_n \ni \xi \colon (u, v, y) \mapsto (\xi u, \xi^{-1} v, \xi^a y).$$

In fact, denote by

$$X = \mathbb{C}^2 / \mu_{dn^2}, \qquad \mu_{dn^2} \ni \xi \colon (z_1, z_2) \mapsto (\xi z_1, \xi^{dna-1} z_2)$$

and let $0 \longrightarrow \mu_{dn} \longrightarrow \mu_{dn^2} \overset{p}{\longrightarrow} \mu_n \longrightarrow 0$ be the natural exact sequence, being $p(\xi) = \xi^{dn}$. We can write $X = Y/\mu_n$, where

$$Y = \mathbb{C}^2 / \mu_{dn}, \qquad \mu_{dn} \ni \xi \colon (z_1, z_2) \mapsto (\xi z_1, \xi^{dna-1} z_2) = (\xi z_1, \xi^{-1} z_2).$$

The analytic algebra of $(Y, 0)$ is

$$\mathbb{C}\{z_1^{dn}, z_2^{dn}, z_1 z_2\} = \mathbb{C}\{u, v, y\}/(uv - y^{dn}),$$

where $u = z_1^{dn}$, $v = z_2^{dn}$, $y = z_1 z_2$. The action of μ_n on $\mathcal{O}_{Y,0}$ is

$$\mu_n \ni \xi : (u, v, y) \mapsto (\xi u, \xi v, \xi^a y).$$

Since a, n are relatively prime, this action is free outside 0.

Theorem 4.1. *Let $a, d, n > 0$ be integers with a, n relatively prime and consider the hypersurface $\mathcal{Y} \subset \mathbb{C}^3 \times \mathbb{C}^d$ of equation $uv - y^{dn} = \sum_{k=0}^{d-1} t_k y^{kn}$, where $t_0, \ldots, t_{d-1}$ are linear coordinates over $\mathbb{C}^d$. The group μ_n acts on $\mathcal{Y}$ by the law*

$$\mu_n \ni \xi : (u, v, y, t_0, \ldots, t_{d-1}) \mapsto (\xi u, \xi^{-1} v, \xi^a y, t_0, \ldots, t_{d-1}).$$

Denote $\mathcal{X} = \mathcal{Y}/\mu_n$ and $\pi : \mathcal{X} \to \mathbb{C}^d$ the factorization to the quotient of the projection $\mathcal{Y} \to \mathbb{C}^d$. Then:

1. *$\pi : (\mathcal{X}, 0) \to (\mathbb{C}^d, 0)$ is a $\mathbb{Q}$-Gorenstein deformation of the cyclic singularity $(X_0, 0)$ of type $\dfrac{1}{dn^2}(1, dna - 1)$.*
2. *Every $\mathbb{Q}$-Gorenstein deformation $(X, 0) \to (\mathbb{C}, 0)$ of $(X_0, 0)$ is isomorphic to the pull-back of π for some germ of holomorphic map $(\mathbb{C}, 0) \to (\mathbb{C}^d, 0)$.*

Proof. [77, 3.9.i, 3.17 and 3.18]. The same result for the cyclic singularity of type $\dfrac{1}{4}(1, 1)$ was previously proved by Riemenschneider [117]. □

In the notation of Theorem 4.1, let $\Delta(t_0, \ldots, t_{d-1})$ be the discriminant of the polynomial $y^{dn} + \sum_{k=0}^{d-1} t_k y^{kn}$. Then the fibre of π over $(t_0, \ldots, t_{d-1})$ is singular if and only if $t_0 \Delta(t_0, \ldots, t_{d-1}) = 0$. In particular the set of point $t \in \mathbb{C}^d$ such that $\pi^{-1}(t)$ is smooth is a Zariski open subset and π is a $\mathbb{Q}$-Gorenstein smoothing.

If $d = 1$ the fibre $\pi^{-1}(t_0)$ is singular if and only if $t_0 = 0$ and then we have:

Corollary 4.12 *Let $f : (X, 0) \to (\mathbb{C}, 0)$ be a $\mathbb{Q}$-Gorenstein deformation of a cyclic singularity of type $\dfrac{1}{n^2}(1, na - 1)$, g.c.d.$(a, n) = 1$. Then f is either a smoothing or it is the trivial deformation.*

Example 4.13 *The Milnor fibre of a $\mathbb{Q}$-Gorenstein smoothing $(X, 0) \to (\mathbb{C}, 0)$ of the cyclic singularity of type $\dfrac{1}{4}(1, 1)$ is isomorphic to the complement in $\mathbb{P}^2$ of a tubular neighbourhood of a smooth conic. In fact such a smoothing is isomorphic to $\{uv - y^2 = -g(t)/2\}/\mu_2$ for some analytic function g such that $g(0) = 0$. The algebra of invariant functions*

$$\frac{\mathbb{C}\{u, v, y, t\}^{\mu_2}}{uv - y^2 + g(t)/2}$$

is generated by

$$\begin{cases} z = u^2 + v^2 \\ x = 2uv = 2y^2 - g(t) \\ w_1 = u^2 - v^2 \\ w_2 = \sqrt{2}y(u - v) \\ w_3 = \sqrt{2}y(u + v) \\ t = t \end{cases}$$

Setting $f_1 = x + g(t)$, $f_2 = z + x$ *and* $f_3 = z - x$ *the smoothing* $(X, 0)$ *is embedded in* $(\mathbb{C}^5 \times \mathbb{C})$, *with coordinates* x, z, w_1, w_2, w_3, t, *and defined by the six equations*

$$\begin{cases} w_i^2 = f_j f_k \\ w_j w_k = w_i f_i \end{cases} \qquad \{i, j, k\} = \{1, 2, 3\} \qquad (1)$$

The same equations can be displayed as

$$\mathrm{rank} \begin{pmatrix} z - x & w_1 & w_2 \\ w_1 & z + x & w_3 \\ w_2 & w_3 & x + g(t) \end{pmatrix} \le 1.$$

We then see that the fibre X_t *of the smoothing is contained in a hyperplane section of the affine cone over the Veronese surface.*

We are now able to study more closely the Milnor fibre of such smoothings.

Proposition 4.14 *Let* F *be the Milnor fibre of a* $\mathbb{Q}$-*Gorenstein smoothing* $(X, 0) \to (\mathbb{C}, 0)$ *of a cyclic singularity* $(X_0, 0)$ *of type* $\dfrac{1}{dn^2}(1, dna - 1)$ *with* g. c. d. $(n, a) = 1$, *then:*

1. $b_2(F) = d - 1$, $\pi_1(F) = \mathbb{Z}_n$ *and the intersection form* $H_2(F, \mathbb{R}) \times H_2(F, \mathbb{R}) \to \mathbb{R}$ *is negative definite.*
2. $\pi_1(\partial F) = \mathbb{Z}_{dn^2}$.
3. *The torsion subgroup of the Picard group* $\mathrm{Pic}(F)$ *of* F *is cyclic of order* n *and it is generated by the canonical bundle* K_F.
4. *The orientation preserving diffeomorphism class of* F *is independent from the smoothing.*
5. *Every orientation preserving diffeomorphism of* ∂F *extends to an orientation preserving diffeomorphism of* $\overline{F}$

Proof. We only give a sketch: full details can de found in the paper [90] and [96].
[1] From Theorem 4.1 it follows that F has an unramified connected covering $F' \to F$ of degree n, where F' is the Milnor fibre of a smoothing of the rational double point A_{dn-1}. Simultaneous resolution implies that F' has the homotopy type of a bouquet of $dn-1$ spheres S^2, hence $\pi_1(F) = \mathbb{Z}_n$, $b_1(F) = 0$ and $e(F) = 1 + b_2(F)$, where e denotes the topological Euler characteristic. Since $ne(F) = e(F') = 1 + b_2(F') = dn$ it follows the equality $b_2(F) = d - 1$.

[2] It follows from the fact that ∂F is diffeomorphic to the link of the cyclic singularity $(X_0, 0)$ and then it is diffeomorphic to the lens space $L(dn^2, 1 - dna)$.

[3] We have seen that the inclusion $F \subset X - 0$ induces an isomorphism $\mathbb{Z}_n = \pi_1(F) \simeq \pi_1(X - 0)$. The universal coefficient theorem gives an isomorphism $Tors H^2(X - 0, \mathbb{Z}) = Tors H^2(F, \mathbb{Z}) = \mathbb{Z}_n$. Since $(X, 0)$ is $\mathbb{Q}$-Gorenstein of index n, the canonical bundle is a torsion element of $Pic(X-0)$ of order exactly n and then the canonical class is a generator of $Tors H^2(X-0, \mathbb{Z})$. According to adjunction formula the natural morphism $H^2(X-0, \mathbb{Z}) \to H^2(F, \mathbb{Z})$ preserves the canonical class.

[4] This follows from Ehresmann fibration theorem and from the fact that, in the notation of Theorem 4.1, the set of points $t \in \mathbb{C}^d$ such that the fibre $\pi^{-1}(t)$ is smooth, is an open connected subset.

[5] Since ∂F has a collar in $\overline{F}$, every diffeomorphism of ∂F isotopic to identity extends to $\overline{F}$. Let $r \in \mathbb{R}$, $r \neq 0$, $|r| << 1$; then F is the quotient G/μ_n, where G is the intersection of the hypersurface $\{uv - y^{dn} = r\}$ with a ball in $\mathbb{C}^3$ of sufficiently small radius $s > 0$.

The diffeomorphisms

$$\tau(u, v, y) = (\overline{u}, \overline{v}, \overline{y}), \qquad \sigma_+(u, v, y) = (v, u, y)$$

commute with the μ_n action and therefore induce orientation preserving diffeomorphism of $\overline{F}$. According to Bonahon [16] the group of orientation preserving diffeomorphisms of the lens space ∂F is generated by the the diffeomorphisms isotopic to the identity and by:

$$\begin{cases} \tau \text{ and } \sigma_+, & \text{if } a = 1, n = 2, \\ \tau, & \text{otherwise.} \end{cases}$$

Full details are in [96, Prop. 1.8]. □

4.4 T-Deformation Equivalence of Surfaces

In this section we introduce the notion of deformation T-equivalence of algebraic surfaces. It will be clear from the definition that deformation equivalence implies deformation T-equivalence.

Definition 4.15 *A surface of class T is a normal projective surfaces with at most rational double points and cyclic singularities of class T.*

Definition 4.16 *The deformation T-equivalence is the equivalence relation, in the set of isomorphism classes of surfaces of class T, generated by the following relation $\sim$:*
Given two surfaces of class T, S_1 and S_2, we set $S_1 \sim S_2$ if they are fibres of a proper flat analytic family $f : S \to C$ such that C is a smooth irreducible affine curve, every fibre of f is of class T and S is $\mathbb{Q}$-Gorenstein.

Theorem 4.2. *Let S_1, S_2 be two smooth compact complex surfaces. If they are deformation T-equivalent then they are orientedly diffeomorphic.*

Proof. The proof is divided in two steps: in the first we associate to every surface S of class T an oriented smooth four-manifold $\widehat{S}$ which is equal to S when S has no singular points; in the second we prove that the diffeomorphism type of $\widehat{S}$ is invariant under $\mathbb{Q}$-Gorenstein deformation over a smooth curve.

1. We first define for every surface of class T, $\widehat{S} = \widehat{S'}$, where $S' \to S$ is the minimal resolution of all rational double points of S. Thus we may suppose that S is a surface with at most cyclic singularities of class T, say at the points $p_1, \ldots, p_n$. Choose closed embeddings $\phi_i \colon (S, p_i) \to (\mathbb{C}^{N_i}, 0)$, fix a sufficiently small real number $r > 0$ and denote $V_i = \phi_i^{-1}(\{z \mid \|z\| < r\})$. If F_i is the Milnor fibre of a $\mathbb{Q}$-Gorenstein smoothing of (S, p_i) we define

$$\widehat{S} = (S - \cup_i V_i) \cup (\cup_i F_i),$$

where the pasting is made by choosing for every i an orientation preserving diffeomorphism $\partial V_i \to \partial F_i$. By Proposition 4.14 $\widehat{S}$ is a well defined smooth oriented four manifold.

If $V \to S$ is the minimal resolution of a surface of class T then, in the terminology of [46], $\widehat{S}$ is obtained by rationally blowing down V.

2. Let $f \colon \mathcal{S} \to \Delta = \{t \in \mathbb{C} \mid |t| < 1\}$ be a $\mathbb{Q}$-Gorenstein deformation of a surface S_0 of class T, we want to prove that $\widehat{S_0} = \widehat{S_t}$ for $|t| \ll 1$.

For simplicity of exposition we prove this under the assumption that S_0 has at most rational double points and cyclic $(1, na - 1)/n^2$ singularities. We refer to [96] for the proof in the general case. By using the simultaneous resolution of rational double points we can assume without loss of generality that S_0 has at most cyclic singularities at points $p_1, \ldots, p_n$.

There exists $0 \le m \le n$ such that the deformation $(\mathcal{S}, p_i) \to (\mathbb{C}, 0)$ is a smoothing for $i \le m$ and trivial for $i > m$.

We can find an integer $N \gg 0$ and closed embeddings $\phi_i \colon (\mathcal{S}, p_i) \to (\mathbb{C}^N \times \Delta, 0)$, $i \le n$. Denoting by $B_r = \{(z, t) \in \mathbb{C}^N \times \Delta \mid |z| < r\}$, we consider, for $r > 0$ sufficiently small, $V_i = S_0 \cap \phi_i^{-1}(B_r)$, $G_i = S_t \cap \phi_i^{-1}(B_r)$, $t \ne 0$ and F_i the Milnor fibre of a $\mathbb{Q}$-Gorenstein smoothing of (S_0, p_i). By construction we have $G_i = F_i$ for $i \le m$, while $G_i = V_i$ and then F_i is the Milnor fibre of a $\mathbb{Q}$-Gorenstein smoothing of (S_t, p_i) for every $i > m$.

The differentiable map of manifold with boundary $f \colon (\mathcal{S} - \cup_i \phi^{-1}(B_r)) \to \Delta$ is proper and has no critical values in a neighbourhood of 0; then, by Ehresmann's fibration theorem, there exists for $0 < |t| \ll 1$, a diffeomorphism $(S_0 - \cup_i V_i) \to (S_t - \cup_i F_i)$ sending ∂V_i into ∂F_i and then there exists a chain of diffeomorphisms

$$\widehat{S_0} \simeq (S_0 - \cup_i V_i) \cup (\cup_i F_i) \simeq (S_t - \cup_i F_i) \cup (\cup_i F_i) \simeq \widehat{S_t}.$$

$\square$

4.5 A Non Trivial Example of T-Deformation Equivalence

Let $Q = \mathbb{P}^1 \times \mathbb{P}^1$ be the quadric and let $H = \mathcal{O}_Q(a, b)$ be a fixed ample line bundle over Q with $a, b \geq 3$; we have $H^2 = 2ab$.

For every integer $0 \leq n \leq 2ab$, let $M_{H,n}$ be the subset of $|H|^3 \times Q^n$ consisting of the elements $(C_1, C_2, C_3, p_1, \ldots, p_n)$ such that:

1. the curves C_1, C_2, C_3 are smooth
2. for every $i \neq j$, C_i intersects transversally C_j
3. $p_i \in C_j$ for every i, j; i.e. the points p_i are contained in the base locus of the linear system generated by C_1, C_2, C_3
4. $p_i \neq p_j$ if $i \neq j$

The set $M_{H,n}$ carries a natural structure of locally closed subscheme of $|H|^3 \times Q^n$.

We denote by $M_{H,n}^0 \subset M_{H,n}$ the (possibly empty) Zariski open subset of elements $(C_1, C_2, C_3, p_1, \ldots, p_n)$ such that $C_1 \cap C_2 \cap C_3 = \{p_1, \ldots, p_n\}$.

The structure of $M_{H,n}^0$ and $M_{H,n}$ is investigated in [96]. In particular it is proved (Lemma 2.1 and Corollary 2.12) the following result.

Lemma 4.17 *In the notation above:*

1. *The scheme $M_{H,n}$ is connected*
2. *If $b = la$ and $n = la(2a - c)$ for some pair of integers $l \geq 2$ and $0 < c < a/2$, then $M_{H,n}^0$ contains at least two connected components*

Assume that the line bundle pair H and the integer n satisfy the condition of Lemma 4.17. For every $m = (C_1, C_2, C_3, p_1, \ldots, p_n) \in M_{H,n}$, let us denote by $p \colon S_m \to Q$ the blow up at $p_1, \ldots, p_n$ and by $D_i \subset S_m$, $i = 1, 2, 3$, the strict transform of C_i: notice that $D_1 \cap D_2 \cap D_3 = \emptyset$ if and only if $m \in M_{H,n}^0$. The divisors D_i are linearly equivalent and then they are the divisors of three sections f_1, f_2, f_3 of a line bundle L. Denote by $\pi \colon V = L \oplus L \oplus L \to S_m$ and by $w_1, w_2, w_3 \in H^0(V, \pi^*L)$ the tautological sections $w_i(x_1, x_2, x_3) = x_i$. Finally let $X_m \subset V$ the subscheme defined by the six equations

$$\begin{cases} w_i^2 = f_j f_k \\ w_j w_k = w_i f_i \end{cases} \qquad \{i, j, k\} = \{1, 2, 3\} \qquad (2)$$

It is easy to see that the isomorphism class of X_m is independent of the choice of the sections f_i. If $x \in X_m$ and $f_1(x) = f_2(x) = f_3(x) = 0$, then we have already seen that (X_m, x) is a cyclic singularity of type $(1, 1)/4$. If otherwise $f_i(x) \neq 0$ for some i, the Jacobian criterion shows that X_m is smooth at the point x. Therefore we have proved that

1. if $m \in M_{H,n}^0$ then X_m is a smooth surface
2. if $m \in M_{H,n}$ then X_m has at most cyclic singularities of type $\frac{1}{4}(1, 1)$

In particular X_m is a surface of class T for every $m \in M_{H,n}$.

If $\phi \colon (\mathbb{C}, 0) \to (M_{H,n}, m)$, $\phi(t) = (C_i(t), p_j(t))$, is any germ of morphism then there exists a $\mathbb{Q}$-Gorenstein deformation $q \colon \mathcal{X} \to (\mathbb{C}, 0)$ such that $q^{-1}(t) = X_{\phi(t)}$. In fact it is sufficient to take $p \colon \mathcal{S} \to Q \times (\mathbb{C}, 0)$ the blow up along the sections $p_j(t)$ and $\mathcal{X} \to \mathcal{S}$ defined by the equations

$$w_i^2 = f_j(t)f_k(t), \qquad w_j w_k = w_i f_i(t)$$

where $f_i(t)$ is a section whose divisor is the strict transform of $C_i(t)$. The computation made in Example 4.13 shows that $\mathcal{X}$ is $\mathbb{Q}$-Gorenstein.

Since $M_{H,n}$ is connected we have that X_m is T-deformation equivalent to X_l for every $l, m \in M_{H,n}$. It is an open question to determine whether the smooth surfaces X_l, X_m are deformation equivalent whenever l, m belong to different connected components of $M_{H,n}^0$.

Remark 4.18 *Other interesting examples of $\mathbb{Q}$-Gorenstein smoothings of T-surfaces have been recently considered in [82] and [115].*

5 Double and Multidouble Covers of Normal Surfaces

5.1 Flat Abelian Covers

Let X be a normal projective irreducible variety and G a finite abelian group acting on X. Denote by $Y = X/G$ and by $\pi \colon X \to Y$ the projection to the quotient. The group G acts on the direct image sheaf $\pi_* \mathcal{O}_X$ and then gives a character decomposition [110]

$$\pi_* \mathcal{O}_X = \bigoplus_\chi \pi_* \mathcal{O}_X^\chi, \qquad \chi \in G^*, \qquad \pi_* \mathcal{O}_X^\chi = \{f \mid \sigma f = \chi(\sigma)f \ \forall \sigma \in G\}.$$

Since there exists a Zariski open subset $U \subset Y$ such that $\pi^{-1}(U) \to U$ is a regular unramified cover with group G, every sheaf $\pi_* \mathcal{O}_X^\chi$ is a line bundle on U [104, p. 70].

Since X, Y are normal, the sheaf $\pi_* \mathcal{O}_X$ is torsion free and reflexive; therefore every $\pi_* \mathcal{O}_X^\chi$, being a direct summand of $\pi_* \mathcal{O}_X$, is a reflexive sheaf of rank 1. In other words, $\pi_* \mathcal{O}_X^\chi$ is the sheaf associated to a Weil divisor on Y. In particular we have that

Lemma 5.1 *In the notation above:*

1. *If Y is locally factorial (e.g. if Y is smooth) then every $\pi_* \mathcal{O}_X^\chi$ is an invertible sheaf.*
2. *π is a flat morphism if and only if every $\pi_* \mathcal{O}_X^\chi$ is an invertible sheaf.*

Proof. The morphism π is flat if and only if $\pi_* \mathcal{O}_X$ is locally free and any direct summand of a locally free sheaf is locally free. □

From now on we consider only flat abelian covers. Therefore every $\pi_* \mathcal{O}_X^\chi$ is an invertible sheaf and there exist line bundles $L_\chi \to Y$ such that $\pi_* \mathcal{O}_X^\chi = \mathcal{O}_Y(-L_\chi)$ for every character χ.

Lemma 5.2 *Let* $\pi \colon X \to Y$ *be a flat G-cover with character decomposition* $\pi_* \mathcal{O}_X = \bigoplus_\chi \mathcal{O}_Y(-L_\chi)$. *Then for every coherent sheaf $\mathcal{F}$ on Y and every* $i \geq 0$ *there exist an isomorphism*

$$\mathrm{Ext}^i_{\mathcal{O}_X}(\pi^* \mathcal{F}, \mathcal{O}_X) = \mathrm{Ext}^i_{\mathcal{O}_Y}(\mathcal{F}, \pi_* \mathcal{O}_X) = \bigoplus_\chi \mathrm{Ext}^i_{\mathcal{O}_Y}(\mathcal{F}, -L_\chi).$$

Proof. We first note (see, e.g. [62, p. 110]) that there exists a natural isomorphism $\mathrm{Hom}_{\mathcal{O}_X}(\pi^* \mathcal{F}, \mathcal{G}) = \mathrm{Hom}_{\mathcal{O}_Y}(\mathcal{F}, \pi_* \mathcal{G})$ for every $\mathcal{O}_X$ module $\mathcal{G}$. This formula, together the exactness of π^* implies that if $\mathcal{I}$ is an injective $\mathcal{O}_X$-module, then $\pi_* \mathcal{I}$ is an injective $\mathcal{O}_Y$-module.

The functor $\mathcal{G} \mapsto \mathrm{Hom}_{\mathcal{O}_X}(\pi^* \mathcal{F}, \mathcal{G})$ is the composition of $\mathcal{G} \mapsto \pi_* \mathcal{G} \mapsto \mathrm{Hom}_{\mathcal{O}_Y}(\mathcal{F}, \pi_* \mathcal{G})$ and then there exists a convergent spectral sequence

$$E_2^{p,q} = \mathrm{Ext}^p_{\mathcal{O}_Y}(\mathcal{F}, R^q \pi_* \mathcal{G}) \quad \Rightarrow \quad \mathrm{Ext}^{p+q}_{\mathcal{O}_X}(\pi^* \mathcal{F}, \mathcal{G}).$$

Since π is finite, $R^q \pi_* \mathcal{G} = 0$ for every coherent $\mathcal{O}_X$-module $\mathcal{G}$ and every $q > 0$. $\square$

Example 5.3 (Simple cyclic covers) *Let* $\pi \colon L \to Y$ *be a line bundle on a normal variety Y and n a positive integer. For every $f \in H^0(Y, \mathcal{O}_Y(nL))$ we define the n-root cover of f as the branched covering $X = Y(\sqrt[n]{f}, L) \longrightarrow Y$ defined by the hypersurface in L of equation $z^n = f$, where $z \in H^0(L, \pi^* L)$ is the tautological section.*

The group $G = \mu_n$ of nth roots of 1 acts on X in the obvious way and the character decomposition is $\pi_* \mathcal{O}_X = \bigoplus_{i=0}^{n-1} \mathcal{O}_Y(-iL)z^i$.

We note that X is normal if and only if the divisor of f is reduced. By definition [110, p. 195] every simple cyclic cover is the nth root cover of some section f as above.

Not every cyclic cover is simple. For example, the triple cyclic cover $\pi \colon \mathbb{P}^1 \to \mathbb{P}^1$, $\pi([x_0, x_1]) = [x_0^3, x_1^3]$ is not simple.

5.2 Flat Double Covers

From now on by a surface we mean a complex projective surface.

Lemma 5.4 *Let X be a normal surface and let $\pi \colon X \to Y$ be the quotient of X by an involution τ. The following conditions are equivalent:*

1. π is flat.

2. *The double cover is simple, i.e. there exists a line bundle $\pi \colon L \to Y$ and a section $f \in H^0(Y, 2L)$ such that X is isomorphic to the subvariety of L defined by the equation $z^2 = f$, where $z \in H^0(L, \pi^*L)$ is the tautological section, and the involution τ corresponds to multiplication by -1 in the fibres of L.*
3. *The fixed subvariety $\mathrm{Fix}(\tau)$ is a Cartier divisor.*

Moreover if X is smooth, then π is flat if and only if Y is smooth.

Proof. [1 ⇒ 2] If π is flat then the group $G = \{1, \tau\}$ acts on the rank 2 locally free sheaf $\pi_*\mathcal{O}_X$ and yields a character decomposition $\pi_*\mathcal{O}_X = \mathcal{O}_Y \oplus \mathcal{O}_Y(-L)$ for some $L \in \mathrm{Pic}(Y)$. X depends only by the $\mathcal{O}_Y$ algebra structure of $\pi_*\mathcal{O}_X$ which is uniquely determined by a map $f \colon \mathcal{O}_Y(-2L) \to \mathcal{O}_Y$, $f \in H^0(Y, 2L)$, and then X is isomorphic to the simple double cover $Y(\sqrt{f}, L) \to Y$.
[2 ⇒ 3] Clear, because $\mathrm{Fix}(\tau) = \{z = 0\}$ is the divisor of a section of π^*L.
[3 ⇒ 1] Let p be a fixed point of τ, then G acts on the local $\mathbb{C}$-algebra $B = \mathcal{O}_{X,p}$. Let us denote by $A = B^G$ the subring of invariant functions and by $I \subset B$ the stalk at p of the ideal sheaf of the closed subscheme $\mathrm{Fix}(\tau)$: by definition I is the ideal generated by all $\tau f - f$, $f \in B$.
Assume that I is a principal ideal; then there exists a generator h of I such that $\tau h = -h$. We claim that B is the free A-module generated by $1, h$. In fact if $a + bh = 0$, with $a, b \in A$, then $a - bh = 0$ which implies $a = bh = 0$. Since h is not a zero divisor $b = 0$. If $f \in B$ and $\tau f = -f$ then $f = (f - \tau f)/2 \in I$ and there exists $b \in B$ such that $f = bh$, $f = (\tau b)h$ and then $b = \tau b \in A$.
If X is smooth, by [1 ⇔ 3] it follows that π is flat if and only if τ has not isolated fixed point, i.e. if and only if Y is smooth. (recall that if Y is smooth then π is always flat). □

Definition 5.5 *Let $\pi \colon X \to Y$ be a flat double cover of normal surfaces. We have seen that $\pi_*\mathcal{O}_X = \mathcal{O}_Y \oplus \mathcal{O}_Y(-L)$ for some line bundle L and X is isomorphic to the subvariety $\{z^2 = f\} \subset L$. Let $R = \mathrm{div}(z) \subset X$ and $D = \mathrm{div}(f) \subset Y$. Note that $\pi^*D = 2R$. The pair L, D is called the* building data *of the flat double cover $X \to Y$.*

The building data determine the cover up to isomorphism.

Lemma 5.6 *In the notation of Definition 5.5:*

1. $\pi_*\mathcal{O}_R = \mathcal{O}_D$
2. *There exists natural isomorphisms*

$$\mathrm{Ext}^1_{\mathcal{O}_X}(\mathcal{O}_R(-R), \mathcal{O}_X) = H^0(X, \mathcal{O}_R(2R)) = H^0(Y, \mathcal{O}_D(D))$$

3. *There exists an exact sequence of $\mathcal{O}_X$-modules*

$$0 \longrightarrow \pi^*\Omega^1_Y \longrightarrow \Omega^1_X \longrightarrow \mathcal{O}_R(-R) \longrightarrow 0$$

Proof. Let $U \subset Y$ be an open affine subset, then

$$\pi_* \mathcal{O}_R(U) = \mathcal{O}_R(\pi^{-1}(U)) = \frac{\mathcal{O}_Y(U)[z]}{(z, z^2 - f)} = \frac{\mathcal{O}_Y(U)}{(f)} = \mathcal{O}_D(U).$$

Applying the functor $\mathcal{H}om_{\mathcal{O}_X}(-, \mathcal{O}_X)$ to the exact sequence

$$0 \longrightarrow \mathcal{O}_X(-2R) \longrightarrow \mathcal{O}_X(-R) \longrightarrow \mathcal{O}_R(-R) \longrightarrow 0$$

we get

$$0 = \mathcal{H}om_{\mathcal{O}_X}(\mathcal{O}_R(-R), \mathcal{O}_X) \to \mathcal{O}_X(R) \to \mathcal{O}_X(2R) \to \mathcal{E}xt^1_{\mathcal{O}_X}(\mathcal{O}_R(-R), \mathcal{O}_X) \to 0$$

and then $\mathcal{E}xt^1_{\mathcal{O}_X}(\mathcal{O}_R(-R), \mathcal{O}_X) = \mathcal{O}_R(2R)$. The Ext spectral sequence gives

$$\begin{aligned}
\mathrm{Ext}^1_{\mathcal{O}_X}(\mathcal{O}_R(-R), \mathcal{O}_X) &= H^0(\mathcal{E}xt^1_{\mathcal{O}_X}(\mathcal{O}_R(-R), \mathcal{O}_X)) = \\
&= H^0(X, \mathcal{O}_R(2R)) = H^0(Y, \pi_* \mathcal{O}_R(\pi^* D)) = H^0(Y, \mathcal{O}_D(D)).
\end{aligned}$$

Let $i \colon X \to L$ be the inclusion. Since $L \xrightarrow{\pi} Y$ is locally a product there exists an obvious inclusion of sheaves $\pi^{-1} \Omega^1_Y \subset i^{-1} \Omega^1_L$; tensoring with the flat module $\mathcal{O}_X$ we get an injection $\pi^* \Omega^1_Y \longrightarrow \Omega^1_L \otimes \mathcal{O}_X$.
The sheaf $\Omega^1_{L/Y}$ is locally free: it is the $\mathcal{O}_L$-dual of the sheaf of vertical vector fields and therefore it is naturally isomorphic to $\pi^*(-L)$.
We have the following standard exact sequences of differentials

$$0 \longrightarrow \pi^* \Omega^1_Y \longrightarrow \Omega^1_L \otimes \mathcal{O}_X \longrightarrow \Omega^1_{L/Y} \otimes \mathcal{O}_X = \mathcal{O}_X(-L) \longrightarrow 0,$$

$$0 \longrightarrow \mathcal{O}_X(-\pi^* D) = \mathcal{O}_X(-X) \longrightarrow \Omega^1_L \otimes \mathcal{O}_X \longrightarrow \Omega^1_X \longrightarrow 0$$

and (3) is obtained by applying the snake lemma to

$$\begin{array}{ccccccc}
0 \longrightarrow & \mathcal{O}_X(-\pi^* D) & \longrightarrow & \Omega^1_L \otimes \mathcal{O}_X & \longrightarrow & \Omega^1_X & \longrightarrow 0 \\
& \downarrow & & \downarrow & & & \\
& \mathcal{O}_X(-L) & = & \mathcal{O}_X(-L) & & &
\end{array}$$

□

If $\pi \colon X \to Y$ is a flat double cover of normal surfaces with building data L, D and ramification divisor $R \subset X$, then Hurwitz' formula gives $K_X = \pi^* K_Y + R$ and therefore $K_X = \pi^*(K_Y + L)$. In particular, assuming X, Y with at most rational double points as singularities, we get that K_X is ample if and only if $K_Y + L$ is ample.

5.3 Automorphisms of Generic Flat Double Covers

If $X \to X/\tau$, is a flat double cover then the group $\mathrm{Aut}(X)$ of biregular automorphisms of X in never trivial since it contains the involution τ. The aim of this section is to prove that if the building data are sufficiently ample and generic then $\mathrm{Aut}(X) = \{1, \tau\}$. We consider for simplicity only the case of surfaces, noticing that the same ideas can be used to prove similar results in any dimension.

Consider the following situation: Y a smooth projective surface, L a line bundle on Y such that $K_Y + L$ is ample and $V \subset \mathbb{P}(H^0(Y, 2L))$ a base point free linear system.

By Bertini's theorem the generic divisor $D \in V$ is smooth and then the double cover $X \to Y$ with building data L, D is a smooth surface of general type. Denote by $\mathrm{Aut}(\pi)$ the (finite) group of automorphisms of the morphism π, i.e. the group whose elements are the commutative diagrams

$$
\begin{array}{ccc}
X & \xrightarrow{g} & X \\
\downarrow{\scriptstyle \pi} & & \downarrow{\scriptstyle \pi} \\
Y & \xrightarrow{\bar{g}} & Y
\end{array}
$$

The involution of the cover τ belongs to the center of $\mathrm{Aut}(\pi)$, the homomorphism $\mathrm{Aut}(\pi) \to \mathrm{Aut}(X)$ is injective, the kernel of $\mathrm{Aut}(\pi) \to \mathrm{Aut}(Y)$ is the subgroup generated by τ and the image of $\mathrm{Aut}(\pi) \to \mathrm{Aut}(Y)$ is contained in the subgroup $\mathrm{Aut}(Y, D)$ of automorphisms $\bar{g}$ such that $\bar{g}(D) \subset D$.

Moreover $\mathrm{Aut}(\pi) \to \mathrm{Aut}(X)$ is bijective if and only if τ belongs to the center of $\mathrm{Aut}(X)$. The next theorem asserts that if D is sufficiently ample and generic then $\mathrm{Aut}(\pi) = \mathrm{Aut}(X)$.

Theorem 5.1. *In the notation above assume that:*

1. *$2L = L_1 + L_2 + L_3$, where $L_i^2 > 0$ for every i.*
2. *There exist base point free linear systems $V_i \subset \mathbb{P}(H^0(Y, L_i))$ such that $D_1 + D_2 + D_3 \in V$ for every $D_i \in V_i$.*
3. *There exists a finite group $G \subset \mathrm{Aut}(Y)$ such that $g(D) = D$ for every $g \in G$, $D \in V$ and $\mathrm{Aut}(Y, D) = G$ for generic D.*

Then for generic $D \in V$ we have $\mathrm{Aut}(\pi) = \mathrm{Aut}(X)$.

Proof. It is not restrictive to assume $\dim V_i = 2$ for every i, denote by $\phi_i \colon Y \to V_i^\vee$ the associated morphism.

There exists a Zariski open G-stable subset $U \subset Y$ such that:

1. G acts freely on U
2. Every ϕ_i has maximal rank on U

Let $p \in U - \cup_i \phi_i^{-1}(\phi_i(Y - U))$ be a fixed point and let $W \subset V$ the linear subsystem of divisors having multiplicity at least 3 at p. The assumption of the theorem implies that W is not empty and if D_0 is the generic member of W then D is smooth on $Y - U$, has an ordinary triple point at p and has at most ordinary double and triple points as singularities. In particular the group G acts freely in the (nonempty) set of ordinary triple points of D_0.

Extend D_0 to a family of divisors $D_t \in V$, $t \in \Delta \subset \mathbb{C}$, such that D_t is smooth and $\mathrm{Aut}(Y, D_t) = G$ for every $t \neq 0$: denote by $\mathcal{X} \to \Delta$ the flat family of surfaces such that X_t is the double cover of Y with building data L, D_t for every t. Note that X_t is smooth for $t \neq 0$ and X_0 has at most rational double points of type A_1 and D_4 as singularities.

Up to a possible base change and shrinking of Δ, there exists a group H of fibre preserving automorphisms of $\mathcal{X}$ such that $H \to \mathrm{Aut}(X_t)$ is injective for every t and an isomorphism for every $t \neq 0$. Given $g \in H$ consider $h = g\tau g^{-1}$. Let $x \in X_0$ be a singular point, then $x \in R_0$ and then $g(x) \in R_0$, $h(x) = x$. According to Lemma 2.7 $h\tau = \tau h$ and then h induces a family of automorphisms $\overline{h_t} \in \mathrm{Aut}(Y)$ such that $\overline{h_t} D_t = D_t$ for every t. By assumption $h_t \in G$ for every t and h_0 acts trivially on the singular points of D_0. This implies that $h_t = Id$ for every t, $h = \tau$ and then τ belongs to the center of H.quad $\square$

Corollary 5.7 *Let Y be a smooth surface of general type, L an ample line bundle on Y and $V \subset \mathbb{P}(H^0(Y, 2L))$ a linear system. If*

1. *$2L = L_1 + L_2 + L_3$, where $L_i^2 > 0$ for every i.*
2. *There exist base point free linear systems $V_i \subset \mathbb{P}(H^0(Y, L_i))$ such that $D_1 + D_2 + D_3 \in V$ for every $D_i \in V_i$.*
3. *V separates points.*

Then for generic $D \in V$ the group of automorphisms of the double cover $X \to Y$ of building data L, D is generated by the involution τ.

Proof. $\mathrm{Aut}(Y)$ is finite and since V separates points, for generic $D \in V$, $\mathrm{Aut}(Y, D) = 0$. We can apply Theorem 5.1 with $G = 0$. $\square$

5.4 Example: Automorphisms of Simple Iterated Double Covers

Definition 5.8 *A finite map between normal algebraic surfaces $p \colon X \to Y$ is called a* simple iterated double cover *associated to a sequence of line bundles $L_1, \ldots, L_n \in \mathrm{Pic}(Y)$ if the following conditions hold:*

1. *There exist $n+1$ normal surfaces $X = X_0, \ldots, X_n = Y$ and n flat double covers $\pi_i \colon X_{i-1} \to X_i$ such that $p = \pi_n \circ \ldots \circ \pi_1$.*
2. *If $p_i \colon X_i \to Y$ is the composition of π_j's, for $j > i$, then we have for every $i = 1, \ldots, n$ the eigensheaves decomposition $\pi_{i*}\mathcal{O}_{X_{i-1}} = \mathcal{O}_{X_i} \oplus p_i^*(-L_i)$.*

In the notation of the definition, we can embed the surface X in the vector bundle $V = L_1 \oplus \cdots \oplus L_n \xrightarrow{p} Y$ by the equations

$$z_i^2 = f_i \qquad i = 1, \ldots, n,$$

where every $z_i \colon V \to p^* L_i$ is the tautological section and $f_i \in H^0(X_i, p_i^* 2L_i)$, where X_i is the surface in $L_{i+1} \oplus \cdots \oplus L_n$ of equations $z_j^2 = f_j$, $j > i$, and π_i is the restriction to X_{i-1} of the natural projection $L_i \oplus \cdots \oplus L_n \to L_{i+1} \oplus \cdots \oplus L_n$. There exists a natural identification of vector spaces

$$H^0(X_i, p_i^* 2L_i) = \bigoplus_{h=0}^{n-i} \bigoplus_{\{j_1,\ldots,j_h\} \subset \{i+1,\ldots,n\}} z_{j_1} \ldots z_{j_h} H^0(Y, 2L_i - L_{j_1} - \ldots - L_{j_h})$$

and then X is defined in V by the equations

$$z_i^2 = \sum_{h=0}^{n-i} \sum_{\{j_1,\ldots,j_h\} \subset \{i+1,\ldots,n\}} z_{j_1} \ldots z_{j_h} g_{j_1,\ldots,j_h},$$

where $g_{j_1,\ldots,j_h} \in H^0(Y, 2L_i - L_{j_1} - \ldots - L_{j_h})$. According to Hurwitz' formula $K_X = p^*(K_Y + L_1 + \ldots + L_n)$.

Proposition 5.9 *Let Y be a smooth surface and $L_1, \ldots, L_n \in \mathrm{Pic}(Y)$ line bundles such that:*

1. $K_Y + L_2 + \ldots + L_n$ *is ample.*
2. *There exists a very ample line bundle H on Y such that $|2L_1 - 3H|$ is base point free.*
3. *The linear systems $|2L_i|$ and $|2L_1 - L_i|$ are base point free for every $i > 1$.*

Then the automorphism group $\mathrm{Aut}(X)$ of a generic simple iterated double cover $p \colon X \to Y$ associated to $L_1, \ldots, L_n$ is isomorphic to $\mathbb{Z}/2$.

Proof. Let X_1 be a generic iterated double cover of Y associated to the sequence $L_2, \ldots, L_n$; since every linear system $|2L_i|$ is base point free, X_1 is a smooth surface with ample canonical bundle.

According to Corollary 5.7 it is sufficient to show that the linear system $|p_1^* 2L_1|$ separates points on X_1. This is clear if $n = 1$, $X_1 = Y$; by induction we may assume that $|p_2^* 2L_1|$ separates points on X_2.

Let $x, y \in X_1$, if $\pi_2(x) \neq \pi_2(y) \in X_2$ then x and y are separated by $|p_1^* 2L_1|$. Otherwise $\pi_2(x) = \pi_2(y)$, $p_1(x) = p_1(y)$ and let $g \in H^0(Y, 2L_1 - L_2)$ such that $g(p_1(x)) \neq 0$. The section $z_2 g \in H^0(X_1, p_1^* 2L_1)$ takes different values in x, y. $\square$

5.5 Flat Multidouble Covers

Here we illustrate the structure of flat abelian covers with group $(\mathbb{Z}/2)^r$ for any $r > 0$. Unless otherwise stated, from now on we always assume $G = (\mathbb{Z}/2)^r$.

It is notationally convenient to consider the group G as a vector space of dimension r over $\mathbb{Z}/2$; there exists a natural isomorphism between the dual vector space $G^\vee$ and the group of characters G^*

$$G^\vee \ni \chi \quad \leftrightarrow \quad (-1)^\chi \in G^*.$$

Definition 5.10 *Given any abelian group Λ, a pair of maps $D: G \to \Lambda$, $L: G^\vee \to \Lambda$ satisfies the* cover condition *if:*

1. $D_0 = L_0 = 0$.
2. *For every $\chi, \eta \in G^\vee$,*

$$L_\chi + L_\eta = L_{\chi+\eta} + \sum_{\chi(\sigma)=\eta(\sigma)=1} D_\sigma.$$

Let $\pi: X \to Y$ be a normal flat G-cover. For every $\sigma \in G - \{0\}$ let $\mathrm{Fix}(\sigma) \subset X$ be the (possibly empty) closed subscheme of fixed points of the involution σ.

For $\sigma \neq 0$ we denote by R_σ the Weil divisorial part of $\mathrm{Fix}(\sigma)$; by convention we set $R_0 = \emptyset$.

We denote $D_\sigma = \pi(R_\sigma) \subset Y$ with the reduced structure.

According to Cartan's Lemma, R_σ is smooth at every smooth point of X and then R_σ is reduced: every R_σ is G-stable and if $\tau \neq \sigma$ then, according to Lemma 2.6, R_σ and R_τ have no common components.

Notice that $\pi^* D_\sigma = 2R_\sigma$ and the Weil divisors R_σ and D_σ are not Cartier in general.

As in the general case, the locally free sheaf $\pi_* \mathcal{O}_X$ admits a character decomposition

$$\pi_* \mathcal{O}_X = \bigoplus_{\chi \in G^\vee} \mathcal{O}_Y(-L_\chi),$$

where $\mathcal{O}_Y(-L_\chi) = \{f \in \pi_* \mathcal{O}_X \mid \sigma f = (-1)^{\chi(\sigma)} f, \forall \sigma \in G\}$. Since $Y = X/G$ we have $\mathcal{O}_Y(-L_0) = \mathcal{O}_Y$.

The line bundles L_χ and the divisors D_σ form a pair $(L, D) \in \mathrm{Pic}(Y)^{G^\vee} \oplus \mathrm{Div}(Y)^G$ which is called the *building data* of the cover $\pi: X \to Y$.

For every pair $\chi, \eta \in G^\vee$, let $\beta_{\chi,\eta}: L_\chi^{-1} \otimes L_\eta^{-1} \to L_{\chi+\eta}^{-1}$ be the multiplication map; we can interpret $\beta_{\chi,\eta}$ as a section of the line bundle $L_\chi \otimes L_\eta \otimes L_{\chi+\eta}^{-1}$. Since the finite morphism $X \to Y$ is uniquely determined by the $\mathcal{O}_Y$-algebra $\pi_* \mathcal{O}_X$, X is isomorphic to the subvariety of the global space of the vector bundle $\pi: V = \bigoplus_{\chi \in G^\vee} L_\chi \to Y$ defined by the equations

$$w_\chi w_\eta = \beta_{\chi,\eta} w_{\chi+\eta}, \qquad \chi, \eta \in G^\vee,$$

where $w_\chi \in H^0(V, \pi^* L_\chi)$ is the tautological section.

A local computation (see [21], [110, proof of Thm. 2.1]) shows that

$$\mathrm{div}(\beta_{\chi,\eta}) = \sum_{\chi(\sigma)=\eta(\sigma)=1} D_\sigma.$$

Therefore, over the open set $U \subset Y$ of regular points the divisors D_σ are Cartier and the pair $(L, \mathcal{O}_U(D)) \in \mathrm{Pic}(U)^{G^\vee} \oplus \mathrm{Pic}(U)^G$ satisfies the cover condition 5.10.

Proposition 5.11 *Let Y be a complete normal variety and denote by $U \subset Y$ the open subset of smooth points. Then every flat normal G-cover $X \to Y$ is uniquely determined, up to G-isomorphism, by the normal variety Y and by the building data (L, D).*
Conversely, given line bundles $L_\chi \in \mathrm{Pic}(Y)$, $\chi \in G^\vee$, and reduced effective Weil divisors $D_\sigma \in \mathrm{Div}(Y)$, $\sigma \in G$, without common components such that the pair $(L, \mathcal{O}_U(D))$ satisfies the cover condition in $\mathrm{Pic}(U)$, then there exists a normal flat G-cover $X \to Y$ with building data (L, D).

Proof. This is Theorem 2.1 of [110], being the hypothesis Y smooth used in [110] easily relaxed to Y normal. □

It is interesting to recall how to construct a G-cover with preassigned building data (L_χ, D_σ); for simplicity we assume that the divisors D_σ are Cartier, pointing out that this construction works for every variety Y.
For every $\sigma \in G$ let $f_\sigma \in \mathcal{O}_Y(D_\sigma)$ be a section defining D_σ. The cover condition implies that there exist isomorphisms of line bundles

$$L_\chi \otimes L_\eta = L_{\chi+\eta} \otimes \mathcal{O}_X\Big(\sum_{\chi(\sigma)=\eta(\sigma)=1} D_\sigma \Big), \qquad \forall \chi, \eta \in G^\vee.$$

The morphisms

$$\prod_{\chi(\sigma)=\eta(\sigma)=1} f_\sigma \colon \mathcal{O}_Y(-L_\chi) \otimes \mathcal{O}_Y(-L_\eta) \to \mathcal{O}_Y(-L_{\chi+\eta})$$

induce a structure of commutative associative $\mathcal{O}_Y$-algebra on $\mathcal{A} = \bigoplus_{\chi \in G^\vee} \mathcal{O}_Y(-L_\chi)$ and then we can take X as the spectrum over Y of $\mathcal{A}$. In more concrete terms, X is the subvariety of $\pi \colon V = \bigoplus_{\chi \in G^\vee} L_\chi \to Y$ defined by the equations

$$w_\chi w_\eta - w_{\chi+\eta} \prod_{\chi(\sigma)=\eta(\sigma)=1} f_\sigma = 0, \qquad \forall \chi, \eta \in G^\vee,$$

where $w_\chi \in H^0(V, \pi^* L_\chi)$ is the tautological section.

Definition 5.12 *Denote:*

1. *For every $x \in X$, $J_x = \{\sigma \neq 0 \mid x \in R_\sigma\}$.*
2. *$J = \cup J_x = \{\sigma \neq 0 \mid R_\sigma \neq \emptyset\}$*

Lemma 5.13 *For every $x \in X$, J_x generates the stabilizer of x. In particular $\cup R_\sigma$ is the locus where π is ramified.*

Proof. It is clear that J_x is contained in the stabilizer of the point $x \in X$. Conversely let $H_x \subset G$ be the subspace generated by J_x; let $\sigma \notin H_x$ be an involution and choose $\chi \in G^\vee$ such that $\chi(\sigma) = 1$ and $\chi(\tau) = 0$ for every $\tau \in J_x$.
Since $w_\chi^2 = \prod_{\chi(\tau)=1} f_\tau$ we have $w_\chi(x) \neq 0$ and then $w_\chi(\sigma(x)) = -w_\chi(x) \neq w_\chi(x)$. □

Definition 5.14 *In the above notation the cover* $\pi\colon X \to Y$ *is called:*

1. Totally ramified: *if J generates G*
2. Simple: *if J is a set of linearly independent vectors of G*
3. Locally simple: *if J_x is a set of linearly independent vectors for every $x \in X$*

Remark 5.15 *Let* $\pi\colon X \to Y$ *be a normal flat* $(\mathbb{Z}/2)^r$*-cover with building data* L_χ, D_σ. *By Hurwitz formula* $K_X = \pi^* K_Y + \sum_\sigma R_\sigma$ *and then* $2K_X = \pi^*(2K_Y + \sum_\sigma D_\sigma)$. *If Y is Cohen–Macaulay then X is Cohen–Macaulay (cf. [97, §23]) and, if Y is smooth, then X is* $\mathbb{Q}$*-Gorenstein.*
If Y is a smooth surface, then X has at most rational double points as singularities and the line bundle $2K_Y + \sum_\sigma D_\sigma$ *is ample then also* K_X *is ample and then X is the canonical model of a surface of general type.*

Remark 5.16 *If* π *is locally simple then the divisors* D_σ *are Cartier; in fact if* $J_x = \{\sigma_1, \dots, \sigma_s\}$ *is a set of linearly independent vectors then there exists* $\chi_1, \dots, \chi_s \in G^\vee$ *such that* $\chi_i(\sigma_j) = \delta_{ij}$ *and therefore, by the cover condition,* $2L_{\chi_i} = \mathcal{O}(D_{\sigma_i})$ *in some Zariski neighbourhood of* $y = \pi(x)$.
For a deeper study of the local properties of abelian covers we refer to the paper of D. Iacono [65].

6 Stability Criteria for Flat Double Covers

The aim of this lecture is to illustrate some of the main ideas used in the paper [91, 93, 95, 96] in order to produce examples of connected components of the moduli space of surfaces of general type $\mathcal{M}$. As a working example we consider the (a, b, c)-surfaces.

Definition 6.1 *Consider an ordered triple* (a, b, c) *of positive integers. An* (a, b, c)*-surface is a surface that is a simple iterated double cover of* $\mathbb{P}^1 \times \mathbb{P}^1$ *associated to the sequence of line bundle* $L_1 = \mathcal{O}(a, b)$ *and* $L_2 = \mathcal{O}(c, b)$.

By definition of simple iterated double cover, every (a, b, c)-surface X is defined in the global space of $L_1 \oplus L_2$ by the equations (see Sect. 5.4).

$$\begin{cases} z_1^2 = f + z_2 h \\ z_2^2 = g \end{cases}$$

where f, g and h are bihomogeneous polynomials of respective bidegrees $(2a, 2b)$, $(2c, 2b)$ and $(2a - c, b)$.

Theorem 6.1. *For every triple* a, b, c *of integers greater than or equal to 4, denote by* $N(a, b, c) \subset \mathcal{M}$ *the subset of canonical surfaces that are* (a, b, c)*-surfaces.*

1. If $a > 2c$, $a \geq b + 2$, $b \geq c + 2$ and b, c are even, then $N(a, b, c)$ is a connected component of the moduli space.
2. If (a, b, c) and (a', b', c') both satisfy the condition (1), then $N(a, b, c) = N(a', b', c')$ if and only if $(a, b, c) = (a', b', c')$.

Denote by $\pi \colon X \to X_1 \to \mathbb{P}^1 \times \mathbb{P}^1$ the projection, where X_1 is the surface of equation $z_2^2 = g$.

We have $K_X = \pi^*(K_Y + L_1 + L_2) = \pi^* \mathcal{O}(a + c - 2, 2b - 2)$ and then $K_X^2 = 8(2b - 2)(a + c - 2)$.

Since $\pi_* \mathcal{O}_X = \mathcal{O}(0, 0) \oplus \mathcal{O}(-c, -b) \oplus \mathcal{O}(-a, -b) \oplus \mathcal{O}(-a - c, -2b)$ we have $\chi(\mathcal{O}_X) = \chi(\pi_* \mathcal{O}_X) = 1 + (c - 1)(b - 1) + (a - 1)(b - 1) + (a + c - 1)(2b - 1) = (a + c)(3b - 2) - 4b + 4$. We note in particular that the numerical invariants of X depend uniquely by $a + c$ and b.

According to the results of Sect. 5.4, if $a, b, c \geq 4$ and $2a > c$, then for the generic $X \in N(a, b, c)$ we have $X / \operatorname{Aut}(X) = X_1$ and therefore the numerical invariants of X_1 are determined by the subset $N(a, b, c) \subset \mathcal{M}$. Since the integer c, b are determined, up to the order, by the numerical invariants K^2 and χ of the surface X_1, the proof of item 2 of Theorem 6.1 follows easily.

The proof of Item 1 is, as expected, more complicated and requires several preliminary results, many of them we consider of independent interest. Denote by $H(a, b, c)$ the (Zariski open) subset of triples

$$(f, g, h) \in H^0(\mathcal{O}(2a, 2b)) \oplus H^0(\mathcal{O}(2c, 2b)) \oplus H^0(\mathcal{O}(2a - c, 2b))$$

such that the surface X of equation

$$z_1^2 = f + z_2 h, \qquad z_2^2 = g$$

is an irreducible surface with at most rational double points as singularities. There exists a natural flat family of canonical surfaces $\mathcal{X} \to H(a, b, c)$ and $N(a, b, c)$ is the image of the induced morphisms $H(a, b, c) \to \mathcal{M}$.

Our goal is to apply the general criteria of Lemma 3.3 and Lemma 3.4 to prove that, under the assumption of the Theorem 6.1, the irreducible subset $N(a, b, c)$ is open and closed in the moduli space.

6.1 Restricted Natural Deformations of Double Covers

Let $\pi \colon X \to Y$ be a flat double cover, say $X = \{z^2 = f\} \subset L$ for some line bundle $L \to Y$ and some section $f \in H^0(Y, 2L)$. Denote by $\tau : z \mapsto -z$ the involution of the cover.

Definition 6.2 A restricted natural deformation of X is a deformation obtained by a small perturbation of f in $H^0(Y, 2L)$.

A Galois deformation of X is a deformation where the action of the involution of the cover τ extends.

Clearly every restricted natural deformation is Galois, while the converse is generally false.

It is clear from the definition that the space of first order Galois deformations is the τ-invariant subspace $\operatorname{Ext}^1_X(\Omega^1_X, \mathcal{O}_X)^\tau$. The τ-equivariant exact sequence

$$0 \longrightarrow \pi^* \Omega^1_Y \longrightarrow \Omega^1_X \longrightarrow \mathcal{O}_R(-R) \longrightarrow 0$$

induces an exact sequence

$$\operatorname{Ext}^1_{\mathcal{O}_X}(\mathcal{O}_R(-R), \mathcal{O}_X) \xrightarrow{\ \epsilon\ } \operatorname{Ext}^1_X(\Omega^1_X, \mathcal{O}_X) \longrightarrow \operatorname{Ext}^1_X(\pi^* \Omega^1_Y, \mathcal{O}_X).$$

According to Lemma 5.6 and Lemma 5.2, the above exact sequence is isomorphic to

$$H^0(Y, \mathcal{O}_D(D)) \xrightarrow{\ \epsilon\ } \operatorname{Ext}^1_X(\Omega^1_X, \mathcal{O}_X) \longrightarrow \operatorname{Ext}^1_Y(\Omega^1_Y, \mathcal{O}_Y \oplus \mathcal{O}_Y(-L)).$$

Taking the τ-invariant part we obtain

$$H^0(Y, \mathcal{O}_D(D)) \xrightarrow{\ \epsilon\ } \operatorname{Ext}^1_X(\Omega^1_X, \mathcal{O}_X)^\tau \longrightarrow \operatorname{Ext}^1_Y(\Omega^1_Y, \mathcal{O}_Y).$$

Lemma 6.3 *In the notation above*

1. *If $\operatorname{Ext}^1_Y(\Omega^1_Y, \mathcal{O}_Y(-L)) = 0$ then every first order deformations is Galois.*
2. *If $H^1(\mathcal{O}_Y) = 0$ then the image of ϵ is the vector space of first order restricted natural deformations.*

Proof (Sketch of proof). The first item is clear. The computation made in [93, p. 751] show that the space of first order restricted natural deformations is the image of the composite morphism

$$H^0(\mathcal{O}_Y(2L)) \to H^0(\mathcal{O}_D(D)) \xrightarrow{\ \epsilon\ } \operatorname{Ext}^1_X(\Omega^1_X, \mathcal{O}_X)$$

and the first morphism is surjective whenever $H^1(\mathcal{O}_Y) = 0$. $\square$

We are now ready to prove the following

Theorem 6.2. *In the above notation let $\mathcal{X} \to \mathcal{Y} \to H$ be a deformation of the map π parametrized by a smooth germ $(H, 0)$ and let $r_X: (H, 0) \to \operatorname{Def}(X)$, $r_Y: (H, 0) \to \operatorname{Def}(Y)$ be the induced maps. Assume:*

1. *r_Y is smooth (and then $\operatorname{Def}(Y)$ is smooth)*
2. *The image of r_X contains the restricted natural deformations*
3. *$\operatorname{Ext}^1_{\mathcal{O}_Y}(\Omega^1_Y, -L) = 0$, $H^1(\mathcal{O}_Y) = 0$*

Then r_X is smooth.

Proof. The differential of the morphisms r_X and r_Y fit into the commutative diagram

$$T_0 H \xrightarrow{\quad dr_X \quad} \mathrm{Ext}^1_{\mathcal{O}_X}(\Omega^1_X, \mathcal{O}_X)$$

$$\Big\downarrow dr_Y \qquad\qquad \Big\downarrow \alpha$$

$$\mathrm{Ext}^1_{\mathcal{O}_Y}(\Omega^1_Y, \mathcal{O}_Y) \xrightarrow{\Phi^{-1}\beta} \mathrm{Ext}^1_{\mathcal{O}_X}(\pi^*\Omega^1_Y, \mathcal{O}_X)$$

where

$$\beta \colon \mathrm{Ext}^1_{\mathcal{O}_Y}(\Omega^1_Y, \mathcal{O}_Y) \to \mathrm{Ext}^1_{\mathcal{O}_Y}(\Omega^1_Y, \mathcal{O}_Y \oplus \mathcal{O}_Y(-L))$$

is the inclusion and

$$\Phi \colon \mathrm{Ext}^1_{\mathcal{O}_X}(\pi^*\Omega^1_Y, \mathcal{O}_X) \to \mathrm{Ext}^1_{\mathcal{O}_Y}(\Omega^1_Y, \mathcal{O}_Y \oplus \mathcal{O}_Y(-L))$$

is the natural isomorphism.

The assumption $\mathrm{Ext}^1_{\mathcal{O}_Y}(\Omega^1_Y, -L) = 0$ implies that $\Phi^{-1}\beta$ is an isomorphism. Since $H^1(\mathcal{O}_Y) = 0$, the kernel of α is the set of restricted natural deformations and by assumption 2 it is contained in the image of dr_X. It is now easy to observe that dr_Y surjective implies dr_X surjective and, since H is smooth, this is sufficient to ensure that r_X is smooth and $\dim \mathrm{Def}(X) = \dim \mathrm{Def}(Y) + \dim \mathrm{Im}\,\epsilon$. $\square$

Example 6.4 Let $X \to \mathbb{P}^2$ be a normal flat double cover branched over a curve $D \subset \mathbb{P}^2$ of degree $\neq 6$. Then the family of restricted natural deformations of X is complete. In fact $\mathbb{P}^2$ is rigid and from Euler exact sequence we get

$$\mathrm{Ext}^1_{\mathbb{P}^2}(\Omega^1_{\mathbb{P}^2}, \mathcal{O}_{\mathbb{P}^2}(-h)) = H^1(\mathbb{P}^2, T(-h)) = 0, \qquad \text{for } h \neq 3,$$

$$\mathrm{Ext}^1_{\mathbb{P}^2}(\Omega^1_{\mathbb{P}^2}, \mathcal{O}_{\mathbb{P}^2}(-3)) = H^1(\mathbb{P}^2, T(-3)) = \mathbb{C}.$$

Example 6.5 Denoting by $Q = \mathbb{P}^1 \times \mathbb{P}^1$ we have

$$\mathrm{Ext}^1_Q(\Omega^1_Q, \mathcal{O}_Q(-a, -b)) = \max(0, (a-3)(1-b)) + \max(0, (b-3)(1-a)),$$

and then $\mathrm{Ext}^1_Q(\Omega^1_Q, \mathcal{O}_Q(-a, -b)) = 0$ whenever $a, b \geq 3$.

In fact $\mathrm{Ext}^1_Q(\Omega^1_Q, \mathcal{O}_Q(-a, -b)) = H^1(T_Q(-a, -b))$, where $T_Q = \mathcal{O}_Q(2, 0) \oplus \mathcal{O}_Q(0, 2)$ is the tangent bundle. For the reader's comfort we recall that for every pair of integers a, b we have

$$\chi = \chi(\mathcal{O}_Q(a, b)) = (a+1)(b+1) = \begin{cases} h^0(\mathcal{O}_Q(a, b)) & \text{if } \chi \geq 0, a + b \geq -2, \\ -h^1(\mathcal{O}_Q(a, b)) & \text{if } \chi \leq 0, \\ h^2(\mathcal{O}_Q(a, b)) & \text{if } \chi \geq 0, a + b \leq -2. \end{cases}$$

Let $X \to Q$ be a normal flat double cover branched over a curve $D \subset Q$ of bidegree $(2a, 2b)$. If $a, b \geq 3$, then the family of restricted natural deformations of X is complete. In fact Q is rigid and $\mathrm{Ext}^1_Q(\Omega^1_Q, \mathcal{O}_Q(-a, -b)) = 0$.

6.2 Openess of $N(a, b, c)$

The notion of restricted natural deformations extends without substantial changes to simple iterated double covers. For example if $p\colon X \xrightarrow{\pi_1} X_1 \xrightarrow{\pi_2} Y$ is a simple iterated double cover associated to $L_1, L_2 \in \mathrm{Pic}(Y)$ and embedded in $L_1 \oplus L_2$ by equations

$$\begin{cases} z_1^2 = f + z_2 h \\ z_2^2 = g, \end{cases}$$

then the restricted natural deformations of $X \to Y$ are obtained by taking small perturbations of f, g, h inside the vector spaces $H^0(Y, 2L_1)$, $H^0(Y, 2L_2)$ and $H^0(Y, 2L_1 - L_2)$ respectively.

Since $H^0(X_1, \pi_2^* 2L_1) = H^0(Y, 2L_1) \oplus H^0(Y, 2L_1 - L_2)$, the restricted natural deformations of the iterated double cover induce the restricted natural deformations of both the flat double covers $X_1 \to Y$ and $X \to X_1$. It is therefore natural to look for inductive applications of Theorem 6.2.

For instance, if

1. $\mathrm{Ext}_Y^1(\Omega_Y^1, \mathcal{O}_Y) = 0$
2. $H^1(\mathcal{O}_Y) = \mathrm{Ext}_Y^1(\Omega_Y^1, -L_2) = 0$

then the restricted natural deformations induce a complete family of deformations of X_1. If moreover

3. $H^1(Y, -L_1) = H^1(Y, -L_2) = 0$
4. $H^0(Y, 2L_2 - L_1) = 0$
5. $\mathrm{Ext}_Y^1(\Omega_Y^1, -L_1) = \mathrm{Ext}_Y^1(\Omega_Y^1, -L_1 - L_2) = 0$

then the restricted natural deformations induce a complete family of deformations of X. In fact applying $\mathrm{Hom}_{X_1}(\ , -L_1)$ to the exact sequence

$$0 \longrightarrow \pi_2^* \Omega_Y^1 \longrightarrow \Omega_{X_1}^1 \longrightarrow \mathcal{O}_{R_2}(-R_2) \longrightarrow 0$$

and setting $D_2 = \{g = 0\} \subset Y$ we get

$$H^0(\mathcal{O}_{D_2}(2L_2 - L_1)) = \mathrm{Ext}_{X_1}^1(\mathcal{O}_{R_2}(-R_2), -L_1) \longrightarrow \mathrm{Ext}_{X_1}^1(\Omega_{X_1}^1, -L_1) \longrightarrow$$
$$\longrightarrow \mathrm{Ext}_{X_1}^1(\pi_2^* \Omega_Y^1, -L_1) = \mathrm{Ext}_Y^1(\Omega_Y^1, -L_1) \oplus \mathrm{Ext}_Y^1(\Omega_Y^1, -L_1 - L_2)$$

and the vanishing of the vector space on the left can be deduced from the exact sequence

$$H^0(Y, 2L_2 - L_1) \longrightarrow H^0(\mathcal{O}_{D_2}(2L_2 - L_1)) \longrightarrow H^1(Y, -L_1).$$

If $a, b, c \geq 3$ and $a > 2c$, then the two line bundles $L_1 = \mathcal{O}_Q(a, b)$ and $L_2 = \mathcal{O}_Q(c, b)$ satisfy the above conditions and therefore, according to Lemma 3.3, $N(a, b, c)$ is open in the moduli space.

The same argument was used in [93] to prove the following theorem:

Theorem 6.3. *Let $L_i = \mathcal{O}_Q(a_i, b_i)$, for $i = 1, \ldots, n$, be a sequence of line bundle on the quadric $Q = \mathbb{P}^1 \times \mathbb{P}^1$ and let $X \to Q$ be a simple iterated double cover associated to $L_1, \ldots, L_n$.*
If $a_i, b_i \geq 3$ for every i and $\min(2a_i - a_j, 2b_i - b_j) < 0$ for every $j < i$, then the restricted natural deformations induce a complete family of deformations of X.
In particular the set of canonical surfaces that are simple iterated double covers of Q associated to $L_1, \ldots, L_n$ is an irreducible open subset of the moduli space.

6.3 RDP-Degenerations of Double Covers

By an RDP-degeneration of surfaces we intend a proper flat family $f \colon \mathcal{X} \to \Delta = \{t \in \mathbb{C} \mid |t| < 1\}$ with the property that every fibre $X_t = f^{-1}(t)$ is a normal surface with at most RDP as singularities.

With the term *RDP-degeneration of double covers* we intend an RDP-degeneration as above endowed of an involution $\tau \colon \mathcal{X} \to \mathcal{X}$ preserving every fibre of f. Moreover, if $\pi \colon \mathcal{X} \to \mathcal{Y} = \mathcal{X}/\tau$ is the projection to quotient, we assume that $\pi_t \colon X_t \to Y_t$ is a flat double cover for every $t \neq 0$.

In general $\pi_0 \colon X_0 \to Y_0$ is not flat; in this section we prove the following theorem which gives a sufficient condition for the flatness of π_0.

Theorem 6.4. *In the above situation suppose that:*

1. *X_t and Y_t are smooth surfaces for every $t \neq 0$.*
2. *The divisibility of the canonical class of Y_t is even for every $t \neq 0$.*

Then Y_0 has at most RDPs and the map $\pi \colon \mathcal{X} \to \mathcal{Y}$ is flat.

Theorem 6.4 follows easily from the following proposition.

Proposition 6.6 *Let $f \colon (\mathcal{X}, 0) \to (\mathbb{C}, 0)$ be a smoothing of a RDP $(X_0, 0)$ and let $f' \colon (\mathcal{Y}, 0) \to (\mathbb{C}, 0)$ be the quotient of $(\mathcal{X}, 0)$ by an involution τ preserving f.*
Assume that $(\mathcal{Y}, 0)$ is a smoothing of the normal singularity $(Y_0, 0)$ and let $F_t \subset Y_t$ be the associated Milnor fibre. Then either one of the following possibilities holds:

1. *$(Y_0, 0)$ is a rational double point and the quotient projection $\pi \colon (\mathcal{X}, 0) \to (\mathcal{Y}, 0)$ is flat.*

2. *$(Y_0, 0)$ is cyclic quotient of type $\dfrac{1}{2d+1}(1, 2d-1)$ and the intersection form on $H_2(F_t, \mathbb{Z})$ is odd and negative definite.*

3. *f' is a $\mathbb{Q}$-Gorenstein smoothing of the cyclic singularity of type $\dfrac{1}{4d}(1, 2d-1)$, the torsion subgroup of $H^2(F_t, \mathbb{Z})$ has order 2 and is generated by the canonical class.*

We now prove Theorem 6.4 as a consequence of Proposition 6.6. It is enough to prove that the map $\mathcal{Y} \to \Delta$ cannot be locally of type (2) or (3) described in the Proposition. Let $p \in \mathcal{Y}$ be a singular point: $(\mathcal{Y}, p)$ cannot be of type (2) above since the inclusion $F_t \subset Y_t$ induces an isometry $H_2(F_t, \mathbb{Z}) \to H_2(Y_t, \mathbb{Z})$ with respect the intersection forms and the intersection form of Y_t is even by Wu's formula.

If $(\mathcal{Y}, p)$ is of type (3) above and if $r \colon H^2(Y_t, \mathbb{Z}) \to H^2(F_t, \mathbb{Z})$ is the natural restriction then $r(c_1(K_Y))$ generates the torsion subgroup of $H^2(F_t, \mathbb{Z})$ which is $\mathbb{Z}/2\mathbb{Z}$ but this gives a contradiction since $c_1(K_X)$ is 2-divisible.

Example 6.7 *The condition $r(Y_t)$ even is essential in order to have Y_0 with at most rational double points. Consider for instance $\mathcal{X} \subset \mathbb{P}^3 \times \Delta$ the hypersurface of equation $x_1 x_2 + x_3^2 + t x_0^2 = 0$ and $\tau \colon \mathcal{X} \to \mathcal{X}$ the involution $\tau \colon x_0 \mapsto -x_0$. For $t \neq 0$ we have $Y_t = \mathbb{P}^2$ and $X_t \to Y_t$ is the simple double cover branched over a smooth conic, while Y_0 is the cone over the nondegenerate rational curve of degree 4 in $\mathbb{P}^4$ and has a cyclic singularity of type $(1, 1)/4$ at its vertex.*

Our strategy of proof of Proposition 6.6 divides in two steps. The first step is the classification of all conjugacy classes of involutions acting on a rational double point; this computation is already done by Catanese and the result is illustrated in the next two tables.

Table 1. Equations of RDP's in $\mathbb{C}^3$

E_8	$z^2 + x^3 + y^5 = 0$
E_7	$z^2 + x(y^3 + x^2) = 0$
E_6	$z^2 + x^3 + y^4 = 0$
$D_n, n \geq 4$	$z^2 + x(y^2 + x^{n-2}) = 0$
A_n	$z^2 + x^2 + y^{n+1} = 0$ or $uv + y^{n+1} = 0$
Smooth	$x = 0$

Table 2. ([23, Th. 2.1]) Conjugacy classes of involutions acting on the RDP's defined as in Table 1

(a)	$y \to -y$	E_6, D_n, A_{2n+1}
(b)	$y \to -y, z \to -z$	Smooth, E_6, D_n, A_{2n+1}
(c)	$(u, v, y) \to (-u, v, -y)$	A_{2n}
(d)	$x \to -x, z \to -z$	A_n
(e)	$(u, v, y) \to (-u, -v, -y)$	A_{2n+1}
(f)	$z \to -z$	All RDP's

Corollary 6.8 *Let $X \to Y$ be a flat double cover of normal surfaces. If X is smooth then Y is smooth. If X has at most RDP's then Y has at most RDP's.*

Proof. According to Table 2 the only involutions whose fixed locus is a Cartier divisor are the ones of types (a) and (f). □

The second step in the proof of Proposition 6.6 is to give a (very rough) classification of the smoothing of the involutions of Table 2 according to the following definition.

Definition 6.9 *Let $(X_0, 0)$ be a singularity and $g_0 \in Aut(X_0, 0)$. A smoothing of g_0 is the data of a smoothing $(\mathcal{X}, 0) \xrightarrow{\ t\ } (\mathbb{C}, 0)$ of $(X_0, 0)$ and an automorphism g of $(\mathcal{X}, 0)$ preserving the map t such that g_0 is the restriction of g to X_0 and the quotient $(\mathcal{Y}, 0) = (\mathcal{X}/g, 0) \xrightarrow{\ t\ } (\mathbb{C}, 0)$ is a smoothing of (X_0/g_0).*

The following Cartan-type Lemma will be very useful for our purposes.

Lemma 6.10 *Let $(\mathcal{X}, 0) \xrightarrow{\ t\ } (\mathbb{C}, 0)$ be a morphism of germs of analytic singularities and let $G \subset Aut(\mathcal{X}, 0)$ be a finite subgroup leaving the morphism t G-invariant.*
Assume that G acts linearly on a finite dimensional $\mathbb{C}$-vector space V and let $i_0 : (X_0, 0) \to (V, 0)$ be an analytic G-embedding.
Then there exists a G-embedding $i : (\mathcal{X}, 0) \to (V \times \mathbb{C}, 0)$ extending i_0 and such that $t = p \circ i$ where p is the projection on the second factor.

Proof (Sketch of proof). We can assume without loss of generality that V is G-isomorphic to $(m_0/m_0^2)^\vee$, the Zariski tangent space of X_0 at 0.
Let $z_1, \ldots, z_n$ be a basis of $V^\vee$, the map i_0 gives a G-equivariant morphism of analytic algebras $i_0^* : \mathbb{C}\{z_1, \ldots, z_n\} \to \mathcal{O}_{X_0}$.
Since G is finite there exists a G-lifting of i_0^*, say $\eta^* : \mathbb{C}\{z_1, \ldots, z_n\} \to \mathcal{O}_{\mathcal{X}}$.
The map $i : (\mathcal{X}, 0) \to (V \times \mathbb{C}, 0)$, induced by the morphism of analytic algebras

$$i^* : \mathbb{C}\{z_1, \ldots, z_n, t\} \to \mathcal{O}_{\mathcal{X}}, \qquad i^*(t) = t, \ i^*(z_i) = \eta^*(z_i),$$

is the desired embedding. □

Lemma 6.11 *The involutions of type (b) and (d) are not smoothable.*

Proof. There are several cases to investigate, here we made only a particular case for illustrating the idea, for the other cases the proof is similar. Let $X_0 = D_n$ and τ be the involution of type (b); assume moreover that the action of τ extends to a smoothing $(\mathcal{X}, 0) \xrightarrow{\ t\ } (\mathbb{C}, 0)$. By Lemma 6.10 we can assume that $(\mathcal{X}, 0)$ is defined in $\mathbb{C}^4$ by the equation

$$z^2 + x(y^2 + x^{n-2}) + t\varphi(x, y, z, t) = 0$$

$\tau(x, y, z, t) = (x, -y, -z, t)$ and φ is τ-invariant.
The fixed locus of τ is the germ of curve of equation $x^{n-1} + t\varphi(x, 0, 0, t)$ contained in the plane $y = z = 0$ and then for $|t| \ll 1$ τ has a finite number of fixed points on X_t and then the quotient X_t/τ is singular. □

Lemma 6.12 *Let* $(\mathcal{X}, 0) \xrightarrow{\ t\ } (\mathbb{C}, 0)$ *be a smoothing of a RDP and let* τ *be an involution of* $(\mathcal{X}, 0)$ *preserving* t. *If* $\tau_{|X_0}$ *is of type (a) or (f) then* X_0/τ *is a RDP and the projection to* $(Y, 0) = (X/\tau, 0)$ *is flat.*

Proof. In case (a), according to Lemma 6.10, we can assume $(\mathcal{X}, 0) \subset (\mathbb{C}^4, 0)$ defined by the equation

$$f(x, y^2, z) + t\varphi(x, y^2, z, t) = 0$$

and $\tau(x, y, z, t) = (x, -y, z, t)$. Thus the equation of $(\mathcal{Y}, 0)$ is

$$f(x, s, z) + t\varphi(x, s, z, t) = 0$$

and $(\mathcal{X}, 0)$ is defined in $(\mathcal{Y} \times \mathbb{C}_y, 0)$ by the equation $y^2 = s$. The case of involution of case (f) is similar. $\square$

Proof (Proof of Proposition 6.6.). By Lemma 6.11 the restriction of τ to X_0 can be only of type (a),(c),(e) or (f). In cases (a) and (f), by Lemma 6.12 the first item of 6.6 holds.

In case (c) by [23, Thms. 2.4 and 3.1] the second case of 6.6 holds.

In case (e) $(\mathcal{Y}, 0)$ is a smoothing of a cyclic singularity of type $\frac{1}{4d}(1, 2d-1)$ [23, Th. 2.5]. Since $\mathcal{Y} - \{0\}$ is smooth τ must act freely on $\mathcal{X} - \{0\}$ and then $\mathcal{Y}$ is $\mathbb{Q}$-Gorenstein of order 2. The statement about the Milnor fibre is proved in Proposition 4.14. $\square$

6.4 RDP-Degenerations of $\mathbb{P}^1 \times \mathbb{P}^1$

We need first to recall some facts about the Segre-Hirzebruch surfaces $\mathbb{F}_d$, for $d \geq 0$ [12, p. 140].

By definition $\mathbb{F}_d \to \mathbb{P}^1$ is the projectivization of the rank 2 vector bundle $p \colon \mathcal{O}(0) \oplus \mathcal{O}(d) \to \mathbb{P}^1$. Clearly $\mathbb{F}_0 = \mathbb{P}^1 \times \mathbb{P}^1$, while for $d > 0$ the surface $\mathbb{F}_d$ contains an unique irreducible curve σ_∞ with negative selfintersection. Moreover σ_∞ is a section of p and $\sigma_\infty^2 = -d$. There exists a section σ_0, unique up to linear equivalence such that $\sigma_0^2 = d$ and $\sigma_0 \cap \sigma_\infty = \emptyset$. If F denotes the fibre of p, then the divisors F and σ_0 are free generators of $\mathrm{Pic}(\mathbb{F}_d)$; the divisor σ_∞ is linearly equivalent to $\sigma_0 - dF$.

If an effective divisor $D \subset \mathbb{F}_d$ is linearly equivalent to $a\sigma_0 + bF$, then $a \geq 0$ and $ad + b \geq 0$. If $b < 0$, then D contains σ_∞ as a component, while if $b < -d$ then D is not reduced since it contains σ_∞ with multiplicity ≥ 2.

Proposition 6.13 *Let* $\mathcal{Y} \to \Delta$ *be a proper flat family of surfaces such that* Y_t *is isomorphic to* $\mathbb{P}^1 \times \mathbb{P}^1$ *for every* $t \neq 0$ *and* Y_0 *has at most RDP as singularities. Then either one of the following situations hold:*

1. *Y_0 is a surface $\mathbb{F}_{2k}$ for some $k \geq 0$ and the pair of line bundles $\mathcal{O}(\sigma_0), \mathcal{O}(F)$ deforms to either $\mathcal{O}(1, k), \mathcal{O}(0, 1)$ or $\mathcal{O}(k, 1), \mathcal{O}(1, 0)$ for $t \neq 0$.*

2. Y_0 is the singular irreducible quadric in $\mathbb{P}^3$, every effective Cartier divisor on Y_0 is linearly equivalent to a multiple of the hyperplane bundle $\mathcal{O}(1)$ and $\mathcal{O}(1)$ deforms to $\mathcal{O}(1,1)$ for $t \neq 0$.

Proof. The case Y_0 smooth is well known (see, e.g. [12, VI.7.1]) and corresponds to case 1.

Assume therefore Y_0 singular and let $\delta \colon Z_0 \to Y_0$ be its minimal resolution; according to Brieskorn-Tyurina simultaneous resolution $Z_0 = \mathbb{F}_{2k}$ for some $k \geq 0$ and contains irreducible rational curves with selfintersection -2. This forces $k = 1$, Y_0 is the singular irreducible quadric and its Picard group is generated by $\mathcal{O}(1) = \delta_* \mathcal{O}(\sigma_0)$. $\square$

6.5 Proof of Theorem 6.1

Proposition 6.14 *Let $f \colon \mathcal{X} \to \Delta$ be a flat family of canonical surfaces; denote $X_t = f^{-1}(t)$ and $\mathcal{X}^* = \mathcal{X} - X_0$. Assume that X_t is smooth for every $t \neq 0$ and that there exists an involution $\tau \colon \mathcal{X}^* \to \mathcal{X}^*$ commuting with f. Assume moreover that $X_t/\tau = \mathbb{P}^1 \times \mathbb{P}^1$ for every $t \neq 0$ and the projection $X_t \to \mathbb{P}^1 \times \mathbb{P}^1$ is isomorphic to a flat double cover ramified over a curve of bidegree $(2c, 2b)$ with $b \geq c + 2$.*

Then X_0 is isomorphic to a flat double cover of a Segre-Hirzebruch surface $\mathbb{F}_{2k}$ for some $k \leq b/(c-1)$.

Proof. By Theorem 2.5 the involution τ extends to an involution of $\mathcal{X}$; denote by $\mathcal{Y} = \mathcal{X}/\tau$ its quotient and by $\pi \colon \mathcal{X} \to \mathcal{Y}$ the projection map. By assumption $Y_t = X_t/\tau$ is isomorphic to $\mathbb{P}^1 \times \mathbb{P}^1$.

According to Theorem 6.4 the surface Y_0 has at most RDP's and the map $\pi \colon \mathcal{X} \to \mathcal{Y}$ is a flat double cover. We have $\pi_* \mathcal{O}_{\mathcal{X}} = \mathcal{O}_{\mathcal{Y}} \oplus \mathcal{L}$ for some line bundle $\mathcal{L}$ such that $L_t = \mathcal{L}_{|Y_t}$ is isomorphic to $g^* \mathcal{O}(c, b)$, for some isomorphism $g \colon Y_t \to \mathbb{P}^1 \times \mathbb{P}^1$.

If Y_0 is singular then it is irreducible singular quadric in $\mathbb{P}^3$ and then the restriction of $\mathcal{L}$ to Y_0 is a multiple of the hyperplane section and then $L_t = \mathcal{O}_{Y_t}(n, n)$ contrary to the assumption.

If Y_0 is smooth then it is isomorphic to $\mathbb{F}_{2k}$ and the branching divisor $D \subset Y_0$ of the flat double cover $X_0 \to Y_0$ extends to a divisor in $\mathbb{P}^1 \times \mathbb{P}^1$ of bidegree $(2c, 2b)$. Thus necessarily D is linearly equivalent to either

1. $2c\sigma_0 + (2b - 2ck)F$ or
2. $2b\sigma_0 + (2c - 2bk)F$.

Since D is reduced and $b \geq c + 2$ it must be $2b - 2ck \geq -2k$ and then $k(c-1) \leq b$. $\square$

We are now ready to prove that the set $N(a, b, c)$ is closed in the moduli space provided that $a > b > c \geq 4$ and b, c are even integers. Denote by $N \subset H(a, b, c)$ the (Zariski open) set of triples

$(f, g, h) \in H(a, b, c) \subset H^0(\mathcal{O}(2a, 2b)) \oplus H^0(\mathcal{O}(2c, 2b)) \oplus H^0(\mathcal{O}(2a - c, 2b))$

such that the bidouble cover $X \to \mathbb{P}^1 \times \mathbb{P}^1$ of equation

$$\begin{cases} z_1^2 = f + z_2 h \\ z_2^2 = g \end{cases}$$

is smooth.

According to Lemma 3.4 we have to show that if $f \colon \mathcal{X} \to \Delta$ is a proper flat family of canonical surfaces such that $\mathcal{X}^* \to \Delta^* = \Delta - \{0\}$ is a family of smooth simple iterated double covers of $\mathbb{P}^1 \times \mathbb{P}^1$ induced by a morphism $\Delta^* \to N$, then also X_0 is a simple iterated double cover associated to $L_1 = \mathcal{O}(a, b), L_2 = \mathcal{O}(c, b)$.

The involution acting of $\mathcal{X}^*$ extends to an involution of $\mathcal{X}$, denote by $\pi_1 \colon \mathcal{X} \to \mathcal{Z}$ the quotient map. For every $t \neq 0$ the surface Z_t is a smooth double cover of $\mathbb{P}^1 \times \mathbb{P}^1$ branched over a curve of bidegree $(2c, 2b)$. In particular K_{Z_t} is the pull back of the line bundle

$$K_{\mathbb{P}^1 \times \mathbb{P}^1} + \mathcal{O}(c, b) = \mathcal{O}(c - 2, b - 2) = 2\mathcal{O}(c/2 - 1, b/2 - 1)$$

and then the canonical divisor of Z_t is 2-divisible. This implies that Z_0 is a surface with at most rational double points and the morphism π_1 is flat.

Denote by $W \to \Delta$ the relative canonical model of $\mathcal{Z}$, we want to prove that the canonical morphism $\delta \colon \mathcal{Z} \to W$ is an isomorphism.

By normality of Z_0 and W_0 the fibres of δ are connected. Assume that there exists an irreducible curve $C \subset Z_0$ contracted by δ and let $D \subset X_0$ be the strict transform of C.

Since π_1 is flat we have $\pi_{1*}\mathcal{O}_{\mathcal{X}} = \mathcal{O}_{\mathcal{Z}} \oplus \mathcal{M}$ for a line bundle $\mathcal{M}$ such that for $t \neq 0$ $\mathcal{M}_t = \delta^* \pi_2^* \mathcal{O}(c, b)$. Since $H^1(\mathcal{O}_{\mathcal{Z}}) = 0$, the Picard group of $\mathcal{Z}$ is discrete and then $\mathcal{M} = \delta^* \pi_2^* \mathcal{O}(c, b)$ and $C \cdot \mathcal{M} = 0$.

Using adjunction formula $K_{X_0} = \pi_1^*(K_{Z_0} + \mathcal{M}_0)$ and then $D \cdot K_{X_0} = 0$ which is impossible since K_{X_0} is ample.

Thus $\mathcal{Z}$ is a family of canonical surfaces and then Z_0 is a flat double cover of a Segre-Hirzebruch surface $\mathbb{F}_{2k}$ branched over a reduced divisor linearly equivalent to $2(c\sigma_0 + (b - 2kc)F)$. The surface X_0 is a simple iterated double cover of $\mathbb{F}_{2k}$ associated to the line bundles $L_1 = a\sigma_0 + (b - ka)F$, $L_2 = c\sigma_0 + (b - kc)F$; the equation of X_0 are

$$z_1^2 = f + z_2 h, \qquad z_2^2 = g$$

and, since X_0 is normal, at least one of the sections f, h vanishes along σ_∞ with multiplicity ≤ 1. This implies that $2b - 2ka \geq -2k$ or $b - k(2a - c) \geq -2k$. Since $a > b + 1 > c + 2$, both inequalities implies $k = 0$.

6.6 Moduli of Simple Iterated Double Covers

We have already noted that for every integer $d \leq c - 4$ the surfaces belonging to $N(a, b, c)$ and $N(a + d, b, c - d)$ have the same numerical invariant K^2, χ and the results of [21] imply also that they are homeomorphic, simply connected and with the same divisibility of the canonical class.

As a consequence of Theorem 6.1 we have that the number of connected components of $\mathcal{M}_{K^2,\chi}$ can be arbitrarily large. Using the same ideas used in this section, it is proved in [93] the following theorem:

Theorem 6.15 *For any sequence* $L_1,\dots,L_n \in \text{Pic}(\mathbb{P}^1 \times \mathbb{P}^1)$ *define* $N(L_1,\dots,L_n)$ *as the image in the moduli space of the set of surfaces of general type whose canonical model is a simple iterated double cover of* $\mathbb{P}^1 \times \mathbb{P}^1$ *associated to* $L_1,\dots,L_n$. *Assume that* $L_i = \mathcal{O}(a_i,b_i)$ *and*

1. $a_i, b_i \geq 3$ *for every* $i = 1,\dots,n$
2. $\min(2a_i - a_j, 2b_i - b_j) < 0$ *for every* $j < i$
3. $a_n \geq b_n + 2$ *and* $b_{n-1} \geq a_{n-1} + 2$
4. a_i, b_i *are even for* $i = 2,\dots,n$
5. $2a_i - a_{i+1} \geq 2$, $2b_i - b_{i+1} \geq 2$ *for every* $i < n$

Then $N(L_1,\dots,L_n)$ *is a nonempty connected component of the moduli space,* $N(L_1,\dots,L_n)$ *is reduced, irreducible and unirational and if* $M_1,\dots,M_m$ *is another sequence satisfying conditions (1),...,(5) and* $N(L_1,\dots,L_n) = N(M_1,\dots,M_m)$ *then* $n = m$ *and* $L_i = M_i$ *for every* $i = 1,\dots,n$.

Another similar result, proved in [95] is the following,

Theorem 6.16 *For any sequence* $L_1,\dots,L_n \in \text{Pic}(\mathbb{P}^2)$ *define* $N(L_1,\dots,L_n)$ *as the image in the moduli space of the set of surfaces of general type whose canonical model is a simple iterated double cover of* $\mathbb{P}^2$ *associated to* $L_1,\dots,L_n$. *Assume that* $L_i = \mathcal{O}(l_i)$ *and write* $\overline{N(L_1,\dots,L_n)}$ *for the closure of* $N(L_1,\dots,L_n)$ *in the moduli space:*

1. *If* $l_n \geq 4$ *and* $l_i > 2l_{i+1}$ *for* $i = 1,\dots,n-1$ *then* $N(L_1,\dots,L_n)$ *is an open subset of the moduli space.*
2. *If* $l_n \geq 5$ *is odd,* l_i *is even and* $l_i > 2l_{i+1}$ *for every* $i = 1,\dots,n-1$ *then* $\overline{N(L_1,\dots,L_n)}$ *is a connected component of the moduli space* $\mathcal{M}$.
3. *If* $M_1,\dots,M_m$ *is another sequence of line bundles satisfying conditions (1) and (2) and* $\overline{N(L_1,\dots,L_n)} = \overline{N(M_1,\dots,M_m)}$ *then* $n = m$ *and* $L_i = M_i$ *for every* $i = 1,\dots,n$.

Theorems 6.15 and 6.16 both imply, with some easy computation (see, e.g. [93]) that does not exist any polynomial $P(K^2,\chi)$ which bounds the number of connected components of $\mathcal{M}_{K^2,\chi}$.

Acknowledgements

I thank Fabrizio Catanese, Gang Tian and the Fondazione C.I.M.E. for their kind invitation to speak in the Session "Symplectic 4-Manifolds And Algebraic Surfaces".

References

1. V. Alexeev: *Moduli spaces $M_{g,n}(W)$ for surfaces.* In: (M. Andreatta and T. Peternell eds.) Proceedings of the international conference, Trento 1994, Walter de Gruyter (1996).

2. A. Andreotti: *Sopra le superficie algebriche che posseggono trasformazioni birazionali in sé.* Rendiconti di Matematica e delle sue applicazioni **V, IX** Roma (1950).

3. E. Arbarello, M. Cornalba, P. Griffiths, J. Harris: *Geometry of algebraic curves, I.* Springer-Verlag (1984).

4. M. Artin: *Deformations of singularities.* Tata Institute of Fundamental Research, Bombay (1976).

5. M. Artin: *Some numerical criteria for contractibility of curves on algebraic surfaces.* Am. J. Math. **84** (1962) 485–496.

6. M. Artin: *On isolated rational singularities of surfaces.* Am. J. Math. **88** (1966) 129–136.

7. M. Artin: *Algebraic approximation of structures over complete local ring.* Publ. Math. IHES **36** (1969) 23–58.

8. M. Artin: *Algebraic construction of Brieskorn's resolutions.* J. Algebra **29** (1974) 330–348.

9. M. Atiyah: *On analytic surfaces with double points.* Proc. Roy. Soc. A **247** (1958) 237–244.

10. L. Badescu: *Suprafete Algebrice.* Editura Academiei Republicii Socialiste Romania (1981). English translation *Algebraic surfaces.* Springer-Verlag Universitext (2001).

11. C. Banica, O. Stanasila: *Méthodes algébrique dans la théorie globale des espaces complexes.* Gauthier-Villars (1977).

12. W. Barth, C. Peters, A. van de Ven: *Compact complex surfaces.* Springer-Verlag, Berlin (1984).

13. W. Barth, K. Hulek, C. Peters, A. van de Ven: *Compact complex surfaces. Second edition.* Springer-Verlag, Berlin (2004).

14. A. Beauville: *Complex algebraic surfaces.* London Math. Soc. LNS **68** (1983).

15. E. Bombieri: *Canonical models of surfaces of general type.* IHES Publ. Sci. **42** (1973) 171–219.

16. F. Bonahon: *Difféotopies des espaces lenticulaires.* Topology **22** (1983) 305–314.

17. E. Brieskorn: *Rationale Singularitäten komplexer Flächen.* Invent. Math. **4** (1967/1968) 336–358.

18. D. Burns, J. Wahl: *Local contribution to global deformations of surfaces.* Invent. Math. **26** (1974) 67–88.

19. H. Cartan: *Quotient d'un espace analytique par un groupe d'automorphismes.* Algebraic geometry and topology: A symposium in honour of S. Lefschetz. Princeton Math. Series **12** (1957) 90–102.

20. F. Catanese: *Moduli of surfaces of general type.* In: *Algebraic geometry: open problems. Proc. Ravello 1982* Springer L.N.M. **997** (1983) 90–112.

21. F. Catanese: *On the moduli spaces of surfaces of general type.* J. Differential Geom. **19** (1984) 483–515.

22. F. Catanese: *Connected components of moduli spaces.* J. Differ. Geom. **24** (1986) 395–399.

23. F. Catanese: *Automorphisms of rational double points and moduli spaces of surfaces of general type.* Compositio Math. **61** (1987) 81–102.

24. F. Catanese: *Moduli of algebraic surfaces.* In C.I.M.E. course *Theory of muduli, Montecatini Terme, 1985* Springer-Verlag LNM **1337** (1988) 1–83.

25. F. Catanese: *Chow varieties, Hilbert schemes and moduli spaces of surfaces of general type.* J. Alg. Geometry **1** (1992) 561–595.

26. F. Catanese: *private communication* (1992).

27. F. Catanese: *(Some) Old and new results on algebraic surfaces.* Proc. I European congress of Math. Paris 1992, Birkhauser (1994).

28. F. Catanese: *Singular bidouble covers and the construction of interesting algebraic surfaces.* in *Algebraic geometry: Hirzebruch 70 (Warsaw, 1998)*, 97–120, Contemp. Math. **241**, Am. Math. Soc., Providence, RI, (1999).

29. F. Catanese: *Moduli spaces of surfaces and real structures.* Ann. of Math. **158** (2003) 577–592.

30. F. Catanese: *Symplectic structures of algebraic surfaces and deformation.* arXiv:math/0207254v1 [math.AG].

31. F. Catanese, B. Wajnryb: *Diffeomorphism of simply connected algebraic surfaces.* J. Differential Geom. **76** (2007), no. 2, 177–213.

32. C. Chevalley: *Invariants of finite groups generated by reflections.* Am. J. Math. **77** (1955), 778-782.

33. C. Ciliberto, E. Sernesi: *Families of varieties and the Hilbert scheme.* Lecture Notes (1990).

34. A. Corti: *Polynomial bounds for automorphisms of surfaces of general type.* Ann. E.N.S. IV, **24** (1991) 113–137.

35. S.K. Donaldson: *Gauge theory and four-manifold topology.* Proc. I Eureopean congress of Math. Paris 1992, Birkhauser (1994).

36. S.K. Donaldson, P.B. Kronheimer: *The geometry of four-manifolds.* Oxford University Press (1990).

37. S. Donaldson: *The Seiberg-Witten equations and 4-manifold topology.* Bull. A.M.S. **33** (1996) 45–70.

38. A. Douady, J.L. Verdier: *Séminaire de géometrie analytique.* Astérisque **16–17** Soc. Math. France (1976).

39. A. Durfee: *Fifteen characterizations of rational double points and simple critical points.* Ens. Math. **25** (1979) 131–163.

40. P. Du Val: *On isolated singularities of surfaces which do not affect the conditions of adjunction.* Proc. Camb. Phil. Soc. **30** 453–459.

41. C. Ehresmann: *Sur les espaces fibrés differentiables.* C. R. Acad. Sci. **224** (1947) 1611–1612.

42. R. Elkik: *Singularites rationelles et deformations.* Invent. Math. **47** (1978) 139–147.

43. H. Esnault, E. Viehweg: *Two dimensional quotient singularities deform to quotient singularities.* Math. Ann. **271** (1985) 439–449.

44. B. Fantechi, M. Manetti: *Obstruction calculus for functors of Artin rings, I.* J. Algebra **202** (1998) 541–576.

45. B. Fantechi, R. Pardini: *Automorphism and moduli spaces of varieties with ample canonical class via deformations of abelian covers.* Comm. in Algebra **25** (1997) 1413–1441.

46. R. Fintushel, R. Stern: *Rational Blowdowns of smooth 4-manifolds.* J. Differ. Geom. **46** (1997) 181–235.

47. G. Fischer: *Complex analytic geometry.* Springer-Verlag LNM **538** (1976).

48. H. Flenner: *Über Deformationen holomorpher Abbildungen.* Habilitationsschrift, Osnabruck 1978.

49. O. Forster, K. Knorr: *Konstruction verseller Familien kompakter komplexer Raüme.* Springer Lect. Notes Math. **705** (1979).

50. M. Freedman: *The topology of four-dimensional manifolds.* J. Diff. Geometry **17** (1982) 357–454.

51. R. Friedman: *Donaldson and Seiberg-Witten invariants of algebraic surfaces.* Algebraic geometry-Santa Cruz 1995, 85–100, Proc. Sympos. Pure Math., **62**, Part 1, Am. Math. Soc., Providence, RI, (1997).

52. R. Friedman, B. Moishezon, J.W. Morgan: *On the C^∞ invariance of the canonical classes of certain algebraic surfaces.* Bull. Am. Math. Soc. (N.S.) **17** (1987) 283–286.

53. R. Friedman, J.W. Morgan: *Algebraic surfaces and four-manifolds: some conjectures and speculations.* Bull. Am. Math. Soc. (N.S.) **18** (1988) 1–19.

54. R. Friedman, J.W. Morgan: *Smooth four-manifolds and complex surfaces.* Ergebnisse der Mathematik **27** Springer-Verlag (1994).

55. R. Friedman, J.W. Morgan: *Algebraic surfaces and Seiberg-Witten invariants.* J. Algebraic Geom. **6** (1997), no. 3, 445–479.

56. D. Gieseker: *Global moduli for surfaces of general type.* Invent. Math. **43** (1977) 233–282.

57. D. Gieseker: *Geometric invariant theory and applications to moduli problems.* Springer lectures notes **996**, Ed. Gherardelli (1982) 45–73.

58. H. Grauert: *Der satz von Kuranishi für kompakte komplexe Raüme.* Invent. Math. **25** (1974) 107–142.

59. G.M. Greuel, U. Karras: *Families of varieties with prescribed singularities.* Comp. Math. **69** (1989) 83–110.

60. G.M. Greuel, J. Steenbrink: *On the topology of smoothable singularities.* Proc. Symp. Pure Math. **40** part 1 (1983) A.M.S. 535–545.

61. A. Grothendieck: *Local Cohomology.* Springer L.N.M. **41** (1967).

62. R. Hartshorne: *Algebraic geometry.* Springer Verlag GTM **52** (1977).

63. M.W. Hirsch: *Differential Topology.* Springer-Verlag GTM **33** (1976).

64. E. Horikawa: *Deformations of holomorphic maps I.* J. Math. Soc. Japan **25** (1973) 372–396; II J. Math. Soc. Japan **26** (1974) 647–667; III Math. Ann. **222** (1976) 275-282.

65. D. Iacono: *Local structure of abelian covers.* J. Algebra **301** (2006), no. 2, 601–615.

66. D. Iacono: *Differential Graded Lie Algebras and Deformations of Holomorphic Maps.* Tesi di Dottorato, Roma (2006). arXiv:math/0701091v1 [math.AG].

67. S. Iitaka: *Deformations of compact complex surfaces II.* J. Math. Soc. Jpn. **22** (1970) 247–261.

68. S. Iitaka: *On D-dimension of algebraic varieties.* J. Math. Soc. Jpn. **23** (1971).

69. V.M. Kharlamov, V.S. Kulikov: *On real structures of real surfaces.* Izv. Ross. Akad. Nauk Ser. Mat. **66** no. 1, (2002), 133–152; translation in Izv. Math. **66** no. 1, (2002), 133–150.

70. A. Kas: *Ordinary double points and obstructed surfaces.* Topology **16** (1977) 51–64.

71. K. Kodaira: *On stability of compact submanifolds of complex manifolds.* Am. J. Math. **85** (1963) 79–94.

72. K. Kodaira: *Complex manifold and deformation of complex structures.* Springer-Verlag (1986).

73. K. Kodaira, L. Nirenberg, D.C. Spencer: *On the existence of deformations of complex analytic structures.* Ann. Math. **68** (1958) 450–459.

74. K. Kodaira, D.C. Spencer: *On the variation of almost complex structures.* In *Algebraic geometry and topology,* Princeton Univ. Press (1957) 139–150.

75. K. Kodaira, D.C. Spencer: *A theorem of completeness for complex analytic fibre spaces.* Acta Math. **100** (1958) 281–294.

76. K. Kodaira, D.C. Spencer: *On deformations of complex analytic structures, I-II, III.* Ann. Math. **67** (1958) 328–466; **71** (1960) 43–76.

77. J. Kollár, N.I. Shepherd-Barron: *Threefolds and deformations of surface singularities.* Invent. Math. **91** (1988) 299–338.

78. M. Kuranishi: *New Proof for the existence of locally complete families of complex structures.* In: Proc. Conf. Complex Analysis (Minneapolis 1964) Springer-Verlag (1965) 142–154.

79. H.B. Laufer: *Taut Two-Dimensional Singularities.* Math. Ann. **205** (1973) 131–164.

80. H.B. Laufer: *Normal twodimensional singularities.* Princeton University press (1971).

81. F. Lazzeri: *Analytic singularities.* C.I.M.E. 1974, editore Cremonese, Roma.

82. Y. Lee, J. Park: *A simply connected surface of general type with $p_g = 0$ and $K^2 = 2$.* Invent. Math. **170** (2007), no. 3, 483–505.

83. A.S. Libgober, J.W. Wood: *Differentiable structures on complete intersections I,II.* I Topology **21** (1982) 469–482. II Proc. Symp. Pure Math. **40** part 2, (1983) 123–133.

84. J. Lipman: *Introduction to resolution of singularities.* Proc. Symp. Pure Math. **29** (Arcata 1974) 187–230.

85. J. Lipman: *Double point resolutions of deformations of rational singularities.* Comp. Math. **38** (1079) 37–43.

86. E.J.N. Looijenga: *Riemann-Roch and smoothings of singularities.* Topology **25** (1986) 293–302.

87. E.J.N. Looijenga: *Isolated singular points on complete intersection.* London Math. Soc. L.N.S. **77** (1983).

88. E.J.N. Looijenga, J. Wahl: *Quadratic functions and smoothing surface singularities.* Topology **25** (1986) 261–291.

89. M. Manetti: $\mathbb{Q}$-*Gorenstein smoothings of quotient singularities.* Preprint Scuola Normale Superiore Pisa (1990). Electronic version available at author's URL.

90. M. Manetti: *Normal degenerations of the complex projective plane.* J. Reine Angew. Math. **419** (1991) 89–118.

91. M. Manetti: *On Some Components of Moduli Space of Surfaces of General Type.* Comp. Math. **92** (1994) 285–297.

92. M. Manetti: *Degenerations of algebraic surfaces and applications to moduli problems.* Tesi di Perfezionamento, Scuola Normale Superiore, Pisa (1995).

93. M. Manetti: *Iterated double covers and connected components of moduli spaces.* Topology **36** (1996) 745-764.

94. M. Manetti: *Automorphisms of generic cyclic covers.* Revista Matemática de la Universidad Complutense de Madrid **10** (1997) 149–156.

95. M. Manetti: *Degenerate double covers of the projective plane.* In: (Catanese, Hulek, Peters and Reid eds.) *New trends in algebraic geometry.* Cambridge Univ. Press (1999).

96. M. Manetti: *On the moduli space of diffeomorphic algebraic surfaces.* Invent. Math. **143** (2001) 29–76.
97. H. Matsumura: *Commutative Ring Theory.* Cambridge University Press (1986).
98. H. Matsumura: *On algebraic groups of birational transformations.* Rend. Accad. Lincei Ser. 8 **34** (1963) 151–155.
99. J. Milnor: *Morse theory.* Ann. Math. Studies **51**, Princeton Univ. Press, Princeton (1963).
100. J. Milnor: *Singular points of complex hypersurfaces.* Ann. Math. Studies **61**, Princeton Univ. Press, Princeton (1968).
101. J. Milnor, J. Stasheff: *Characteristic classes.* Ann. Math. Studies **76**, Princeton Univ. Press, Princeton (1974).
102. B.G. Moishezon: *Analogs of Lefschetz Theorems for linear systems with isolated singularities.* J. Diff. Geometry **31** (1990) 47–72.
103. D. Mumford: *The topology of normal singularities of an algebraic surface and a criterion for simplicity.* Publ. Math. IHES **9** (1961) 5–22.
104. D. Mumford: *Abelian varieties.* Oxford Univ. Press (1970).
105. M. Namba: *Families of meromorphic functions on compact Riemann surfaces.* Lecture Notes in Mathematics **767** Springer-Verlag, New York/Berlin, (1979).
106. P.E. Newstead: *Introduction to moduli problems and orbit spaces.* Tata Institute Fundamenta Research (1978).
107. C. Okonek, A. Teleman: *Les invariants de Seiberg-Witten et la conjecture de van de Ven.* C. R. Acad. Sci. Paris Sér. I Math. **321** (1995), no. 4, 457–461.
108. V.P. Palamodov: *Deformations of complex spaces.* Uspekhi Mat. Nauk. **31:3** (1976) 129–194. Transl. Russian Math. Surveys **31:3** (1976) 129–197.
109. V.P. Palamodov: *Deformations of complex spaces.* In: *Several complex variables IV.* Encyclopaedia of Mathematical Sciences **10**, Springer-Verlag (1986) 105–194.
110. R. Pardini: *Abelian covers of algebraic varieties.* J. Reine Angew. Math. **417** (1991) 191–213.
111. U. Persson: *An introduction to the geography of surfaces of general type.* Proc. Sympos. Pure Math. **46** (1987) 195–218.
112. H. Pinkham: *Singularités rationelles des surfaces (avec un appendice).* Lectures Notes in Math. **777** Springer Verlag (1980) 147–178.
113. H. Pinkham: *Some local obstructions to deforming global surfaces.* Nova acta Leopoldina NF **52** Nr. 240 (1981) 173–178.
114. D. Prill: *Local classification of quotients of complex manifolds by discontinuous groups.* Duke Math. J. **34** (1967) 375–386.
115. R. Răsdeaconu, I. Şuvaina: *The algebraic rational blow-down.* arXiv:math/0601270v1 [math.SG].
116. M. Reid: *Young person's guide to canonical singularities.* Proc. Symp. Pure Math. **46** part 1 (1987) A.M.S. 345–414.
117. O. Riemenschneider: *Deformationen von quotientensingularitäten (nach zyklischen gruppen).* Math. Ann. **209** (1974) 211–248.
118. M. Salvetti: *On the number of non-equivalent differentiable structures on 4-manifolds.* Man. Math. **63** (1989) 157–171.
119. M. Salvetti: *A lower bound for the number of differentiable structures on 4-manifolds.* Bollettino U.M.I. **(7) 5-A** (1991) 33–40.
120. M. Schlessinger: *Functors of Artin rings.* Trans. Am. Math. Soc. **130** (1968) 208–222.

121. M. Schlessinger: *Rigidity of quotient singularities*. Invent. Math. **14** (1971) 17–26.

122. C.S. Seshadri: *Quotient spaces modulo reductive algebraic groups*. Ann. of Math. (2) **95** (1972), 511–556; errata, ibid. (2) 96 (1972), 599.

123. J.P. Serre: *Cours d'arithmétique*. P.U.F. Paris (1970).

124. J. Steenbrink: *Mixed Hodge structures associated with isolated singularities*. Proc. Symp. Pure Math. **40** part 2 (1983) A.M.S. 513–536.

125. S.G. Tankeev: *On a global theory of moduli of algebraic surfaces of general type*. Math. USSR Izvestija **6** (1972) 1200–1216.

126. B. Teissier: *The hunting invariants in the geometry of discriminants*. In: (P. Holm ed.) *Real and complex singularities*. Oslo (1976) 565–678.

127. A.N. Tyurin: *Six lectures on four manifolds*. In: C.I.M.E. lectures *Transcendental methods in algebraic geometry, 1994*. Springer-Verlag LNM **1646** (1996).

128. G.N. Tyurina: *On the tautness of rationally contractible curves on a surface*. Math. USSR Izvestija **2** (1968) 907–934.

129. G.N. Tyurina: *Resolutions of singularities of plane deformations of double rational points*. Funk. Anal. i. Priloz. **4** (1970) 77–83.

130. E. Viehweg: *Quasi-projective moduli for polarized manifolds*. Springer-Verlag Ergebnisse Math. Grenz. **30** (1995).

131. J. Wahl: *Smoothings of normal surface singularities*. Topology **20** (1981) 219–246.

132. J. Wahl: *Elliptic deformations of minimally elliptic singularities*. Math. Ann. **253** (1980) 241–262.

133. J. Wahl: *Simultaneous resolution of rational singularities*. Comp. Math. **38** (1979) 43–54.

134. J. Wahl: *Equisingular deformations of normal surface singularities*. Ann. Math. **104** (1976) 325–356.

135. C.T.C. Wall: *On the orthogonal groups of unimodular quadratic forms*. Math. Ann. **147** (1962) 328–338.

136. J.J. Wavrik: *Obstruction to the existence of a space of moduli*. In *Global Analysis* Princeton University Press (1969), 403–414.

137. E. Witten: *Monopoles and four-manifolds*. Math. Res. Lett. **1** (1994) 769–796.

138. O. Zariski: *Sur la normalité analytique des variétés normales*. Ann. Inst. Fourier **2** (1950) 161–164.

Lectures on Four-Dimensional Dehn Twists

Paul Seidel

Department of Mathematics, 5734 S. University Avenue, Chicago, IL 60637, USA
seidel@math.uchicago.edu

1 Introduction

Let M be a closed symplectic manifold, with symplectic form ω. A symplectic automorphism is a diffeomorphism $\phi : M \to M$ such that $\phi^*\omega = \omega$. We equip the group $Aut(M) = Aut(M, \omega)$ of all such maps with the C^∞-topology. Like the whole diffeomorphism group, this is an infinite-dimensional Lie group in a very loose sense: it has a well-defined Lie algebra, which consists of closed one-forms on M, but the exponential map is not locally onto. We will be looking at the homotopy type of $Aut(M)$, and in particular the *symplectic mapping class group* $\pi_0(Aut(M))$.

Remark 1.1 *If $H^1(M; \mathbb{R}) \neq 0$, the C^∞-topology is is many respects not the right one, and should be replaced by the* Hamiltonian topology, *denoted by $Aut^h(M)$. This is defined by taking a basis of neighbourhoods of the identity to be the symplectic automorphisms generated by time-dependent Hamiltonians $H : [0; 1] \times M \to \mathbb{R}$ with $||H||_{C^k} < \epsilon$ for some k, ϵ. A smooth isotopy is continuous in the Hamiltonian topology iff it is Hamiltonian. The relation between $\pi_0(Aut(M))$ and $\pi_0(Aut^h(M))$ is determined by the image of the flux homomorphism, which we do not discuss since it is thoroughly covered elsewhere [28]. In fact, for simplicity we will mostly use $Aut(M)$, even when this restricts us to manifolds with $H^1(M; \mathbb{R}) = 0$ (if this irks, see Remark 3.12).*

When M is two-dimensional, Moser's lemma tells us that $Diff^+(M)$ retracts onto $Aut(M)$, so $\pi_0(Aut(M))$ is the ordinary mapping class group, which leaves matters in the hands of topologists. Next, suppose that M is a four-manifold. Diffeomorphism groups in four dimensions are not well understood, not even the local case of $\mathbb{R}^4$. Contrarily to what this seems to indicate, the corresponding symplectic problem is far easier: this was one of Gromov's [18] original applications of the pseudo-holomorphic curve method. In extreme simplification, the strategy is to fibre a symplectic four-manifold by a family of such curves, and thereby to reduce the isotopy question to a fibered version of the two-dimensional case. For instance, it turns out that the compactly supported symplectic automorphism group $Aut^c(\mathbb{R}^4)$ is weakly contractible. Here are some more of Gromov's results:

Theorem 1.2 *(1) $Aut(\mathbb{CP}^2)$ is homotopy equivalent to $PU(3)$. (2) For a monotone symplectic structure, $Aut(S^2 \times S^2)$ is homotopy equivalent to $(SO(3) \times SO(3)) \rtimes \mathbb{Z}/2$. (3) (not actually stated in [18], but follows by the same method) for a monotone symplectic structure, $Aut(\mathbb{CP}^2 \# \overline{\mathbb{CP}^2})$ is homotopy equivalent to $U(2)$.*

Recall that a symplectic manifold is monotone if $c_1(M) = r[\omega] \in H^2(M; \mathbb{R})$ for some $r > 0$. (Our formulation is slightly anachronistic: it is true that symplectic forms on $\mathbb{CP}^2$, $S^2 \times S^2$, and $\mathbb{CP}^2 \# \overline{\mathbb{CP}^2}$ are determined up to isomorphism by their cohomology classes, but this is a more recent result, whose proof depends on Seiberg–Witten invariants and Taubes' work; originally, Theorem 1.2 would have been formulated in terms of monotone Kähler forms, which obviously give rise to unique symplectic structures.) Note that in all cases, the result says that $Aut(M)$ is homotopy equivalent to the group of holomorphic automorphisms. One can average the Kähler form with respect to a maximal compact subgroup of this, and then $Aut(M)$ becomes homotopy equivalent to the Kähler isometry group.

After surmonting considerable difficulties, Abreu and McDuff [1, 2] (see also [5]) extended Gromov's method to non-monotone symplectic forms. Their results show that the symplectic automorphism group changes radically if one varies the symplectic class. Moreover, it is not typically homotopy equivalent to any compact Lie group, so that Kähler isometry groups are no longer a good model. Nevertheless, they obtained an essentially complete understanding of the topology of $Aut(M)$, in particular:

Theorem 1.3 *Suppose that M is either $S^2 \times S^2$ or $\mathbb{CP}^2 \# \overline{\mathbb{CP}^2}$, with a non-monotone symplectic form. Then $\pi_0(Aut(M))$ is trivial.*

Now we bring a different source of intuition into play. Let B be a connected pointed manifold. A symplectic fibration with fibre M and base B is a smooth proper fibration $\pi : E \to B$ together with a family of symplectic forms $\{\Omega_b\}$ on the fibres such that $[\Omega_b] \in H^2(E_b; \mathbb{R})$ is locally constant, and a preferred isomorphism between the fibre over the base point and M. There is a universal fibration in the homotopy theory sense, whose base is the classifying space $BAut(M)$. The main advantage of the classifying space viewpoint is that it provides a link with algebraic geometry. Namely, let $\mathcal{E}$, $\mathcal{B}$ be smooth quasi-projective varieties, and $\pi : \mathcal{E} \to \mathcal{B}$ a proper smooth morphism, with a line bundle $\mathbb{L} \to \mathcal{E}$ which is relatively very ample. This means that the sections of $\mathbb{L}|\mathcal{E}_b$ define an embedding of $\mathcal{E}$ into a projective bundle over $\mathcal{B}$. From this embedding one can get a family of (Fubini-Study) Kähler forms on the fibres, so $\mathcal{E}$ becomes a symplectic fibration, classified by a map

$$\mathcal{B} \longrightarrow BAut(\mathcal{M}) \tag{1}$$

where $\mathcal{M}$ is the fibre over some base point, equipped with its Kähler form. In some cases, one can construct a family which is universal in the sense of moduli theory, and then the associated map (1) is the best of its kind. More generally, one needs to consider versal families together with the automorphism groups of their fibres (this is very much the case in the situation studied by Abreu and McDuff; it would be nice to have a sound stack-theoretic formulation, giving the right generalization of the universal base space at least as a homotopy type). Of course, there is no a priori guarantee that algebraic geometry comes anywhere near describing the whole topology of the symplectic automorphism group, or vice versa, that symplectic topology detects all of the structure of algebro-geometric moduli spaces.

Example 1.4 *Suppose that some $\mathcal{M}$ is a double cover of $\mathbb{CP}^1 \times \mathbb{CP}^1$ branched along a smooth curve of bidegree (6,6), and ι the corresponding involution. Let $\mathcal{E} \to \mathcal{B}$ be any algebraic family, with connected $\mathcal{B}$ of course, such that $\mathcal{M}$ is one of the fibres. By looking at the canonical linear systems, one can show that the fibres of $\mathcal{E}$ over a Zariski-open subset of $\mathcal{B}$ are also double covers of smooth quadrics. Suppose that we take the line bundle $\mathbb{L} \to \mathcal{B}$ which is some high power of the fibrewise canonical bundle; it then follows that the image of the monodromy homomorphism $\pi_1(\mathcal{B}) \to \pi_0(Aut(\mathcal{M}))$ consists of elements which commute with $[\iota]$. Donaldson asked whether all symplectic automorphisms act on $H_*(\mathcal{M})$ in a ι-equivariant way; this remains an open question.*

As this concrete example suggests, there are currently no tools strong enough to compute symplectic mapping class groups for algebraic surfaces (or general symplectic four-manifolds) which are not rational or ruled. However, the relation between $\pi_1(\mathcal{B})$ and $\pi_0(Aut(\mathcal{M}))$ can be probed by looking at the behaviour of some particularly simple classes of symplectic automorphisms, and one of these will be the subject of these lectures.

Namely, let M be a closed symplectic four-manifold, and $L \subset M$ an embedded Lagrangian two-sphere. One can associate to this a Dehn twist or Picard–Lefschetz transformation, which is an element $\tau_L \in Aut(M)$ determined up to isotopy. The definition is a straightforward generalization of the classical Dehn twists in two dimensions. However, the topology turns out to be rather different: because of the Picard–Lefschetz formula

$$(\tau_L)_*(x) = \begin{cases} x + (x \cdot l)\, l & x \in H_2(M; \mathbb{Z}), \\ x & x \in H_k(M; \mathbb{Z}), \ k \neq 2 \end{cases} \tag{2}$$

where $l = \pm[L]$ satisfies $l \cdot l = -2$, the square τ_L^2 acts trivially on homology, and in fact it is isotopic to the identity in $Diff(M)$. The obvious question is whether the same holds in $Aut(M)$ as well. The first case which comes to mind is that of the anti-diagonal in $M = S^2 \times S^2$ with the monotone symplectic structure, and there τ_L^2 is indeed symplectically isotopic to the identity. But this is a rather untypical situation: we will show that under fairly weak conditions on a symplectic four-manifold, $[\tau_L^2] \in \pi_0(Aut(M))$ is nontrivial whatever the choice of L. To take a popular class of examples,

Theorem 1.5 *Let $M \subset \mathbb{CP}^{n+2}$ be a smooth complete intersection of complex dimension two. Suppose that M is neither $\mathbb{CP}^2$ nor $\mathbb{CP}^1 \times \mathbb{CP}^1$, which excludes the multidegrees $(1,\dots,1)$ and $(2,1,\dots,1)$. Then the homomorphism*

$$\pi_0(Aut(M)) \longrightarrow \pi_0(Diff(M)) \tag{3}$$

induced by inclusion is not injective.

There is in fact a slightly subtler phenomenon going on, which has to do with the change in topology of $Aut(M)$ as the symplectic structure varies. Let ϕ be a symplectic automorphism with respect to the given symplectic form ω. We say that ϕ is *potentially fragile* if there is a smooth family ω^s of symplectic forms, $s \in [0; \epsilon)$ for some $\epsilon > 0$, and a smooth family ϕ^s of diffeomorphisms such that $(\phi^s)^*\omega^s = \omega^s$, with the following properties: (1) $(\phi^0, \omega^0) = (\phi, \omega)$; (2) for all $s > 0$, ϕ^s is isotopic to the identity inside $Aut(M, \omega^s)$. If in addition, (3) ϕ is not isotopic to the identity in $Aut(M, \omega)$, we say that ϕ is *fragile*. It is a basic fact that squares of Dehn twists are always potentially fragile, and so we have:

Corollary 1.6 *Every two-dimensional complete intersection other than $\mathbb{CP}^2$, $\mathbb{CP}^1 \times \mathbb{CP}^1$ admits a fragile symplectic automorphism.*

As suggested by their alternative name, Dehn twists do occur as monodromy maps in families of algebraic surfaces, so the nontriviality of τ^2 proves that symplectic mapping class groups do detect certain kinds of elements of π_1 of a moduli space, which are hidden from ordinary topology. Moreover, the fragility phenomenon has a natural interpretation in these terms.

Theorem 1.5 and Corollary 1.6 are taken from the author's Ph.D. thesis [45]. Time and [48] have made many of the technical arguments standard, and that frees us to put more emphasis on examples and motivation, but otherwise the structure and limitations of the original exposition have been preserved. However, it seems reasonable to point out some related results that have been obtained since then. For $K3$ and Enriques surfaces containing two disjoint Lagrangian spheres L_1, L_2, it was shown in [47] that $[\tau_{L_1}] \in \pi_0(Aut(M))$ has infinite order, and therefore that the map (3) has infinite kernel. Reference [23] proves that for the noncompact four-manifold M given by the equation $xy + z^{m+1} = 1$ in $\mathbb{C}^3$, there is a commutative diagram

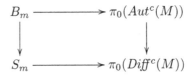

where the upper $\rightarrow$ is injective. In particular, the kernel of the right $\downarrow$ contains a copy of the pure braid group PB_m. Similar phenomena happen for closed four-manifolds: for instance, for a suitable symplectic form on the $K3$ surface, one can show using [23, 49] that the kernel of (3) contains a copy of PB_m for at least $m = 15$. In fact, a simplified version of the same phenomenon (with a more direct proof) already occurs for the del Pezzo surface $\mathbb{CP}^2 \# 5\overline{\mathbb{CP}^2}$, see Example 2.13 below. All of this fits in well with the idea that maps (1) should be an important ingredient in understanding symplectic mapping class groups of algebraic surfaces.

2 Definition and First Properties

(2a) The construction of four-dimensional Dehn twists is standard [6, 46, 48], but we will need the details as a basis for further discussion. Consider T^*S^2 with its standard symplectic form ω, in coordinates

$$T^*S^2 = \{(u, v) \in \mathbb{R}^3 \times \mathbb{R}^3 \ : \ \langle u, v \rangle = 0, ||v|| = 1\}, \quad \omega = du \wedge dv.$$

This carries the $O(3)$-action induced from that on S^2. Maybe less obviously, the function $h(u, v) = ||u||$ induces a Hamiltonian circle action σ on $T^*S^2 \setminus S^2$,

$$\sigma_t(u, v) = \left(\cos(t)u - \sin(t)||u||v, \cos(t)v + \sin(t)\frac{u}{||u||} \right).$$

σ_π is the antipodal map $A(u, v) = (-u, -v)$, while for $t \in (0; \pi)$, σ_t does not extend continuously over the zero-section. Geometrically with respect to the round metric on S^2, σ is the *normalized geodesic flow*, transporting each tangent vector at unit speed (irrespective of its length) along the geodesic emanating from it. Thus, the existence of σ is based on the fact that all geodesics on S^2 are closed. Now take a function $r : \mathbb{R} \to \mathbb{R}$ satisfying $r(t) = 0$ for $t \gg 0$ and $r(-t) = r(t) - t$. The Hamiltonian flow of $H = r(h)$ is $\phi_t(u, v) = \sigma_{t\,r'(||u||)}(u, v)$, and since $r'(0) = 1/2$, the time 2π map can be extended continuously over the zero-section as the antipodal map. The resulting compactly supported symplectic automorphism of T^*S^2,

$$\tau(u, v) = \begin{cases} \sigma_{2\pi\,r'(||u||)}(u, v) & u \neq 0, \\ (0, -v) & u = 0 \end{cases}$$

is called a model Dehn twist. To implant this local model into a given geometric situation, suppose that $L \subset M$ is a Lagrangian sphere in a closed symplectic four-manifold, and choose an identification $i_0 : S^2 \to L$. The Lagrangian tubular neighbourhood theorem [8, Theorem 1.5] tells us that i_0 extends to a symplectic embedding

$$i : T^*_{\leq \lambda} S^2 \longrightarrow M$$

of the space $T^*_{\leq \lambda} S^2 \subset T^*S^2$ of cotangent vectors of length $\leq \lambda$, for some small $\lambda > 0$. By choosing $r(t) = 0$ for $t \geq \lambda/2$, one gets a model Dehn twist τ supported inside that subspace, and then one defines the Dehn twist τ_L to be

$$\tau_L(x) = \begin{cases} i\tau i^{-1}(x) & x \in im(i), \\ x & \text{otherwise.} \end{cases}$$

The construction is not strictly unique, but it is unique up to symplectic isotopy. The only choice that carries any topology is the identification i_0, but this can be dealt with by observing that τ is $O(3)$-equivariant, and $\mathit{Diff}(S^2) \simeq O(3)$ by Smale's theorem. In particular, τ_L does not depend on a choice of orientation of L.

If the circle action σ extended smoothly over the zero-section, then we could write down a compactly supported symplectic isotopy between τ^2 and the identity by moving along the orbits,

$$\psi_t(u,v) = \sigma_{4\pi t \, r'(||u||)}(u,v). \tag{1}$$

This may seem a pointless remark, since σ does not extend over S^2, but it comes into its own after a perturbation of the symplectic structure. Take the standard symplectic form on S^2, $\beta_v(X,Y) = \langle v, X \times Y \rangle$, and pull it back to T^*S^2. Then $\omega^s = \omega + s\beta$, $s \in \mathbb{R}$, is still an $SO(3)$-invariant symplectic form.

Proposition 2.1 *There is a smooth family (ϕ^s) of compactly supported diffeomorphisms of T^*S^2, with the following properties: (1) ϕ^s is symplectic for ω^s; (2) for all $s \neq 0$, ϕ^s is isotopic to the identity by an isotopy in $\mathit{Aut}^c(T^*S^2, \omega^s)$; (3) ϕ^0 is the square τ^2 of a model Dehn twist.*

We begin with an elementary general fact. For concreteness, we will identify $\mathfrak{so}_3^* \cong \mathfrak{so}_3 \cong \mathbb{R}^3$ by using the cross-product and the standard invariant pairing.

Lemma 2.2 *Let M be a symplectic manifold, carrying a Hamiltonian $SO(3)$-action ρ with moment map μ. Then $h = ||\mu||$ is the Hamiltonian of a circle action on $M \setminus \mu^{-1}(0)$.*

Proof. h Poisson-commutes with all components of μ (since this is true for the Poisson bracket on $\mathfrak{so}_3^*$, a well-known fact from mechanics), so its flow maps each level set $\mu^{-1}(w)$ to itself. The associated vector field X satisfies

$$X|\mu^{-1}(w) = K_{w/||w||}|\mu^{-1}(w)$$

where K are the Killing vector fields, which is clearly a circle action (the quotient $\mu^{-1}(w)/S^1$ can be identified with the symplectic quotient $M//SO(3)$ with respect to the coadjoint orbit of w). $\square$

The moment map for the $SO(3)$-action on T^*S^2 is $\mu(u,v) = -u \times v$, so the induced circle action is just σ. With respect to the deformed symplectic structures ω^s, the $SO(3)$-action remains Hamiltonian but the moment map is $\mu^s(u,v) = -sv - u \times v$, which is nowhere zero and hence gives rise to a circle action σ^s on the whole cotangent space. As $r \to 0$, σ^s converges on compact subsets of $T^*S^2 \setminus S^2$ to σ. For simplicity, assume that our model Dehn twist τ is defined using a function h which satisfies $h'(t) = 1/2$ for small t. Then

$$\phi^s(u, v) = \sigma^s_{4\pi h'(\|v\|)}(u, v)$$

for $s \neq 0$ defines a family of compactly supported ω_s-symplectic automorphisms. These are all equal to the identity in a neighbourhood of the zero section, hence they match up smoothly with $\phi^0 = \tau_L^2$. By replacing σ with σ^s in (1) one finds ω^s-symplectic isotopies between each ϕ^s, $s \neq 0$, and the identity. This concludes the proof of Proposition 2.1. It is no problem to graft this local construction into any Dehn twist, which yields:

Corollary 2.3 *For any Lagrangian sphere L in a closed symplectic four-manifold M, the square τ_L^2 of the Dehn twist is potentially fragile.*

(2b) It is easy to see that any compactly supported $O(3)$-equivariant symplectic automorphism of T^*S^2 has the form

$$\phi(x) = \begin{cases} \sigma_{2\pi r'(\|x\|)}(x) & x \notin S^2, \\ A^k(x) & x \in S^2 \end{cases}$$

for some $k \in \mathbb{Z}$, and where $r : \mathbb{R} \to \mathbb{R}$ is a function with $r(t) = 0$ for $r \gg 0$, and $r(-t) = r(t) - kt$ everywhere. There is no topologically nontrivial information in this data except for k, so the space of such automorphisms is weakly homotopy equivalent to the discrete set $\mathbb{Z}$ (the adjective weakly is a technical precaution, since there are several slightly different choices for the topology on compactly supported diffeomorphism groups, which however all have the same spaces of continuous maps from finite-dimensional manifolds). We will now see that the topology does not change if the equivariance condition is dropped:

Proposition 2.4 *The compactly supported automorphism group $\mathrm{Aut}^c(T^*S^2)$ is weakly homotopy equivalent to the discrete set $\mathbb{Z}$, with $1 \in \mathbb{Z}$ mapped to the model Dehn twist.*

In particular $[\tau^k] \in \pi_0(\mathrm{Aut}^c(T^*S^2))$ is nontrivial for all $k \neq 0$. The result also says that up to isotopy and iterating, a Dehn twist is the only construction of a symplectic automorphism that can be done locally near a Lagrangian sphere.

Proof. This is an easy consequence of Gromov's work. Take $M = S^2 \times S^2$ with the standard product symplectic form (in which both factors have the same volume), $L = \{x_1 + x_2 = 0\}$ the antidiagonal, and $\Delta = \{x_1 = x_2\}$ the diagonal. Consider the groups

$$\begin{aligned} \mathcal{G}_1 &= \{\phi \in \mathrm{Aut}(M) \; : \; \phi(\Delta) = \Delta\}, \\ \mathcal{G}_2 &= \{\phi \in \mathcal{G}_1 \; : \; \phi|\Delta = id\}, \\ \mathcal{G}_3 &= \{\phi \in \mathcal{G}_2 \; : \; \phi|U = id \text{ for some open } U \supset \Delta\}. \end{aligned}$$

First of all, $M \setminus \Delta$ is isomorphic to $T^*_{<\lambda} S^2$ some λ, with L corresponding to the zero section. Therefore we have a weak homotopy equivalence $\mathcal{G}_3 \simeq Aut^c(T^*S^2)$. Next, there is a weak fibration

$$\mathcal{G}_3 \longrightarrow \mathcal{G}_2 \xrightarrow{D} Map(S^2, S^1)$$

where $Map(S^2, S^1)$ is thought of as the group of unitary gauge transformations of the normal bundle to Δ, and D essentially the map which associates to each automorphism its derivative in normal direction. It is an easy observation that $Map(S^2, S^1) \simeq S^1$. Third, we have a weak fibration

$$\mathcal{G}_2 \longrightarrow \mathcal{G}_1 \longrightarrow Diff^+(S^2),$$

with $Diff^+(S^2) \simeq SO(3)$. Finally

$$\mathcal{G}_1 \longrightarrow \mathcal{G}_0 \longrightarrow \mathcal{S}_\Delta,$$

where $\mathcal{S}_\Delta$ is the space of embedded symplectic two-spheres in $S^2 \times S^2$ which can be mapped to Δ by a symplectic automorphism. Gromov's theorem says that $Aut(M) \simeq (SO(3) \times SO(3)) \rtimes \mathbb{Z}/2$, and a variation of another of his basic results is that $\mathcal{S}_\Delta \simeq SO(3)$. Appying these sequences in the reverse order, one finds that $\mathcal{G}_1$ is homotopy equivalent to $SO(3) \times \mathbb{Z}/2$, and that $\mathcal{G}_2 \simeq \mathbb{Z}/2$, so the higher homotopy groups of $\mathcal{G}_3$ vanish while $\pi_0(\mathcal{G}_3)$ sits in a short exact sequence

$$1 \to \mathbb{Z} \longrightarrow \pi_0(\mathcal{G}_3) \xrightarrow{\alpha} \mathbb{Z}/2 \to 1, \tag{2}$$

where α assigns to each symplectomorphism the sign $\phi_*[L] = \pm[L]$. The last step yields the following additional information: take a map $\phi \in \mathcal{G}_3$ which preserves the orientation of L, and let ϕ_t be a homotopy from it to the identity inside $\mathcal{G}_2$. Then the element of $ker(\alpha)$ represented by ϕ is the degree of $S^1 \to Sp_4(\mathbb{R})$, $t \mapsto D_x\phi_t$ at any point $x \in \Delta$. By applying this to the isotopy $\tau_L^2 \simeq id$ constructed in Example 2.9 below, one sees that (2) does not split, and that $[\tau_L]$ is a generator of $\pi_0(Aut^c(T^*S^2)) \cong \mathbb{Z}$. $\square$

It is an interesting exercise to see how the above argument changes if one passes to the symplectic form $\omega^s = \omega + s\beta$ for small s.

(2c) Corollary 2.3 is too essential to pass it off as the result of some ad hoc local construction. A proper understanding involves looking at the real nature of Dehn twists as monodromy maps.

Definition 2.5 *Let S be an oriented surface, possibly non-compact or with boundary. A (six-dimensional) symplectic Lefschetz fibration over S is a six-manifold E with a proper map $\pi : E \to S$, $\pi^{-1}(\partial S) = \partial E$, a closed two-form $\Omega \in \Omega^2(E)$, a complex structure J_E defined on a neighbourhood of the set of critical points E^{crit}, and a positively oriented complex structure j_S defined on a neighbourhood of the set of critical values S^{crit}. The requirements are:*

- *Near the critical points, π is a holomorphic map with respect to J_E and j_S, and the critical points themselves are nondegenerate. Moreover, E^{crit} is disjoint from ∂E, and $\pi|E^{crit}$ is injective.*

- Ω is a Kähler form for J_E in a neighbourhood of E^{crit}. For any point $x \notin E^{crit}$, the restriction of Ω_x to $TE_x^v = \ker(D\pi_x)$ is nondegenerate.

The geometry of these fibrations is not very different from the familiar four-dimensional case treated in [8]; one possible reference for the results stated below is [48]. Away from the critical fibres Lefschetz fibrations are clearly symplectic fibrations and in fact carry a preferred Hamiltonian connection TE^h, the Ω-orthogonal complement to TE^v (the word "Hamiltonian" refers to the structure group Aut^h from Remark 1.1, and does *not* mean that the monodromy consists of maps Hamiltonian isotopic to the identity). Hence, for any smooth path $\gamma : [0; 1] \to S \setminus S^{crit}$ we have a canonical parallel transport map $P_\gamma : E_{\gamma(0)} \to E_{\gamma(1)}$. Given a path $\gamma : [0; 1] \to S$ with $\gamma^{-1}(S^{crit}) = \{1\}$, $\gamma'(1) \neq 0$, one can look at the limit of $P_{\gamma|[0;t]}$ as $t \to 1$, and this gives rise to a Lagrangian two-sphere $V_\gamma \subset E_{\gamma(0)}$, which is the vanishing cycle of γ. The Picard–Lefschetz theorem says that if λ is a loop in $S \setminus S^{crit}$ with $\lambda(0) = \lambda(1) = \gamma(0)$, winding around γ in positive sense, its monodromy is the Dehn twist around the vanishing cycle, at least up to symplectic isotopy:

$$P_\lambda \simeq \tau_{V_\gamma} \in Aut(E_{\gamma(0)}).$$

Let's pass temporarily to algebro-geometric language, so $\pi : \mathcal{E} \to \mathcal{S}$ is a proper holomorphic map from a threefold to a curve, with the same kind of critical points as before, and $\mathbb{L} \to \mathcal{E}$ is a relatively very ample line bundle. Atiyah [7] (later generalized by Brieskorn [11]) discovered the phenomenon of *simultaneous resolution*, which can be formulated as follows: let $r : \hat{\mathcal{S}} \to \mathcal{S}$ be a branched covering which has double ramification at each preimage of points in $\mathcal{S}^{crit}$. Then there is a commutative diagram

$$
\begin{array}{ccc}
\hat{\mathcal{E}} & \xrightarrow{\ R\ } & \mathcal{E} \\
\downarrow{\scriptstyle \hat{\pi}} & & \downarrow{\scriptstyle \pi} \\
\hat{\mathcal{S}} & \xrightarrow{\ r\ } & \mathcal{S}
\end{array}
$$

where $\hat{\pi}$ has no critical points (proper smooth morphism), and the restriction of R gives an isomorphism

$$\hat{\mathcal{E}} \setminus R^{-1}(\mathcal{E}^{crit}) \xrightarrow{\ \cong\ } r^*(\mathcal{E} \setminus \mathcal{E}^{crit}).$$

In particular, away from the singular fibres $\hat{\mathcal{E}}$ is just the pullback of $\mathcal{E}$. If λ is a small loop in $\mathcal{S} \setminus \mathcal{S}^{crit}$ going once around a critical value, then its iterate λ^2 can be lifted to $\hat{\mathcal{S}}$, which means that the monodromy around it must be isotopic to the identity as a diffeomorphism. Of course, by the Picard–Lefschetz formula $P_{\lambda^2} \simeq \tau_V^2$ for the appropriate vanishing cycle V. The preimage of each critical point $x \in \mathcal{E}$ is a rational curve $C_x \subset \hat{\mathcal{E}}$ with normal bundle $\mathcal{O}(-2)$ in its fibre. Suppose that there is a line bundle $\Lambda \to \hat{\mathcal{E}}$ such that $\Lambda|C_x$ has positive degree for each x (this may or may not exist, depending on the choice of resolution).

Then $\hat{\mathbb{L}} = \mathbb{L}^{\otimes d} \otimes \Lambda^{\otimes e}$ is relatively very ample for $d \gg e \gg 0$. This shows that the monodromy around λ^2 becomes symplectically trivial after a change of the symplectic form, which is essentially the same property as potential fragility of τ_V^2 except that algebraic geometry does not actually allow us to see this change as a continuous deformation. However, one can easily copy the local construction of the simultaneous resolution in the symplectic setting, and this gives an alternative proof of Corollary 2.3 avoiding any explicit computation.

Remark 2.6 *More generally, potential fragility occurs naturally in situations involving hyperkähler quotients. Let X be a hyperkähler manifold, and pick a preferred complex structure on it. Suppose that it carries a hyperkähler circle action with moment map $h = (h_{\mathbb{R}}, h_{\mathbb{C}}) : X \to \mathbb{R} \times \mathbb{C}$, and a connected component of the fixed point set on which $h \equiv 0$. For simplicity we will assume that the action is otherwise free, and ignore problems arising from the non-compactness of X (so the following statements are not entirely rigorous). If one fixes $s \in \mathbb{R}$ then*

$$X^s_{\mathbb{C}^*} = (h_{\mathbb{R}}^{-1}(s) \setminus h_{\mathbb{C}}^{-1}(0))/S^1 \xrightarrow{h_{\mathbb{C}}} \mathbb{C}^*$$

is a holomorphic map, and the total space carries a natural quotient Kähler form. One can therefore define the monodromy around a circle of some radius $\epsilon > 0$ in the base $\mathbb{C}^*$, which is a symplectic automorphism ϕ^s of the hyperkähler quotient $X^s_\epsilon = (h_{\mathbb{R}}^{-1}(s) \cap h_{\mathbb{C}}^{-1}(\epsilon))/S^1$. Varying s does not affect the complex structure on X^s_ϵ, but the Kähler class varies, so one can consider the ϕ^s as a family of automorphisms for a corresponding family ω^s of symplectic forms on a fixed manifold. For $s = 0$ there is a singular fibre X^0_0 at the center of the circle, and one would hope that ϕ^0 reflects this fact; in contrast, for $s \neq 0$ we have that X^s_z is smooth for all z, so ϕ^s is symplectically isotopic to the identity. The case of squared Dehn twists is a particularly simple example of this, with $X = \mathbb{H}^2$; see [26]. A straightforward generalization leads to analogues of τ^2 on $T^*\mathbb{CP}^n$, which were discussed in [47].

(2d) Donaldson's theory of almost holomorphic functions is an attempt to reduce all questions about symplectic four-manifolds to two-dimensional ones, and hence to combinatorial group theory. The paper [14] achieves this for the fundamental classification problem, but the wider program also embraces symplectic mapping groups. The relevant deeper results are still being elaborated, but the elementary side of the theory is sufficient to understand the potential fragility of squared Dehn twists. The following discussion is due to Donaldson (except possibly for mistakes introduced by the author). Compared to the exposition in [8], to which the reader is referred for the basic theory of Lefschetz pencils, we will just need to exercise a little more care concerning the definition of symplectic forms on the total spaces.

Let S be a closed oriented surface, equipped with a symplectic form η and a finite set of marked points $\Sigma = \{z_1, \ldots, z_p\}$, which may be empty. We assume that the Euler characteristic $\chi(S \setminus \Sigma) < 0$. Denote by $Aut^h(S, \Sigma)$ the group of symplectic automorphisms of S which are the identity in a neighbourhood of Σ, with the Hamiltonian topology. For any simple closed curve $\gamma \subset S \setminus \Sigma$, we have the (classical) Dehn twist t_γ, which is an element of $Aut^h(S, \Sigma)$ unique up to isotopy within that topological group (note that if γ, γ' are nonseparating curves which are isotopic to each other, but not Hamiltonian isotopic, then t_γ and $t_{\gamma'}$ have different classes in $\pi_0(Aut^h)$). Choose a small loop ζ_k around each z_k. Take a finite ordered family $(\gamma_1, \ldots, \gamma_m)$ of simple closed non-contractible curves in $S \setminus \Sigma$, such that

$$t_{\gamma_1} \ldots t_{\gamma_m} \simeq t_{\zeta_1} \ldots t_{\zeta_p} \tag{3}$$

in $Aut^h(S, \Sigma)$. From this one constructs a four-manifold M together with a family ω^s of closed forms, which are symplectic for $s \gg 0$. For brevity, we will call this an *asymptotically symplectic manifold*. The first step is take the (four-dimensional topological) Lefschetz fibration $\tilde{M} \to S^2$ with smooth fibre S and vanishing cycles $\gamma_1, \ldots, \gamma_m$. Using a suitable Hamiltonian connection, one can define a closed two-form $\tilde{\omega}$ on $\tilde{M}$ whose restriction to each smooth fibre is symplectic. The family $\tilde{\omega}^s = \tilde{\omega} + s\beta$, where β is the pullback of a positive volume form on S^2, consists of symplectic forms for $s \gg 0$. Each base point z_k will give rise to a section, whose image is a symplectic sphere with self-intersection -1. Blowing down these spheres completes the construction of $(M, \{\omega^s\})$. Of course there is some choice in the details, but the outcome is unique up to asymptotic symplectic isomorphism, which is the existence of a family of diffeomorphisms $\{\phi^s\}$ which are symplectic for $s \gg 0$; and moreover, this family is canonical up to asymptotically symplectic isotopy, which is enough for our purpose. For later reference, we note the following fact about the cohomology class of ω^s. The primitive part $H^2(M; \mathbb{R})^{prim}$, which is just the quotient of $H^2(M; \mathbb{R})$ by the Poincaré dual of the fibre $S \subset M$, can be described as the middle cohomology group of a complex [15]

$$H^1(S \setminus \Sigma; \mathbb{R}) \xrightarrow{a} \mathbb{R}^m \xrightarrow{a'} H^1(S; \mathbb{R}) \tag{4}$$

where a is given by integrating over the γ_k, and a' involves a certain dual set of vanishing cycles γ_k'. The class of ω^s in $H^2(M; \mathbb{R})^{prim}$ is independent of s, and is represented by a vector in $\mathbb{R}^m$ in (4) defined by choosing a one-form θ on $S \setminus \Sigma$ with $d\theta = \eta$, and integrating that over the γ_k. In particular, if θ can be chosen in such a way that $\int_{\gamma_k} \theta = 0$ for all k, then all ω^s are multiples of $PD([S])$, which is the case of a Lefschetz *pencil*.

If one replaces the γ_k by curves Hamiltonian isotopic to them, M remains the same, up to the same kind of isomorphism as before. We call the equivalence class of $(\gamma_1, \ldots, \gamma_m)$ under this relation a *Lefschetz fibration datum*; this will be denoted by Γ, and the associated manifold by $(M_\Gamma, \{\omega_\Gamma^s\})$. More interestingly, there are two nontrivial modifications of a Lefschetz fibration datum which do not change M; together they amount to an action of $G = \pi_0(Aut^h(S, \Sigma)) \times B_m$ on the set of such data. The first factor acts by applying a symplectic automorphism ϕ to all of the γ_k, and the generators of the braid group B_m act by elementary Hurwitz moves

$$(\gamma_1, \ldots, \gamma_m) \longmapsto (\gamma_1, \ldots, \gamma_{k-1}, t_{\gamma_k}(\gamma_{k+1}), \gamma_k, \gamma_{k+2}, \ldots, \gamma_m). \qquad (5)$$

Roughly speaking, what the two components of the G-action do is to change the way in which the fibre of $\tilde{M}_\Gamma$ is identified with S, respectively the way in which its base is identified with S^2. By uniqueness, we have for every $g \in G$ such that $g(\Gamma) = \Gamma$ an induced asymptotically symplectic automorphism $\{\phi^s\}$ of M_Γ. Denoting by $G_\Gamma \subset G$ the subgroup which stabilizes Γ, and by $Aut(M_\Gamma, \{\omega_\Gamma^s\})$ the group of asymptotically symplectic automorphisms, we therefore have a canonical map

$$G_\Gamma \longrightarrow \pi_0(Aut(M_\Gamma, \{\omega_\Gamma^s\})). \qquad (6)$$

(in the case of a Lefschetz pencil, the right hand side reduces to $Aut(M_\Gamma, \omega_\Gamma^\sigma)$ for some fixed $\sigma \gg 0$). Usually (6) is not injective. For instance, consider the situation where two subsequent curves γ_k, γ_{k+1} are disjoint. Applying (5) just exchanges the curves; the square of this operation is a nontrivial element of G_Γ, but the associated asymptotically symplectic automorphism is isotopic to the identity. This can be most easily seen by thinking of families of Lefschetz fibrations: in our case, we have a family parametrized by S^1 in which two critical values in S^2 rotate around each other, and whose monodromy is the image of our Hurwitz move in (6); but since the vanishing cycles are disjoint, we can move the two critical points into the same fibre, and so the family can be extended over D^2, which trivializes the monodromy.

Suppose that we are in the Lefschetz pencil situation where $\int_{\gamma_k} \theta = 0$, and that two subsequent curves γ_k, γ_{k+1} agree. One can then use their bounding "Lefschetz thimbles" to construct a Lagrangian sphere $L \subset M$, and its inverse Dehn twist τ_L^{-1} is the image of the elementary Hurwitz move (5) under (6). Now move γ_k, γ_{k+1} away from each other in a non-Hamiltonian way, by an opposite amount of area. The resulting new configuration of curves $\gamma_1', \ldots, \gamma_m'$ still satisfies the basic equation (3), and defines the same four-manifold $M_{\Gamma'} = M_\Gamma$ with a different symplectic form: an argument using (4) shows that $\omega_{\Gamma'}^s$ differs from ω_Γ^s by a multiple of $PD(L)$, which becomes comparatively small as $s \to \infty$. Since γ_k', γ_{k+1}' are disjoint, the element of G_Γ which led to τ_L^2 now becomes an element of $G_{\Gamma'}'$ inducing a trivial asymptotically symplectic automorphism, which is the statement of potential fragility in this framework.

Remark 2.7 *We should briefly mention the expected deeper results concerning the map (6) (these were first stated by Donaldson, and their proof is the subject of ongoing work of Auroux–Munoz–Presas). The main idea is that the image of (6) for Lefschetz pencils should ultimately exhaust the symplectic automorphism group as the degree of the pencil goes to ∞. More precisely, given a symplectic manifold and integral symplectic form ω, and an arbitrary symplectic automorphism ϕ, there should be a Lefschetz pencil whose fibres lie* in the class $k[\omega]$ for $k \gg 0$, and an element of the resulting G_Γ which maps to $[\phi]$. There is also a list of relations for the kernel of (6) which is conjectured to be complete in a suitable $k \to \infty$ sense, but a rigorous formulation of that would be quite complicated since it involves "degree doubling".

(2e) As usual, let $L \subset M$ be a Lagrangian sphere in a closed symplectic four-manifold. Having considered the fragility of τ_L^2 from different points of view, we now turn to the main question, which is whether it is isotopic to the identity in $Aut(M)$. We know that this is a nontrivial question because the answer for the corresponding local problem is negative, by Proposition 2.4, and as mentioned in the Introduction this answer carries over to the vast majority of closed four-manifolds. For now, however, the discussion will start from the opposite direction, as we try to accumulate examples where τ_L^2 is symplectically isotopic to the identity, and then probe the line where something nontrivial happens.

First of all, there is an elementary construction based directly on the circle action σ used in the definition of the Dehn twist.

Lemma 2.8 *Suppose that there is a Hamiltonian circle action $\bar\sigma$ on $M\backslash L$ and a Lagrangian tubular neighbourhood $i : T_{<\lambda}^* S^2 \to M$ of L which is equivariant with respect to σ, $\bar\sigma$. Then τ_L^2 is isotopic to the identity in $Aut(M)$.*

The proof is straightforward, and we leave it to the reader.

Example 2.9 *As in the proof of Proposition 2.4 take $M = S^2 \times S^2$ with the monotone symplectic form, and $L = \{x_1 + x_2 = 0\}$ the antidiagonal. The diagonal $SO(3)$-action has moment map $\mu(x) = -x_1 - x_2 \in \mathbb{R}^3$, and from Lemma 2.2 above we know that $\bar h(x) = ||x_1 + x_2||$ is the moment map for a circle action $\bar\sigma$ on $M \backslash L$. This has the desired property with respect to any $SO(3)$-equivariant Lagrangian tubular neighbourhood for L. A slight refinement of Lemma 2.8 shows that τ_L itself is symplectically isotopic to the involution $(x_1, x_2) \mapsto (x_2, x_1)$. Somewhat less transparently, this could also be derived from Gromov's Theorem 1.2.*

Example 2.10 *A related case is the "regular pentagon space", a manifold often used as a basic example in the theory of symplectic quotients [39, Chap. 4, Sect. 5] [24, Chap. 16.1] [38, Chap. 8] (incidentally, it is also the same as the Deligne-Mumford space $\overline{\mathcal{M}}_{0,5}$). Take S^2 with its standard symplectic form, and consider the diagonal action of $SO(3)$ on $(S^2)^5$ with moment map $\mu(x) = -(x_1 + \cdots + x_5)$. The symplectic quotient $M = \mu^{-1}(0)/SO(3)$ is the space of quintuples of vectors of unit length in $\mathbb{R}^3$ which add up to zero, up to simultaneous rotation. This is a compact symplectic four-manifold, and it contains a natural Lagrangian sphere*

$$L_1 = \{x_1 + x_2 = 0\}. \tag{7}$$

$M \setminus L_1$ carries a Hamiltonian circle action $\bar{\sigma}_1$, given by rotating x_1 around the axis formed by $x_1 + x_2$ while leaving $x_1 + x_2$, x_3, x_4, x_5 fixed. The relevant moment map is $\bar{h}_1(x) = ||x_1 + x_2||$ as before, which already looks much like our standard circle action on $T^*S^2 \setminus S^2$. Indeed, one can find a tubular neighbourhood of L_1 satisfying the conditions of Lemma 2.8, so $\tau_{L_1}^2$ is symplectically isotopic to the identity. In fact, by cyclically permuting coordinates, one finds a configuration of Lagrangian spheres $L_1, \ldots, L_5$ whose intersections are indicated by a pentagon graph

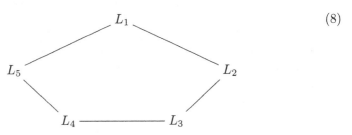

(8)

Because of the resulting braid relations [46, Appendix], $\tau_{L_1}, \ldots, \tau_{L_4}$ generate a homomorphism $B_5 \to \pi_0(Aut(M))$; on the other hand, we have the additional relation $[\tau_{L_k}^2] = 1$, so this actually factors through the symmetric group S_5.

It is worth while to identify M more explicitly. Take the maps induced by inclusion j and projection p,

$$H^2((S^2)^5; \mathbb{R}) \xrightarrow{j^*} H^2(\mu^{-1}(0); \mathbb{R}) \xleftarrow{p^*} H^2(M; \mathbb{R}).$$

Our group being $SO(3)$, a look at the standard spectral sequence shows that cohomology and equivariant cohomology coincide in degree two. This implies that p^* is an isomorphism. Now, the pullback of the symplectic form on M via p agrees with the restriction of the symplectic form on $(S^2)^5$ via j, and the same holds for the first Chern classes of their respective tangent bundles. We conclude that M is monotone, so by general classification results [27] it must be either $\mathbb{CP}^1 \times \mathbb{CP}^1$ or $\mathbb{CP}^2$ blown up at $0 \le k \le 8$ points. The same consideration with equivariant cohomology as before, together with Kirwan's surjectivity theorem, shows that j^* is onto, so $b_2(M) \le 5$. On the other hand, by looking at the intersection matrix of the configuration (8) one sees that the part of $H^2(M; \mathbb{R})$ orthogonal to the symplectic class has at least dimension 4. Therefore $b_2(M) = 5$ and so

$$M \cong \mathbb{CP}^2 \# 4\overline{\mathbb{CP}^2}.$$

A more elementary approach is to observe that $M \setminus (L_1 \cup L_3)$ carries a T^2-action with three fixed points, which directly yields $\chi(M) = 7$. Finally, one can vary this example by considering quintuples of vectors of different lengths, see [19, 20, 25]. This yields examples of Lagrangian spheres on $\mathbb{CP}^2 \# 2\overline{\mathbb{CP}^2}$ and $\mathbb{CP}^2 \# 3\overline{\mathbb{CP}^2}$ with τ^2 symplectically isotopic to the identity, however the relevant symplectic forms are not monotone.

Another way of finding examples of $\tau^2 \simeq id$ is based on the connection with algebraic geometry and monodromy, which means on the construction of suitable families of algebraic surfaces together with the Picard–Lefschetz theorem.

Lemma 2.11 *Let $\pi : \mathcal{E} \to \mathcal{B}$ be a proper smooth map between quasi-projective varieties of relative dimension 2, and $\mathbb{L} \to \mathcal{E}$ a relatively very ample line bundle. Suppose that there is a partial compactification $\bar{\pi} : \bar{\mathcal{E}} \to \bar{\mathcal{B}}$ where the total space and base are still smooth, and with the following properties: (1) The discriminant $\Delta = \bar{\mathcal{B}} \setminus \mathcal{B}$ is a hypersurface, and the fibre over a generic point $\delta \in \Delta$ is reduced and has a single ordinary double point singularity; (2) the meridian around δ is an element of order 2 in $\pi_1(\mathcal{B})$; (3) $\mathbb{L}$ extends to a relatively very ample line bundle $\bar{\mathbb{L}} \to \bar{\mathcal{E}}$. If these conditions hold, the smooth fibre $\mathcal{E}_b$, $b \in \mathcal{B}$, with its induced Kähler structure, contains a Lagrangian sphere L such that τ_L^2 is symplectically isotopic to the identity.*

Proof. Take a generic point $\delta \in \Delta$, a neighbourhood $U \subset \bar{\mathcal{B}}$ of δ, and local holomorphic coordinates $(\zeta, y_1, \ldots, y_n) : U \to \mathbb{C}^{n+1}$ such that $\Delta = \{\zeta = 0\}$. Take a small generic value of the map $y \circ \bar{\pi} : \bar{\mathcal{E}}|U \to U \to \mathbb{C}^n$. The preimage of that is a smooth threefold $\bar{\mathcal{E}}_y$ with a holomorphic map $\bar{\pi}_y = \zeta \circ \bar{\pi} : \bar{\mathcal{E}}_y \to U_y \subset \mathbb{C}$, such that $\bar{\pi}_y^{-1}(0)$ has a single ordinary double point. This implies that $\bar{\pi}_y$ has a nondegenerate critical point. After using $\bar{\mathbb{L}}$ to put a suitable Kähler form on $\bar{\mathcal{E}}_y$, we find that the monodromy of $\bar{\pi}_y$ around 0 is the Dehn twist along some Lagrangian sphere. On the other hand, the monodromy is the image of the meridian around δ by the map (1), so it must be of order two. $\square$

Example 2.12 *A classical case is that of del Pezzo surfaces with small rank. The necessary algebro-geometric background can be found in the first few lectures of [13]. Fix $2 \le k \le 4$. Consider a configuration $b = \{b_1, \ldots, b_k\}$ of k unordered distinct points in $\mathbb{CP}^2$ which are in general position, meaning that no three of them are collinear. If that is the case, the anticanonical bundle on the blowup $Bl_b(\mathbb{CP}^2)$ is ample, in fact very ample. Over the space $\mathcal{B} \subset Conf_k(\mathbb{CP}^2)$ of configurations c in general position, there is a natural family of blowups $\mathcal{E} \to \mathcal{B}$, and the anticanonical bundle is relatively very ample. The action of $PSL_3(\mathbb{C})$ on $\mathcal{B}$ is transitive, and using that, it is easy to determine*

$$\pi_1(\mathcal{B}) \cong S_k. \tag{9}$$

What this says is that a symplectic automorphism obtained as monodromy map from the family of blowups is symplectically isotopic to the identity iff it acts trivially on homology.

To partially compactify $\mathcal{B}$, we will now relax the genericity conditions by allowing two points to collide. Let $b \subset \mathcal{O}_{\mathbb{CP}^2}$ be an ideal sheaf of length k. This means that it is a configuration of points with multiplicities, which add up to k, and additional infinitesimal information at the multiple points. We say that *b is in almost general position if any point occurs at most with multiplicity two, and its restriction to any line in $\mathbb{C}P^2$ has length at most two. The space of such ideals is a partial compactification $\bar{\mathcal{B}}$ of $\mathcal{B}$, and the discriminant Δ is a smooth divisor. For $\delta \in \Delta$, $Bl_\delta(\mathbb{C}P^2)$ is a surface with an ordinary double point. This means that $\mathcal{E}$ extends to a family $\bar{\mathcal{E}} \to \bar{\mathcal{B}}$ such that the fibres over Δ have an ordinary double point. It is not difficult to show that $\bar{\mathcal{E}}$ is smooth, and that the anticanonical bundle is still fibrewise very ample. In view of (9), Lemma 2.11 implies that there is a Lagrangian sphere L in the smooth fibre, which is $\mathbb{C}P^2 \# k\overline{\mathbb{C}P^2}$ with a monotone symplectic form, such that τ_L^2 is symplectically isotopic to the identity (of course, for $k = 4$ we already know this from Example 2.10, but the cases $k = 2, 3$ are new).*

The structure of the moduli space changes for del Pezzo surfaces of rank $5 \le k \le 8$, where the action of $PSL_3(\mathbb{C})$ on the corresponding space of generic configurations is no longer transitive. As one would expect from the general philosophy, this also affects the structure of symplectic mapping class groups. The first case $k = 5$ can be treated by elementary means, and we will do so now.

Example 2.13 *Take $\mathcal{B}^{ord} \subset Conf_5^{ord}(\mathbb{C}P^2)$ to be the space of ordered quintuples of points $b = (b_1, \ldots, b_5)$ in the projective plane which are in general position, in the same sense as before. $PSL_3(\mathbb{C})$ acts freely on this, and the quotient $\mathcal{B}^{ord}/PSL_3(\mathbb{C})$ is isomorphic to the moduli space of ordered quintuples of points on the line, $Conf_5^{ord}(\mathbb{C}P^1)/PSL_2(\mathbb{C})$. One can see this by direct computation: each $PSL_3(\mathbb{C})$-orbit on $\mathcal{B}^{ord}$ contains exactly one point of the form*

$$b_1 = [1:0:0], \; b_2 = [0:1:0], \; b_3 = [0;0;1],$$
$$b_4 = [1:1:1], \; b_5 = [z/w; (1-z)/(1-w); 1] \tag{10}$$

with $z, w \ne \{0, 1\}$ and $z \ne w$; and correspondingly, each $PSL_2(\mathbb{C})$-orbit on $\mathbb{C}P^1$ contains a unique configuration of the form $(0, 1, \infty, z, w)$. A more geometric construction goes as follows: there is a unique (necessarily nonsingular, by the general position condition) conic Q which goes through the points $b_1, \ldots, b_5$. One can identify $Q \cong \mathbb{C}P^1$ and then the $b_k \in Q$ become a configuration of points on the line. A straightforward computation shows that this gives back our previous identification; the requirement that this should work out explains the strange coordinates used in (10).

This approach can be imitated on a symplectic level. Let $M = \mathbb{C}P^2 \# 5\overline{\mathbb{C}P^2}$ with its monotone symplectic structure. Let $E_1, \ldots, E_5$ be the homology classes of the exceptional curves in the $\overline{\mathbb{C}P^2}$ summands, and L the homology class of the line in $\mathbb{C}P^2$. Take an arbitrary (not generic in any sense) compatible almost complex structure J. Each of the classes

$$E_1, \ldots, E_5, 2L - E_1 - \cdots - E_5 \tag{11}$$

is minimal, in the sense that it cannot be written as the sum of two classes of positive symplectic area, hence the moduli space of J-holomorphic spheres in *that class is compact. The adjunction formula [34, Theorem 2.2.1] proves that this space consists of embedded spheres, and the regularity theorem from [21] implies that it is smooth. By deforming to the standard complex structure, one sees that each class (11) is represented by a unique embedded J-holomorphic sphere. The multiplicity theorem [34, Theorem 2.1.1] shows that the sphere representing $2L - E_1 - \cdots - E_5$ intersects each E_k sphere transversally in a single point. Hence, by identifying that sphere with $\mathbb{CP}^1$ one gets an element of $Conf_5^{ord}(\mathbb{CP}^1)/PSL_2(\mathbb{C})$. This can be done on the fibres of a symplectic fibration, as long as the homological monodromy is trivial (allowing one to identify the homology classes (11) in different fibres), so one gets a map*

$$\beta : BAut^0(M) \longrightarrow Conf_5^{ord}(\mathbb{CP}^1)/PSL_2(\mathbb{C})$$

unique up to homotopy, where $Aut^0(M)$ is the subgroup of symplectic automorphisms acting trivially on homology. The monodromy of the universal family of blowups over $\mathcal{B}^{ord}/PSL_3(\mathbb{C})$ gives a map

$$\alpha : \mathcal{B}^{ord}/PSL_3(\mathbb{C}) \longrightarrow BAut^0(M)$$

such that $\beta \circ \alpha$ is homotopy equivalent to the previous isomorphism of spaces. Hence α induces an injective homomorphism from $\Gamma_5^{ord} = \pi_1(\mathcal{B}^{ord}/PSL_3(\mathbb{C}))$ to $\pi_0(Aut(M))$. By taking up the discussion from the previous example, one can see that the image of that homomorphism is generated by squared Dehn twists, hence maps trivially to $\pi_0(Diff(M))$. To take into account maps which act nontrivially on homology, one should introduce an extension Γ_5 of Γ_5^{ord} by the Weyl group $W(D_5)$, which is the automorphism group of the lattice $H_2(M)$ preserving c_1 and the intersection product (this is slightly larger than the extension $\pi_1^{orb}(\mathcal{B}/PSL_3(\mathbb{C}) \times S_5)$ that one gets from passing to unordered configurations). There is a corresponding extended map $\Gamma_5 \to \pi_0(Aut(M))$, which is obviously also injective, and which fits into a commutative diagram

$$\begin{array}{ccc} \Gamma_5 & \hookrightarrow & \pi_0(Aut(M)) \\ \downarrow & & \downarrow \\ W(D_5) & \longrightarrow & \pi_0(Diff(M)). \end{array} \qquad (12)$$

$M = \mathbb{CP}^2 \# 5\overline{\mathbb{CP}^2}$ also occurs as space of parabolic rank two odd degree bundles with fixed determinant and weights $1/2$ on the five-pointed sphere (this gives another explanation for the isomorphism of configuration spaces in Example 2.13). In terms of flat connections, one can write it as

$$M = \{A_1 \ldots A_5 \in C_{1/2} \; : \; A_1 \ldots A_5 = I\}/PU(2). \qquad (13)$$

where $S^2 \cong C_{1/2} \subset SU(2)$ is the conjugacy class of $diag(i, -i)$, and $PU(2)$ acts by simultaneous conjugation. There is an obvious action of the mapping class group of the five-pointed sphere on M. The gauge-theoretic definition shows that the action is by symplectomorphisms, and up to symplectic isotopy one can identify it with the top $\to$ from (12) restricted to

$\pi_1^{orb}(\mathcal{B}/PSL_3(\mathbb{C}) \times S_5) \subset \Gamma_5$. The injectivity of this map is interesting because of the (conjectural) relation between symplectic Floer homology and certain gauge theoretic invariants of knots [12]. As a final remark, note that there is a striking similarity between (13) and the definition of the regular pentagon space: indeed, if one replaces $C_{1/2}$ with the conjugation class of $diag(e^{\pi i \alpha}, e^{-\pi i \alpha})$ for some $\alpha < 2/5$, GIT arguments show that the resulting space is symplectically deformation equivalent to the pentagon space (as one passes the critical weight $2/5$, the space undergoes a single blowup). This makes the difference between the behaviour of squared Dehn twists even more remarkable.

If one goes further to $k = 6$, where the blowup is a cubic surface in $\mathbb{CP}^3$, the situation becomes considerably more complicated, mainly because the notion of general position involves an additional condition on conics. Take the space $\mathcal{B}^{ord}/PSL_3(\mathbb{C})$ of ordered configurations of points in general position, which is the same as the moduli space of marked cubic surfaces. A theorem of Allcock [3] says that this is a $K(\Gamma_6^{ord}, 1)$, and the group Γ_6^{ord} is quite large: it contains infinitely generated normal subgroups [4]. For purposes of comparison with $\pi_0(Aut)$, the right group Γ_6 is an extension of Γ_6^{orb} by $W(E_6)$, which is the orbifold fundamental group of the moduli space of cubic surfaces. Libgober [30] proved that Γ_6 is a quotient of the generalized braid group $B(E_6)$, and Looijenga [33] has given an explicit presentation of it. The last-mentioned paper also contains a discussion of the $k = 7$ case, in which Γ_7 is the orbifold fundamental group of the moduli space of *non-hyperelliptic* genus three curves. We will return to these del Pezzo surfaces in Example 3.10 below.

Example 2.14 *Here is another, even simpler, application of Lemma 2.11. For any algebraic surface $\mathcal{M}$, there is a smooth family $\mathcal{E} \to \mathcal{B}$ over the configuration space $\mathcal{B} = Conf_2(\mathcal{M})$, whose fibre at b is the blowup $Bl_b(\mathcal{M})$. Take an ample line bundle Λ on $\mathcal{M}$, and equip each blowup with $\mathbb{L}_b = \Lambda^{\otimes d} \otimes \mathcal{O}(-E_1 - E_2)^{\otimes e}$ for some $d \gg e \gg 0$. Both the family and the line bundle extend to the compactification $\overline{\mathcal{B}} = Hilb_2(\mathcal{M})$ where the two points are allowed to come together, and the fibres over the discriminant Δ have ordinary double points. The monodromy around the meridian is a Dehn twist along a Lagrangian sphere in the class $E_1 - E_2$, and using the short exact sequence*

$$1 \to \pi_1(\mathcal{M})^2 \to \pi_1(\mathcal{B}) \to \mathbb{Z}/2 \to 1$$

one sees that the square of this Dehn twist is isotopic to the identity. This is actually a local phenomenon: $\mathbb{C}^2 \# 2\overline{\mathbb{CP}^2}$, with the two exceptional divisors having equal area, contains a Lagrangian sphere whose squared Dehn twist is isotopic to the identity in the compactly supported symplectic automorphism group. This can be implanted into $M \# 2\overline{\mathbb{CP}^2}$ for any closed symplectic four-manifold M, as long as the area of the exceptional divisors remains equal and sufficiently small.

3 Floer and Quantum Homology

(3a) Fix a closed symplectic four-manifold M with $H^1(M; \mathbb{R}) = 0$, and a coefficient field $\mathbb{K}$ ($\mathbb{K} = \mathbb{Q}$ will do in all the basic examples, but including positive characteristic fields gives slightly sharper general results). The universal Novikov field Λ over $\mathbb{K}$ is the field of formal series

$$f(q) = \sum_{d \in \mathbb{R}} a_d q^d$$

with coefficients $a_d \in \mathbb{K}$, with the following one-sided growth condition: for any $D \in \mathbb{R}$ there are at most finitely many $d \leq D$ such that $a_d \neq 0$. Floer homology associates to any $\phi \in Aut(M)$ a finite-dimensional $\mathbb{Z}/2$-graded Λ-vector space, the Floer homology group

$$HF_*(\phi) = HF_0(\phi) \oplus HF_1(\phi),$$

and these groups come with the following additional structure:

- There is a distinguished element $e \in HF_0(id)$ and a distinguished linear map $p : HF_0(id) \to \Lambda$.

- For any ϕ, ψ there is a canonical product, the so-called pair-of-pants product
$$* = *_{\phi,\psi} : HF_*(\phi) \otimes_\Lambda HF_*(\psi) \longrightarrow HF_*(\phi\psi)$$

- For any ϕ, ψ there is a conjugation isomorphism
$$c_{\phi,\psi} : HF_*(\phi) \xrightarrow{\cong} HF_*(\psi\phi\psi^{-1})$$

- For any smooth path $\lambda : [0; 1] \to Aut(M)$ there is a canonical continuation element $I_\lambda \in HF_0(\lambda_0^{-1}\lambda_1)$.

We now write down a rather long list of axioms satisfied by Floer homology theory. The aim is partly pedagogical, since this compares unfavourably with the later formulation in terms of a topological quantum field theory.

- $*$ is associative, in the sense that the two possible ways of bracketing give the same trilinear map $HF_*(\phi) \otimes HF_*(\psi) \otimes HF_*(\eta) \to HF_*(\phi\psi\eta)$. It is commutative, which means that the following diagram commutes:

$$
\begin{array}{ccc}
HF_*(\phi) \otimes HF_*(\psi) & \xrightarrow{\text{(signed) exchange}} & HF_*(\psi) \otimes HF_*(\phi) \\
\downarrow{*} & & \downarrow{*} \\
HF_*(\phi\psi) & \xrightarrow{\quad c_{\phi\psi,\phi^{-1}} \quad} & HF_*(\psi\phi)
\end{array}
$$

$e \in HF_*(id)$ is a two-sided unit for $*$, and for any ϕ we get a nondegenerate pairing between $HF_*(\phi)$ and $HF_*(\phi^{-1})$ by setting $\langle x, y \rangle = p(x * y)$.

- $c_{\phi,id}$ is the identity for any ϕ, and so is self-conjugation $c_{\phi,\phi}$ for any ϕ. Conjugation isomorphisms are well-behaved under composition, $c_{\psi\phi\psi^{-1},\eta} \circ c_{\phi,\psi} = c_{\phi,\eta\psi}$. They are compatible with pair-of-pants products, $c_{\phi,\eta}(x) * c_{\psi,\eta}(y) = c_{\phi\psi,\eta}(x * y)$. Moreover, conjugation $c_{id,\phi} : HF_*(id) \to HF_*(id)$ for any ϕ leaves e and p invariant.

- Any constant path λ gives rise to the element $I_\lambda = e \in HF_*(id)$. Two paths which are homotopic rel endpoints have the same continuation elements. Concatenation of paths corresponds to product of continuation elements, $I_{\lambda \circ \mu} = I_\lambda * I_\mu$. Next, if we compose a path λ with a fixed map ϕ, more precisely if $(L_\phi\lambda)_t = \phi\lambda_t$ and $(R_\phi\lambda)_t = \lambda_t\phi$, then

$$I_{L_\phi\lambda} = I_\lambda, \qquad I_{R_\phi\lambda} = c_{\lambda_0^{-1}\lambda_1,\phi^{-1}}(I_\lambda).$$

Remark 3.1 *We cannot pass this monument to abstract nonsense without lifting our hat to gerbes. For simplicity we consider only finite cyclic gerbes, so suppose that X is a connected topological space carrying a bundle of projective spaces $\mathbb{CP}^n \to E \to X$ with a $PU(n+1)$-connection, and ΩX the based loop space. To any $\phi \in \Omega X$ one can associate the monodromy $m_\phi \in PU(n+1)$. Take the set of all preimages of m_ϕ in $U(n+1)$, and let $I(\phi)$ be the $\mathbb{C}$-vector space freely generated by this set. (1) $I(constant\ path)$ is the group ring $\mathbb{C}[\mathbb{Z}/(n+1)]$, and we can define a canonical element e and linear map p as usual. (2) Since $m_{\phi\psi} = m_\phi m_\psi$, multiplication in $U(n+1)$ defines a composition map $I(\phi) \otimes I(\psi) \to I(\phi\psi)$. (3) Conjugation with m_ψ gives rise to an isomorphism $I(\phi) \to I(\psi\phi\psi^{-1})$. (4) For any homotopy λ_t in ΩX one can define a preferred element of $I(\lambda_1^{-1}\lambda_0)$ by deforming $\lambda_1^{-1}\lambda_0$ to the constant path, and taking e there. This satisfies all the properties stated above.*

The first consequence of the axioms is that $HF_*(id)$ is a graded commutative algebra with unit e. Actually, the trace p makes it into a Frobenius algebra. The conjugation maps $c_{id,\phi}$ define an action of $Aut(M)$ on $HF_*(id)$ by Frobenius algebra automorphisms, and this descends to an action of $\pi_0(Aut(M))$. To see that, note that for any $x \in HF_*(id)$ and any path λ starting at $\lambda_0 = id$, with corresponding reversed path $\bar\lambda$, we have $c_{\lambda_1,\lambda_1} = id$ and $I_\lambda * I_{\bar\lambda} = e$, hence

$$c_{id,\lambda_1}(x) = c_{id,\lambda_1}(x) * c_{\lambda_1,\lambda_1}(I_\lambda) * I_{\bar\lambda} = c_{\lambda_1,\lambda_1}(x * I_\lambda) * I_{\bar\lambda} = x * I_\lambda * I_{\bar\lambda} = x.$$

$HF_*(id)$ acts on each $HF_*(\phi)$ by left pair-of-pants product (one could equally use the product on the right, since for $x \in HF_*(id)$ and $y \in HF_*(\phi)$, $y * x = (-1)^{deg(x)deg(y)}c_{\phi,id}(x * y) = (-1)^{deg(x)deg(y)}x * y$). Here are some simple properties of the module structure, directly derived from the axioms:

Lemma 3.2 *(1) $x, c_{id,\phi}(x) \in HF_*(id)$ act in the same way on $y \in HF_*(\phi)$. (2) Up to isomorphism of $HF_*(id)$-modules, $HF_*(\phi)$ is an invariant of $[\phi] \in \pi_0(Aut(M))$. (3) There is a nondegenerate pairing $HF_*(\phi) \otimes HF_*(\phi^{-1}) \to \Lambda$ satisfying $\langle x*y, z\rangle = (-1)^{deg(x)deg(y)}\langle y, x*z\rangle$ for all $x \in HF_*(id)$, $y \in HF_*(\phi)$, $z \in HF_*(\phi^{-1})$.*

Proof. (1) $x*y = c_{\phi,\phi}(x*y) = c_{id,\phi}(x)*c_{\phi,\phi}(y) = c_{id,\phi}(x)*y$. (2) For any path λ, right multiplication with I_λ is an isomorphism $HF_*(\lambda_0) \to HF_*(\lambda_1)$ which commutes with left multiplication by elements of $HF_*(id)$. (3) The pairing is defined as $\langle y, z \rangle = p(y * z)$, and obviously has the desired properties. $\square$

(3b) By a theorem of Piunikhin–Salamon–Schwarz [41], Ruan–Tian [42], and Liu–Tian [31], $HF_*(id)$ is canonically isomorphic to the (small) *quantum homology ring* $QH_*(M)$. As a vector space, this is simply $H_*(M; \Lambda)$ with the grading reduced to $\mathbb{Z}/2$. The identity $e \in HF_*(id)$ is the fundamental class $[M]$, and the linear map p is induced from collapse $M \to point$. The action of $\pi_0(Aut(M))$ is the obvious action of symplectomorphisms on the homology of our manifold. The only non-topological element is the quantum intersection product, which corresponds to the pair-of-pants product in Floer cohomology, hence will be denoted by the same symbol $*$. It is defined by

$$(x_0 q^0 * y_0 q^0) \cdot z_0 q^0 = \sum_{A \in H_2(M; \mathbb{Z})} \Phi_{3,A}(x_0, y_0, z_0)\, q^{\omega(A)}$$

for $x_0, y_0, z_0 \in H_*(M; \mathbb{K})$, where $\cdot$ is the ordinary intersection pairing with Λ-coefficients, and $\Phi_{3,A}(x_0, y_0, z_0) \in \mathbb{K}$ the simplest kind of genus zero Gromov invariant, counting pseudo-holomorphic spheres in class A with three marked points lying on suitable representatives of x_0, y_0, z_0 respectively. Note that since symplectic four-manifolds are *weakly monotone*, we can (and will) use the older approach of Ruan–Tian [43] and McDuff–Salamon [35] to define Gromov invariants with coefficients in an arbitrary field $\mathbb{K}$. The leading term $\Phi_{3,0}(x_0, y_0, z_0)q^0$ counting constant pseudo-holomorphic curves is the ordinary triple intersection pairing, so the leading term in the quantum product is the ordinary intersection product.

Proposition 3.3 ([36, Corollary 1.6], largely based on results of [32]) *Let M be a closed symplectic four-manifold which is minimal, and not rational or ruled. Then $\Phi_{3,A} = 0$ for all $A \neq 0$.*

Among the cases not covered by the Proposition, rational surfaces are of primary interest because of the connection to classical enumerative problems in projective geometry. Here is a very simple example:

Example 3.4 *We will be using some representation theory of finite groups, so let $char(\mathbb{K}) = 0$ throughout the following computation. Take $M = \mathbb{CP}^2 \# k\overline{\mathbb{CP}^2}$, $5 \leq k \leq 8$, equipped with its monotone symplectic structure, normalized to $[\omega] = c_1$. Monotonicity simplifies the structure of the quantum product considerably: in the expansion*

$$x * y = (x \cap y) + (x *_1 y)q + (x *_2 y)q^2 + \dots$$

the q^d term has degree $2d-4$ with respect to the ordinary grading of $H_*(M; \Lambda)$, in particular the terms $q^5, q^6, \dots$ all disappear. Fixing some compatible (integrable) complex structure, one finds that the only holomorphic spheres with

$c_1(A) = 1$ *are the exceptional divisors, of which there is precisely one for each element of* $\mathcal{E} = \{A \in H_2(M; \mathbb{Z}) \ : \ c_1(A) = 1, \ A \cdot A = -1\}$. *By the divisor axiom for Gromov invariants, these classes A satisfy* $\Phi_{3,A}(x, y, z) = (x \cdot A)(y \cdot A)(z \cdot A)$ *for* $x, y, z \in H_2(M; \mathbb{K})$, *and hence*

$$x *_1 y = \sum_{A \in \mathcal{E}} (x \cdot A)(y \cdot A)A. \tag{1}$$

Let K be the Poincaré dual of $-c_1$, *and* $K^\perp \subset H_2(M; \mathbb{K})$ *its orthogonal complement with respect to the intersection form. Fact: for all* $x, y \in K^\perp$, $x *_1 y$ *is a multiple of K. This follows from* (1) *by explicit computation [9, Proposition 3.5.5]. For the most complicated cases* $k = 7, 8$ *one can also use a trick from [13, p. 33]:* $\bar{A} = (k - 6)(A \cdot K)K - A$ *is an involution of* $H_2(M; \mathbb{Z})$ *preserving K and the intersection form. It acts freely on* $\mathcal{E}$, *and the contributions of A and* $\bar{A}$ *to* $x *_1 y$ *add up to a multiple of K. Next, let W be the group of linear automorphisms of* $H_2(M; \mathbb{Z})$ *which preserve the intersection form, and leave K fixed. This is a reflection group of type* D_5, E_6, E_7 *or* E_8 *and it acts irreducibly on* $K^\perp$. *Moreover, each element of W can be realized by a symplectic automorphism of M, and so the quantum product is W-equivariant (this can also be checked by a direct computation of Gromov invariants, without appealing to the* $Aut(M)$*-action). Therefore, both* $*_1 : (K^\perp)^{\otimes 2} \to \mathbb{K}K \subset H_2(M; \mathbb{K})$ *and* $*_2 : (K^\perp)^{\otimes 2} \to H_4(M; \mathbb{K}) = \mathbb{K}$ *must be scalar multiples of the intersection form. We record this for later use, Fact: There is a* $z \in QH_*(M)$ *of the form* $z = [point] + \alpha_1 K q + \alpha_2 [M]q^2$ *for some* $\alpha_1, \alpha_2 \in \mathbb{K}$, *such that for all* $x, y \in K^\perp$, $x * y = (x \cdot y)z$.

(3c) The case $\phi = id$ is misleading in so far as for a general symplectic automorphism ϕ, $HF_*(\phi)$ has no known interpretation in terms of topology or Gromov–Witten invariants, and is hard or impossible to compute. Our insight into Dehn twists and their squares depends entirely on the following result:

Proposition 3.5 *For any Lagrangian sphere* $L \subset M$, *there is a long exact sequence*

$$H_*(S^2; \Lambda) \xrightarrow{\quad\quad} QH_*(M) \xrightarrow{\ G\ } HF_*(\tau_L)$$
$$\underset{\partial}{\xleftarrow{\hspace{3cm}}}$$

where the grading of $H_*(S^2; \Lambda)$ *is reduced to a* $\mathbb{Z}/2$*-grading,* ∂ *has odd degree, and G is a map of* $QH_*(M)$*-modules.*

The origins of this will be discussed extensively later, but for now let's pass directly to applications. Let $I_l \subset QH_*(M)$ be the ideal generated by $l = [L]q^0$.

Lemma 3.6 $dim_\Lambda I_l = 2$, *and moreover* I_l *is contained in* $QH_0(M)$.

Proof. Assume first that $char(\mathbb{K}) \neq 2$. Since $L \cdot L = -2$, we know that l is nontrivial and linearly independently from $l * l = -2[point] + \ldots$, so $dim_\Lambda I_l \geq 2$. The other half uses the Picard–Lefschetz formula (2). Since $(\tau_L)_*(l) = -l$, multiplication with l is an endomorphism of $QH_*(M)$ which exchanges the ± 1 eigenspaces of $(\tau_L)_*$. The $+1$ eigenspace has codimension one, and the -1 eigenspace has dimension one, and so the kernel of the multiplication map has codimension at most two, which means that its image has dimension at most two.

Without assumptions on the characteristic, one has to argue slightly more carefully as follows. We know that $[L] \in H_2(M; \mathbb{Z})$ is nontrivial and primitive, so there is a $w \in H_2(M; \mathbb{Z})$ with $w \cdot [L] = 1$. Denote the induced element of $H_2(M; \mathbb{K})$ equally by w. Then $l * w = [point] + \ldots$, from which it follows as before that $dim_\Lambda I_l \geq 2$. From the Picard–Lefschetz formula one gets

$$w * l + ((w * l) \cdot l)l = (\tau_L)_*(w * l) = (\tau_L)_*(w) * (\tau_L)_*(l) = -w * l - l * l,$$

which shows that $l * l$ lies in the linear subspace generated by l and $w * l$; and similarly for any $x \in QH_*(M)$,

$$w * x + ((w * x) \cdot l)l = (\tau_L)_*(w * x) = w * x + l * x + (x \cdot l)(w * l + l * l)$$

which shows that $l * x$ lies in the subspace generated by l and $w * l$. $\square$

Lemma 3.7 *The kernel of any $QH_*(M)$-module map $G : QH_*(M) \to HF_*(\tau_L)$ must contain I_l.*

Proof. Let w be as in the proof of the previous Lemma. From Lemma 3.2(1) we know that for any $y \in HF_*(\tau_L)$, $l * y = (\tau_L)_*(w) * y - w * y = 0$. Hence $G(l) = G(l * e) = l * G(e) = 0$, and therefore also $G(x * l) = 0$ for any x. $\square$

For the long exact sequence from Proposition 3.5, this means that the kernel of G is precisely I_l and that the differential δ is zero, showing that

$$HF_*(\tau_L) \cong QH_*(M)/I_l$$

as a $QH_*(M)$-module. Now suppose that τ_L^2 is symplectically isotopic to the identity. By Lemma 3.2(2) we have an isomorphism $HF_*(\tau_L^{-1}) \cong HF_*(\tau_L)$ of $QH_*(M)$-modules, and part (3) of the same Lemma shows that there is a nondegenerate pairing on $QH_*(M)/I_l$ which satisfies $\langle x * y, z \rangle = \pm \langle y, x * z \rangle$. Taking $y = e$ shows that $\langle x, z \rangle = \langle e, x * z \rangle$, so the pairing comes from the linear map $\langle e, - \rangle$ and the quantum product on $QH_*(M)/I_l$.

Corollary 3.8 *If τ_L^2 is symplectically isotopic to the identity, the quotient algebra $QH_*(M)/I_l$ is Frobenius. In particular, any linear subspace $W \subset QH_0(M)/I_l$ which satisfies $x \cdot y = 0$ for all $x, y \in W$ must satisfy $dim_\Lambda W \leq \frac{1}{2} dim_\Lambda QH_0(M)/I_l$.*

The first part is just the outcome of the preceding discussion, and the second part is an elementary fact about Frobenius algebras: W is an isotropic subspace with respect to the pairing, whence the bound on the dimension.

Corollary 3.9 *Let* M *be a closed minimal symplectic four-manifold with* $H^1(M; \mathbb{R}) = 0$, *and not rational or ruled. Suppose that* $\dim H_2(M; \mathbb{K}) \geq 3$. *Then for every Lagrangian sphere* $L \subset M$, τ_L^2 *is not symplectically isotopic to the identity, hence fragile.*

Proof. From Proposition 3.3,

$$QH_*(M)/I_l = H_*(M; \Lambda)/(\Lambda l \oplus \Lambda[point])$$

with the algebra structure induced by the ordinary intersection product. In particular, $W = H_2(M; \Lambda)/\Lambda l$ is a subspace satisfying the conditions of Corollary 3.8, and $\dim W = \dim H_2(M; \Lambda) - 1 > \frac{1}{2}\dim H_2(M; \Lambda) = \frac{1}{2}\dim QH_*(M)/I_l$. $\square$

Example 2.14 shows that the minimality assumption cannot be removed. The condition that M should not be rational excludes the case of $S^2 \times S^2$ discussed in Example 2.9. As for the final assumption $\dim H_2(M; \mathbb{K}) \geq 3$, a lack of suitable examples makes it hard to decide whether it is strictly necessary. In the algebro-geometric world, there are minimal surfaces of general type with Betti numbers $b_1(M) = 0$, $b_2(M) = 2$ exist, but the Miyaoka inequality $\chi - 3\sigma \geq \frac{9}{2}\#\{nodes\}$ [37] implies that they do not admit degenerations to nodal ones, thereby barring the main route to constructing Lagrangian spheres in them. Moreover, the most common explicit examples in the literature are uniformized by a polydisc, so they cannot contain any embedded spheres with nonzero selfintersection.

Example 3.10 *Take* $M = \mathbb{CP}^2 \# k\overline{\mathbb{CP}^2}$, $5 \leq k \leq 8$, *with a monotone symplectic form. As in Example 3.4 we use a coefficient field with* $char(\mathbb{K}) = 0$. *The computation carried out there shows that for any* $x, y \in K^\perp$, $x * y = -\frac{1}{2}(x \cdot y)l * l \in I_l$. *Hence, the image of* $K^\perp$ *in* $QH_*(M)/I_l$, *which is of dimension*

$$k - 1 > \frac{1}{2}(k + 1) = \frac{1}{2}\dim QH_*(M)/I_l, \tag{2}$$

violates the conditions of Corollary 3.8. It follows that in contrast with the situation for $k \leq 4$, *squared Dehn twists are never symplectically isotopic to the identity. For* $k = 5$, *we already saw some cases of this phenomenon in Example 2.13, and as explained there, this goes well with the intuition provided by the topology of moduli spaces. In a slightly different direction, one should note that the nontriviality of* τ^2 *has implications for the* π_1 *of spaces of symplectic embeddings of* k *balls into* $\mathbb{CP}^2$, *via the symplectic interpretation of blowup, see e.g. [10].*

It would be interesting to extend the entire discussion to arbitrary (not monotone) symplectic forms on rational four-manifolds. Although the Gromov invariants are constant under deformations of the symplectic class, the exponents $q^{\omega(A)}$ change, which affects the algebraic structure of the quantum homology ring, and thereby the criterion which we have used to explore the nature of squared Dehn twists. As a sample question, take a Lagrangian sphere L on, say, the cubic surface, and then perturb the symplectic class in a generic way subject only to the condition that L continues to be Lagrangian. Is it true that then, $QH_*(M)/I_l$ becomes semisimple? This is relevant because semisimple algebras are obviously Frobenius (see [9] for a proof of the generic semisimplicity of $QH_*(M)$ itself).

Finally, we turn to the proof of Theorem 1.5 stated in the introduction (together with Corollary 2.3, this also proves Corollary 1.6). Let $M \subset \mathbb{CP}^{n+2}$ be a nontrivial complete intersection of degrees $\mathbf{d} = (d_1, \ldots, d_n)$, $n \geq 1$ and $d_k \geq 2$, with the symplectic structure ω induced by the Fubini-Study form ω_{FS}, which we normalize to $\omega_{FS}^{n+2} = 1$. Each such M contains a Lagrangian sphere, which can be obtained as vanishing cycle in a generic pencil of complete intersections. Moreover,

$$\pi_1(M) = 1,$$

$$\chi(M) = \frac{1}{2}\Big(\prod_k d_k\Big)\Big[\Big(\sum_k d_k - (n+3)\Big)^2 + \sum_k d_k^2 - (n+3)\Big],$$

$$c_1(M) = \Big(n + 3 - \sum_k d_k\Big)[\omega].$$

With the exception of six choices of degrees $\mathbf{d} = (2), (3), (4), (2,2), (2,3)$, $(2,2,2)$, $c_1(M)$ is a negative multiple of $[\omega]$, so M is minimal and of general type, and $\chi(M) > \sum_k d_k(d_k - 1) \geq 6$, which means $b_2(M) \geq 4$, so Corollary 3.9 applies. Out of the remaning cases, three are $K3$ surfaces, $\mathbf{d} = (4), (2,3), (2,2,2)$, to which Corollary 3.9 also applies. The other three are $\mathbf{d} = (2)$ which is the quadric $\mathbb{CP}^1 \times \mathbb{CP}^1$, hence excluded from the statement of Theorem 1.5, and $\mathbf{d} = (3), (2,2)$ which are the del Pezzo surfaces of rank $k = 6, 5$ respectively, and therefore fall under Example 3.10.

(3d) As promised, we will now present Floer homology theory as a TQFT in $1 + 1$ dimensions "coupled with" symplectic fibrations. This is a generalization of the setup from [41] where only the trivial fibration was allowed (Lalonde has recently introduced a very similar generalization, but his intended applications are quite different). Throughout the following discussion, all symplectic fibrations have fibres isomorphic to M, without any specific choice of isomorphism. The basic data are:

- For any symplectic fibration $F \to Z$ over an oriented circle Z, we have a Floer homology group $HF_*(Z, F)$.
- For any isomorphism $\Gamma : F_1 \to F_2$ between such fibrations covering an orientation-preserving diffeomorphism $\gamma : Z_1 \to Z_2$, there is an induced canonical isomorphism $C(\gamma, \Gamma) : HF_*(Z_1, F_1) \to HF_*(Z_2, F_2)$.
- Let S be a connected compact oriented surface with $p+q$ boundary circles. We arbitrarily divide the circles into positive and negative ones, and reverse the natural induced orientation of the latter, so that

$$\partial S = \bar{Z}_1^- \cup \cdots \cup \bar{Z}_p^- \cup Z_{p+1}^+ \cup \cdots \cup Z_{p+q}^+.$$

Given a symplectic fibration $E \to S$, with restrictions $F_k^{\pm} = E|Z_k^{\pm}$, we have a relative Gromov invariant

$$G(S, E) : \bigotimes_{k=1}^{p} HF_*(Z_k^-, F_k^-) \longrightarrow \bigotimes_{k=p+1}^{p+q} HF_*(Z_k^+, F_k^+).$$

This is independent of the way in which the Z^-, Z^+ are numbered, up to the usual signed interchange of factors in the tensor product.

The maps $C(\gamma, \Gamma)$, sometimes omitted from more summary expositions, are a natural part of the theory: after all, the "cobordism category" is more properly a 2-category [50], and the algebraic framework should reflect this. The TQFT axioms are:

- The identity automorphism of each (Z, F) induces the identity $C(id_Z, id_F)$ on Floer homology. The maps $C(\gamma, \Gamma)$ are well-behaved under composition of isomorphisms. Moreover, if (γ^t, Γ^t), $t \in [0; 1]$ is a smooth family of isomorphisms $F_1 \to F_2$, then $C(\gamma^0, \Gamma^0) = C(\gamma^1, \Gamma^1)$.

- Let $\xi : S_1 \to S_2$ be an orientation-preserving diffeomorphism of surfaces, which respects the decomposition of the boundary into positive and negative circles, and suppose that this is covered by an isomorphism $\Xi : E_1 \to E_2$ of symplectic fibrations. Let $\gamma_k^{\pm}$, $\Gamma_k^{\pm}$ be the restriction of ξ, Ξ to the boundary components. Then the following diagram commutes:

$$
\begin{array}{ccc}
\bigotimes_{k=1}^{p} HF_*(Z_{1,k}^-, F_{1,k}^-) & \xrightarrow{G(S_1, E_1)} & \bigotimes_{k=p+1}^{p+q} HF_*(Z_{1,k}^+, F_{1,k}^+) \\
\Big\downarrow{\scriptstyle \bigotimes_{k=1}^{p} C(\gamma_k^-, \Gamma_k^-)} & & \Big\downarrow{\scriptstyle \bigotimes_{k=p+1}^{p+q} C(\gamma_k^+, \Gamma_k^+)} \\
\bigotimes_{k=1}^{p} HF_*(Z_{2,k}^-, F_{2,k}^-) & \xrightarrow{G(S_2, E_2)} & \bigotimes_{k=p+1}^{p+q} HF_*(Z_{2,k}^+, F_{2,k}^+)
\end{array}
$$

- Take any $F \to Z$ and pull it back by projection to a fibration $E \to S = [1; 2] \times Z$. (a) If we take $Z_1 = \{1\} \times Z$ negative and $Z_2 = \{2\} \times Z$ positive, the relative Gromov invariant $HF_*(Z, F) \to HF_*(Z, F)$ is the identity. (b) Take both Z_1, Z_2 to be negative. Then the relative Gromov invariant $HF_*(Z, F) \otimes HF_*(\bar{Z}, \bar{F}) \to \Lambda$, where $(\bar{Z}, \bar{F})$ denotes orientation-reversal on the base, is a nondegenerate pairing.

- The gluing or cut-and-paste axiom. Let S_1, S_2 be two surfaces carrying symplectic fibrations E_1, E_2, and suppose that we have an isomorphism (γ, Γ) between the induced fibrations over the mth positive boundary circle of S_1 and the nth negative one of S_2. One can glue together the two boundary components to form a surface $S = S_1 \cup S_2$ and a symplectic fibration E over it, and the associated relative invariant $G(S, E)$ is the composition

$$\left(\bigotimes_k HF_*(Z_{1,k}^-, F_{1,k}^-) \right) \otimes \left(\bigotimes_{l \neq n} HF_*(Z_{2,l}^-, F_{2,l}^-) \right)$$

$$\Big\downarrow G(S_1, E_2) \otimes id$$

$$\left(\bigotimes_k HF_*(Z_{1,k}^+, F_{1,k}^+) \right) \otimes \left(\bigotimes_{l \neq n} HF_*(Z_{2,l}^-, F_{2,l}^-) \right)$$

$$\cong \Big\downarrow \text{exchange and } C(\gamma, \Gamma)$$

$$\left(\bigotimes_{k \neq m} HF_*(Z_{1,k}^+, F_{1,k}^+) \right) \otimes \left(\bigotimes_l HF_*(Z_{2,l}^-, F_{2,l}^-) \right)$$

$$\Big\downarrow id \otimes G(S_2, E_2)$$

$$\left(\bigotimes_{k \neq m} HF_*(Z_{1,k}^+, F_{1,k}^+) \right) \otimes \left(\bigotimes_l HF_*(Z_{2,l}^+, F_{2,l}^+) \right)$$

How does this set of axioms for Floer homology imply the previously used one? To any $\phi \in Aut(M)$ one can associate the mapping torus $F_\phi = \mathbb{R} \times M/(t,x) \sim (t-1, \phi(x))$, which is naturally a symplectic fibration over $S^1 = \mathbb{R}/\mathbb{Z}$. Set $HF_*(\phi) = HF_*(S^1, F_\phi)$. Given an isotopy (λ_t) in $Aut(M)$ (with λ_t constant for t close to the endpoints $0, 1$), one can define an isomorphism $\Gamma_\lambda : F_{\lambda_0} \to F_{\lambda_1}$ by $[0;1] \times M \to [0;1] \times M$, $(t,x) \mapsto (t, \lambda_t^{-1}\lambda_0(x))$, and the corresponding map $C(id_{S^1}, \Gamma_\lambda)$ is our previous I_λ. For any ψ there is a canonical isomorphism $\Gamma_{\phi,\psi} : F_\phi \to F_{\psi\phi\psi^{-1}}$, $(t,x) \mapsto (t, \psi(x))$, and we correspondingly define the conjugation isomorphism $c_{\phi,\psi} = C(id_{S^1}, \Gamma_{\phi,\psi})$. In the case where $\phi = \psi$, the isomorphism $\Gamma_{\phi,\phi}$ can be deformed to the identity by rotating the base once, $(t,x) \mapsto (t-\tau, \phi(x))$, and this explains the previously stated property that $c_{\phi,\phi} = id$. Extracting the remaining structure, such as the pair-of-pants and its properties, is staple TQFT fare, which can be found in any expository account such as [44].

Having come this far, we can make a straightforward extension to the formalism, which is to replace symplectic fibrations by Lefschetz fibrations in the sense of Definition 2.5. This requires some small modifications of the axioms, since even in the absence of critical points, our definition of Lefschetz fibrations contains more information (the two-form Ω on the total space) than that of symplectic fibration. Essentially, one has to add another property saying that relative Gromov invariants are unchanged under deformation of a Lefschetz fibration. But we have spent enough time with exercises in axiomatics, so we leave the precise formulation to the reader. The essential new ingredient that comes from Lefschetz fibrations is this: given any Lagrangian sphere $L \subset M$, one can construct a Lefschetz fibration E' over a disc S' with a single critical point, whose associated vanishing cycle is $L \subset M$. By the Picard–Lefschetz theorem, the monodromy around the boundary is isotopic to the Dehn twist $\tau_L \in Aut(M)$, and the associated relative Gromov invariant provides a distinguished element

$$\theta_L \overset{\text{def}}{=} G(S', E') \in HF_*(\tau_L). \tag{3}$$

One can show that E' is unique up to deformation, so this class is independent of the details of the construction. Pair-of-pants product with θ_L yields for any $\phi \in Aut(M)$ a canonical homomorphism

$$HF_*(\phi) \longrightarrow HF_*(\phi \circ \tau_L),$$

and the special case $\phi = id_M$ is the map G from Proposition 3.5. The fact that this map is a homomorphism of $QH_*(M)$-modules follows from associativity of the pair-of-pants product.

Remark 3.11 *One expects that Proposition 3.5 is a special case of a more general long exact sequence, of the form*

$$HF_*(L, \phi(L)) \longrightarrow HF_*(\phi) \longrightarrow HF_*(\phi \circ \tau_L) \qquad (4)$$

for an arbitrary $\phi \in Aut(M)$. The appearance of Lagrangian intersection Floer homology means that in order to understand this sequence, our framework should be further extended to an "open-closed" string theory, where the symplectic fibrations are allowed to carry Lagrangian boundary conditions, see [48]. In cases where Lagrangian Floer homology is well-behaved, such as when M is an exact symplectic manifold with boundary, the sequence (4) can be readily proved by adapting of the argument from [48]. When M is a closed four-manifold, $HF_*(L_0, L_1)$ is not always defined, but this should not be an issue for the case for the group in (4), since the obstructions in the sense of [17, 40] coming from L and $\phi(L)$ ought to cancel out. With this in mind, the proof should go through much as before, but there are still some technical points to be cleared up, so we will stop short of claiming it as a theorem.

Remark 3.12 *The condition $H^1(M; \mathbb{R}) = 0$ can be removed from the whole section, at the cost of replacing $Aut(M)$ by $Aut^h(M)$. The Dehn twist along a Lagrangian sphere is unique up to Hamiltonian isotopy; the axioms for Floer homology remain the same except that the elements I_λ exist only for Hamiltonian isotopies; and since the basic Proposition 3.5 continues to hold, so do all its consequences. In the construction of the TQFT, one has to replace symplectic fibrations by Hamiltonian fibrations (Lefschetz fibrations as we defined them are already Hamiltonian).*

A more interesting question is whether for $H^1(M; \mathbb{R}) \neq 0$, it can happen that τ_L^2 is symplectically and not Hamiltonian isotopic to the identity. Assuming some unproved but quite likely statements, one can give a negative answer to this at least in the case when $c_1 = \lambda[\omega]$ for $\lambda < 0$ and the divisibility of c_1 is $N \geq 2$. Suppose that $\phi = \tau_L^{-2}$ is symplectically, but not Hamiltonian, isotopic to the identity. There is a theorem of Lê-Ono [29] which determines $HF_*(\phi)$ in the opposite sign case where $\lambda > 0$. It seems reasonable to expect this to hold in our case too, so that

$$HF_*(\phi) \cong H_*(M; \underline{\Lambda}) \qquad (5)$$

where $\underline{\Lambda}$ is a nontrivial local system of Λ-coefficients determined by the flux of ϕ. We also assume the long exact sequence (4) for ϕ, which would be

$$H_*(L; \underline{\Lambda}) \longrightarrow HF_*(\phi) \longrightarrow HF_*(\phi \circ \tau_L) \qquad (6)$$
$$\underbrace{\qquad\qquad\qquad\qquad}_{\delta \ (\text{of degree } -1)}$$

Floer homology groups are now $\mathbb{Z}/2N$-graded, and by combining (5) with (6) and standard facts about Novikov homology, one sees that $HF_*(\phi \circ \tau_L)$ is concentrated in three adjacent degrees. On the other hand, we still have $HF_*(\tau_L) \cong QH_*(M)/I_l$, which is nonzero in four degrees, hence $HF_*(\tau_L^{-1}) \not\cong HF_*(\phi \circ \tau_L)$.

4 Pseudo-Holomorphic Sections and Curvature

(4a) The aim of this section is to explain the proof of Proposition 3.5, but we start on a much more basic level with the definition of Floer homology according to Hofer–Salamon [22], recast in fibre bundle language. Let

$$p : F \longrightarrow S^1 = \mathbb{R}/\mathbb{Z}$$

be a smooth proper fibration with four-dimensional fibres, equipped with a closed two-form Ω whose restriction to each fibre is symplectic. We have the corresponding symplectic connection $TF = TF^h \oplus TF^v$. Let $\mathcal{S}(S^1, F)$ be the space of all smooth sections of p, and $\mathcal{S}^h(S^1, F)$ the subspace of horizontal sections, which are those with $d\sigma/dt \in TF^h$. To $\sigma \in \mathcal{S}^h(S^1, F)$ one can associate a linear connection ∇^σ on the pullback bundle $\sigma^* TF^v$,

$$\nabla^\sigma_{\partial_t} X = [d\sigma/dt, X].$$

We say that F has *nondegenerate horizontal sections* if for every σ, there are no nonzero solutions of $\nabla^\sigma X = 0$. We will assume from now on that this is the case; then $\mathcal{S}^h(S^1, F)$ is finite, and one defines the Floer chain group as

$$CF_*(S^1, F) = \bigoplus_{\sigma \in \mathcal{S}^h(S^1, F)} \Lambda\langle\sigma\rangle.$$

The $\mathbb{Z}/2$ degree of a generator $\langle\sigma\rangle$ is determined by the sign of $\det(1 - R^\sigma)$, where R^σ is the monodromy of ∇^σ around S^1. There is a closed action one-form da on $\mathcal{S}(S^1, F)$ whose critical point set is $\mathcal{S}^h(S^1, F)$, namely

$$da(\sigma)X = \int_{S^1} \Omega(d\sigma/dt, X)\, dt$$

for $X \in T_\sigma \mathcal{S}(S^1, F) = C^\infty(\sigma^* TF^v)$. Nondegeneracy of the horizontal sections corresponds to the Morse nondegeneracy of a local primitive a. Now take a smooth family of $\Omega|F_t$-compatible almost complex structures J_{F_t} on the fibres. The negative gradient flow lines of da with respect to the resulting L^2 metric are the solutions of Floer's equation. In view of later developments, we find it convenient to write the equation as follows. Take $\pi = id_{\mathbb{R}} \times p :$ $E = \mathbb{R} \times F \to S = \mathbb{R} \times S^1$. Equip S with its standard complex structure j, and E with the almost complex structure J characterized by the following properties: (1) π is (J,j)-holomorphic; (2) The restriction of J to any fibre $E_{s,t}$ is equal to J_{F_t}; (3) J preserves the splitting of TE into horizontal and vertical parts induced by the pullback of Ω. Then Floer's equation translates into the pseudo-holomorphic section equation

$$\begin{cases} u : S \longrightarrow E, \quad \pi \circ u = id_S \\ Du \circ j = J \circ Du, \\ \lim_{s \to \pm\infty} u(s,\cdot) = \sigma_{\pm}, \end{cases} \tag{1}$$

where $\sigma_{\pm} \in \mathcal{S}^h(S^1, F)$. Note that for any $\sigma \in \mathcal{S}^h(S^1, F)$ there is a trivial or stationary solution $u(s,t) = (s, \sigma(t))$ of (1). We denote by $\mathcal{M}^*(S, E; \sigma_-, \sigma_+)$ the space of all *other* solutions, divided by the free $\mathbb{R}$-action of translation in s-direction; and by $\mathcal{M}_0^*(S, E; \sigma_-, \sigma_+)$ the subspace of those solutions whose virtual dimension is equal to zero. The Floer differential on $CF_*(S^1, F)$ is defined by

$$\partial\langle\sigma_-\rangle \;=\; \sum_{\substack{\sigma_+ \in \mathcal{S}^h(S^1,F) \\ u \in \mathcal{M}_0^*(S,E;\sigma_-,\sigma_+)}} \pm q^{\epsilon(u)}\langle\sigma_+\rangle$$

where the energy is $\epsilon(u) = \int_S u^*\Omega \in (0; \infty)$, and the sign is determined by coherent orientations which we will not explain further. For this to actually work and give the correct Floer homology $HF_* = H_*(CF_*, \partial)$, the J_{F_t} need to satisfy a number of generic "transversality" properties:

- There are no non-constant J_{F_t}-holomorphic spheres of Chern number ≤ 0.
- If $v : S^2 \to F_t$ is a J_{F_t}-holomorphic map with Chern number one, the image of v is disjoint from $\sigma(t)$ for all $\sigma \in \mathcal{S}^h(S^1, F)$.
- The linearized operator $D\bar\partial_u$ attached to any solution of Floer's equation is onto. This means that the spaces $\mathcal{M}^*(S, E; \sigma_-, \sigma_+)$ are all smooth of the expected dimension.

The space of pseudo-holomorphic spheres with Chern number ≤ 0 in a four-manifold has virtual dimension ≤ -2, so that even in a one-parameter family of manifolds the virtual dimension remains negative. As for the images of pseudo-holomorphic spheres with Chern number one, they form a codimension 2 subset in a four-manifold, and the same thing holds in a family, so they should typically avoid the image of any fixed finite set of sections, which is one-dimensional. In both cases, the fact that the condition is actually generic is proved by appealing to the theory of somewhere injective pseudo-holomorphic curves, see for instance [35]. The last requirement is slightly more tricky because of the $\mathbb{R}$-symmetry on the moduli space; see [16] for a proof.

We now introduce the second ingredient of the TQFT, the relative Gromov invariants. As a basic technical point, the surfaces with boundary which we used to state the axioms must be replaced by noncompact surfaces with a boundary at infinity. For ease of formulation, we will consider only the case of $S = \mathbb{R} \times S^1$, which is the one relevant for our applications. Let $\pi : E \to S$ be a smooth proper fibration with four-dimensional fibres, equipped with a closed two-form Ω whose restriction to any fibre is symplectic. The behaviour of E over the two ends of our surface is governed by the following "tubular ends" assumptions: there are fibrations $p^\pm : F^\pm \to S^1$ with two-forms $\Omega^\pm$ as before, with the property that the horizontal sections are nondegenerate, and fibered diffeomorphisms $\Psi^- : E|(-\infty; s_-] \longrightarrow (-\infty; s_-] \times F^-$, $\Psi^+ : E|[s_+; \infty) \longrightarrow [s_+; \infty) \times F^+$ for some $s_- < s_+$, such that $(\Psi^\pm)^* \Omega^\pm = \Omega$.

Take a positively oriented complex structure j on S. We say that an almost complex structure J on E is *semi-compatible* with Ω if π is (J, j)-holomorphic, and the restriction of J to each fibre is compatible with the symplectic form in the usual sense. With respect to the splitting $TE_x = TE_x^h \oplus TE_x^v$, this means that

$$J = \begin{pmatrix} j & 0 \\ J^{vh} & J^{vv} \end{pmatrix} \tag{2}$$

where J^{vv} is a family of compatible almost complex structures on the fibres, and J^{vh} is a $\mathbb{C}$-antilinear map $TE^h \to TE^v$ (this corresponds to the "inhomogeneous term" in the theory of pseudoholomorphic maps). We also need to impose some conditions at infinity. Choose families of almost complex structures $J_{F_t^-}$, $J_{F_t^+}$ on the fibres of $F^\pm$ which are admissible for Floer theory, meaning that they satisfy the transversality properties stated above and can therefore be used to define $HF_*(S^1, F^\pm)$. These give rise to almost complex structures $J^\pm$ on the products $\mathbb{R} \times F^\pm$, and the requirements are that j is standard on $(-\infty; s_-] \times S^1$ and $[s_+; \infty) \times S^1$, and $\Psi^\pm$ is $(J, J^\pm)$-holomorphic. We denote the space of such pairs (j, J), for a fixed choice of $J^\pm$, by $\mathcal{J}(S, E)$.

For $\sigma_- \in \mathcal{S}^h(S^1, F^-)$, $\sigma_+ \in \mathcal{S}^h(S^1, F^+)$, consider the space $\mathcal{M}(S, E; \sigma_-, \sigma_+)$ of sections $u : S \to E$ satisfying the same (1) as before, where the convergence conditions should be more properly formulated as $\Psi^\pm(u(s, t)) = (s, u^\pm(s, t))$ with $u^\pm(s, \cdot) \to \sigma_\pm$ in $\mathcal{S}(S^1, F^\pm)$. Writing $\mathcal{M}_0(S, E; \sigma_-, \sigma_+)$ for the subspace where the virtual dimension is zero, one defines a chain homomorphism $CG(S, E) : CF_*(S^1, F^-) \to CF_*(S^1, F^+)$ by

$$CG(S, E)\langle \sigma_- \rangle = \sum_{\substack{\sigma_+ \in \mathcal{S}^h(S^1, F^+) \\ u \in \mathcal{M}_0(S, E; \sigma_-, \sigma_+)}} \pm q^{\epsilon(u)} \langle \sigma_+ \rangle \tag{3}$$

The relative Gromov invariant is the induced map on homology. As before, there are a number of conditions that J has to satisfy, in order for (3) to be a well-defined and meaningful expression:

- There are no J-holomorphic spheres in any fibre of E with strictly negative Chern number.
- If $v : S^2 \to E_{s,t}$ is a non-constant J-holomorphic sphere with Chern number zero, its image does not contain $u(s,t)$ for any $u \in \mathcal{M}_0(S, E; \sigma_-, \sigma_+)$.
- The linearized operator $D\bar{\partial}_u$ associated to any $u \in \mathcal{M}(S, E; \sigma_-, \sigma_+)$ is onto.

Note that because our fibration is a two-parameter family of symplectic four-manifolds, pseudo-holomorphic spheres in the fibres with Chern number zero can no longer be avoided, even though one can always achieve that a particular fixed fibre contains none of them. The proof that the above conditions are generic is standard; for details consult [22, 35].

There is little difficulty in replacing our symplectic fibration with a Lefschetz fibration $\pi : E \to S$, having the same kind of behaviour at infinity. In this case, the definition of $\mathcal{J}(S, E)$ includes the additional requirements that $j = j_S$ in a neighbourhood of the critical values, and $J = J_E$ near the critical points. A smooth section cannot pass through any critical point, so the analytic setup for the moduli spaces $\mathcal{M}_0(S, E; \sigma_-, \sigma_+)$ remains the same as before. Of course, pseudo-holomorphic spheres in the singular fibres appear in the Gromov–Uhlenbeck compactification of the space of sections, and to avoid potential problems with them one has to impose another condition on J:

- If $(s,t) \in S^{crit}$ and $v : S^2 \to E$ is a nonconstant J-holomorphic map with image in $E_{s,t}$, then $\langle c_1(E), [v] \rangle > 0$.

To prove genericity of this, one considers the minimal resolution $\hat{E}_{s,t}$ of $E_{s,t}$, which is well-defined because our complex structure J is integrable near the singularities. It is a feature of ordinary double points in two complex dimensions (closely related to simultaneous resolutions) that $c_1(\hat{E}_{s,t})$ is the pullback of $c_1(E)|E_{s,t}$. By a small perturbation of the almost complex structure on the resolution, supported away from the exceptional divisor, one can achieve that there are no pseudo-holomorphic curves $\hat{v} : S^2 \to \hat{E}_{s,t}$ with $\langle c_1(\hat{E}_{s,t}), [\hat{v}] \rangle \le 0$ except for the exceptional divisor itself and its multiple covers. The desired result follows by lifting pseudoholomorphic spheres from $E_{s,t}$ to the resolution.

(4b) Solutions of Floer's equation have two properties not shared by more general pseudoholomorphic sections: (1) there is an $\mathbb{R}$-action by translations; (2) the energy of any pseudoholomorphic section is $\epsilon(u) \ge 0$, and those with zero energy are horizontal sections of the symplectic connection. While (1) is characteristic of Floer's equation, (2) can be extended to a wider class of geometric situations, as follows. Let $\pi : E \to S = \mathbb{R} \times S^1$ be a Lefschetz fibration with the same "tubular end" structure as before. We say that E has *nonnegative (Hamiltonian) curvature* if for any point $x \notin E^{crit}$, the restriction of Ω to TE_x^h is nonnegative with respect to the orientation induced from $TS_{\pi(x)}$. A pair $(j, J) \in \mathcal{J}(S, E)$ is *fully compatible* if $\Omega(\cdot, J\cdot)$ is symmetric, or equivalently $J(TE_x^h) \subset TE_x^h$ for all $x \notin E^{crit}$. With respect to the decomposition (2) this means that $J^{vh} = 0$. The following result is straightforward:

Lemma 4.1 *Suppose that E has nonnegative curvature, and that J is fully compatible. Then any $u \in \mathcal{M}(S, E; \sigma_-, \sigma_+)$ satisfies $\epsilon(u) \geq 0$. Any horizontal (covariantly constant) section is automatically J-holomorphic; in the converse direction, any $u \in \mathcal{M}(S, E; \sigma_-, \sigma_+)$ with $\epsilon(u) = 0$ must necessarily be horizontal.*

To take advantage of this, one would like to make the spaces of pseudo-holomorphic sections regular by choosing a generic J within the class of fully compatible almost complex structures. It is easy to see that all non-horizontal $u \in \mathcal{M}(S, E; \sigma_-, \sigma_+)$ can be made regular in this way, but the horizontal sections persist for any choice of J, so we have to enforce their regularity by making additional assumptions. The following Lemma is useful for that purpose:

Lemma 4.2 *In the situation of Lemma 4.1, let u be a horizontal section. Suppose that $\epsilon(u) = 0$, and that the associated linearized operator $D\bar{\partial}_u$ has index zero. Then u is regular, which is to say that $D\bar{\partial}_u$ is onto.*

This is an easy consequence of a Weitzenböck argument, see [48, Lemmas 2.11 and 2.27]. Hence, if any horizontal u satisfies the condition of the Lemma, one can indeed choose a fully compatible J which makes the moduli spaces of pseudo-holomorphic sections regular. Full compatibility does not restrict the behaviour of J on the fibres, so we can also achieve all the other conditions needed to make the relative Gromov invariant well-defined. After expanding the resulting chain homomorphism into powers of q,

$$CG(S, E) = \sum_{d \geq 0} CG(S, E)_d q^d \tag{4}$$

one finds that the leading term $CG(S, E)_0$ counts only horizontal sections.

Lemma 4.3 *Suppose that E has nonnegative curvature, and that for any $\sigma_+ \in \mathcal{S}^h(S^1, F^+)$ there is a horizontal section u of E with $\lim_{s \to +\infty} u(s, \cdot) = \sigma_+$, such that $\epsilon(u) = 0$ and $D\bar{\partial}_u$ has index zero. Then, for a suitable choice of fully compatible almost complex structure J, the cochain level map $CG(S, E) :$ $CF_*(S^1, F^-) \to CF_*(S^1, F^+)$ is surjective.*

Proof. Since horizontal sections are determined by their value at any single point, it follows that for any σ_+ there is a unique horizontal section $u = u_{\sigma_+}$ approaching it, and that these are all horizontal sections. Consider the map $R : CF_*(S^1, F^+) \to CF_*(S^1, F^-)$ which maps $\langle \sigma_+ \rangle$ to the generator $\langle \sigma_- \rangle$ associated to the negative limit of u_{σ_+}. From (4) one sees that $(CG(S, E) \circ R)\langle \sigma_+ \rangle = \pm \langle \sigma_+ \rangle +$ (strictly positive powers of q), which clearly shows that $CG(S, E)$ is onto. $\square$

We now turn to the concrete problem posed by a Lagrangian sphere L in a symplectic four-manifold M. Choose a symplectic embedding $i : T^*_{<\lambda} S^2 \to M$

with $i(S^2) = L$. Take a model Dehn twist τ defined using a function r which satisfies

$$r'(t) \begin{cases} \in [1/4; 3/4] & t \in [0; \mu), \\ \in [0; 1/2] & t \in [\mu; \lambda/2), \\ = 0 & t \geq \lambda/2 \end{cases} \tag{5}$$

for some $\mu < \lambda/2$, and transplant it to a Dehn twist τ_L using i. Next, choose a a Morse function $H : M \to \mathbb{R}$ with the properties that (1) $H(i(u,v)) = ||u||$ for all $\mu \leq ||u|| < \lambda$; (2) h has precisely two critical points in $im(i)$, both of which lie on L (their Morse indices will of course be 0 and 2). Let ϕ be the Hamiltonian flow of H for small positive time $\delta > 0$, and $\tilde{\tau}_L = \tau_L \circ \phi$. By construction, $\tilde{\tau}_L = \phi$ outside $i(T^*_{<\lambda/2} S^2)$.

Lemma 4.4 *If δ is sufficiently small, $\tilde{\tau}_L$ has no fixed points inside $im(i)$.*

Proof. As δ becomes small, the fixed points of $\tilde{\tau}_L$ accumulate at the fixed points of τ_L, hence they will lie outside $i(T^*_{\leq\mu} S^2)$. Recalling the definition of the model Dehn twist, we have that

$$(i^{-1}\tilde{\tau}_L i)(u,v) = \sigma_{2\pi(r'(||u||)+\delta)}(u,v)$$

for $||u|| \geq \mu$, and since $r'(||u||) + \delta \in [\delta; 1/2 + \delta]$ cannot be an integer, $i(u,v)$ cannot be a fixed point. $\square$

Let $S' \subset S = \mathbb{R} \times S^1$ be the disc of radius $1/4$ around $(s,t) = (0,0)$, and $S'' = S \setminus int(S')$. There is a Lefschetz fibration $E' \to S'$ with fibre M, whose monodromy around ∂D is a Dehn twist τ_L defined using a function r that satisfies (5). This is explicitly constructed in [48, Sect. 1.2], where properties somewhat stricter to (5) are subsumed under the notion of "wobblyness". To complement this, there is a fibration $E'' \to S''$ with a two-form Ω'' such that the monodromy of the resulting symplectic connection is ϕ around the loop $\{-1\} \times S^1$, $\tau_L \circ \phi = \tilde{\tau}_L$ around $\{+1\} \times S^1$, and τ_L around $\partial S''$. This is actually much simpler to write down:

$$E'' = \frac{\{z = s + it \in \mathbb{R} \times [0; 1] \ : \ |z| \geq 1/4, |z - i| \geq 1/4\}}{(s, 1, x) \sim (s, 0, \phi(x)) \text{ for } s < 0, \ (s, 1, x) \sim (s, 0, \tilde{\tau}_L(x)) \text{ for } s > 0}$$

One can glue together the two pieces along $\partial S' = \partial S''$ to a Lefschetz fibration $E \to S$, and the resulting chain level map is

$$CG(S, E) : CF_*(\phi) \longrightarrow CF_*(\tilde{\tau}_L). \tag{6}$$

It can be arranged that E' has nonnegative curvature, and that E'' is flat (zero curvature), so the curvature of E is again nonnegative. Actually, the construction of E', like that of τ_L itself, is based on the local model of $T^*_{\leq\lambda} S^2$, so that E' contains a trivial piece $S' \times (M \setminus im(i))$. Using this and Lemma 4.4 one sees that for any fixed point x of $\tilde{\tau}_L$, which the same as a critical point of H lying outside $im(i)$, there is a horizontal section u of E such that $u(s,t) = (s,t,x)$ for $s > 1/2$, and that these sections satisfy the conditions of Lemma 4.3. Hence (6) is onto for a suitable choice of almost complex structure; but from the definition of H, we know that its kernel is two-dimensional and concentrated in $CF_{even}(\phi)$, which implies that the induced map $G = G(S, E)$ on Floer homology fits into a long exact sequence as stated in Proposition 3.5. On the other hand, a gluing argument which separates the two pieces in our construction of E shows that one can indeed write G as pair-of-pants product with an element θ_L as in (3).

Acknowledgments

Obviously, this work owes enormously to Simon Donaldson, whose student I was while carrying it out originally. Peter Kronheimer pointed out simultaneous resolution, which was a turning-point. I have profited from discussions with Norbert A'Campo, Denis Auroux, Mike Callahan, Allen Knutson, Anatoly Libgober, Dusa McDuff, Leonid Polterovich, Ivan Smith, Dietmar Salamon, Richard Thomas, Andrey Todorov and Claude Viterbo. Finally, I'd like to thank the organizers and audience of the C.I.M.E. summer school for giving me the opportunity to return to the subject.

References

1. M. Abreu, *Topology of symplectomorphism groups of $S^2 \times S^2$*, Invent. Math. **131** (1998), 1–24.
2. M. Abreu and D. McDuff, *Topology of symplectomorphism groups of rational ruled surfaces*, J. Am. Math. Soc. **13** (2000), 971–1009.
3. D. Allcock, *Asphericity of moduli spaces via curvature*, J. Differ. Geom. **55** (2000), 441–451.
4. D. Allcock, J. Carlson, and D. Toledo, *A complex hyperbolic structure for moduli of cubic surfaces*, C. R. Acad. Sci. Ser. I **326** (1998), no. 1, 49–54.
5. S. Anjos, *Homotopy type of symplectomorphism groups of $S^2 \times S^2$*, Geom. Topol. **6** (2002), 195–218.
6. V. I. Arnol'd, *Some remarks on symplectic monodromy of Milnor fibrations*, The Floer Memorial Volume (H. Hofer, C. Taubes, A. Weinstein, and E. Zehnder, eds.), Progress in Mathematics, vol. 133, Birkhäuser, 1995, pp. 99–104.
7. M. F. Atiyah, *On analytic surfaces with double points*, Proc. Roy. Soc. Ser. A **247** (1958), 237–244.
8. D. Auroux and I. Smith, *Lefschetz pencils, branched covers and symplectic invariants*, Symplectic 4-manifolds and algebraic surfaces (F. Catanese and G. Tian, eds.), Lect. Notes Math. vol. 1938, Springer, 2008, pp. 1–54.
9. A. Bayer and Yu. Manin, *(semi)simple exercises in quantum cohomology*, Preprint math.AG/0103164.
10. P. Biran, *Connectedness of spaces of symplectic embeddings*, Int. Math. Res. Notices (1996), 487–491.

11. E. Brieskorn, *Die Auflösung rationaler Singularitäten holomorpher Abbildungen*, Math. Ann. **178** (1968), 255–270.

12. O. Collin and B. Steer, *Instanton homology for knots via 3-orbifolds*, J. Differ. Geom. **51** (1999), 149–202.

13. M. Demazure, H. Pinkham, and B. Teissier (eds.), *Séminaire sur les singularités des surfaces (séminaire Palaiseau)*, Lecture Notes in Math., vol. 777, Springer, 1980.

14. S. K. Donaldson, *Lefschetz pencils on symplectic manifolds*, J. Differ. Geom. **53** (1999), 205–236.

15. ———, *Polynomials, vanishing cycles, and Floer homology*, Mathematics: frontiers and perspectives, Am. Math. Soc., 2000, pp. 55–64.

16. A. Floer, H. Hofer, and D. Salamon, *Transversality in elliptic Morse theory for the symplectic action*, Duke Math. J. **80** (1995), 251–292.

17. K. Fukaya, Y.-G. Oh, H. Ohta, and K. Ono, *Lagrangian intersection Floer theory – anomaly and obstruction*, Preprint, 2000.

18. M. Gromov, *Pseudoholomorphic curves in symplectic manifolds*, Invent. Math. **82** (1985), 307–347.

19. J.-C. Hausmann and A. Knutson, *Polygon spaces and Grassmannians*, Enseign. Math. **43** (1997), 173–198.

20. ———, *The cohomology ring of polygon spaces*, Ann. Inst. Fourier (Grenoble) **48** (1998), 281–321.

21. H. Hofer and V. Lizan J.-C. Sikorav, *On genericity for holomorphic curves in four-dimensional almost-complex manifolds*, J. Geom. Anal. **7** (1997), 149–159.

22. H. Hofer and D. Salamon, *Floer homology and Novikov rings*, The Floer memorial volume (H. Hofer, C. Taubes, A. Weinstein, and E. Zehnder, eds.), Progress in Mathematics, vol. 133, Birkhäuser, 1995, pp. 483–524.

23. M. Khovanov and P. Seidel, *Quivers, Floer cohomology, and braid group actions*, J. Am. Math. Soc. **15** (2002), 203–271.

24. F. Kirwan, *Cohomology of quotients in symplectic and algebraic geometry*, Mathematical Notes, vol. 31, Princeton University Press, 1984.

25. A. Klyachko, *Spatial polygons and stable configurations of points in the projective line*, Algebraic geometry and its applications (Yaroslavl, 1992), Vieweg, 1994, pp. 67–84.

26. P. Kronheimer, *The construction of ALE spaces as hyper-kähler quotients*, J. Differ. Geom. **29** (1989), 665–683.

27. F. Lalonde and D. McDuff, *J-curves and the classification of rational and ruled symplectic 4-manifolds*, Contact and symplectic geometry (C. B. Thomas, ed.), Cambridge Univ. Press, 1996, pp. 3–42.

28. F. Lalonde, D. McDuff, and L. Polterovich, *On the flux conjectures*, Geometry, topology, and dynamics (Montreal, PQ, 1995) (F. Lalonde, ed.), Amer. Math. Soc., 1998, pp. 69–85.

29. Le Hong Van and K. Ono, *Cup-length estimate for symplectic fixed points*, Contact and symplectic geometry (C. B. Thomas, ed.), Cambridge Univ. Press, 1996, pp. 268–295.

30. A. Libgober, *On the fundamental group of the moduli space of cubic surfaces*, Math. Z. **162** (1978), 63–37.

31. G. Liu and G. Tian, *On the equivalence of multiplicative structures in Floer homology and quantum homology*, Acta Math. Sin. (Engl. Ser.) **15** (1999), 53–80.

32. A. Liu, *Some new applications of general wall crossing formula, Gompf's conjecture and its applications*, Math. Res. Lett. **3** (1996), 569–585.

33. E. Looijenga, *Affine Artin groups and the fundamental groups of some moduli spaces*, Preprint math.AG/98001117.

34. D. McDuff, *Positivity of intersections*, Chapter 6 in *Holomorphic curves in symplectic geometry* (M. Audin and J. Lafontaine, eds.), Birkhäuser, 1994.

35. D. McDuff and D. Salamon, *J-holomorphic curves and quantum cohomology*, University Lecture Notes Series, vol. 6, Am. Math. Soc., 1994.

36. _____, *A survey of symplectic 4-manifolds with $b^+ = 1$*, Turkish J. Math. **20** (1996), 47–60.

37. Y. Miyaoka, *The maximal number of quotient singularities on a surface with given numerical invariants*, Math. Ann. **268** (1984), 159–171.

38. D. Mumford, J. Fogarty, and F. Kirwan, *Geometric invariant theory, 3rd edition*, Springer, 1994.

39. P. Newstead, *Introduction to moduli problems and orbit spaces*, Tata Institute Lectures, vol. 51, Springer, 1978.

40. Y.-G. Oh, *Addendum to: Floer cohomology of Lagrangian intersections and pseudo-holomorphic discs I*, Comm. Pure Appl. Math. **48** (1995), 1299–1302.

41. S. Piunikhin, D. Salamon, and M. Schwarz, *Symplectic Floer-Donaldson theory and quantum cohomology*, Contact and symplectic geometry (C. B. Thomas, ed.), Cambridge Univ. Press, 1996, pp. 171–200.

42. Y. Ruan and G. Tian, *Bott-type symplectic Floer cohomology and its multiplication structures*, Math. Res. Lett. **2** (1995), 203–219.

43. _____, *A mathematical theory of Quantum Cohomology*, J. Differ. Geom. **42** (1995), 259–367.

44. G. Segal, *Lectures on topological field theory (Stanford notes)*, Unfinished manuscript available from http://www.cgtp.duke.edu/ITP99/segal.

45. P. Seidel, *Floer homology and the symplectic isotopy problem*, Ph.D. thesis, Oxford University, 1997.

46. _____, *Lagrangian two-spheres can be symplectically knotted*, J. Differ. Geom. **52** (1999), 145–171.

47. _____, *Graded Lagrangian submanifolds*, Bull. Soc. Math. France **128** (2000), 103–146.

48. _____, *A long exact sequence for symplectic Floer cohomology*, Topology **42** (2003), 1003–1063.

49. P. Seidel and R. Thomas, *Braid group actions on derived categories of coherent sheaves*, Duke Math. J. **108** (2001), 37–108.

50. U. Tillmann, *s-structures for k-linear categories and the definition of a modular functor*, J. Lond. Math. Soc. **58** (1998), 208–228.

Lectures on Pseudo-Holomorphic Curves and the Symplectic Isotopy Problem

Bernd Siebert[1] and Gang Tian[2]

[1] Mathematisches Institut, Albert-Ludwigs-Universität Freiburg, Eckerstr. 1,
79104 Freiburg, Germany
`bernd.siebert@math.uni-freiburg.de`
[2] Department of Mathematics, MIT, Cambridge, MA 02139-4307, USA
`tian@math.mit.edu`

Summary. The purpose of these notes is a more self-contained presentation of the results of the authors in Siebert and Tian (Ann Math 161:955–1016, 2005). Some applications are also given.

1 Introduction

This text is an expanded version of the lectures delivered by the authors at the CIME summer school "Symplectic 4-manifolds and algebraic surfaces," Cetraro (Italy), September 2–10, 2003. The aim of these lectures were mostly to introduce graduate students to pseudo-holomorphic techniques for the study of isotopy of symplectic submanifolds in dimension four. We tried to keep the style of the lectures by emphasizing the basic reasons for the correctness of a result rather than by providing full details.

Essentially none of the content claims any originality, but most of the results are scattered in the literature in sometimes hard-to-read locations. For example, we give a hands-on proof of the smooth parametrization of the space of holomorphic cycles on a complex surface under some positivity assumption. This is usually derived by the big machinery of deformation theory together with Banach-analytic methods. For an uninitiated person it is hard not only to follow the formal arguments needed to deduce the result from places in the literature, but also, and maybe more importantly, to understand *why* it is true. While our treatment here has the disadvantage to consider only a particular situation that does not even quite suffice for the proof of the Main Theorem (Theorem 9.1) we hope that it is useful for enhancing the understanding of this result outside the community of hardcore complex analysts and algebraic geometers.

One lecture was devoted to the beautiful theorem of Micallef and White on the holomorphic nature of pseudo-holomorphic curve singularities. The original paper is quite well written and this might be the reason that a proof of this theorem has not appeared anywhere else until very recently, in the excellent monograph [McSa2]. It devotes an appendix of 40 pages length to a careful derivation of this theorem and of most of the necessary analytical tools. Following the general principle of these lecture notes our purpose here is not to give a complete and necessarily technical proof, but to condense the original proof to the essentials. We tried to achieve this goal by specializing to the paradigmical case of tacnodal singularities with a "half-integrable" complex structure.

Another section treats the compactness theorem for pseudo-holomorphic curves. A special feature of the presented proof is that it works for sequences of almost complex structures only converging in the $\mathscr{C}^0$-topology. This is essential in the application to the symplectic isotopy problem.

We also give a self-contained presentation of Shevchishin's study of moduli spaces of equisingular pseudo-holomorphic maps and of second variations of pseudo-holomorphic curves. Here we provide streamlined proofs of the two results from [Sh] by only computing the second variation in the directions actually needed for our purposes.

The last section discusses the proof of the main theorem, which is also the main result from [SiTi3]. The logic of this proof is a bit difficult, involving several reduction steps and two inductions, and we are not sure if the current presentation really helps in understanding what is going on. Maybe somebody else has to take this up again and add new ideas into streamlining this piece.

Finally there is one section on the application to symplectic Lefschetz fibrations. This makes the link to the other lectures of the summer school, notably to those by Auroux and Smith.

2 Pseudo-Holomorphic Curves

2.1 Almost Complex and Symplectic Geometry

An *almost complex structure* on a manifold M is a bundle endomorphism $J : T_M \to T_M$ with square $-\operatorname{id}_{T_M}$. In other words, J makes T_M into a complex vector bundle and we have the canonical decomposition

$$T_M \otimes_{\mathbb{R}} \mathbb{C} = T_M^{1,0} \oplus T_M^{0,1} = T_M \oplus \overline{T_M}$$

into real and imaginary parts. The second equality is an isomorphism of complex vector bundles and $\overline{T_M}$ is just another copy of T_M with complex structure $-J$. For switching between complex and real notation it is important to write down the latter identifications explicitly:

$$T_M \longrightarrow T_M^{1,0}, \quad X \longmapsto \tfrac{1}{2}(X - iJX),$$

and similarly with $X + iJX$ for $T_M^{0,1}$. Standard examples are complex manifolds with $J(\partial_{x_\mu}) = \partial_{y_\mu}$, $J(\partial_{y_\mu}) = -\partial_{x_\mu}$ for holomorphic coordinates $z_\mu = x_\mu + iy_\mu$.

Then the above isomorphism sends $\partial_{x_\mu}, \partial_{y_\mu} \in T_M$ to $\partial_{z_\mu}, i\partial_{z_\mu} \in T_M^{1,0}$ and to $\partial_{\bar{z}_\mu}, i\partial_{\bar{z}_\mu} \in T_M^{0,1}$ respectively. Such *integrable* almost complex structures are characterized by the vanishing of the Nijenhuis tensor, a $(2,1)$-tensor depending only on J. In dimension 2 an almost complex structure is nothing but a conformal structure and hence it is integrable by classical theory. In higher dimensions there are manifolds having almost complex structures but no integrable ones. For example, any symplectic manifold (M, ω) possesses an almost complex structure, as we will see instantly, but there are symplectic manifolds not even homotopy equivalent to a complex manifold, see e.g. [OzSt].

The link between symplectic and almost complex geometry is by the notion of *tameness*. An almost complex structure J is *tamed* by a symplectic form ω if $\omega(X, JY) > 0$ for any $X, Y \in T_M \setminus \{0\}$. The space $\mathscr{J}^\omega$ of ω-tamed almost complex structures is contractible. In fact, one first proves this for the space of *compatible* almost complex structures, which have the additional property $\omega(JX, JY) = \omega(X, Y)$ for all X, Y. These are in one-to-one correspondence with Riemannian metrics g via $g(X, Y) = \omega(X, JY)$, and hence form a contractible space. In particular, a compatible almost complex structure J_0 in (M, ω) exists. Then the generalized Cayley transform

$$ J \longmapsto (J + J_0)^{-1} \circ (J - J_0) $$

produces a diffeomorphism of $\mathscr{J}^\omega$ with the space of J_0-antilinear endomorphisms A of T_M with $\|A\| < 1$ (this is the mapping norm for $g_0 = \omega(\,.\,, J_0\,.\,)$).

A differentiable map $\varphi : N \to M$ between almost complex manifolds is *pseudo-holomorphic* if $D\varphi$ is complex linear as map between complex vector bundles. If φ is an embedding this leads to the notion of pseudo-holomorphic submanifold $\varphi(N) \subset M$. If the complex structures are integrable then pseudo-holomorphicity specializes to holomorphicity. However, there are many more cases:

Proposition 2.1 *For any symplectic submanifold $Z \subset (M, \omega)$ the space of $J \in \mathscr{J}^\omega(M)$ making Z into a pseudo-holomorphic submanifold is non-empty and contractible.*

The proof uses the same arguments as for the contractibility of $\mathscr{J}^\omega$ outlined above. Another case of interest for us is the following, which can be proved by direct computation.

Proposition 2.2 *[SiTi3, Proposition 1.2] Let (M, ω) be a closed symplectic 4-manifold and $p : M \to B$ a smooth fiber bundle with all fibers symplectic. Then for any symplectic form ω_B on B and any almost complex structure J on M making the fibers of p pseudo-holomorphic, $\omega_k := \omega + k\, p^*(\omega_B)$ tames J for $k \gg 0$.*

The Cauchy–Riemann equation is over-determined in dimensions greater than two and hence the study of pseudo-holomorphic maps $\varphi : N \to M$

promises to be most interesting for $\dim N = 2$. Then N is a (not necessarily connected) Riemann surface, that we write Σ with almost complex structure j understood. The image of φ is called *pseudo-holomorphic curve*, or *J-holomorphic curve* if one wants to explicitly refer to an almost complex structure J on M. A pseudo-holomorphic curve is *irreducible* if Σ is connected, otherwise *reducible*, and its *irreducible components* are the images of the connected components of Σ. If φ does not factor non-trivially over a holomorphic map to another Riemann surface we call φ *reduced* or *non-multiply covered*, otherwise *non-reduced* or *multiply covered*.

2.2 Basic Properties of Pseudo-Holomorphic Curves

Pseudo-holomorphic curves have a lot in common with holomorphic curves:

(1) *Regularity.* If $\varphi : \Sigma \to (M, J)$ is of Sobolev class $W^{1,p}$, $p > 2$ (one weak derivative in L^p) and satisfies the Cauchy–Riemann equation $\frac{1}{2}(D\varphi + J \circ D\varphi \circ j) = 0$ weakly, then φ is smooth ($\mathscr{C}^\infty$; we assume J smooth). Note that by the Sobolev embedding theorem $W^{1,p}(\Sigma, M) \subset C^0(\Sigma, N)$, so it suffices to work in charts.

(2) *Critical points.* The set of critical points $\mathrm{crit}(\varphi) \subset \Sigma$ of a pseudo-holomorphic map $\varphi : \Sigma \to M$ is discrete.

(3) *Intersections and identity theorem.* Two different irreducible pseudo-holomorphic curves intersect discretely and, if $\dim M = 4$, with positive, finite intersection indices.

(4) *Local holomorphicity.* Let $C \subset (M, J)$ be a pseudo-holomorphic curve with finitely many irreducible components and $P \in C$. Then there exists a neighborhood $U \subset M$ of P and a C^1-diffeomorphism $\Phi : U \to \Phi(U) \subset \mathbb{C}^n$ such that $\Phi(C)$ is a holomorphic curve near $\Phi(P)$.

This is the content of the theorem of Micallef and White that we discuss in detail in Sect. 4. Note that this implies (2) and (3).

(5) *Removable singularities.* Let $\Delta^* \subset \mathbb{C}$ denote the pointed unit disk and $\varphi : \Delta^* \to (M, J)$ a pseudo-holomorphic map. Assume that φ has *bounded energy*, that is $\int_{\Delta^*} |D\varphi|^2 < \infty$ for any complete Riemannian metric on M. Then φ extends to a pseudo-holomorphic map $\Delta \to M$.

If ω tames the almost complex structure the energy can be bounded by the symplectic area: $\int_{\Delta^*} |D\varphi|^2 < c \cdot \int_\Sigma \varphi^* \omega$. Note that $\int_\Sigma \varphi^* \omega$ is a topological entity provided Σ is closed.

(6) *Local existence.* For any $X \in T_M$ of sufficiently small length there exists a pseudo-holomorphic map $\varphi : \Delta \to M$ with $D\varphi_{|0}(\partial_t) = X$. Here t is the standard holomorphic coordinate on the unit disk.

The construction is by application of the implicit function theorem to appropriate perturbations of the exponential map. Therefore it also works in families. In particular, any almost complex manifold can locally be fibered into pseudo-holomorphic disks. In dimension 4 this implies the local existence of complex (-valued) coordinates z, w such that $z = $ const is a pseudo-holomorphic disk with w restricting to a holomorphic coordinate. There exist then complex functions a, b with

$$T_M^{0,1} = \mathbb{C} \cdot (\partial_{\bar{z}} + a\partial_z + b\partial_w) + \mathbb{C} \cdot \partial_{\bar{w}}.$$

Conversely, any choices of a, b lead to an almost complex structure with z, w having the same properties. This provides a convenient way to work with almost complex structures respecting a given fibration of M by pseudo-holomorphic curves.

2.3 Moduli Spaces

The real use of pseudo-holomorphic curves in symplectic geometry comes from looking at whole spaces of them rather than single elements. There are various methods to set up the analysis to deal with such spaces. Here we follow the treatment of [Sh], to which we refer for details. Let $\mathbb{T}_g$ be the Teichmüller space of complex structures on a closed oriented surface Σ of genus g. The advantage of working with $\mathbb{T}_g$ rather than with the Riemann moduli space is that $\mathbb{T}_g$ parametrizes an actual family of complex structures on a fixed closed surface Σ. Let G be the holomorphic automorphism group of the identity component of any $j \in \mathbb{T}_g$, that is $G = \mathrm{PGL}(2, \mathbb{C})$ for $g = 0$, $G = U(1) \times U(1)$ for $g = 1$ and $G = 0$ for $g \geq 2$. Then $\mathbb{T}_g$ is an open ball in $\mathbb{C}^{3g-3+\dim_{\mathbb{C}} G}$, and it parametrizes a family of G-invariant complex structures. Let $\mathscr{J}$ be the Banach manifold of almost complex structures on M, of class C^l for some integer $l > 2$ fixed once and for all. The particular choice is insignificant. The *total moduli space* $\mathscr{M}$ of pseudo-holomorphic maps $\Sigma \to M$ is then a quotient of a subset of the Banach manifold

$$\mathscr{B} := \mathbb{T}_g \times W^{1,p}(\Sigma, M) \times \mathscr{J}.$$

The local shape of this Banach manifold is exhibited by its tangent spaces

$$T_{\mathscr{B},(j,\varphi,J)} = H^1(T_\Sigma) \times W^{1,p}(\varphi^* T_M) \times \mathscr{C}^l(\mathrm{End}\, T_M).$$

Here $H^1(T_\Sigma)$ is the cohomology with values in T_Σ, viewed as holomorphic line bundle over Σ. In more classical notation $H^1(T_\Sigma)$ may be replaced by the space of holomorphic quadratic differentials. In any case, the tangent space to $\mathbb{T}_g$ is also a subspace of $\mathscr{C}^\infty(\mathrm{End}(T_\Sigma))$ via variations of j and this is how we are going to represent its elements. To describe $\mathscr{M}$ consider the Banach bundle $\mathscr{E}$ over $\mathscr{B}$ with fibers

$$\mathscr{E}_{(j,\varphi,J)} = L^p(\Sigma, \varphi^*(T_M, J) \otimes_{\mathbb{C}} \Lambda^{0,1}),$$

where $\Lambda^{0,1}$ is our shorthand notation for $(T_\Sigma^{0,1})^*$ and where we wrote (T_M, J) to emphasize that T_M is viewed as a complex vector bundle via J. Consider the section $s : \mathscr{B} \to \mathscr{E}$ defined by the condition of complex linearity of $D\varphi$:

$$s(j, \varphi, J) = D\varphi + J \circ D\varphi \circ j.$$

Thus $s(j, \varphi, J) = 0$ iff $\varphi : (\Sigma, j) \to (M, J)$ is pseudo-holomorphic. We call the operator defined by the right-hand side the *nonlinear $\bar{\partial}$-operator*. If $M = \mathbb{C}^n$ with its standard complex structure this is just twice the usual $\bar{\partial}$-operator applied to the components of φ. Define $\hat{\mathcal{M}}$ as the zero locus of s minus those (j, φ, J) defining a multiply covered pseudo-holomorphic map. In other words, we consider only generically injective φ. This restriction is crucial for transversality to work, see Lemma 2.4. There is an obvious action of G on $\mathcal{B}$ by composing φ with biholomorphisms of (Σ, j). The moduli space of our interest is the quotient $\mathcal{M} := \hat{\mathcal{M}}/G$ of the induced action on $\hat{\mathcal{M}} \subset \mathcal{B}$.

Proposition 2.3 *$\hat{\mathcal{M}} \subset \mathcal{B}$ is a submanifold and G acts properly and freely.*

Sketch of proof. A torsion-free connection ∇ on M induces connections on $\varphi^* T_M$ (also denoted ∇) and on $\mathscr{E}$ (denoted $\nabla^{\mathscr{E}}$). For $(j', v, J') \in T_{(j,\varphi,J)}$ and $w \in T_\Sigma$ a straightforward computation gives, in real notation

$$\left(\nabla^{\mathscr{E}}_{(j',v,J')}s\right)w = \nabla_w v + J \circ \nabla_{j(w)} v + \nabla_v J \circ D_{j(w)}\varphi + J' \circ D_{j(w)}\varphi + J \circ D_{j'(w)}\varphi.$$

Replacing w by $j(w)$ changes signs, so $\nabla^{\mathscr{E}}_{(j',v,J')}s$ lies in $L^p(\varphi^* T_M \otimes_{\mathbb{C}} \Lambda^{0,1}) = \mathscr{E}_{(j,\varphi,J)}$ as it should. The last two terms treat the variations of J and j respectively. The first three terms compute the derivative of s for fixed almost complex structures. They combine to a first order differential operator on $\varphi^* T_M$ denoted by

$$D_{\varphi,J}v = \nabla v + J \circ \nabla_{j(.)}v + \nabla_v J \circ D_{j(.)}\varphi. \tag{2.1}$$

This operator is not generally J-linear, but has the form

$$D_{\varphi,J} = 2\bar{\partial}_{\varphi,J} + R,$$

with a J-linear differential operator $\bar{\partial}_{\varphi,J}$ of type $(0,1)$ and with the J-antilinear part R of order 0. (With our choices $R = N_J(., D\varphi \circ j)$ for N_J the Nijenhuis tensor of J.) Then $\bar{\partial}_{\varphi,J}$ *defines* a holomorphic structure on $\varphi^* T_M$, and this ultimately is the source of the holomorphic nature of pseudo-holomorphic maps. It is then standard to deduce that $D_{\varphi,J}$ is Fredholm as map from $W^{1,p}(\varphi^* T_M)$ to $L^p(\varphi^* T_M \otimes \Lambda^{0,1})$. To finish the proof apply the implicit function theorem taking into account the following Lemma 2.4, whose proof is an exercise.

The statements on the action of G are elementary to verify.

Lemma 2.4 *If $\varphi : \Sigma \to M$ is injective over an open set in Σ, then coker $D_{\varphi,J}$ can be spanned by terms of the form $J' \circ D\varphi \circ j$ for $J' \in T_J \mathcal{J} = \mathscr{C}^l(\mathrm{End}(T_M))$.*

The proof of the proposition together with the Riemann–Roch count (index theorem) for the holomorphic vector bundle $(\varphi^* T_M, \bar{\partial}_{\varphi,J})$ give the following.

Corollary 2.5 *The projection* $\pi : \hat{\mathscr{M}} \to \mathbb{T}_g \times \mathscr{J}$ *is Fredholm of index*

$$\operatorname{ind}(D_{\varphi,J}) = \operatorname{ind}(\bar{\partial}_{\varphi,J}) = 2 \left(\deg \varphi^* T_M + \dim_{\mathbb{C}} M \cdot (1 - g) \right)$$
$$= 2 \left(c_1(M) \cdot \varphi_*[\Sigma] + \dim_{\mathbb{C}} M \cdot (1 - g) \right)$$

A few words of caution are in order. First, for $g = 0, 1$ the action of G on $\mathscr{B}$ is *not differentiable*, although it acts by differentiable transformations. The reason is that differentiating along a family of biholomorphisms costs one derivative and hence leads out of any space of finite differentiability. A related issue is that the differentiable structure on $\mathscr{B}$ depends on the choice of map from $\mathbb{T}_g$ to the space of complex structures on Σ. Because of elliptic regularity all of these issues disappear after restriction to $\hat{\mathscr{M}}$, so we may safely take the quotient by G there. One should also be aware that there is still the mapping class group of Σ acting on $\mathscr{M}$. Only the quotient is the set of isomorphism classes of pseudo-holomorphic curves on M. However, this quotient does not support a universal family of curves anymore, at least not in the naïve sense. As our interest in $\mathscr{M}$ is for Sard-type results it is technically simpler to work with $\mathscr{M}$ rather than with this discrete quotient.

Moreover, for simplicity we henceforth essentially ignore the action of G. This merely means that we drop some correction terms in the dimension counts for $g = 0, 1$.

Remark 2.6 (1) The derivative of a section of a vector bundle $\mathscr{E}$ over a manifold $\mathscr{B}$ does not depend on the choice of a connection *after restriction to the zero locus*. In fact, if $v \in \mathscr{E}_p$ lies on the zero locus, then $T_{\mathscr{E},v} = T_{M,p} \oplus \mathscr{E}_p$ canonically, and this decomposition is the same as induced by any connection. Thus Formula (2.1) has intrinsic meaning along $\hat{\mathscr{M}}$. In particular, Ds defines a section of $\operatorname{Hom}(T_{\mathscr{B}}, \mathscr{E})|_{\hat{\mathscr{M}}}$ that we need later on.

(2) The projection $\pi : \hat{\mathscr{M}} \to \mathbb{T}_g \times \mathscr{J}$ needs not be proper – sequences of pseudo-holomorphic maps can have reducible or lower genus limits. Here are two typical examples in the integrable situation.

(a) *A family of plane quadrics degenerating to two lines.*
   Let $\Sigma = \mathbb{CP}^1$, $M = \mathbb{CP}^2$ and $\varphi_\varepsilon([t, u]) = [\varepsilon t^2, tu, \varepsilon u^2]$. Then $\operatorname{im}(\varphi_\varepsilon)$ is the plane algebraic curve $V(xz - \varepsilon^2 y^2) = \{[x, y, z] \in \mathbb{CP}^2 \mid xz - \varepsilon^2 y^2 = 0\}$. For $\varepsilon \to 0$ the image Hausdorff converges to the union of two lines $xz = 0$. Hence φ_ε can not converge in any sense to a pseudo-holomorphic map $\Sigma \to M$.

(b) *A drop of genus by the occurrence of cusps.*
   For $\varepsilon \neq 0$ the cubic curve $V(x^2 z - y^3 - \varepsilon z^3) \subset \mathbb{CP}^2$ is a smooth holomorphic curve of genus 1, hence the image of a holomorphic map φ_ε from $\Sigma = S^1 \times S^1$. For $\varepsilon \to 0$ the image of φ_ε converges to the cuspidal cubic $V(x^2 z - y^3)$, which is the bijective image of

$$\mathbb{CP}^1 \to \mathbb{CP}^2, \quad [t, u] \to [t^3, t^2 u, u^3].$$

The *Gromov compactness theorem* explains the nature of this non-compactness precisely. In its modern form it states the compactness of a Hausdorff enlargement of the space of isomorphism classes of pseudo-holomorphic curves over any compact subset of the space of almost complex structures $\mathscr{J}$. The elements of the enlargement are so-called *stable maps*, which are maps with domains nodal Riemann surfaces. For a detailed discussion see Sect. 6.

2.4 Applications

I. *Ruled surfaces.* In complex geometry a ruled surface is a holomorphic $\mathbb{CP}^1$-bundle. They are all Kähler. A ruled surface is rational (birational to $\mathbb{CP}^2$) iff the base of the fibration is $\mathbb{CP}^1$. A symplectic analogue are S^2-bundles with a symplectic structure making all fibers symplectic. The fibers are then symplectic spheres with self-intersection 0. Conversely, a result of McDuff says that a symplectic manifold with a symplectic sphere $C \subset M$ with $C \cdot C \geq 0$ is either $\mathbb{CP}^2$ with the standard structure (and C is a line or a conic) or a symplectic ruled surface (and C is a fiber or, in the rational case, a positive section), up to symplectic blowing-up [MD]. The proof employs pseudo-holomorphic techniques similar to what follows.

Now let $p : (M^4, I) \to (S^2, i)$ be a holomorphic rational ruled surface. In the notation we indicated the complex structures by I and i. Let ω be a Kähler form on M and $\mathscr{J}^\omega$ the space of ω-tamed almost complex structures. The following result was used in [SiTi3] to reduce the isotopy problem to a fibered situation.

Proposition 2.7 *For any $J \in \mathscr{J}^\omega$, M arises as total space of an S^2-bundle $p' : M \to S^2$ with all fibers J-holomorphic. Moreover, p' is homotopic to p through a homotopy of S^2-bundles.*

Sketch of proof. By connectedness of $\mathscr{J}^\omega$ there exists a path $(J_t)_{t \in [0,1]}$ connecting $I = J_0$ with $J = J_1$. By a standard Sard-type argument

$$\mathscr{M}_{(J_t)} := [0,1] \times_{\mathscr{J}} \mathscr{M} = \{(t, j, \varphi, J) \in [0,1] \times \mathscr{M} \mid J = J_t\}$$

is a manifold for an appropriate ("general") choice of (J_t). Let $\mathscr{M}_{F,J_t} \subset \mathscr{M}_{(J_t)}$ be the subset of J_t-holomorphic curves homologous to a fiber $F \subset M$ of p. The exact sequence of complex vector bundles over the domain $\Sigma = S^2$ of such a J_t-holomorphic curve (see Sect. 3.2)

$$0 \longrightarrow T_\Sigma \longrightarrow T_M|_\Sigma \longrightarrow N_{\Sigma|M} \longrightarrow 0$$

gives $c_1(M) \cdot [F] = c_1(T_\Sigma) \cdot [\Sigma] + F \cdot F = 2$. Then the dimension formula from Corollary 2.5 shows

$$\dim_{\mathbb{C}} \mathscr{M}_{F,(J_t)} = c_1(M) \cdot [F] + \dim_{\mathbb{C}} M \cdot (1 - g) - \dim G = 2 + 2 - 3 = 1.$$

Moreover, $\mathscr{M}_{F,J_t}$ is compact by the Gromov compactness theorem since $[F]$ is a primitive class in $\{A \in H_2(M, \mathbb{Z}) \mid \int_{[A]} \omega > 0\}$. In fact, by primitivity any pseudo-holomorphic curve C representing $[F]$ has to be irreducible. Moreover, the genus formula (Proposition 3.2 below) implies that any irreducible pseudo-holomorphic curve representing $[F]$ has to be an embedded sphere. We will see in Proposition 3.4 that then the deformation theory of any $C \in \mathcal{M}_{F,J_t}$ is unobstructed.

Next, the positivity of intersection indices of pseudo-holomorphic curves implies that any two curves in $\mathcal{M}_{F,J_t}$ are either disjoint or equal. Together with unobstructedness we find that through any point $P \in M$ passes exactly one J_t-holomorphic curve homologous to F. Define

$$p_t : M \to \mathcal{M}_{F,J_t}, \quad P \longmapsto C, \quad C \text{ the curve passing through } P.$$

Since $\mathcal{M}_{F,J_0} \simeq S^2$ via $C \leftrightarrow p^{-1}(x)$ we may identify $p_0 = p$. A computation on the map of tangent spaces shows that p_t is a submersion for any t. Finally, for homological reasons $\mathcal{M}_{F,J_t} \simeq S^2$ for any t. The proof is finished by setting $p' = p_1$.

II. *Isotopy of symplectic surfaces.* The main topic of these lectures is the isotopy classification of symplectic surfaces. We are now ready to explain the relevance of pseudo-holomorphic techniques for this question. Let (M^4, I, ω) be a Kähler surface. We wish to ask the following question.

If $B \subset M$ is a symplectic surface then is B isotopic to a holomorphic curve?

By *isotopy* we mean connected by a path inside the space of smooth symplectic submanifolds. In cases of interest the space of smooth holomorphic curves representing $[B] \in H_2(M, \mathbb{Z})$ is connected. Hence a positive answer to our question shows uniqueness of symplectic submanifolds homologous to B up to isotopy.

The use of pseudo-holomorphic techniques is straightforward. By the discussion in Sect. 2.2 there exists a tamed almost complex structure J making B a pseudo-holomorphic curve. As in 2.4,I choose a generic path $(J_t)_{t \in [0,1]}$ in $\mathcal{J}^\omega$ connecting J with the integrable complex structure I. Now try to deform B as pseudo-holomorphic curve together with J. In other words, we want to find a family $(B_t)_{t \in [0,1]}$ of submanifolds with B_t pseudo-holomorphic for J_t and with $B_0 = B$.

There are two obstructions to carrying this through. Let $\mathcal{M}_{B,(J_t)}$ be the moduli space of pseudo-holomorphic submanifolds $C \subset M$ homologous to B and pseudo-holomorphic for some J_t. The first problem arises from the fact that the projection $\mathcal{M}_{B,(J_t)} \to [0,1]$ may have critical points. Thus if $(B_t)_{t \in [0,t_0]}$ is a deformation of B with the requested properties over the interval $[0, t_0]$ with t_0 a critical point, it might occur that B_{t_0} does not deform to a J_t-holomorphic curve for any $t > t_0$. We will see in Sect. 3 that this phenomenon does indeed not occur under certain positivity conditions on M. The second reason is non-properness of the projection $\mathcal{M}_{B,(J_t)} \to [0,1]$. It might happen that a family (B_t) exists on $[0, t_0)$, but does not extend to t_0. In view of the

Gromov compactness theorem a different way to say this is to view $\tilde{\mathcal{M}}_{B,(J_t)}$ as an open subset of a larger moduli space $\tilde{\mathcal{M}}_{B,(J_t)}$ of pseudo-holomorphic cycles. Then the question is if the closed subset of singular cycles does locally disconnect $\tilde{\mathcal{M}}_{B,(J_t)}$ or not. Thus for this second question one has to study deformations of singular pseudo-holomorphic cycles.

III. *Pseudo-holomorphic spheres with prescribed singularities.* Another variant of the above technique allows the construction of pseudo-holomorphic spheres with prescribed singularities. The following is from [SiTi3], Proposition 7.1.

Proposition 2.8 *Let* $p : (M, I) \to \mathbb{CP}^1$ *be a rational ruled surface and*

$$\varphi : \Delta \longrightarrow M$$

an injective holomorphic map, $\Delta \subset \mathbb{C}$ *the unit disk. Let* J *be an almost complex structure on* M *making* p *pseudo-holomorphic and agreeing with* I *in a neighborhood of* $\varphi(0)$.

Then for any $k > 0$ *there exists a* J-*holomorphic sphere*

$$\psi_k : \mathbb{CP}^1 \longrightarrow M$$

approximating φ *to* kth *order at* 0:

$$d_M(\varphi(\tau), \psi_k(\tau)) = o(|\tau|^k).$$

Here d_M is the distance function for some Riemannian metric on M. This result says that any plane holomorphic curve singularity arises as the singularity of a J-holomorphic sphere, for J with the stated properties. The proof relies heavily on the fact that M is a rational ruled surface. Note that the existence of a J-holomorphic sphere not homologous to a fiber excludes ruled surfaces over a base of higher genus.

Sketch of proof. It is not hard to see that J can be connected to I by a path of almost complex structures with the same properties as J. Therefore the idea is again to start with a holomorphic solution to the problem and then to deform the almost complex structure. Excluding the trivial case $p \circ \varphi = \text{const}$ write

$$\varphi(\tau) = (\tau^m, h(\tau))$$

in holomorphic coordinates on $M \setminus (F \cup H) \simeq \mathbb{C}^2$, F a fiber and H a positive holomorphic section of p ($H \cdot H \geq 0$). Then h is a holomorphic function. Now consider the space of pseudo-holomorphic maps $\mathbb{CP}^1 \to M$ of the form

$$\tau \longmapsto (\tau^m, h(\tau) + o(|\tau|^l)).$$

For appropriate l the moduli space of such maps has expected dimension 0. Then for a generic path $(J_t)_{t \in [0,1]}$ of almost complex structures the union of such moduli spaces over this path is a differentiable one-dimensional manifold $q : \mathcal{M}_{\varphi,(J_t)} \to [0,1]$ without critical points over $t = 0, 1$. By a straightforward dimension estimate the corresponding moduli spaces of reducible pseudo-holomorphic curves occurring in the Gromov compactness theorem are empty. Hence the projection $q : \mathcal{M}_{\varphi,(J_t)} \to [0,1]$ is proper. Thus $\mathcal{M}_{\varphi,(J_t)}$ is a compact one-dimensional manifold with boundary and all boundary points lie over $\{0,1\} \subset [0,1]$. The closed components of $\mathcal{M}_{\varphi,(J_t)}$ do not contribute to the moduli spaces for $t = 0, 1$. The other components have two ends each, and they either map to the same boundary point of $[0,1]$ or to different ones. In any case the parity of the cardinality of $q^{-1}(0)$ and of $q^{-1}(1)$ are the same, as illustrated in the following Fig. 2.1. Finally, an explicit computation shows that in the integrable situation the moduli space has exactly one element. Therefore $q^{-1}(1)$ can not be empty either. An element of this moduli space provides the desired J-holomorphic approximation of φ.

2.5 Pseudo-Analytic Inequalities

In this section we lay the foundations for the study of critical points of pseudo-holomorphic maps. As this is a local question we take as domain the unit disk, $\varphi : \Delta \to (M, J)$. The main point of this study is that any singularity bears a certain kind of holomorphicity in itself, and the amount of holomorphicity indeed increases with the complexity of the singularity. The reason for this to happen comes from the following series of results on differential inequalities for the $\bar{\partial}$-operator. For simplicity of proof we formulate these only for functions with values in $\mathbb{C}$ rather than $\mathbb{C}^n$ as needed, but comment on how to generalize to $n > 1$.

Lemma 2.9 *Let $f \in W^{1,2}(\Delta)$ fulfill $|\partial_{\bar{z}} f| \leq \phi \cdot |f|$ almost everywhere for $\phi \in L^p(\Delta)$, $p > 2$. Then either $f = 0$ or there exist a uniquely determined integer μ and $g \in W^{1,p}(\Delta)$, $g(0) \neq 0$ with*

$$f(z) = z^\mu \cdot g(z) \quad \text{almost everywhere.}$$

Proof. A standard elliptic bootstrapping argument shows $f \in W^{1,p}(\Delta)$, see for example [IvSh1], Lemma 3.1.1,(i). This step requires $p > 2$. Next

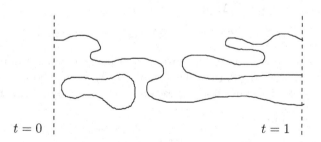

$t = 0$ $t = 1$

Fig. 2.1. A one-dimensional cobordism

comes a trick attributed to Carleman to reduce to the holomorphic situation: By hypothesis $\left|\frac{\partial_{\bar{z}} f}{f}\right| \leq \phi$. We will recall in Proposition 3.1 below that $\partial_{\bar{z}} : W^{1,p}(\Delta) \to L^p(\Delta)$ is surjective. Hence there exists $\psi \in W^{1,p}(\Delta)$ solving $\partial_{\bar{z}} \psi = \frac{\partial_{\bar{z}} f}{f}$. Then

$$\partial_{\bar{z}}(e^{-\psi} f) = e^{-\psi}(-\partial_{\bar{z}} \psi) f + e^{-\psi} \partial_{\bar{z}} f = 0$$

shows that $e^{-\psi} f$ is a holomorphic function (*Carleman similarity principle*). Now complex function theory tells us that $e^{-\psi} f = z^\mu \cdot h$ for $h \in \mathcal{O}(\Delta) \cap W^{1,p}(\Delta)$, $h(0) \neq 0$. Putting $g = e^\psi \cdot h$ gives the required representation of f. □

Remark 2.10 (1) As for an intrinsic interpretation of μ note that it is the intersection multiplicity of the graph of h with the graph of the zero function inside $\Delta \times \mathbb{C}$ at $(0,0)$. Multiplication by e^ψ induces a homeomorphism of $\Delta \times \mathbb{C}$ and transforms the graph of h into the graph of f. Hence μ is a topologically defined entity depending only on f.

(2) The Carleman trick replaces the use of a general removable singularities theorem for solutions of differential inequalities due to Harvey and Polking that was employed in [IvSh1], Lemma 3.1.1. Unlike the Carleman trick this method generalizes to maps $f : \Delta \to \mathbb{C}^n$ with $n > 1$. Another possibility that works also for $n > 1$ is to use the Hartman-Wintner theorem on the polynomial behavior of solutions of certain partial differential equations in two variables, see e.g. [McSa2]. A third approach appeared in the printed version [IvSh2] of [IvSh1]; here the authors noticed that one can deduce a Carleman similarity principle also for maps $f : \Delta \to \mathbb{C}^n$ by viewing f as a holomorphic section of $\Delta \times \mathbb{C}^n$, viewed as holomorphic vector bundle with non-standard $\bar{\partial}$-operator. This is arguably the easiest method to deduce the result for all n.

A similar looking lemma of quite different flavor deduces holomorphicity up to some order from a polynomial estimate on $|\partial_{\bar{z}} f|$. Again we took this from [IvSh1], but the proof given there makes unnecessary use of Lemma 2.9.

Lemma 2.11 *Let $f \in L^2(\Delta)$ fulfill $|\partial_{\bar{z}} f| \leq \phi \cdot |z|^\nu$ almost everywhere for some $\phi \in L^p(\Delta)$, $p > 2$ and $\nu \in \mathbb{N}$. Then either $f = 0$ or there exists $P \in \mathbb{C}[z]$, $\deg P \leq \nu$ and $g \in W^{1,p}(\Delta)$, $g(0) = 0$ with*

$$f(z) = P(z) + z^\nu \cdot g(z) \quad \text{almost everywhere.}$$

Proof. By induction over ν, the case $\nu = 0$ being trivial. Assume the case $\nu - 1$ is true. Elliptic regularity gives $f \in W^{1,p}(\Delta)$. By the Sobolev embedding theorem f is Hölder continuous of exponent $\alpha = 1 - \frac{2}{p} \in (0,1)$. Hence $f_1 = \frac{f - f(0)}{z}$ is L^2 and

$$|\partial_{\bar{z}} f_1| = \left|\frac{\partial_{\bar{z}} f}{z}\right| \leq \phi \cdot |z|^{\nu-1}.$$

Therefore induction applies to f_1 and we see $f_1 = P_1 + z^{\nu-1} \cdot g$ with g of the required form. Plugging in the definition of f_1 gives $f = (f(0) + zP_1) + z^\nu \cdot g$, so $P = f(0) + zP_1$ is the correct definition of P. □

Remark 2.12 The lemma generalizes in a straightforward manner to maps $f : \Delta \to \mathbb{C}^n$. In this situation the line $\mathbb{C} \cdot P(0)$ has a geometric interpretation as complex tangent line of $\mathrm{im}(f)$ in $f(0)$, which by definition is the limit of lines $\mathbb{C} \cdot (f(z) - f(0)) \in \mathbb{CP}^{n-1}$ for $z \neq 0$, $z \to 0$.

Combining the two lemmas gives the following useful result, which again generalizes to maps $f : \Delta \to \mathbb{C}^n$.

Proposition 2.13 Let $f \in L^2(\Delta)$ fulfill $|\partial_{\bar{z}} f| \leq \phi |z|^\nu |f|$ almost everywhere for $\phi \in L^p(\Delta)$, $p > 2$ and $\nu \in \mathbb{N}$. Then either $f = 0$ or there exist uniquely determined $\mu, \nu \in \mathbb{N}$ and $P \in \mathbb{C}[z]$, $\deg P \leq \nu$, $P(0) \neq 0$, $g \in W^{1,p}(\Delta)$, $g(0) = 0$ with

$$f(z) = z^\mu \big(P(z) + z^\nu \cdot g(z) \big) \quad \text{almost everywhere.}$$

Proof. Lemma 2.9 gives $f = z^\mu g$. Now by hypothesis g also fulfills the stated estimate:

$$|\partial_{\bar{z}} g| = \left| \frac{\partial_{\bar{z}} f}{z^\mu} \right| \leq \phi |z|^\nu \left| \frac{f}{z^\mu} \right| = \phi |z^\nu| \cdot |g|.$$

Thus replacing f by g reduces to the case $f(0) \neq 0$ and $\mu = 0$. The result then follows by Lemma 2.11 applied to f with ϕ replaced by $\phi \cdot |f|$. □

3 Unobstructedness I: Smooth and Nodal Curves

3.1 Preliminaries on the $\bar{\partial}$-Equation

The crucial tool to study the $\bar{\partial}$-equation analytically is the inhomogeneous Cauchy integral formula. It says

$$f = Hf + T(\partial_{\bar{z}} f) \tag{3.1}$$

for all $f \in \mathscr{C}^1(\overline{\Delta})$ with integral operators

$$Hf(z) = \frac{1}{2\pi i} \int_{\partial \Delta} \frac{f(w)}{w - z} \, dw, \quad Tg(z) = \frac{1}{2\pi i} \int_\Delta \frac{g(w)}{w - z} \, dw \wedge d\bar{w}.$$

(All functions in this section are $\mathbb{C}$-valued.) The first operator H maps continuous functions defined on $S^1 = \partial \Delta$ to holomorphic functions on Δ. Continuity of Hf along the boundary is not generally true if f is just continuous. To understand this note that any $f \in \mathscr{C}^0(S^1)$ can be written as a Fourier series $\sum_{n \in \mathbb{Z}} a_n z^n$ and then $Hf = \sum_{n \in \mathbb{N}} a_n z^n$ is the projection to the space of functions spanned by non-negative Fourier modes. This function needs not be continuous.

The integrand of the second integral operator T looks discontinuous, but in fact it is not as one sees in polar coordinates with center $w = z$. For differentiability properties one computes $\partial_{\bar{z}} T = \mathrm{id}$ from (3.1), while $\partial_z T = S$ with S the singular integral operator

$$Sg(z) = \frac{1}{2\pi i} \lim_{\varepsilon \to 0} \int_{\Delta \backslash B_\varepsilon(z)} \frac{g(w)}{(w-z)^2} \, dw \wedge d\bar{w} \, .$$

The Calderon–Zygmund theorem says that S is a continuous map from $L^p(\Delta)$ to itself for $1 < p < \infty$. Recall also that the Sobolev space $W^{1,p}(\Delta)$ consists of L^p-functions with weak partial derivatives in L^p too. For $p > 2$ it holds $1 - \frac{2}{p} > 0$, so the Sobolev embedding theorem implies that any $f \in W^{1,p}(\Delta)$ has a continuous representative. Moreover, the map $W^{1,p}(\Delta) \to C^0(\overline{\Delta})$ thus defined is continuous. Thus (3.1) holds for $f \in W^{1,p}(\Delta)$ for any $p > 2$. Summarizing the discussion, the Cauchy integral formula induces the following remarkable direct sum decomposition of $W^{1,p}(\Delta)$.

Proposition 3.1 *Let $2 < p < \infty$. Then $(H, \bar{\partial}) : W^{1,p}(\Delta) \longrightarrow (\mathcal{O}(\Delta) \cap W^{1,p}(\Delta)) \times L^p(\Delta)$ is an isomorphism. In particular, for any $g \in L^p(\Delta)$ there exists $f \in W^{1,p}(\Delta)$ with $\partial_{\bar{z}} f = g$.*

Thus any $f \in W^{1,p}(\Delta)$ can be written in the form $h + T(\partial_{\bar{z}} f)$ with h holomorphic in Δ and continuously extending to $\overline{\Delta}$ and $T(\partial_{\bar{z}} f)|_{\partial \Delta}$ gathering all negative Fourier coefficients of $f|_{\overline{\Delta}}$.

3.2 The Normal $\bar{\partial}$-Operator

We have already described pseudo-holomorphic maps by a non-linear PDE. One trivial variation of a pseudo-holomorphic map is by reparametrization. It is sometimes useful to get rid of this part, especially if one is interested in pseudo-holomorphic curves rather than pseudo-holomorphic maps. This is achieved by the normal $\bar{\partial}$-operator that we now introduce.

Recall that for a pseudo-holomorphic map $\varphi : \Sigma \to M$ the operator $\bar{\partial}_{\varphi,J}$ from (2.2) in Sect. 2.3 defines a natural holomorphic structure on $\varphi^* T_M$ compatible with the holomorphic structure on T_Σ. In fact, a straightforward computation shows

$$D\varphi \circ \bar{\partial}_{T_\Sigma} = \bar{\partial}_{\varphi,J} \circ D\varphi.$$

If φ is an immersion we thus obtain a short exact sequence of holomorphic vector bundles over Σ

$$0 \longrightarrow T_\Sigma \longrightarrow \varphi^*(T_M) \longrightarrow N \longrightarrow 0.$$

This sequence *defines* the normal bundle N along φ. If φ has critical points it is still possible to define a normal bundle as follows. For a complex vector bundle V denote by $\mathcal{O}(V)$ the sheaf of holomorphic sections of V. (See Sect. 5.4 for a short reminder of sheaf theory.) While at critical points $D\varphi : T_\Sigma \to \varphi^* T_M$ is not injective the map of sheaves $\mathcal{O}(T_\Sigma) \to \mathcal{O}(\varphi^* T_M)$ still is. As an example consider $\varphi(t) = (t^2, t^3)$ as map from Δ to $\mathbb{C}^2$ with standard complex structure. Then

$$D\varphi(\partial_t) = 2t\,\partial_z + 3t^2\,\partial_w, \tag{3.2}$$

and as germ of holomorphic function the right-hand side is non-zero. Thus in any case we obtain a short exact sequence of sheaves of $\mathcal{O}_\Sigma$-modules

$$0 \longrightarrow \mathcal{O}(T_\Sigma) \overset{D\varphi}{\longrightarrow} \mathcal{O}(\varphi^* T_M) \longrightarrow \mathcal{N} \longrightarrow 0. \tag{3.3}$$

From the definition, $\mathcal{N}$ is just some coherent sheaf on Σ. But on a Riemann surface any coherent sheaf splits uniquely into a skyscraper sheaf (discrete support) and the sheaf of sections of a holomorphic vector bundle. Thus we may write

$$\mathcal{N} = \mathcal{N}^{\text{tor}} \oplus \mathcal{O}(N)$$

for some holomorphic vector bundle N. We call $\mathcal{N}$ the *normal sheaf* along φ and N the *normal bundle*. The skyscraper sheaf $\mathcal{N}^{\text{tor}}$ is the subsheaf of $\mathcal{N}$ generated by sections that are annihilated by multiplication by some non-zero holomorphic function ("torsion sections"). In our example $\varphi(t) = (t^2, t^3)$ the section $v = 2\,\partial_z + 3t\,\partial_w$ of $\varphi^* T_M$ is contained in the image of $\mathcal{O}(T_\Sigma)$ for $t \neq 0$, but not at $t = 0$, while $tv = D\varphi(\partial_t)$. In fact, $\mathcal{N}^{\text{tor}}$ is isomorphic to a copy of $\mathbb{C}$ over $t = 0$ generated by the germ of v at 0. Then N is the holomorphic line bundle generated by any $a(t)\partial_z + b(t)\partial_w$ with $b(0) \neq 0$.

As a simple but very powerful application of the normal sequence (3.3) we record the genus formula for pseudo-holomorphic curves in dimension four.

Proposition 3.2 *Let (M, J) be an almost complex 4-manifold and $C \subset M$ an irreducible pseudo-holomorphic curve. Then*

$$2g(C) - 2 \leq c_1(M) \cdot C + C \cdot C,$$

with equality if and only if C is smooth.

Proof. Let $\varphi : \Sigma \to M$ be the pseudo-holomorphic map with image C. Since $\deg T_\Sigma = 2g(C) - 2$ the normal sequence (3.3) shows

$$2g(C) - 2 \;=\; \deg \varphi^* T_M + \deg N + \lg \mathcal{N}^{\text{tor}} \;=\; c_1(M) \cdot C + \deg N + \lg \mathcal{N}^{\text{tor}}.$$

Here $\lg \mathcal{N}^{\text{tor}}$ is the sum of the $\mathbb{C}$-vector space dimensions of the stalks of $\mathcal{N}$. This term vanishes iff $\mathcal{N}^{\text{tor}} = 0$, that is, iff φ is an immersion. The degree of N equals $C \cdot C$ if C is smooth and drops by the self-intersection number of φ in the immersed case. By the PDE description of the space of pseudo-holomorphic maps from a unit disk to M, it is not hard to show that locally in Σ any pseudo-holomorphic map $\Sigma \to M$ can be perturbed to a pseudo-holomorphic immersion. At the expense of changing J away from the singularities of C slightly this statement globalizes. This process does not change any of $g(C)$, $c_1(M) \cdot C$ and $C \cdot C$. Hence the result for general C follows from the immersed case. $\square$

To get rid of $\mathcal{N}^{\text{tor}}$ it is convenient to go over to meromorphic sections of T_Σ with poles of order at most $\text{ord}_P D\varphi$ in a critical point P of $D\varphi$. In fact, the sheaf of such meromorphic sections is the sheaf of holomorphic sections of a line bundle that we conveniently denote $T_\Sigma[A]$, where A is the divisor $\sum_{P \in \text{crit}(\varphi)} (\text{ord}_P D\varphi) \cdot P$. Then $T_\Sigma[A] = \text{kern}(\varphi^* T_M \to N)$ and hence we obtain the short exact sequence of holomorphic vector bundles

$$0 \longrightarrow T_\Sigma[A] \xrightarrow{D\varphi} \varphi^* T_M \longrightarrow N \longrightarrow 0.$$

Thus by (3.2) together with $R \circ D\varphi = 0$ the operator $D_{\varphi,J} = \bar{\partial}_{\varphi,J} + R : W^{1,p}(\varphi^* T_M) \to L^p(\varphi^* T_M \otimes \Lambda^{0,1})$ fits into the following commutative diagram with exact rows.

$$
\begin{array}{ccccccccc}
0 & \to & W^{1,p}(T_\Sigma[A]) & \to & W^{1,p}(\varphi^* T_M) & \to & W^{1,p}(N) & \to & 0 \\
& & \bar{\partial}_{T_\Sigma} \downarrow & & \downarrow D_{\varphi,J} & & \downarrow D^N_{\varphi,J} & & \\
0 & \to & L^p(T_\Sigma[A] \otimes \Lambda^{0,1}) & \to & L^p(\varphi^* T_M \otimes \Lambda^{0,1}) & \to & L^p(N \otimes \Lambda^{0,1}) & \to & 0
\end{array}
$$

This defines the *normal $\bar{\partial}_J$-operator* $D^N_{\varphi,J}$. As with $D_{\varphi,J}$ we have the decomposition $D^N_{\varphi,J} = \bar{\partial}_N + R_N$ into complex linear and a zero order complex anti-linear part. By the snake lemma the diagram readily induces the long exact sequence

$$
\begin{array}{c}
0 \to H^0(T_\Sigma[A]) \to \text{kern}\, D_{\varphi,J} \to \text{kern}\, D^N_{\varphi,J} \\
\to H^1(T_\Sigma[A]) \to \text{coker}\, D_{\varphi,J} \to \text{coker}\, D^N_{\varphi,J} \to 0
\end{array}
$$

The cohomology groups on the left are Dolbeault cohomology groups for the holomorphic vector bundle $T_\Sigma[A]$ or sheaf cohomology groups of the corresponding coherent sheaves. Forgetting the twist by A then gives the following exact sequence.

$$
\begin{array}{c}
0 \to H^0(T_\Sigma) \to \text{kern}\, D_{\varphi,J} \to \text{kern}\, D^N_{\varphi,J} \oplus H^0(\mathcal{N}^{\text{tor}}) \\
\to H^1(T_\Sigma) \to \text{coker}\, D_{\varphi,J} \to \quad \text{coker}\, D^N_{\varphi,J} \quad \to 0
\end{array}
\tag{3.4}
$$

The terms in this sequence have a geometric interpretation. Each column is associated to a deformation problem. The left-most column deals with deformations of Riemann surfaces: $H^0(T_\Sigma)$ is the space of holomorphic vector fields on Σ. It is trivial except in genera 0 and 1, where it gives infinitesimal holomorphic reparametrizations of φ. As already mentioned in Sect. 2.3, the

space $H^1(T_\Sigma)$ is isomorphic to the space of holomorphic quadratic differentials via Serre-duality and hence describes the tangent space to the Riemann
or Teichmüller space of complex structures on Σ. Every element of this space
is the tangent vector of an actual one-parameter family of complex structures
on Σ – this deformation problem is unobstructed.

The middle column covers the deformation problem of φ as pseudo-
holomorphic curve *with almost complex structures both on M and on Σ fixed*.
In fact, $D_{\varphi,J}$ is the linearization of the Fredholm map describing the space of
pseudo-holomorphic maps $(\Sigma, j) \to (M, J)$ with fixed almost complex structures (see proof of Proposition 2.3). If coker $D_{\varphi,J} \neq 0$ this moduli space might
not be smooth of the expected dimension $\mathrm{ind}(D_{\varphi,J})$, and this is then an obstructed deformation problem.

The maps from the left column to the middle column also have an interest-
ing meaning. On the upper part, $H^0(T_\Sigma) \to \mathrm{kern}\, D_{\varphi,J}$ describes infinitesimal
holomorphic reparametrizations as infinitesimal deformations of φ. On the
lower part, the image of $H^1(T_\Sigma)$ in coker $D_{\varphi,J}$ exhibits those obstructions of
the deformation problem of φ with fixed almost complex structures that can
be killed by variations of the complex structure of Σ.

The right column is maybe the most interesting. First, there are no obstructions to the deformations of φ as J-holomorphic map iff coker $D^N_{\varphi,J} = 0$,
provided we allow variations of the complex structure of Σ. Thus when it
comes to the smoothness of the moduli space relative $\mathscr{J}$ then coker $D^N_{\varphi,J}$ is
much more relevant than the more traditional coker $D_{\varphi,J}$. Finally, the term on
the upper right corner consists of two terms, with $H^0(\mathcal{N}^{\mathrm{tor}})$ reflecting deformations of the singularities. In fact, infinitesimal deformations with vanishing
component on this part can be realized by sections of $\varphi^* T_M$ with zeros of the
same orders as those of $D\varphi$. This follows directly from the definition of $\mathcal{N}^{\mathrm{tor}}$.
Such deformations are exactly those keeping the number of critical points of
φ. Note that while $\mathcal{N}^{\mathrm{tor}}$ does not explicitly show up in the obstruction space
coker $D^N_{\varphi,J}$, it does influence this space by lowering the degree of N. The exact
sequence also gives the (non-canonical) direct sum decomposition

$$\mathrm{kern}\, D^N_{\varphi,J} \oplus H^0(\mathcal{N}^{\mathrm{tor}}) = \big(\, \mathrm{kern}\, D_{\varphi,J}/H^0(T_\Sigma) \big) \oplus \mathrm{kern}\, \big(H^1(T_\Sigma) \to \mathrm{coker}\, D_{\varphi,J}\big).$$

The decomposition on the right-hand side mixes local and global contributions. The previous discussion gives the following interpretation:
$\mathrm{kern}\, D_{\varphi,J}/H^0(T_\Sigma)$ is the space of infinitesimal deformations of φ as
J-holomorphic map modulo biholomorphisms; $\mathrm{kern}\, \big(H^1(T_\Sigma) \to \mathrm{coker}\, D_{\varphi,J}\big)$
is the tangent space to the space of complex structures on Σ that can be
realized by variations of φ as J-holomorphic map.

Summarizing this discussion, it is the right column that describes the moduli space of pseudo-holomorphic maps for fixed J. In particular, if
coker $D^N_{\varphi,J} = 0$ then the moduli space $\mathscr{M}_J \subset \mathscr{M}$ of J-holomorphic maps
$\Sigma \to M$ for *arbitrary* complex structures on Σ is smooth at (φ, J, j) with
tangent space

$$T_{\mathscr{M}_J,(\varphi,J,j)} = \mathrm{kern}\, D^N_{\varphi,J} \oplus H^0(\mathcal{N}^{\mathrm{tor}}).$$

3.3 Immersed Curves

If $\dim M = 4$ and $\varphi : \Sigma \to M$ is an immersion then N is a holomorphic line bundle and
$$N \otimes T_\Sigma \simeq \varphi^*(\det T_M).$$
Here T_M is taken as complex vector bundle. From this we are going to deduce a cohomological criterion for the surjectivity of
$$D^N_{\varphi,J} = \bar\partial + R : W^{1,p}(N) \longrightarrow L^p(N \otimes \Lambda^{0,1}).$$
By elliptic theory the cokernel of $D^N_{\varphi,J}$ is dual to the kernel of its formal adjoint operator
$$(D^N_{\varphi,J})^* : W^{1,p}(N^* \otimes \Lambda^{1,0}) \longrightarrow L^p(N^* \otimes \Lambda^{1,1}).$$
Note that $(D^N_{\varphi,J})^* = \bar\partial - R^*$ is also of Cauchy–Riemann type. Now in dimension 4 the bundle N is a holomorphic line bundle over Σ. In a local holomorphic trivialization $(D^N_{\varphi,J})^*$ therefore takes the form $f \mapsto \bar\partial f + \alpha f + \beta \bar f$ for some functions α, β. Solutions of such equations are called *pseudo-analytic* [Ve]. While related this notion predates pseudo-holomorphicity and should not be mixed up with it.

Lemma 3.3 *Let $\alpha, \beta \in L^p(\Delta)$ and let $f \in W^{1,p}(\Delta) \setminus \{0\}$ fulfill*
$$\partial_{\bar z} f + \alpha f + \beta \bar f = 0.$$
Then all zeros of f are isolated and positive.

Proof. This is another application of the Carleman trick, cf. the proof of Lemma 2.9. Replacing α by $\alpha + \beta \cdot \bar f/f$ reduces to the case $\beta = 0$. Note that $\bar f/f$ is bounded, so $\beta \cdot \bar f/f$ stays in L^p. By Proposition 3.1 there exists $g \in W^{1,p}(\Delta)$ solving $\partial_{\bar z} g = \alpha$. Then
$$\partial_{\bar z}(e^g f) = e^g(\partial_{\bar z} g \cdot f + \partial_{\bar z} f) = e^g(\alpha f + \partial_{\bar z} f) = 0.$$
Thus the diffeomorphism $\Psi : (z, w) \mapsto (z, e^{g(z)} w)$ transforms the graph of f into the graph of a holomorphic function. $\square$

Here is the cohomological unobstructedness theorem for immersed curves.

Proposition 3.4 *Let (M, J) be a 4-dimensional almost complex manifold, and let $\varphi : \Sigma \to M$ be an immersed J-holomorphic curve with $c_1(M) \cdot [C] > 0$. Then the moduli space $\mathscr{M}_J$ of J-holomorphic maps to M is smooth at (j, φ, J).*

Proof. By the previous discussion the result follows once we show the vanishing of $\mathrm{kern}(D^N_{\varphi,J})^*$. An element of this space is a section of the holomorphic line bundle $N^* \otimes \Lambda^{1,0}$ over Σ of degree
$$\deg(N^* \otimes \Lambda^{1,0}) = \deg(\det T_M^*|_C) = -c_1(M) \cdot [C] < 0.$$
In a local holomorphic trivialization it is represented by a pseudo-analytic function. Thus by Lemma 3.3 and the degree computation it has to be identically zero. $\square$

3.4 Smoothings of Nodal Curves

A pseudo-holomorphic curve $C \subset M$ is a *nodal curve* if all singularities of C are transversal unions of two smooth branches. It is natural to consider a nodal curve as the image of an injective map from a *nodal Riemann surface*. A nodal Riemann surface is a union of Riemann surfaces Σ_i with finitely many disjoint pairs of points identified. The identification map $\hat{\Sigma} := \coprod_i \Sigma_i \to \Sigma$ from the disjoint union is called *normalization*. (This notion has a more precise meaning in complex analysis.) A map $\varphi : \Sigma \to M$ is pseudo-holomorphic if it is continuous and if the composition $\hat{\varphi} : \hat{\Sigma} \to \Sigma \to M$ is pseudo-holomorphic. Analogously one defines $W^{1,p}$-spaces for $p > 2$.

For a nodal curve C it is possible to extend the above discussion to include topology change by smoothing the nodes, as in $zw = t$ for $(z, w) \in \Delta \times \Delta$ and t the deformation parameter. This follows from the by now well-understood gluing construction for pseudo-holomorphic maps. There are various ways to achieve this, see for example [LiTi, Si]. They share a formulation by a family of non-linear Fredholm operators

$$\prod_i H^1(T_{\Sigma_i}) \times W^{1,p}(\varphi^* T_M) \longrightarrow L^p(\varphi^* T_M \otimes \Lambda^{0,1}),$$

parametrized by $l = \sharp$nodes gluing parameters $(t_1, \ldots, t_l) \in \mathbb{C}^l$ of sufficiently small norm. The definitions of $W^{1,p}$ and L^p near the nodes vary from approach to approach. Thus fixed $(t_1, \ldots, t_l)$ gives the deformation problem with given topological type as discussed above, and putting $t_i = 0$ means keeping the ith node. The linearization $D'_{\varphi,J}$ of this operator for $(t_1, \ldots, t_l) = 0$ and fixed almost complex structures fits into a diagram very similar to the one above:

$$
\begin{array}{ccccccccc}
0 & \to & W^{1,p}(T_\Sigma[A]) & \to & W^{1,p}(\varphi^* T_M) & \to & W^{1,p}(N') & \to & 0 \\
 & & \bar{\partial}_{T_\Sigma} \downarrow & & \downarrow D'_{\varphi,J} & & \downarrow D^{N'}_{\varphi,J} & & \\
0 & \to & L^p(T_\Sigma[A] \otimes \Lambda^{0,1}) & \to & L^p(\varphi^* T_M \otimes \Lambda^{0,1}) & \to & L^p(N' \otimes \Lambda^{0,1}) & \to & 0.
\end{array}
$$

In the right column N' denotes the image of N under the normalization map:

$$N' := \bigoplus_i \varphi_i^* T_M / D\varphi_i(T_{\Sigma_i}).$$

Thus N' is a holomorphic line bundle on Σ only away from the nodes, while near a node it is a direct sum of line bundles on each of the two branches. Note that surjectivity of $W^{1,p}(\Sigma, \varphi^* T_M) \to W^{1,p}(\Sigma, N')$ is special to the nodal case in dimension 4 since it requires the tangent spaces of the branches at a node $P \in \varphi(C)$ to generate $T_{M,P}$. A crucial observation then is that the obstructions to this extended deformation problem can be computed on the normalization:

$$\operatorname{coker} D^{N'}_{\varphi,J} = \operatorname{coker} D^N_{\hat{\varphi},J}.$$

This follows by chasing the diagrams. Geometrically the identity can be understood by saying that it is the same to deform φ as pseudo-holomorphic map or to deform each of the maps $\Sigma_i \to M$ separately. In fact, the position of the identification points of the Σ_i are uniquely determined by the maps to M. In view of the cohomological unobstructedness theorem and the implicit function theorem relative $\mathscr{J} \times \mathbb{C}^l$ we obtain the following strengthening of Proposition 3.4.

Proposition 3.5 *[Sk] Let $C \subset M$ be a nodal J-holomorphic curve on an almost complex 4-manifold (M, J), $C = \bigcup C_i$. Assume that $c_1(M) \cdot C_i > 0$ for every i. Then the moduli space of J-holomorphic curves homologous to C is a smooth manifold of real dimension $c_1(M) \cdot C + C \cdot C$. The subset parametrizing nodal curves is locally a transversal union of submanifolds of real codimension 2. In particular, there exists a sequence of smooth J-holomorphic curves $C_n \subset M$ with $C_n \to C$ in the Hausdorff sense (C can be smoothed), and such smoothings are unique up to isotopy through smooth J-holomorphic curves.*

4 The Theorem of Micallef and White

4.1 Statement of Theorem

In this section we discuss the theorem of Micallef and White on the holomorphicity of germs of pseudo-holomorphic curves up to $\mathscr{C}^1$-diffeomorphism. The precise statement is the following.

Theorem 4.1 *(Micallef and White [MiWh], Theorem 6.2.) Let J be an almost complex structure on a neighborhood of the origin in $\mathbb{R}^{2n}$ with $J_{|0} = I$, the standard complex structure on $\mathbb{C}^n = \mathbb{R}^{2n}$. Let $C \subset \mathbb{R}^{2n}$ be a J-holomorphic curve with $0 \in C$.*

Then there exists a neighborhood U of $0 \in \mathbb{R}^{2n}$ and a $\mathscr{C}^1$-diffeomorphism $\Phi : U \to V \subset \mathbb{C}^n$, $\Phi(0) = 0$, $D\Phi_{|0} = \mathrm{id}$, such that $\Phi(C) \subset V$ is defined by complex polynomial equations. In particular, $\Phi(C)$ is a holomorphic curve.

The proof in loc. cit. might seem a bit computational on first reading, but the basic idea is in fact quite elegant and simple. As it is one substantial ingredient in our proof of the isotopy theorem, we include a discussion here. For simplicity we restrict to the two-dimensional, pseudo-holomorphically fibered situation, just as in Sect. 2.4,I. In other words, there are complex coordinates z, w with $(z, w) \mapsto z$ pseudo-holomorphic. Then w can be chosen in such a way that $T_M^{1,0} = \mathbb{C} \cdot (\partial_{\bar{z}} + b\partial_w) + \mathbb{C} \cdot \partial_{\bar{w}}$ for a $\mathbb{C}$-valued function b with $b(0, 0) = 0$. This will be enough for our application and it still captures the essentials of the fully general proof.

4.2 The Case of Tacnodes

Traditionally a *tacnode* is a higher order contact point of a union of smooth holomorphic curves, its branches. The same definition makes sense pseudo-holomorphically. We assume this tangent to be $w = 0$. Then the ith branch of our pseudo-holomorphic tacnode is the image of

$$\Delta \longrightarrow \mathbb{C}^2, \quad t \longmapsto (t, f_i(t)),$$

with $f_i(0) = 0$, $Df_{i|0} = 0$. The pseudo-holomorphicity equation takes the form $\partial_{\bar{t}} f_i = b(t, f_i(t))$. For $i \neq j$ this gives the equation

$$0 = \big(\partial_{\bar{t}} f_j - b(t, f_j)\big) - \big(\partial_{\bar{t}} f_i - b(t, f_i)\big)$$
$$= \partial_{\bar{t}}(f_j - f_i) - \frac{b(t, f_j) - b(t, f_i)}{f_j - f_i} \cdot (f_j - f_i).$$

Now $\big(b(t, f_j) - b(t, f_i)\big)/(f_j - f_i)$ is bounded, and hence $f_j - f_i$ is another instance of a pseudo-analytic function. The Carleman trick in Lemma 3.3 now implies that f_i and f_j osculate only to finite order. (This also follows from Aronszajn's unique continuation theorem [Ar].)

On the other hand, if $f_j - f_i = O(|t|^n)$ then $|\partial_{\bar{t}}(f_j - f_i)| = \big|b(t, f_i(t)) - b(t, f_j(t))\big| = O(|t|^n)$ by pseudo-holomorphicity and hence, by Lemma 2.11

$$f_j(t) - f_i(t) = at^n + o(|t|^n). \tag{4.1}$$

The polynomial leading term provides the handle to holomorphicity. The diffeomorphism Φ will be of the form

$$\Phi(z, w) = (z, w - E(z, w)).$$

To construct $E(z, w)$ consider the approximations

$$f_{i,n} = M.V.\{f_j | f_j - f_i = o(|t|^n)\}$$

to f_i, for every i and $n \geq 1$. $M.V.$ stands for the arithmetic mean. Because we are dealing with a tacnode, $f_{i,1} = f_{j,1}$ for every i, j, and by finiteness of osculation orders, there exists N with $f_{i,n} = f_i$ for every $n \geq N$. Now (4.1) gives $a_{i,n} \in \mathbb{C}$ and functions $E_{i,n}$ with

$$f_{i,n} - f_{i,n-1} = a_{i,n} t^n + E_{i,n}(t), \quad E_{i,n}(t) = o(|t|^n).$$

Summing from $n = 1$ to N shows

$$f_i = \sum_{n=2}^{N} a_{i,n} t^n + f_{i,1}(t) + \sum_{n=2}^{N} E_{i,n}(t). \tag{4.2}$$

The rest is a matter of merging the various $E_{i,n}$ into $E(z, w)$ to achieve $\Phi(t, f_i(t)) = (t, \sum_{n=2}^{N} a_{i,n} t^n)$. We are going to set $E = \sum_{i=1}^{N} E_i$ with

$E_i(z, w) = E_{i,n}(z)$ in a strip osculating to order n to the graph of $f_{i,n}$. More precisely, choose a smooth cut-off function $\rho : \mathbb{R}_{\geq 0} \to [0, 1]$ with $\rho(s) = 1$ for $s \in [0, 1/2]$ and $\rho(s) = 0$ for $s \geq 1$. Then for $n = 1$ define

$$E_1(z, w) = \rho\left(\frac{|w|}{|z|^{3/2}}\right) \cdot f_{i,1}(z).$$

The exponent $3/2$ may be replaced by any number in $(1, 2)$. For $n \geq 2$ take

$$E_n(z, w) = \begin{cases} \rho\left(\dfrac{|w - f_{i,n}(z)|}{\varepsilon|z|^n}\right) \cdot E_{i,n}(z), & |w - f_{i,n}(z)| \leq \varepsilon|z|^n \\ 0, & \text{otherwise.} \end{cases}$$

To see that this is well-defined for ε and $|z|$ sufficiently small, note that by construction $f_{i,n}$ and $f_{j,n}$ osculate polynomially to dominant order. If this order is larger than n then $f_{i,n} = f_{j,n}$. Otherwise $|f_{j,n}(z) - f_{i,n}(z)| > \varepsilon|z|^n$ for ε and $|z|$ sufficiently small. See Fig. 4.1 for illustration. The distinction between the cases $n = 1$ and $n > 1$ at this stage could be avoided by formally setting $E_{i,1} = f_{i,1}$. However, to also treat branches with different tangents later on, we want Φ to be the identity outside a region of the form $|z| \leq |w|^a$ with $a > 1$.

Finally note that by construction $E_n(t, f_i(t)) = E_{i,n}(t)$, and hence

$$\Phi(t, f_i(t)) = \left(t, f_i(t) - \sum_n E_n(t, f_i(t))\right)$$

$$= \left(t, f_i(t) - \sum_{n=2}^{N} E_{i,n}(t) - f_{i,1}(t)\right) \overset{(4.2)}{=} \left(t, \sum_{n=1}^{N} a_{i,n} t^n\right).$$

This finishes the proof of Theorem 4.1 for the case of tacnodes under the made simplifying assumptions.

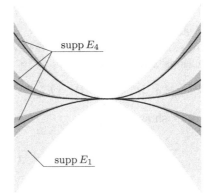

Fig. 4.1. The supports of E_n

As for differentiability it is clear that Φ is smooth away from $(0,0)$. But $\partial_w E_n$ involves the term

$$\left| \partial_{\bar{w}} \rho \left(\frac{|w - f_{i,n}|}{\varepsilon \cdot |z|^n} \right) \right| = \frac{1}{|z|^n} \cdot o(|z|^n) = o(1).$$

Thus φ may only be $\mathscr{C}^1$.

4.3 The General Case

In the general case the branches of C are images of pseudo-holomorphic maps

$$\Delta \longrightarrow \mathbb{C}^2, \quad t \longmapsto (t^{Q_i}, f_i(t)),$$

for some $Q_i \in \mathbb{N}$. Note our simplifying assumption that the projection onto the first coordinate be pseudo-holomorphic. By composing with branched covers $t \mapsto t^{m_i}$ we may assume $Q_i = Q$ for all i. The pseudo-holomorphicity equation then reads $\partial_{\bar{t}} f_i = Q \bar{t}^Q b(t, f_i(t))$. The proof now proceeds as before but we deal with multi-valued functions $E_{i,n}$ of z. A simple way to implement this is by enlarging the set of functions f_i by including compositions with $t \mapsto \zeta t$ for all Qth roots of unity ζ. The definition of $E_n(z, w)$ then reads

$$E_n(z, w) = \begin{cases} \rho \left(\dfrac{|w - f_{i,n}(t)|}{\varepsilon |t|^n} \right) \cdot E_{i,n}(t), & |w - f_{i,n}(t)| \le \varepsilon |z|^n \\ 0, & \text{otherwise}, \end{cases}$$

for any t with $t^Q = z$. This is well-defined as before since the set of functions f_i is invariant under composition with $t \mapsto \zeta t$ whenever $\zeta^Q = 1$.

Finally, if C has branches with different tangent lines do the construction for the union of branches with given tangent line separately. The diffeomorphisms obtained in this way are the identity outside of trumpet-like sets osculating to the tangent lines as in Fig. 4.2. Hence their composition maps each branch to an algebraic curve as claimed in the theorem.

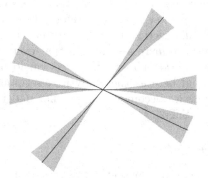

Fig. 4.2. Different tangent lines

5 Unobstructedness II: The Integrable Case

5.1 Motivation

We saw in Sect. 3 that if $c_1(M)$ evaluates strictly positively on a smooth pseudo-holomorphic curve in a four-manifold then this curve has unobstructed deformations. The only known generalizations to singular curves rely on parametrized deformations. These deformations preserve the geometric genus and, by the genus formula, lead at best to a nodal curve. Unobstructedness in this restricted class of deformations is a stronger statement, which thus requires stronger assumptions. In particular, the types of the singular points enter as a condition, and this limits heavily the usefulness of such results for the isotopy problem. Note these problems do already arise in the integrable situation. For example, not every curve on a complex surface can be deformed into a curve of the same geometric genus with only nodes as singularities. This is a fact of life and can not be circumvented by clever arguments.

Thus we need to allow an increase of geometric genus. There are two points of views for this. The first one looks at a singular pseudo-holomorphic curve as the image of a pseudo-holomorphic map from a nodal Riemann surface, as obtained by the Compactness Theorem (Sect. 7). While this is a good point of view for many general problems such as defining Gromov-Witten invariants, we are now dealing with maps from a *reducible domain* to a complex two-dimensional space. This has the effect that (total) degree arguments alone as in Sect. 3 do not give good unobstructedness results. For example, unobstructedness fails whenever the limit stable map contracts components of higher genus. Moreover, it is not hard to show that not all stable maps allowed by topology can arise. There are subtle and largely ununderstood analytical obstructions preventing this. Again both problems are inherited from the integrable situation. A characterization of holomorphic or algebraic stable maps that can arise as limit is an unsolved problem in algebraic geometry. There is some elementary discussion on this in the context of stable reduction in [HaMo]. If a good theory of unobstructedness in the integrable case from the point of view of stable maps was possible it would likely generalize to the pseudo-holomorphic setting. Unfortunately, such theory is not in sight.

The second point of view considers deformations of the limit as a cycle. The purpose of this section is to prove unobstructedness *in the integrable situation* under the mere assumption that every component evaluates strictly positively on $c_1(M)$. This is the direct analogue of Proposition 3.4. Integrability is essential here as the analytic description of deformations of pseudo-holomorphic cycles becomes very singular under the presence of multiple components, see [SiTi2].

5.2 Special Covers

We now begin with the proper content of this section. Let M be a complex surface, not necessarily compact but without boundary. In our application M will be a tubular neighborhood of a limit pseudo-holomorphic cycle in a rational ruled almost complex four-manifold M, endowed with a *different*, integrable almost complex structure. Let $C = \sum_i m_i C_i$ be a compact holomorphic cycle on M of complex dimension one. We assume C to be *effective*, that is $m_i > 0$ for all i. There is a general theory on the existence of moduli spaces of compact holomorphic cycles of any dimension, due to Barlet [Bl]. For the case at hand it can be greatly simplified and this is what we will do in this section.

The approach presented here differs from [SiTi3], Sect. 4 in being strictly local around C. This has the advantage to link the linearization of the modeling PDE to cohomology on the curve very directly. However, such a treatment is not possible if C contains fibers of our ruled surface $M \to S^2$, and then the more global point of view of [SiTi3] becomes necessary.

The essential simplification is the existence of special covers of a neighborhood of $|C| = \bigcup_i C_i$.

Hypothesis 5.1 *There exists an open cover $\mathcal{U} = \{U_\mu\}$ of a neighborhood of C in M with the following properties:*

1. *$\mathrm{cl}(U_\mu) \simeq A_\mu \times \bar{\Delta}$, where A_μ is a compact, connected Riemann surface with non-empty boundary and $\bar{\Delta} \subset \mathbb{C}$ is the closed unit disk.*
2. *$(A_\mu \times \partial\Delta) \cap |C| = \emptyset$.*
3. *$U_\kappa \cap U_\mu \cap U_\nu = \emptyset$ for any pairwise different κ, μ, ν. The fiber structures given by projection to A_μ are compatible on overlaps.*

The symbol "$\simeq$" denotes biholomorphism. The point of these covers is that on U_μ there is a holomorphic projection $U_\mu \to \mathrm{inn}(A_\mu)$ with restriction to $|C|$ proper and with finite fibers, hence a branched covering. Such cycles in $A_\mu \times \mathbb{C}$ of degree b over A_μ have a description by b holomorphic functions via Weierstraß polynomials, see below. Locally this indeed works in any dimension. But it is generally impossible to make the projections match on overlaps as required by (3), not even locally and for smooth C. The reason is that the analytic germ of M along C need not be isomorphic to the germ of the holomorphic normal bundle along the zero section. This almost always fails in the positive ("ample") case that we are interested in, see [La]. For our application to cycles in ruled surfaces, however, we can use the bundle projection, provided C does not contain fiber components, see Lemma 5.1 below. Without this assumption the arguments below still work but require a more difficult discussion of change of projections, as in [Bl]. As the emphasis of these lectures is on explaining why things are true rather than generality we choose to impose this simplifying assumption.

Another, more severe, simplification special to one-dimensional C is that we do not allow triple intersections. This implies that A_μ can not be contractible for all μ unless all components of C are rational, and hence our charts have a certain global flavor. Under the presence of triple intersections it seems impossible to get our direct connection of the modeling of the moduli space with cohomological data on C.

Lemma 5.1 *Assume there exists a map $p : M \to S$ with $\dim S = 1$ such that p is a holomorphic submersion near $|C|$ and no component of $|C|$ contains a connected component of a fiber of p. Then Hypothesis 5.1 is fulfilled.*

Proof. Denote by $Z \subset M$ the finite set of singular points of $|C|$ and of critical points of the projection $|C| \to S$. Add finitely many points to Z to make sure that Z intersects each irreducible component of $|C|$. For $P \in |C|$ let $F = p^{-1}(p(P))$ be the fiber through P. Then $F \cap |C|$ is an analytic subset of F that by hypothesis does not contain a connected component of F. Since F is complex one-dimensional P is an isolated point of $F \cap C$. Because p is a submersion there exists a holomorphic chart $(z, w) : U(P) \to \mathbb{C}^2$ in a neighborhood of P in M with $(z, w)(P) = 0$ and $z = \text{const}$ describing the fibers of p. We may assume $U(P)$ to be so small that $|C|$ is the zero locus of a holomorphic function h defined on all of $U(P)$. Let $\varepsilon > 0$ be such that $F \cap w^{-1}(B_\varepsilon(0)) = \{P\}$. Then $\min\{|f(0, w)| \,|\, |w| = \varepsilon\} > 0$. Hence for $\delta > 0$ sufficiently small $\min\{|f(z, w)| \,|\, |w| = \varepsilon, |z| \leq \delta\}$ remains nonzero. This shows $|C| \cap (z, w)^{-1}(\overline{B}_\delta(0) \times \partial B_\varepsilon(0)) = \emptyset$. This verifies Hypothesis 5.1,1 and 2 for $U(P)$, while the compatibility of fiber structures in 5.1,3 holds by construction. This defines elements $U_1, \ldots, U_{\sharp Z}$ of the open cover $\mathcal{U}$ intersecting Z.

To finish the construction we define one more open set U_0 as follows. This construction relies on Stein theory, see e.g. [GrRe] or [KpKp] for the basics. (With some effort this can be replaced by more elementary arguments, but it does not seem worthwhile doing here for a technical result like this.) Choose a Riemannian metric on M. Let δ be so small that $B_{2\delta}(Z) \subset \bigcup_{\nu \geq 1} U_\nu$. Since we have chosen Z to intersect each irreducible component of $|C|$, the complement $|C| \setminus \text{cl}\, B_\delta(Z)$ is a union of open Riemann surfaces. Thus $|C| \setminus \text{cl}\, B_\delta(Z)$ is a Stein submanifold of $M \setminus \text{cl}\, B_\delta(Z)$, hence has a Stein neighborhood $W \subset M \setminus \text{cl}\, B_\delta(Z)$. Now every hypersurface in a Stein manifold is defined by one global holomorphic function, say f in our case. So f is a global version of the fiber coordinate w before. Then for δ sufficiently small the projection

$$U_0 := \big\{ P \in W \setminus \text{cl}\, B_{3\delta/2}(Z) \,\big|\, |f(P)| < \delta \big\} \xrightarrow{p} S$$

factors holomorphically through $\pi : U_0 \to |C| \setminus \text{cl}\, B_{3\delta/2}(Z)$. This is an instance of so-called Stein factorization, which contracts the connected components of the fibers of a holomorphic map. Choosing δ even smaller this factorization gives a biholomorphism

$$U_0 \xrightarrow{(\pi, f)} |C| \setminus \text{cl}\, B_\delta(Z) \times \Delta$$

extending to $\text{cl}(U_0)$. Hence $U_0, U_1, \ldots, U_{\sharp Z}$ provides the desired open cover $\mathcal{U}$. $\square$

5.3 Description of the Deformation Space

Having a cover fulfilling Hypothesis 5.1 it is easy to describe the moduli space of small deformations of C as the fiber of a holomorphic, non-linear Fredholm map. We use Čech notation $U_{\mu\nu} = U_\mu \cap U_\nu$, $U_{\kappa\mu\nu} = U_\kappa \cap U_\mu \cap U_\nu$. Write $V_\mu = \text{inn}(A_\mu)$. Fix $p > 2$ and denote by $\mathcal{O}^p(V_\mu)$ the space of holomorphic functions on $\text{inn}(V_\mu)$ of class L^p. This is a Banach space with the L^p-norm defined by a Riemannian metric on A_μ, chosen once and for all. Similarly define $\mathcal{O}^p(V_{\mu\nu})$ and $\mathcal{O}^p(V_{\kappa\mu\nu})$.

To describe deformations of C let us first consider the local situation on $V_\mu \times \Delta \subset M$. For this discussion we drop the index μ. Denote by w the coordinate on the unit disk. For any holomorphic cycle C' on $V \times \Delta$ with $|C'| \cap (V \times \partial\Delta) = \emptyset$ the Weierstraß preparation theorem gives bounded holomorphic functions $a_1, \ldots, a_b \in \mathcal{O}^p(V)$ with $w^b + a_1 w^{b-1} + \ldots + a_b = 0$ describing C', see e.g. [GrHa]. Here b is the relative degree of C' over V, and everything takes multiplicities into account. The tuple $(a_1, \ldots, a_b)$ should be thought of as a holomorphic map from V to the b-fold symmetric product $\text{Sym}^b \mathbb{C}$ of $\mathbb{C}$ with itself.

Digression on $\text{Sym}^b \mathbb{C}$. By definition $\text{Sym}^b \mathbb{C}$ is the quotient of $\mathbb{C} \times \cdots \times \mathbb{C}$ by the permutation action of the symmetric group S_b on b letters. Quite generally, if a finite group acts on a complex manifold or complex space X then the topological space X/G has naturally the structure of a complex space by declaring a function on X/G holomorphic whenever its pull-back to X is holomorphic. For the case of the permutation action on the coordinates of $\mathbb{C}^b$ we claim that the map

$$\Phi : \text{Sym}^d \mathbb{C} \longrightarrow \mathbb{C}^d$$

induced by $(\sigma_1, \ldots, \sigma_b) : \mathbb{C}^b \to \mathbb{C}^b$ is a biholomorphism. Here

$$\sigma_k(w_1, \ldots, w_b) = \sum_{1 \leq i_1 < i_2 < \ldots < i_k \leq b} w_{i_1} w_{i_2} \ldots w_{i_b}$$

is the ith elementary symmetric polynomial. In fact, set-theoretically $\text{Sym}^b \mathbb{C}$ parametrizes unordered tuples of b not necessarily disjoint points in $\mathbb{C}$. By the fundamental theorem of algebra there is precisely one monic (leading coefficient equal to 1) polynomial of degree b having this zero set, with multiplicities. The coefficients of this polynomial are the elementary symmetric functions in the zeros. This shows that Φ is bijective. Now any symmetric holomorphic function in $w_1, \ldots, w_b$ is a holomorphic function in $\sigma_1, \ldots, \sigma_b$. Thus Φ is a biholomorphism. By the same token, if w, w' are two holomorphic coordinates on an open set $W \subset \mathbb{C}$ then the induced holomorphic coordinates σ_i, σ'_i, $i = 1, \ldots, b$, are related by a biholomorphic transformation. Note however that something peculiar is happening to the differentiable structure: Not every S_b-invariant smooth function on $\mathbb{C}^b$ leads to a smooth function on $\text{Sym}^b \mathbb{C}$ for the differentiable structure coming from holomorphic geometry. As an example consider $b = 2$ and the function $f(w_1, w_2) = (w_1 - w_2)(\bar{w}_1 - \bar{w}_2) = |w_1 - w_2|^2$. This is the pull-back of $|\sigma_1^2 - \sigma_2|$, which is only a Lipschitz function on $\text{Sym}^2 \mathbb{C}$. Another, evident feature of symmetric products is that a neighborhood of a point $\sum_i m_i P_i$ of $\text{Sym}^b \mathbb{C}$ with the P_i pairwise disjoint, is canonically biholomorphic to an open set in $\prod_i \text{Sym}^{m_i} \mathbb{C}$. End of digression.

From this discussion it follows that we can compare Weierstraß representations with compatible projections. Let $(a_\mu : V_\mu \to \mathrm{Sym}^{b_\mu} \mathbb{C}) \in \mathcal{O}^p(V_\mu)^{b_\mu}$ be the Weierstraß representations for the given cycle C. Let $V^b_{\mu\nu} \subset V_{\mu\nu}$ be the union of connected components where the covering degree is equal to b. Note that $V^b_{\mu\nu} = \emptyset$ whenever $b > \min\{b_\mu, b_\nu\}$. For a'_μ sufficiently close to a_μ in L^p there is a cycle C'_μ in U_μ with Weierstraß representation a'_μ. Denote by $\mathscr{F}_\mu$ a sufficiently small neighborhood of a_μ in $\mathcal{O}^p(V_\mu)^{b_\mu}$ where this is the case. For every μ, ν, b let $\mathscr{F}^b_{\mu\nu} = \mathcal{O}^p(V^b_{\mu\nu})^b$, viewed as space of maps $V^b_{\mu\nu} \to \mathrm{Sym}^b \mathbb{C}$. The above discussion gives comparison maps

$$\Theta^b_{\mu\nu} : \mathscr{F}_\nu \longrightarrow \mathscr{F}^b_{\mu\nu}.$$

Define the gluing map by

$$\Theta : \prod_\mu \mathscr{F}_\mu \longrightarrow \prod_{\mu<\nu, b} \mathscr{F}^b_{\mu\nu}, \quad \Theta(a'_\mu) = (a'_\mu - \Theta^b_{\mu\nu}(a'_\nu))_{\mu\nu}. \tag{5.1}$$

Clearly $(a'_\mu)_\mu$ glues to a holomorphic cycle iff $\Theta(a'_\mu) = 0$. We will see that Θ is a holomorphic Fredholm map with kernel and cokernel canonically isomorphic to the first two cohomology groups of the normal sheaf of C in M. To follow this plan, two more digressions are necessary. One to explain the notion of normal sheaf and one for the needed properties of sheaf cohomology.

5.4 The Holomorphic Normal Sheaf

One definition of tangent vector of a differentiable manifold M at a point P works by emphasizing its property of *derivation*. Let $\mathscr{C}^\infty_{M,P}$ be the space of germs of smooth functions at P. An element of this space is represented by a smooth function defined on a neighborhood of P, and two such functions give the same element if they agree on a common neighborhood. Applying a tangent vector $X \in T_{M,P}$ on representing functions defines an $\mathbb{R}$-linear map $D : \mathscr{C}^\infty_{M,P} \to \mathbb{R}$. For $f, g \in \mathscr{C}^\infty_{M,P}$ Leibniz' rule gives

$$X(fg) = fX(g) + gX(f).$$

In particular, $X(f^2) = 0$ if $f(P) = 0$ and $X(1) = X(1^2) = 2X(1) = 0$. Thus the interesting part of D is the induced map on $\mathfrak{m}^\infty_P/(\mathfrak{m}^\infty_p)^2$ where $\mathfrak{m}^\infty_P = \{f \in \mathscr{C}^\infty_{M,P} \mid f(P) = 0\}$ is the maximal ideal of the ring $\mathscr{C}^\infty_{M,P}$. We claim that the map $T_{M,P} \to \mathrm{Hom}_\mathbb{R}(\mathfrak{m}^\infty_P/(\mathfrak{m}^\infty_p)^2, \mathbb{R})$ is an isomorphism. In fact, let $x_1, \ldots, x_n$ be coordinates of M around P. Then $X = \sum_i a_i \partial_{x_i}$ applied to x_i yields a_i, so the map is injective. On the other hand, if $f \in \mathfrak{m}^\infty_P$ then by the Taylor formula $f - \sum_i \partial_{x_i} f(P) \cdot x_i \in (\mathfrak{m}^\infty)^2$, and hence any linear map $\mathfrak{m}^\infty_P/(\mathfrak{m}^\infty_p)^2 \to \mathbb{R}$ is determined by its values on x_i. This gives the well-known canonical identification

$$T_{M,P} = \mathrm{Hom}_\mathbb{R}(\mathfrak{m}^\infty_P/(\mathfrak{m}^\infty_p)^2, \mathbb{R}).$$

If M is a complex manifold then the holomorphic tangent space at P can similarly be described by considering holomorphic functions and $\mathbb{C}$-linear maps.

Now if $Z \subset M$ is a submanifold the same philosophy applies to the normal bundle. For this and what follows we will inevitably need some elementary sheaf theory that we are also trying to explain briefly.

An (abelian) *sheaf* is an association of an abelian group $\mathcal{F}(U)$ (the *sections* of $\mathcal{F}$ over U) to every open subset $U \subset M$, together with restriction homomorphisms $\rho_{VU} : \mathcal{F}(U) \to \mathcal{F}(V)$ whenever $V \subset U$. The restriction maps must be compatible with composition. Moreover, the following sheaf axioms must hold for every covering $\{U_i\}$ of an open set $U \subset M$. Write $U_{ij} = U_i \cap U_j$.

(S1) (*local-global principle*) If $s \in \mathcal{F}(U)$ and $\rho_{U_i U}(s) = 0$ for all i then $s = 0$.
(S2) (*gluing axiom*) Given $s_i \in \mathcal{F}(U_i)$ with $\rho_{U_{ij}U_i}(s_i) = \rho_{U_{ij}U_j}(s_j)$ for every i,j there exists $s \in \mathcal{F}(U)$ with $s_i = \rho_{U_i U}(s)$.

The following are straightforward examples:

1. $\mathcal{C}_M^\infty : U \mapsto \{f : U \to \mathbb{R} \text{ smooth}\}$ with restriction of functions defining the restriction maps.
2. For $E \downarrow M$ a fiber bundle the sheaf $\mathcal{C}^\infty(E)$ of smooth sections of E.
3. For M a complex manifold, $\mathcal{O}_M : U \mapsto \{f : U \to \mathbb{C} \text{ holomorphic}\}$. This is a subsheaf of the sheaf of complex valued smooth functions on M.
4. For G an abelian group the constant sheaf

$$G_M : U \mapsto \{f : U \to G \text{ locally constant}\},$$

   ρ_{VU} the restriction. The sections of this sheaf over any *connected* open set are identified with elements of G.

A homomorphism of sheaves $\mathcal{F} \to \mathcal{G}$ is a system of homomorphisms $\mathcal{F}(U) \to \mathcal{G}(U)$ compatible with restriction.

Returning to the normal bundle of $Z \subset M$ let $\mathcal{I}_Z^\infty$ be the sheaf of smooth functions vanishing along Z. Then a normal vector field ν along Z on $U \subset Z$ induces a well-defined map

$$\mathcal{I}_Z^\infty(U) \longrightarrow \mathcal{C}_Z^\infty(U), \quad f \longmapsto \left(\tilde{\nu}(f)\right)\big|_Z,$$

where $\tilde{\nu}$ is a lift of ν to a vector field on M defined on a neighborhood of U in M. As we restrict to Z after evaluation and since f vanishes along Z the result does not depend on the choice of lift. Now as with $T_{M,P}$ one checks that $(\mathcal{I}_Z^\infty(U))^2$ maps to zero and that the map of sheaves

$$\mathcal{C}^\infty(N_{Z|M}) \longrightarrow \mathcal{H}om_{\mathcal{C}_Z^\infty}(\mathcal{I}_Z^\infty/(\mathcal{I}_Z^\infty)^2, \mathcal{C}_Z^\infty)$$

is an isomorphism. A section over U of the Hom-sheaf on the right is a $(\mathcal{C}_Z^\infty)|_U$-linear sheaf homomorphism $(\mathcal{I}_Z^\infty/(\mathcal{I}_Z^\infty)^2)|_U \to (\mathcal{C}_Z^\infty)|_U$. Note that multiplication of sections $\mathcal{I}_Z^\infty/(\mathcal{I}_Z^\infty)^2$ by sections of $\mathcal{C}_Z^\infty$ is well-defined, so this makes sense.

Generally one has to be careful here where to put the brackets because the notions of quotient and Hom-sheaves are a bit delicate. For example, if $\mathcal{F} \to \mathcal{G}$ is a sheaf homomorphism the sheaf axioms need not hold for $\mathcal{Q} : U \mapsto \mathcal{G}(U)/\mathcal{F}(U)$. The standard example for this is the inclusion $\mathbb{Z}_M \to \mathscr{C}_M^\infty$ with $M = \mathbb{R}^2 \backslash \{0\}$ or any other non-simply connected space. On the complements of the positive and negative real half-axes $1/2\pi$ of the angle in polar coordinates give sections of $\mathscr{C}_M^\infty$ agreeing on $\mathbb{R}^2 \setminus \mathbb{R}$ modulo integers, but they do not glue to a single-valued function on M.

In any case, there is a canonical procedure to force the sheaf axioms, by taking the sheaf of continuous sections of the space of stalks

$$\text{Ét}(\mathcal{Q}) := \coprod_{P \in M} \varinjlim_{U \ni P} \mathcal{Q}(U)$$

(étale space) of $\mathcal{Q}$. Every $s \in \mathcal{Q}(U)$ induces a section of $\text{Ét}(\mathcal{Q})$ over U, and the images of these sections are taken as basis for the topology of $\text{Ét}(\mathcal{Q})$. In writing down quotient or Hom-sheaves it is understood to apply this procedure. This is the general definition that we will need for the holomorphic situation momentarily, but it is in fact not needed for sheaves that allow partitions of unity (fine sheaves) such as $\mathcal{I}_Z^\infty$. So a section of $\text{Hom}_{\mathscr{C}_Z^\infty}(\mathcal{I}_Z^\infty/(\mathcal{I}_Z^\infty)^2, \mathscr{C}_Z^\infty)$ over U is indeed just a $\mathscr{C}_Z^\infty(U)$-linear map of section spaces $\mathcal{I}_Z^\infty(U)/(\mathcal{I}_Z^\infty(U))^2 \to \mathscr{C}_Z^\infty(U)$.

Now if M is a complex manifold and $Z \subset M$ is a complex submanifold then $N_{Z|M}$ is a holomorphic vector bundle. Let $\mathcal{I}_Z \subset \mathcal{O}_M$ be the sheaf of holomorphic functions on M vanishing along Z. Then the natural map

$$\mathcal{O}(N_{Z|M}) \longrightarrow \text{Hom}_{\mathcal{O}_Z}(\mathcal{I}_Z/\mathcal{I}_Z^2, \mathcal{O}_Z)$$

is an isomorphism. Explicitly, if Z is locally given by one equation $f = 0$ then f generates $\mathcal{I}_Z/\mathcal{I}_Z^2$ as an $\mathcal{O}_Z$-module. Hence a section φ of the Hom-sheaf on the right is uniquely defined by $\varphi(f)$, a holomorphic function on Z. This provides an explicit local identification of $\mathcal{O}(N_{Z|M})$ with $\mathcal{O}_Z$, which is nothing but a local holomorphic trivialization of $N_{Z|M}$.

Now the whole point of this discussion is that it generalizes well to the non-reduced situation. Let us discuss this in the most simple situation of a multiple point mP in $M = \mathbb{C}$. Taking P the origin and w for the coordinate on $\mathbb{C}$ (z will have a different meaning below) we have $\mathcal{I}_{mP} = \mathcal{O}_\mathbb{C} \cdot w^m$ and $\mathcal{O}_{mP} = \mathcal{O}_\mathbb{C}/\mathcal{I}_{mP}$ is an m-dimensional complex vector space with basis $1, w, \ldots, w^{m-1}$. A homomorphism $\mathcal{I}_{mP}/(\mathcal{I}_{mP})^2 \to \mathcal{O}_{mP}$ is uniquely defined by $(\alpha_0, \ldots, \alpha_m) \in \mathbb{C}^m$ via

$$w^m \longmapsto \alpha_0 + \alpha_1 w + \ldots + \alpha_{m-1} w^{m-1}.$$

This fits well with limits as follows. Consider mP as the limit of m pairwise different points $Z_t := \{P_1(t), \ldots, P_m(t)\}$, $t > 0$, given by the vanishing of $f_t := w^m + a_1(t)w^{m-1} + \ldots + a_m(t)$ where $a_i(t) \xrightarrow{t \to 0} 0$. Then $\mathcal{I}_{Z(t)}/(\mathcal{I}_{Z(t)})^2 = \bigoplus_i \mathcal{I}_{P_i(t)}/(\mathcal{I}_{P_i(t)})^2$ is the sheaf with one copy of $\mathbb{C}$ at each point of $Z(t)$ (a "skyscraper sheaf"). Thus $\bigoplus_i T_{P_i(t)} = \text{Hom}(\mathcal{I}_{Z(t)}/(\mathcal{I}_{Z(t)})^2, \mathcal{O}_{Z(T)})$. Now

$\mathcal{I}_{Z(t)}/(\mathcal{I}_{Z(t)})^2$ is globally generated over $\mathcal{O}_{Z(T)}$ by f_t, and $1, w, \ldots, w^{m-1}$ are a basis for the sections of $\mathcal{O}_{Z(T)}$ as a complex vector space. This gives an identification of $\bigoplus_i T_{P_i(t)}$ with polynomials $\alpha_0 + \alpha_1 w + \ldots + \alpha_{m-1} w^{m-1}$, so this is compatible with the description at $t = 0$! Note that a family of vector fields along $Z(t)$ has a limit for $t \to 0$ if and only if it extends to a continuous family of holomorphic vector fields in a neighborhood of P. The limit is then the limit of $(m-1)$-jets of this family.

It therefore makes sense to define, for any subspace $Z \subset M$ defined by an ideal sheaf $\mathcal{I}_Z \subset \mathcal{O}_M$, reduced or not, the *holomorphic normal sheaf*

$$\mathcal{N}_{Z|M} := \mathcal{H}om_{\mathcal{O}_Z}(\mathcal{I}_Z/\mathcal{I}_Z^2, \mathcal{O}_Z),$$

where $\mathcal{O}_Z = \mathcal{O}_M/\mathcal{I}_Z$ is the sheaf of holomorphic functions on Z. In our application we have Z the generally non-reduced subspace of M defined by the one-codimensional cycle C. By abuse of notation we use C both to denote the cycle and this subspace. Explicitly, $\mathcal{I}_C$ is the sheaf locally generated by $\prod_i f_i^{m_i}$ if $C = \sum_i m_i C_i$ and f_i vanishes to first order along C_i. Note that such a choice of generator of $\mathcal{I}_C$ gives a local identification of $\mathcal{N}_{C|M}$ with $\mathcal{O}_C$.

The importance of the normal sheaf for us comes from its relation with local deformations of holomorphic cycles.

Lemma 5.2 *Let V be a complex manifold and consider the open subset $M \subset \mathcal{O}^p(V, \mathrm{Sym}^b \mathbb{C})$ of tuples $(a_1, \ldots, a_b)$ such that the zero set of $f_{(a_1,\ldots,a_b)}(z, w) := w^b + a_1(z) w^{b-1} + \ldots + a_b(z)$ is contained in $V \times \Delta$. Then there is a canonical isomorphism*

$$T_{(a_1,\ldots,a_b)} M \simeq \Gamma(C, \mathcal{N}_{C|V \times \Delta}),$$

for C the holomorphic cycle in $V \times \Delta$ defined by $f_{(a_1,\ldots,a_b)}$.

Proof. The map sends $\frac{d}{dt}\big|_{t=0}(a_1(t), \ldots, a_b(t)) = (\alpha_1, \ldots, \alpha_b)$ to the section

$$f_{(a_1,\ldots,a_b)} \longmapsto w^b + \alpha_1 w^{b-1} + \ldots + \alpha_b.$$

By the above discussion every global holomorphic function on C has a unique representative of the form $w^b + \alpha_1 w^{b-1} + \ldots + \alpha_b$. Hence this map is an isomorphism. $\square$

5.5 Computation of the Linearization

If $\mathcal{F}$ is an abelian sheaf on a topological space X and $\mathcal{U} = \{U_i\}$ is an open cover of X the Čech cohomology groups $\check{H}^k(\mathcal{U}, \mathcal{F})$ are the cohomology groups of the Čech complex $(C^\bullet(\mathcal{U}, \mathcal{F}))$ with cochains

$$C^k(\mathcal{U}, \mathcal{F}) = \prod_{i_0 < i_1 < \ldots < i_k} \Gamma(U_{i_0} \cap \ldots \cap U_{i_k}, \mathcal{F}),$$

and differentials

$$\check{d}(s_{i_0} \ldots s_{i_k})_{i_0 \ldots i_k} = \left(\sum_l (-1)^l s_{i_0 \ldots \hat{i}_l \ldots i_k} \right)_{i_0 \ldots i_k}.$$

By the gluing axiom (S2) it holds $\check{H}^0(\mathcal{U}, \mathcal{F}) = \mathcal{F}(X)$.

Theorem 5.3 *The gluing map Θ from (5.1) is a holomorphic map with* $\mathrm{kern}\, D\Theta = \check{H}^0(\mathcal{V}, \mathcal{N}_{C|M})$ *and* $\mathrm{coker}\, D\Theta = \check{H}^1(\mathcal{V}, \mathcal{N}_{C|M})$.

Proof. The holomorphicity claim is evident. For the linearization we remark that the components of Θ factor through

$$(a'_\mu, a'_\nu) \longmapsto (a'_\mu, \Theta^b_{\mu\nu}(a'_\nu)) \longmapsto (a'_\mu - \Theta^b_{\mu\nu}(a'_\nu)).$$

In view of Lemma 5.2 the linearization at (a_μ, a_ν) of the components of the first map are canonically the restriction maps $\Gamma(C \cap U_\mu, \mathcal{N}_{C|M}) \to \Gamma(C \cap U_{\mu\nu}, \mathcal{N}_{C|M})$, $\Gamma(C \cap U_\nu, \mathcal{N}_{C|M}) \to \Gamma(C \cap U_{\mu\nu}, \mathcal{N}_{C|M})$. Hence $D\Theta$ is canonically isomorphic to the Čech complex

$$\prod_\mu \Gamma(C \cap U_\mu, \mathcal{N}_{C|M}) \longrightarrow \prod_{\mu<\nu} \Gamma(C \cap U_{\mu\nu}, \mathcal{N}_{C|M}), \quad (\alpha_\mu)_\mu \longmapsto (\alpha_\mu - \alpha_\nu)_{\mu\nu}.$$

Note that $C^k(C \cap \mathcal{U}, \mathcal{N}_{C|M}) = 0$ for $k > 1$ because triple intersections in $\mathcal{U}$ are empty.   □

5.6 A Vanishing Theorem

For this paragraph we assume some familiarity with sheaf cohomology and Serre duality on singular curves. So this section will be (even) harder to read for somebody without training in complex geometry. Unfortunately we could not find a more elementary treatment.

It is well-known that the Čech cohomology groups $\check{H}^i(C \cap \mathcal{V}, \mathcal{N}_{C|M})$ are finite dimensional and canonically isomorphic to the *sheaf cohomology groups* $H^i(C, \mathcal{N}_{C|M})$, see [GrHa] and the references given there. In particular, Θ is a non-linear Fredholm map. Our aim in this paragraph is to prove surjectivity of its linearization by the following result.

Proposition 5.4 *Let $C = \sum_{i=0}^r m_i C_i$ be a compact holomorphic cycle on a complex surface M. Assume that $c_1(M) \cdot C_i > 0$ and $C_i \cdot C_i \geq 0$ for every i. Then $H^1(C, \mathcal{N}_{C|M}) = 0$.*

Proof. In view of the identification $\mathcal{N}_{C|M} = \mathcal{O}_C(C)$ a stronger statement is the vanishing of $H^1(C, \mathcal{O}_{C'}(C))$ for every effective subcycle $C' \subset C$. This latter formulation allows an induction over the sum of the multiplicities of C'.

As an auxiliary statement we first show the vanishing of $H^1(\mathcal{O}_{C_i}(C''))$ for every i and every subcycle C'' of C containing C_i. Serre duality (see e.g. [BtPeVe], Theorem II.6.1) shows that $H^1(\mathcal{O}_{C_i}(C''))$ is dual to $H^0(\mathcal{H}om_{\mathcal{O}_{C_i}}(\mathcal{O}_{C_i}(C''), \mathcal{O}_{C_i}) \otimes \omega_{C_i})$. Here ω_{C_i} is the *dualizing sheaf* of C_i. If $\mathcal{K}_M$ denotes the sheaf of holomorphic sections of $\det T_M^*$ then it can be computed as $\omega_{C_i} = \mathcal{K}_M \otimes \mathcal{N}_{C_i|M} = \mathcal{K}_M \otimes \mathcal{O}_{C_i}(C_i)$. Therefore

$$\mathcal{H}om(\mathcal{O}_{C_i}(C''), \mathcal{O}_{C_i}) \otimes \omega_{C_i} \simeq \mathcal{O}_{C_i}(-C'') \otimes \omega_{C_i} \simeq \mathcal{K}_M \otimes \mathcal{O}_{C_i}(C_i - C'').$$

But $\mathcal{K}_M \otimes \mathcal{O}_{C_i}(C_i - C'')$ is the sheaf of sections of a holomorphic line bundle over C_i of degree $c_1(T_M^*) \cdot C_i - (C'' - C_i) \cdot C_i \le -c_1(M) \cdot C_i < 0$. Here we use that C'' contains C_i and $C_i \cdot C_i \ge 0$. Because C_i is reduced and irreducible this implies that any global section of this line bundle is trivial. Hence $H^1(\mathcal{O}_{C_i}(C'')) = 0$.

Setting $C' = C_i$ for some i starts the induction. For the induction step assume $H^1(C, \mathcal{O}_{C'}(C)) = 0$ and let i be such that $C' + C_i$ is still a subcycle of C. Let $\mathcal{I}_{C'|C'+C_i}$ be the ideal sheaf of C' in $C' + C_i$. Because $\mathcal{I}_{C'+C_i} = \mathcal{I}_{C'} \cdot \mathcal{I}_{C_i}$, multiplication induces an isomorphism $\mathcal{O}_{C_i}(-C') = \mathcal{I}_{C'} \otimes (\mathcal{O}_M/\mathcal{I}_{C_i}) \simeq \mathcal{I}_{C'|C'+C_i}$. Thus we have a restriction sequence

$$0 \longrightarrow \mathcal{O}_{C'+C_i}(C) \otimes \mathcal{O}_{C_i}(-C') \longrightarrow \mathcal{O}_{C'+C_i}(C) \longrightarrow \mathcal{O}_{C'}(C) \longrightarrow 0.$$

Observing $\mathcal{O}_{C'+C_i}(C) \otimes \mathcal{O}_{C_i}(-C') \simeq \mathcal{O}_{C_i}(C - C')$ the long exact cohomology sequence reads

$$\ldots \longrightarrow H^1(\mathcal{O}_{C_i}(C - C')) \longrightarrow H^1(\mathcal{O}_{C'+C_i}(C)) \longrightarrow H^1(\mathcal{O}_{C'}(C)) \longrightarrow \ldots$$

The term on the right vanishes by induction hypothesis, while the term on the left vanishes by the auxiliary result applied to $C'' = C - C'$. Hence $H^1(\mathcal{O}_{C'+C_i}(C)) = 0$ proving the induction step. $\quad\square$

5.7 The Unobstructedness Theorem

Under the assumptions of Proposition 5.4 and Hypothesis 5.1 we now have a description of deformations of C by the fiber of a holomorphic map between complex Banach manifolds whose linearization is surjective with finite dimensional kernel. Applying the implicit function theorem gives the main theorem of this lecture.

Theorem 5.5 *Let M be a complex surface and $C = \sum_i m_i C_i$ a compact holomorphic 1-cycle with $c_1(M) \cdot C_i > 0$ and $C_i \cdot C_i \ge 0$ for all i. Assume that a covering $\mathcal{U}$ of a neighborhood of $|C|$ in M exists satisfying Hypotheses 5.1, 1–3. Then the space of holomorphic cycles in M is a complex manifold of dimension $\Gamma(\mathcal{N}_{C|M})$ in a neighborhood of C. Moreover, analogous statements hold for a family of complex structures on M preserving the data described in Hypothesis 5.1.*

Remark 5.6 The hypotheses of the theorem do not imply that the cycle C is the limit of smooth cycles. For example, $\Gamma(\mathcal{N}_{C|M})$ may still be trivial and then C does not deform at all. Smoothability only follows with the additional requirement that $\mathcal{N}_{C|M}$ is *globally generated*. This statement can be checked by a transversality argument inside symmetric products of $\mathbb{C}$, cf. [SiTi3], Lemma 4.8.

The proof of global generatedness at some $P \in M$ in our fibered situation $p : M \to S^2$ follows from comparing dimensions of $\Gamma(\mathcal{N}_{C|M})$ and $\Gamma(\mathcal{N}_{C|M}(-F))$ where $F = p^{-1}(p(P))$. These dimensions differ maximally, namely by the sum of the fibers of $\mathcal{N}_{C|M}$ over $|C| \cap F$ if also $H^1(\mathcal{N}_{C|M}(-F)) = 0$. This is true by the same method as seen because also $(c_1(M) - H) \cdot C_i > 0$. See [SiTi3], Lemma 4.4 for details.

In any case, if C is the limit of smooth curves then the subset of the moduli space parametrizing singular cycles is a proper analytic subset of the moduli space, which is smooth, and hence does not locally disconnect the moduli space at C. In particular, any two smoothings of C are isotopic through a family of smooth holomorphic curves staying close to C.

6 Application to Symplectic Topology in Dimension Four

One point of view on symplectic topology is as an area somewhere between complex geometry and differential topology. On one hand symplectic constructions sometimes have the same or almost the flexibility as constructions in differential topology. As an example think of Gompf's *symplectic normal sum* [Go]. It requires two symplectic manifolds M_1, M_2 with symplectic hypersurfaces $D_1 \subset M_1$, $D_2 \subset M_2$ (real codimension two) and a symplectomorphism $\Phi : D_1 \to D_2$ with a lift to an isomorphism of symplectic line bundles $\tilde{\Phi} : N_{D_1|M_1} \to \Phi^*(N^*_{D_2|M_2})$. The result is a well-defined one-parameter family of symplectic manifolds $M_1 \amalg_{\Phi,\varepsilon} M_2$; each of its elements is diffeomorphic to the union of $M_i \setminus U_i$, where U_i is a tubular neighborhood of D_i and the boundaries ∂U_i are identified via $\tilde{\Phi}$. So the difference to a purely differential-topological construction is that (1) the bundle isomorphism Φ needs to preserve the symplectic normal structure along D_i and (2) there is a finite-dimensional parameter space to the construction.

Compare this with the analogous problem in complex geometry. Here $D_i \subset M_i$ is a divisor and one can form a singular complex space $M_1 \amalg_\Phi M_2$ by gluing M_1 and M_2 via an isomorphism $\Phi : D_1 \simeq D_2$. The singularity looks locally like D_i times the union of coordinate lines $zw = 0$ in $\mathbb{C}^2$. However, even if the holomorphic line bundles $N_{D_1|M_1}$ and $\Phi^* N^*_{D_2|M_2}$ are isomorphic there need not exist a smoothing of this space [PsPi]. A smoothing would locally replace $zw = 0$ by $zw = \varepsilon$ in appropriate holomorphic coordinates. Should this smoothing problem be unobstructed, it has as local parameter space the product of a complex disk for the smoothing parameter ε and some finite-dimensional space dealing with deformations of the singular space $M_1 \amalg_\Phi M_2$. Note how the deformation parameter ε reappears on the symplectic side as gluing parameter. The most essential difference to the symplectic situation is the appearance of obstructions.

The correspondence between complex and symplectic geometry can be expected to be especially interesting in dimension four, where a great deal is known classically on the complex side and where differential topology is so rich. In this context it is quite natural to consider the question when a symplectic submanifold in $\mathbb{CP}^2$ is isotopic to a holomorphic curve, which is the main topic of these lectures. An even stronger motivation is explained in the contribution by Auroux and Smith to this volume [AuSm], where they discuss how closely related the classification of symplectic manifolds is to the classification of symplectic surfaces in $\mathbb{CP}^2$. (These surfaces can have classical singularities, that is, nodes and cusps.)

The purpose of this section is to give a slightly different view on the relation between complex and symplectic geometry via Lefschetz fibrations. We will see that a certain class of Lefschetz fibrations called hyperelliptic arises as two-fold covers of rational ruled surfaces, and that our isotopy theorem for symplectic submanifolds of S^2-bundles over S^2 gives a classification of a subclass of such Lefschetz fibrations. This point of view also has an interpretation via representations of the braid group, a topic of independent interest.

6.1 Monodromy Representations – Hurwitz Equivalence

A *symplectic Lefschetz fibration* of an oriented four-manifold (M, ω) is a proper differentiable surjection $q : M \to S^2$ with only finitely many critical points in pairwise disjoint fibers with local model $\mathbb{C}^2 \to \mathbb{C}$, $(z, w) \mapsto zw$. Here z, w and the coordinate on S^2 are complex-valued and compatible with the orientations. With the famous exception of certain genus-one fiber bundles without sections, for example a Hopf-surface $S^3 \times S^1 \to S^2$, see [McSa1], Expl. 6.5 for a discussion, M then has a distinguished deformation class of symplectic structures characterized by the property that each fiber is symplectic [GoSt]. Note that if ω has this property then this is also the case for $q^* \omega_{S^2} + \varepsilon \omega$ for any $\varepsilon > 0$. In particular, this deformation class of symplectic structures has $q^* \omega_{S^2}$ in its closure.

For a general discussion of Lefschetz fibrations we refer to the lectures of Auroux and Smith. From this discussion recall that any symplectic four-manifold (M, ω) with $[\omega] \in H^2(M, \mathbb{Q})$ arises as total space of a Lefschetz fibration after blowing up finitely many disjoint points and with fibers Poincaré dual to the pull-back of $k[\omega]$ for $k \gg 0$ [Do]. The fibration structure is unique up to isotopy for each $k \gg 0$. In other words, for each ray $\mathbb{Q}_{>0}\omega$ of symplectic structures with rational cohomology one can associate a sequence of Lefschetz fibrations, which is unique up to taking subsequences. However, the sequence depends heavily on the choice of $[\omega]$, and it also seems difficult to control the effect of increasing k on the fibration structure. Conversely, it is also difficult to characterize Lefschetz fibrations arising in this way. Necessary conditions are the existence of sections with self-intersection number -1 and irreducibility of all singular fibers for $k \gg 0$ [Sm2], but these conditions are certainly not sufficient. So at the moment the use of this point of view for an effective classification of symplectic four-manifolds is limited.

On the other hand, in algebraic geometry Lefschetz fibrations have been especially useful for low degrees, that is, for low genus of the fibers. This is the point of view taken up in this section symplectically.

6.2 Hyperelliptic Lefschetz Fibrations

Auroux and Smith explain in their lectures that a symplectic Lefschetz fibration $\pi : M \to S^2$ with singular fibers over $s_1, \ldots, s_\mu \in S^2$ is characterized by its monodromy representation into the mapping class group

$$\rho : \pi_1(S^2 \setminus \{s_1, \ldots, s_\mu\}, s_0) \longrightarrow \pi_0 \operatorname{Diff}^+(\pi^{-1}(\Sigma)).$$

Here $s_0 \in S^2 \setminus \{s_1, \ldots, s_\mu\}$ is some fixed non-critical point in the base and $\Sigma = \pi^{-1}(s_0)$. For each loop running around only one critical point once the monodromy is a Dehn twist. There is even a one-to-one correspondence between isomorphism classes of Lefschetz fibrations with μ singular fibers along with a diffeomorphism $\pi^{-1}(s_0) \simeq \Sigma$, and such representations [Ka].

If one chooses a generating set of μ loops $\gamma_1, \ldots, \gamma_\mu$ intersecting only in s_0 and each encircling one of the critical points then ρ is uniquely determined by the tuple $(\tau_1, \ldots, \tau_\mu)$ of μ Dehn twists $\tau_i = \rho(\gamma_i)$ of Σ. Conversely, any such tuple with the property $\prod_i \tau_i = 1$ arises from such a representation. This gives a description of symplectic Lefschetz fibrations up to isomorphism by finite algebraic data, namely by the word $\tau_1 \ldots \tau_\mu$ of Dehn twists in the genus-g mapping class group $\mathrm{MC}_g \simeq \pi_0 \operatorname{Diff}^+(\pi^{-1}(\Sigma))$. This description is unique up to an overall conjugation (coming from the choice of isomorphism $\mathrm{MC}_g \simeq \pi_0 \operatorname{Diff}^+(\pi^{-1}(\Sigma))$) and up to so-called *Hurwitz equivalence*. The latter accounts for the choice of $\gamma_1, \ldots, \gamma_\mu$. It is generated by transformations of the form

$$\tau_1 \ldots \tau_r \tau_{r+1} \ldots \tau_\mu \longrightarrow \tau_1 \ldots \tau_{r+1} (\tau_r)_{\tau_{r+1}} \ldots \tau_\mu,$$

(Hurwitz move) where $(\tau_r)_{\tau_{r+1}} = \tau_{r+1}^{-1} \tau_r \tau_{r+1}$. Note that the set of Dehn twists is stable under conjugation, and hence the word on the right-hand side still consists of Dehn twists.

We now want to look at a special class of Lefschetz fibrations called *hyperelliptic*. By definition their monodromy representations take values in the hyperelliptic mapping class group $\mathrm{HMC}_g \subset \mathrm{MC}_g$. Recall that a hyperelliptic curve of genus g is an algebraic curve that admits a two-fold cover $\kappa : \Sigma \to \mathbb{CP}^1$ branched in $2g + 2$ points. The hyperelliptic mapping class group is the subgroup of MC_g of isotopy classes of diffeomorphisms of Σ respecting κ. So each $\sigma \in \mathrm{HMC}_g$ induces a diffeomorphism of S^2 fixing the branch set, well-defined up to isotopy. This defines a homomorphism

$$\mathrm{HMC}_g \longrightarrow \mathrm{MC}(S^2, 2g + 2)$$

to the mapping class group of S^2 marked with a set of $2g + 2$ points. The kernel is generated by the hyperelliptic involution that swaps the two points in the fibers of κ. For genus two it happens that $\mathrm{HMC}_g = \mathrm{MC}_g$, otherwise the inclusion $\mathrm{HMC}_g \subset \mathrm{MC}_g$ is strict. For all this a good reference is the book [Bi].

Of course, given a closed surface Σ of genus g there are many involutions with $2g + 2$ fixed points exhibiting Σ as a two-fold cover of S^2, and these give different copies of HMC_g in MC_g. The definition of hyperelliptic Lefschetz fibrations requires that $\mathrm{im}\,\rho \subset \mathrm{HMC}_g$ for one such choice of involution.

One method to produce a hyperelliptic Lefschetz fibration $M \to S^2$ is as composition of a two-fold cover with an S^2-bundles $p : P \to S^2$, with branch locus a so-called *Hurwitz curve* $B \subset P$ of degree $2g+2$ over S^2. Recall from the lectures of Auroux and Smith that a smooth submanifold $B \subset P$ is a Hurwitz curve if near the critical points of the composition $B \to P \to S^2$ there are local complex coordinates (z, w) on P such that $p(z, w) = z$ and B is locally given by $w = z^2$. Then $B \to S^2$ is a branched cover of degree $2g + 2$ with only simple branch points. The critical points of this projection produce singular fibers as follows. In the local coordinates (z, w) introduced above, M is locally the solution set of $v^2 - w^2 + z$. This four-dimensional manifold has complex coordinates $v' = v - w$, $w' = v + w$ because one can eliminate z. The projection to the z-coordinate then has the standard description $(v', w') \mapsto v'w'$ of a Lefschetz fibration near a singular point.

This construction yields hyperelliptic Lefschetz fibrations with only irreducible singular fibers. But there are relatively minimal hyperelliptic Lefschetz fibrations with reducible fibers. (To be relatively minimal means that we have not introduced reducible fibers artificially by blowing up the total space. Technically this leads to spheres contained in fibers with self-intersection number -1, and "relatively minimal" means there are no such spheres.) After a slight perturbation of the Lefschetz fibration any fiber contains at most one critical point (we took this as a condition in the definition of Lefschetz fibrations), and then any reducible fiber is a union of two surfaces, of genera h and $g - h$ for some $0 < h < [g/2]$.

Now how can one construct hyperelliptic Lefschetz fibrations with reducible fibers? Here is the construction of the model for a neighborhood of a singular fiber with irreducible components of genera h and $g - h$. Let $\Delta \subset \mathbb{C}$ be the unit disk, and consider in $\Delta \times \mathbb{CP}^1$ the holomorphic curve $\bar{B}$ given by

$$(w - \alpha_1) \cdot \ldots \cdot (w - \alpha_{2(g-h)+1}) \cdot (w - \beta_1 z^2) \cdot \ldots \cdot (w - \beta_{2h+1} z^2) = 0. \quad (6.1)$$

Here z is the coordinate on Δ, w is the affine coordinate on $\mathbb{C} \subset \mathbb{CP}^1$, and both the α_i and the β_j are pairwise disjoint and non-zero. So $\bar{B}$ consists of $2g + 2$ irreducible components, each projecting isomorphically to Δ. There is one singular point at $(0,0)$, a tacnode with tangent line $w = 0$ and contained in $2h + 1$ branches of $\bar{B}$. Figure 6.1, left, depicts the case $g = 2$, $h = 1$. Now let $\rho_1 : P_1 \to \Delta \times \mathbb{CP}^1$ be the blow up at $(0,0)$. This replaces the fiber F over $z = 0$ by two (-1)-spheres, the strict transform F_1 of F and another one E contracted under ρ_1. By taking the strict transform B_1 of $\bar{B}$ (the closure

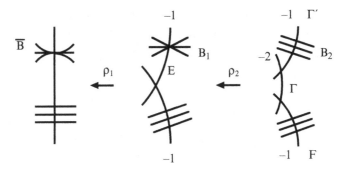

Fig. 6.1. Producing reducible fibers ($g = 2, h = 1$)

of $\rho_1^{-1}(\bar{B}) \setminus E$) the tacnode of $\bar{B}$ transforms to an ordinary singular point of multiplicity $2h + 1$ lying on $E \setminus F$. Another blow up $\rho_2 : \tilde{P} \to P_1$ in this singular point desingularizes B_1. The strict transform Γ of E is a sphere of self-intersection -2. So the fiber over 0 of $\tilde{P} \to \Delta$ is a union of two (-1)-spheres and one (-2)-sphere intersecting as depicted in Fig. 6.1 on the right. Denote by B_2 the strict transform of B_1 under the second blow-up.

It is not hard to check that, viewed as divisor, $B_2 + \Gamma$ is divisible by 2 up to rational equivalence, just as $\bar{B}$. Hence there exists a holomorphic line bundle L on $\tilde{P}$ with a section s with zero locus $B_2 + \Gamma$. The solution set to $u^2 = s$ with $u \in L$ is a two-fold cover of $\tilde{P}$ branched over $B_2 + \Gamma$. This is an instance of the standard construction of cyclic branched covers with given branch locus.

The projection $\tilde{M} \to \Delta$ is a local model for a genus-g Lefschetz fibration. The singular fiber over $0 \in \Delta$ is a chain of three components, of genera $g - h$, 0 and h, respectively. Since Γ is a (-2)-sphere in the branch locus the rational component $\tilde{\kappa}^{-1}(\Gamma)$ has self-intersection number -1 and hence can be contracted. The result is a manifold M with the projection to Δ the desired local model of a relatively minimal Lefschetz fibration with a reducible fiber with components of genera h and $g - h$. It is obviously hyperelliptic by construction.

One interesting remark is that the covering involution of $\tilde{M}$ descends to M. This action has one fixed point at the unique critical point of the fibration $M \to \Delta$. In local holomorphic coordinates it looks like $(u, v) \mapsto (-u, -v)$, and the ring of invariant holomorphic functions is generated by $x = u^2$, $y = v^2$, $z = uv$ (this can be chosen to agree with the z from before). The generators fulfill the relation $z^2 = xy$, so the quotient is isomorphic to the two-fold cover of $\mathbb{C}^2$ branched over the coordinate axes. This is called the A_1-singularity, and we have just verified the well-known fact that this singularity is what one obtains locally by contracting the (-2)-curve Γ on $\tilde{P}$. Alternatively, and maybe more appropriately, one should view the singular space P obtained by this contraction as the orbifold $M/(\mathbb{Z}/2)$.

6.3 Braid Monodromy and the Structure of Hyperelliptic Lefschetz Fibrations

We have now found a way to construct a hyperelliptic Lefschetz fibration starting from a certain branch surface $\bar{B}$ in S^2-bundles $\bar{p} : \bar{P} \to S^2$. This surface may have tacnodal singularities with non-vertical tangent line, and these account for reducible singular fibers in the resulting Lefschetz fibration. Otherwise the projection $\bar{B} \to S^2$ is a simply branched covering with the simple branch points leading to irreducible singular fibers.

Theorem 6.1 *[SiTi1] Any hyperelliptic Lefschetz fibration arises in this way.*

The case of genus-two has also been proved in [Sm1], and a slightly more topological proof is contained in [Fu]. The key ingredients of our proof are to describe B by its monodromy in the braid group $B(S^2, 2g + 2)$ of the sphere on $2g + 2$ strands, and to observe that $B(S^2, 2g + 2)$ is also a $\mathbb{Z}/2$-extension of $\mathrm{MC}(S^2, 2g + 2)$, just as HMC_g. While nevertheless $B(S^2, 2g + 2)$ and HMC_g are not isomorphic, there is a one-to-one correspondence between the set of half-twists in the braid group on one side and the set of Dehn-twists in HMC_g on the other side. This correspondence identifies the two kinds of monodromy representations. We now give some more details.

For a topological space X the *braid group on d strands* $B(X, d)$ can be defined as the fundamental group of the *configuration space*

$$X^{[d]} := (X \times \ldots \times X \setminus \Delta)/S_d.$$

Here $\Delta = \{(x_1, \ldots, x_d) \in X^d \,|\, \exists i \neq j, x_i = x_j\}$ is the generalized diagonal and S^d acts by permutation of the components. So a braid takes a number of fixed points on X, moves them with $t \in [0, 1]$ such that at no time t two points coincide, and such that at $t = 1$ we end up with a permutation of the tuple we started with.

If X is a two-dimensional oriented manifold and $\gamma : [0, 1] \to X$ is an embedded path connecting two different P_j, P_k and disjoint from $\{P_1, \ldots, P_d\}$ otherwise, there is a braid exchanging P_j and P_k as in the following Fig. 6.2. Any such braid is called a *half-twist*. Note that the set of half-twists is invariant under the action of the group of homeomorphisms on X fixing $\{P_1, \ldots, P_k\}$.

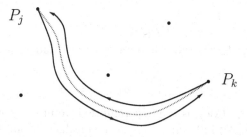

Fig. 6.2. A half-twist

The classical *Artin braid group* $B_d := B(\mathbb{R}^2, d)$ can be explicitly described as follows. Take as points $P_j := e^{2\pi\sqrt{-1}j/d}$ the dth roots of unity. Then define σ_j as the half-twist associated to the line segment connecting P_j and P_{j+1} for $1 \leq j < d$. In the following we take the index j modulo d. Then B_d is the group generated by the σ_j subject to the famous braid relations

$$\sigma_j\sigma_k = \sigma_k\sigma_j \quad \text{for } |j - k| \geq 2,$$
$$\sigma_j\sigma_{j+1}\sigma_j = \sigma_j\sigma_{j+1}\sigma_j \quad \text{for all } j.$$

It is important to note that there are infinitely many different sets of generators such as the σ_j, one for each self-intersection free path running through all the P_j. This is responsible for some of the complications when dealing with the braid group.

Now given one of our branch surfaces $p : \bar{B} \subset \bar{P}$ with critical set $\{s_1, \ldots, s_\mu\} \subset S^2$, a closed path γ in $S^2 \setminus \{s_1, \ldots, s_\mu\}$ defines a braid in S^2 by trivializing $\bar{P}$ over γ and by interpreting the pull-back of B over γ as the strands of the braid. This defines the monodromy representation

$$\rho' : \pi_1(S^2 \setminus \{s_1, \ldots, s_\mu\}, s_0) \longrightarrow B(S^2, 2g + 2),$$

which characterizes $\bar{B}$ uniquely up to isotopy. Note that the braid group on the right is really the braid group of the fiber $\bar{p}^{-1}(s_0)$ with the point set $\bar{B} \cap \bar{p}^{-1}(s_0)$.

The following possibilities arise for the monodromy around a loop γ enclosing only one of the s_i. If $\bar{B}$ is smooth over s_i then $\bar{B} \to S^2$ has a simple branch point over s_i. In this case $\rho'(\gamma)$ is the half-twist swapping the two branches of B coming together at the branch point. Otherwise $\bar{B}$ has a tacnode mapping to s_i. Then the local standard form (6.1) gives the following description of $\rho'(\gamma)$. There is an embedded loop $S^1 \hookrightarrow S^2 = \bar{p}^{-1}(s_0)$ passing through a subset $P_{i_1}, \ldots, P_{i_{2h+1}}$ of $\bar{B} \cap \bar{p}^{-1}(s_0)$ and *not enclosing any other* P_j. Now $\rho'(\gamma)$ is given by a full counterclockwise rotation of these points along the loop, and by the identity on all other points.

The point now is that in the hyperelliptic case any Dehn twist arises as a two-fold cover of a distinguished braid of the described form once a choice of a north pole $\infty \in S^2$ has been made. In fact, the three groups HMC_g, $\mathrm{MC}(S^2, 2g + 2)$ and $B(S^2, 2g + 2)$ all have $2g + 2$ generators $\sigma_1, \ldots, \sigma_{2g+1}$ fulfilling the Artin braid relations, and in addition:

1. $\mathrm{MC}(S^2, 2g + 2)$: $I = 1$, $T = 1$
2. HMC_g: $I^2 = 1$, $T = 1$, and I is central ($I\sigma_i = \sigma_i I$ for all i)
3. $B(S^2, 2g + 2)$: $I = 1$, and this implies $T^2 = 1$ and T central

where

$$I = \sigma_1 \ldots \sigma_{2g+1}\sigma_{2g+1} \ldots \sigma_1, \quad T = (\sigma_1 \ldots \sigma_{2g+1})^{2g+2}$$

Geometrically, $I \in \mathrm{HMC}_g$ is the hyperelliptic involution; it induces the trivial element in $\mathrm{MC}(S^2, 2g+2)$, and it can not be produced from braids via two-fold covers. On the other hand, T is the full-twist along a loop passing through all the points; its square is the trivial braid as one sees by "pulling the bundle of strands across $\infty \in S^2$," and again it induces the trivial element in both $MC(S^2, 2g + 2)$ and in HMC_g.

Thus given a hyperelliptic Lefschetz fibration one can produce a branch surface $\bar{B}$ uniquely up to isotopy by going via braid monodromy. There are two minor global issues with this, one being the extension of $\bar{B}$ over one point at infinity, the other homological two-divisibility of $\bar{B}$. The latter follows by an a priori computation of the possible numbers of singular fibers of each type, of the critical points of $\bar{B}$ and their relation to the homology class of $\bar{B}$. The first issue can be resolved by closing up $\bar{B}$ either in the trivial or in the non-trivial S^2-bundle over S^2.

6.4 Symplectic Noether–Horikawa Surfaces

A simple, but important observation is that any of the branch surfaces $\tilde{B} \subset \tilde{P}$ are symplectic with respect to $\omega_{\tilde{P}} + k\tilde{p}^*\omega_{S^2}$, for $k \gg 0$. Here $\omega_{\tilde{P}}$ and ω_{S^2} are any Kähler structures on $\tilde{P}$ and on S^2, and $\tilde{P} \to P$ resolves the tacnodes of $\bar{B} \subset P$. Thus the question whether a hyperelliptic Lefschetz fibrations is isomorphic (as a Lefschetz fibration) to a holomorphic one, is equivalent to asking if $\tilde{B}$ can be deformed to a holomorphic curve within the class of branch surfaces in $\tilde{P}$.

For the understanding of symplectic Lefschetz fibrations this point of view is certainly limited for the following two reasons. First, it is not true that any hyperelliptic Lefschetz fibration is isomorphic to a holomorphic one. For example, [OzSt] shows that fiber sums of two copies of a certain genus-2 Lefschetz fibration produce infinitely many pairwise non-homeomorphic symplectic 4-manifolds of which only finitely many can be realized as complex manifolds. And second, the general classification of holomorphic branch curves up to isotopy, hence of hyperelliptic holomorphic Lefschetz fibrations, is complicated, see e.g. [Ch].

On the other hand, the complex geometry becomes regular in a certain stable range, when the deformation theory of the branch curve is always unobstructed. This is the case when the total number μ of singular fibers is much larger than the number t of reducible singular fibers. In the genus-2 case, the discussions in [Ch] suggest $\mu > 18t$ as this stable range. The example of [OzSt] has $\mu = 4t$. So in our opinion, the holomorphic point of view is appropriate for a classification in a certain stable range.

Conjecture 6.2 *For any g there exists an integer N_g such that any hyperelliptic symplectic genus-g Lefschetz fibration with μ singular fibers of which t are reducible, and such that*

$$\mu > N_g t,$$

is isomorphic to a holomorphic one.

The holomorphic classification in the stable range should in turn be simple. We expect that there is only a very small number of deformation classes of holomorphic genus-g Lefschetz fibrations with fixed numbers and types of singular fibers (given by the genera of its irreducible components), distinguished by topological invariants of the total space such as Euler characteristic and signature.

The conjecture in particular says that any hyperelliptic Lefschetz fibration with reducible fibers is holomorphic. By the discussion above this is equivalent to saying that each smooth branch surface (no tacnodes) in a rational ruled surface is isotopic as branch surface to a holomorphic curve. The main theorem of these lectures Theorem 9.1 says that this is true for connected $\bar{B}$ provided $\deg(\bar{B} \to S^2) \leq 7$. If $\bar{B}$ is disconnected it is either a product, or it has precisely two components and one of them is a section. In the disconnected case the monodromy representation does not act transitively on the set of strands, while this is true in the case with connected $\bar{B}$. (With hindsight we will even see that in the connected case the monodromy representation is surjective.) In any case, we say the case with connected $\bar{B}$ has *transitive monodromy*. Then we have the following:

Theorem 6.3 *[SiTi3] A Lefschetz fibration with only irreducible singular fibers of genus two, or of genus one with a section, and with transitive monodromy is isomorphic to a holomorphic Lefschetz fibration.*

By the standard technique of degeneration to nodal curves (see [Te] for a survey) it is not hard to compute the braid monodromy for smooth algebraic branch curves in $\mathbb{CP}^1$-bundles over $\mathbb{CP}^1$, i.e. in Hirzebruch surfaces $\mathbb{F}_k$, see [Ch].

Proposition 6.4 *The braid monodromy word of a smooth algebraic curve $\bar{B} \subset \mathbb{F}_k$ of degree d and with μ simple critical points is Hurwitz-equivalent to one of the following:*

1. $(\sigma_1 \dots \sigma_{d-1}\sigma_{d-1} \dots \sigma_1)^{\frac{\mu}{2d-2}}$ *($\bar{B}$ connected and k even)*
2. $(\sigma_1 \dots \sigma_{d-1}\sigma_{d-1} \dots \sigma_1)^{\frac{\mu-d(d-1)}{2d-2}} (\sigma_1 \dots \sigma_{d-1})^d$ *($\bar{B}$ connected and k odd)*
3. $(\sigma_1 \dots \sigma_{d-2})^{\frac{\mu}{d-2}}$ *($\bar{B}$ disconnected; $k = 2d$)*

Taken together this gives a complete classification of symplectic Lefschetz fibrations with only irreducible singular fibers and transitive monodromy in genus two.

In the non-hyperelliptic case it is not clear what an analogue of Conjecture 6.2 should be. Any symplectic manifold arises as total space of a symplectic Lefschetz fibration without reducible singular fibers, after blowing up finitely many points [Do, Sm2]. Thus the absence of reducible singular fibers alone certainly does not suffice as obstruction to holomorphicity.

By purely braid-theoretic methods Auroux very recently achieved the following beautiful stable classification result:

Theorem 6.5 *[Au2] For each g there exists a universal genus-g Lefschetz fibration π_g^0 with the following property: Given two genus-g Lefschetz fibrations $\pi_i : M_i \to S^2$, $i = 1, 2$, with the same numbers of reducible fibers then the fiber connected sums of π_i with sufficiently many copies of π_g^0 are isomorphic Lefschetz fibrations, provided*

1. *there exist sections $\Sigma_i \subset M_i$ with the same self-intersection numbers, and*
2. *M_1 and M_2 have the same Euler number.*

This refines a previous, slightly simpler result by the same author for the genus 2 case [Au1], building on Theorem 6.3.

7 The $\mathscr{C}^0$-Compactness Theorem for Pseudo-Holomorphic Curves

In this section we discuss a compactness theorem for J-holomorphic maps in the case that J is only assumed to be continuous. Such a compactness theorem was first due to Gromov [Gv] and was further discussed by Parker-Wolfson, Pansu, Ye and Ruan-Tian [PrWo, Pn, Ye, RuTi]. For the reader's convenience, we will present a proof of this compactness theorem and emphasize that it depends only on the $\mathscr{C}^0$-norm of the involved almost complex structures. Our proof basically follows [Ti], where further smoothness was discussed. We should point out that the dependence on a weaker norm for the almost complex structures is crucial in our study of the symplectic isotopy problem.

7.1 Statement of Theorem and Conventions

First we note that in this section by a J-holomorphic map we mean a Hölder continuous ($\mathscr{C}^{0,\alpha}$, $0 < \alpha < 1$) map from a Riemann surface Σ into M whose derivative is L^2-bounded and which satisfies the J-holomorphicity equation in the distributional sense. Explicitly, the last phrase says that for any smooth vector field X on M with compact support in a neighborhood of $f(\Sigma)$ and any smooth vector field v with compact support in Σ,

$$\int_\Sigma g\big(X, Df(v) + J \cdot Df(j_\Sigma(v))\big)dz = 0,$$

where j_Σ denotes the conformal structure of Σ. This coincides with the standard J-holomorphicity equation whenever f is smooth. By our assumption on f, any L^2-section of f^*T_M over Σ can be approximated in the L^2-topology by the pull-back of a locally constant vector field on M, so it follows that the above equation for f also holds when X is replaced by any L^2-section of f^*T_M.

As before, we denote by (M, ω) a compact symplectic manifold and by g a fixed Riemannian metric. Let J_i be a sequence of *continuous* almost complex structures on M converging to J_∞ in the $\mathscr{C}^0$-topology and uniformly tamed in the following sense: There exists a constant $c > 0$ such that for any $X \in T_M$ and any i

$$cg(X, X) \leq \omega(X, J_i X) \leq c^{-1} g(X, X). \tag{7.1}$$

Here is the main result of this section.

Theorem 7.1 *Let (M, ω) and g be as above. Assume that Σ_i is a sequence of Riemann surfaces of fixed genus and $f_i : \Sigma_i \to M$ are J_i-holomorphic with uniformly bounded homology classes $f_{i*}[\Sigma_i] \in H_2(M, \mathbb{Z})$.*

Then there is a connected singular Riemann surface Σ_∞ with finitely many irreducible components $\Sigma_{\infty,a}$, and smooth maps $\phi_i : \Sigma_i \to \Sigma_\infty$ such that the following holds:

1. *ϕ_i is invertible on the pre-image of the regular part of Σ_∞*
2. *A subsequence of $f_i \circ \phi_i^{-1}$ converges to a J_∞-holomorphic map f_∞ on the regular part of Σ_∞ in the $\mathscr{C}^0$-topology*
3. *Each $f_\infty|_{\Sigma_{\infty,a}}$ extends to a J_∞-holomorphic map from $\Sigma_{\infty,a}$ into M, and the homology classes $f_{i*}[\Sigma_i]$ converge to $f_{\infty*}[\Sigma_\infty]$ in $H_2(M, \mathbb{Z})$*

The rest of this section is devoted to the proof. We start with the monotonicity formula for pseudo-holomorphic maps.

7.2 The Monotonicity Formula for Pseudo-Holomorphic Maps

For notational simplicity we will denote by J one of the almost complex structures J_i or J_∞. Let I and g_{stan} denote the standard almost complex structure and standard flat Riemannian metric on $\mathbb{R}^{2n}$, respectively. By (7.1), for any $\eta > 0$ there is a uniform δ_η such that for any geodesic ball $B_R(p)$ ($p \in M$ and $R \leq \delta_\eta$), there is a $\mathscr{C}^1$-diffeomorphism $\phi : B_R(p) \to B_R(0) \subset \mathbb{R}^{2n}$ such that

$$\|J - \phi^* I\|_{\mathscr{C}^0(B_R(p))} \leq \eta, \tag{7.2}$$

where norms are taken with respect to g. We may further assume that $\|g - \phi^* g_{\mathrm{stan}}\|_{\mathscr{C}^0} \leq C$ for some uniform constant C.

Denote by Δ_r the disk in $\mathbb{C}$ with center at the origin and radius r, and $\Delta = \Delta_1$. Throughout the proof c will be a uniform constant whose actual value may vary.

Lemma 7.2 *There is an $\epsilon > 0$ such that for any $\alpha \in (0, 1)$ and any J-holomorphic map $f : \Delta_r \to M$ ($r > 0$) with $\int_{\Delta_r} |Df|_g^2 dz \leq \epsilon$, we have*

$$\int_{B_{r'}(q)} |Df|_g^2 dz \leq c_\alpha r'^{2\alpha}, \quad \forall q \in \Delta_{r/2} \text{ and } r' \leq r/4, \tag{7.3}$$

Moreover, we have

$$\mathrm{diam} f(\Delta_{r'/2}) \leq c_\alpha \sqrt{\epsilon} \left(\frac{r'}{r}\right)^\alpha. \tag{7.4}$$

Here c_α is a uniform constant which may depend on α.

Proof. Since all estimates are scaling-invariant, we may assume $r = 1$. Let $\eta > 0$ be a sufficiently small positive number and let δ_η be given as in (7.2). For simplicity, we will write δ for δ_η and identify $B_R(p)$, $p = f(0)$, with $B_R(0) \subset \mathbb{R}^{2n}$ by the diffeomorphism ϕ in (7.2). Because g and $\phi^* g_{\text{stan}}$ are uniformly equivalent we may then also replace $|\,.\,|_g$ in the statement by the standard norm $|\,.\,|$ in $\mathbb{R}^n$. Choose any $\rho_0 \leq \frac{1}{2}$ so that $f(B_{2\rho_0}(0)) \subset B_\delta(p)$. As input to Morrey's Lemma (see e.g. [GiTr], Lemma 12.2) we now derive a growth condition for the local L^2-norm of Df at a fixed $y \in B_{\rho_0}(0)$, see (7.6).

In polar coordinates (r, θ) centered at y, the Cauchy–Riemann equation becomes

$$\frac{\partial f}{\partial r} + \frac{1}{r} J \frac{\partial f}{\partial \theta} = 0,$$

and $|Df|^2 = |\partial_r f|^2 + |r^{-1} \partial_\theta f|^2$. In particular, both $|\partial_r f|^2$ and $|r^{-1} \partial_\theta f|^2$ are close to $\frac{1}{2} |Df|^2$:

$$
\begin{aligned}
\big| 2|\partial_r f|^2 - |Df|^2 \big| &= \big| |I\partial_r f|^2 - |r^{-1} \partial_\theta f|^2 \big| \\
&\leq \big| (J - I)\partial_r f \big|^2 + \big| |J\partial_r f|^2 - |r^{-1} \partial_\theta f|^2 \big| \leq \eta |Df|^2, \quad (7.5)
\end{aligned}
$$
$$\big| 2|r^{-1} \partial_\theta f|^2 - |Df|^2 \big| \leq \eta |Df|^2.$$

We also obtain the pointwise estimate

$$
\begin{aligned}
0 = \big| \partial_r f + r^{-1} J \partial_\theta f \big|^2 &= \big| \partial_r f + r^{-1} I \partial_\theta f + r^{-1} (J - I) \partial_\theta f \big|^2 \\
&= \big| \partial_r f + r^{-1} I \partial_\theta f \big|^2 + 2 \langle \partial_r f + r^{-1} I \partial_\theta f, r^{-1} (J - I) \partial_\theta f \rangle + \big| r^{-1} (J - I) \partial_\theta f \big|^2 \\
&\geq \big| \partial_r f \big|^2 + 2 \langle \partial_r f, r^{-1} I \partial_\theta f \rangle + \big| r^{-1} I \partial_\theta f \big|^2 - 2\eta \cdot \big(|\partial_r f| + |r^{-1} \partial_\theta f| \big) |r^{-1} \partial_\theta f| \\
&\geq (1 - 4\eta) |Df|^2 + 2 \langle \partial_r f, r^{-1} I \partial_\theta f \rangle.
\end{aligned}
$$

Then integrating by parts twice, we have for $\rho \leq \rho_0$ and any constant vector $\lambda \in \mathbb{R}^n$,

$$
\begin{aligned}
0 \geq (1 - 4\eta) \int_{B_\rho(y)} \big| Df \big|^2 r\,dr\,d\theta &+ 2 \int_0^\rho \int_0^{2\pi} \langle \frac{\partial f}{\partial r}, I(\frac{\partial f}{\partial \theta}) \rangle dr\,d\theta \\
= (1 - 4\eta) \int_{B_\rho(y)} \big| Df \big|^2 r\,dr\,d\theta &+ 2 \int_0^{2\pi} \langle f - \lambda, I(\frac{\partial f}{\partial \theta}) \rangle (\rho, \theta)\,d\theta \\
-2 \int_0^\rho \int_0^{2\pi} \langle f - \lambda, I(\frac{\partial^2 f}{\partial r \partial \theta}) \rangle dr\,d\theta \\
= (1 - 4\eta) \int_{B_\rho(y)} \big| Df \big|^2 r\,dr\,d\theta &+ 2 \int_0^{2\pi} \langle f - \lambda, I(\frac{\partial f}{\partial \theta}) \rangle (\rho, \theta)\,d\theta \\
+ 2 \int_0^\rho \int_0^{2\pi} \langle \frac{\partial f}{\partial \theta}, I(\frac{\partial f}{\partial r}) \rangle dr\,d\theta.
\end{aligned}
$$

The last term gives another $(1 - 2\eta) \int_{B_\rho(y)} \big| Df \big|^2 r\,dr\,d\theta$ by the following:

$$2r^{-1}\langle \partial_\theta f, I\partial_r f\rangle = \langle r^{-1}\partial_\theta f, I\partial_r f\rangle - \langle r^{-1}I\partial_\theta f, \partial_r f\rangle$$
$$\geq -2|Df|^2 \cdot \eta + \langle r^{-1}\partial_\theta f, J\partial_r f\rangle + \langle r^{-1}J\partial_\theta f, \partial_r f\rangle$$
$$= -2|Df|^2 \cdot \eta + |r^{-1}\partial_\theta f|^2 + |\partial_r f|^2 = (1-2\eta)|Df|^2.$$

It follows that

$$(1-3\eta)\int_{B_\rho(y)}|Df|^2 r dr d\theta \leq -\int_0^{2\pi}\langle f-\lambda, I(\frac{\partial f}{\partial \theta})\rangle d\theta.$$

$$\leq \Big(\int_0^{2\pi}|f-\lambda|^2 d\theta\Big)^{1/2} \cdot \Big(\int_0^{2\pi}\Big|\frac{\partial f}{\partial \theta}\Big|^2 d\theta\Big)^{1/2}.$$

Now choose

$$\lambda = \frac{1}{2\pi}\int_0^{2\pi} f d\theta.$$

Then by the Poincaré inequality on the unit circle, we have

$$\int_0^{2\pi}|f-\lambda|^2 d\theta \leq \int_0^{2\pi}\Big|\frac{\partial f}{\partial \theta}\Big|^2 d\theta \leq \rho^2 \int_0^{2\pi}\Big|r^{-1}\frac{\partial f}{\partial \theta}\Big|^2 d\theta.$$

Moreover, $|r^{-1}\partial_\theta f|^2 \leq \frac{1+\eta}{2}|Df|^2$ by (7.5) and $\frac{1-3\eta}{1+\eta} \geq 1-4\eta$. Plugging all this into the previous inequality gives

$$(1-4\eta)\int_{B_\rho(y)}|Df|^2 r dr d\theta \leq \frac{\rho^2}{2}\int_0^{2\pi}|Df|^2 d\theta.$$

But

$$\frac{\partial}{\partial \rho}\int_{B_\rho(y)}|Df|^2 r dr d\theta = \rho \int_0^{2\pi}|Df|^2 d\theta,$$

so the above is the same as

$$2(1-4\eta)\int_{B_\rho(y)}|Df|^2 r dr d\theta \leq \rho \frac{\partial}{\partial \rho}\int_{B_\rho(y)}|Df|^2 r dr d\theta,$$

that is,

$$\frac{\partial}{\partial \rho}\Big(\rho^{-2(1-4\eta)}\int_{B_\rho(y)}|Df|^2 r dr d\theta\Big) \geq 0.$$

This implies, for any $\rho < \rho_0$ and $y \in B_{\rho_0}(0)$,

$$\int_{B_\rho(y)}|Df|^2 r dr d\theta \leq c\Big(\frac{\rho}{\rho_0}\Big)^{2(1-4\eta)}\int_\Delta |Df|^2 r dr d\theta, \qquad (7.6)$$

where c is a uniform constant. It follows from this and Morrey's lemma that

$$\sup_{x,y\in\Delta_{\rho_0}}\frac{|f(x)-f(y)|}{|x-y|^{1-4\eta}} \leq c_\eta \rho_0^{-1+4\eta}\Big(\int_\Delta |Df|^2 r dr d\theta\Big)^{1/2},$$

where c_η is some uniform constant depending only on η. In particular, choosing $\eta \leq 1/8$, we obtain for $x, y \in \Delta_{\rho_0}$

$$|f(x) - f(y)| \leq c_\eta \left(\frac{|x-y|}{\rho_0}\right)^{1/2} \cdot \sqrt{\varepsilon} \leq c_\eta \sqrt{2}\sqrt{\varepsilon}.$$

Thus the diameter of $f(\Delta_{\rho_0})$ is bounded by $c\sqrt{\varepsilon}$.

It remains to prove that if ϵ is sufficiently small, then $f(\Delta_{1/2})$ is contained in a ball of radius δ. In fact, we can then set $\rho_0 = 1/2$ above and conclude the desired uniform estimates for $\alpha = 1 - 4\eta$. For any $x \in \Delta$, define

$$t(x) = \sup\left\{t \in [0, 1/2] \,\big|\, \mathrm{diam}(f(B_{t(1-|x|)}(x))) \leq \delta\right\}.$$

If the above claim is false then $t(0) < 1/2$. Let $x_0 \in \Delta_{1/2}$ be such that $t(x_0) = \inf t(x) < 1/2$. Set $a(x) = t(x)(1-|x|)$. Then for any $x \in B_{a(x_0)}(x_0)$, we have

$$a(x) \geq t(x_0)(1-|x|) \geq a(x_0) - t(x_0)|x - x_0| > a(x_0) - t(x_0)a(x_0) > \frac{1}{2}a(x_0).$$

This implies $B_{a(x)}(x) \supset B_{a(x_0)/2}(x)$ and thus, from the above diameter estimate, for any $x \in B_{a(x_0)}(x_0)$, we have

$$\mathrm{diam}(f(B_{a(x_0)/2}(x))) \leq c\sqrt{\epsilon}.$$

It follows that

$$\mathrm{diam}(f(B_{a(x_0)}(x_0))) \leq 2c\sqrt{\epsilon}.$$

Since the constant c here depends only on δ, we get a contradiction if ϵ is sufficiently small. The claim is proved.

7.3 A Removable Singularities Theorem

As an application of the Monotonicity Lemma we derive the following sort of Uhlenbeck removable singularity theorem under the condition that J is only continuous as described above.

Proposition 7.3 *Let (M, ω) and J be as above. If $f : \Delta_{r_0} \backslash \{0\} \to M$ is a J-holomorphic map with $\int_{\Delta_{r_0}} |Df|_g^2 dz < \infty$, then f extends to a Hölder continuous map from Δ into M.*

Proof. Fix any $\alpha \in (0, 1)$. By choosing r_0 smaller, we may assume

$$\int_{\Delta_{r_0}} |Df|_g^2 dz < \epsilon,$$

where ϵ is as in Lemma 7.2.

Let $x, y \in \Delta_{r_0/2}$ with $|y| \leq |x|$, say. If $|x - y| \leq |x|/2$, then by Lemma 7.2 applied to the restriction of f to $B_{r_0/2}(x) \subset \Delta_{r_0}$, we have

$$d(f(x), f(y)) \leq \text{diam}(f(B_{|x-y|}(x))) \leq 2c_\alpha \sqrt{\epsilon} \left(\frac{|x-y|}{r_0} \right)^\alpha.$$

If $|x-y| > |x|/2 \geq |y|/2$, choose z such that it is collinear to x and has $|z| = |y|$. We can cover $\partial B_{|y|}(0)$ by 12 balls of radius $|y|/2$, so applying Lemma 7.2 at most 12 times, we get

$$d(f(z), f(y)) \leq 12c_\alpha \sqrt{\epsilon} \left(\frac{|y|}{r_0} \right)^\alpha \leq 24c_\alpha \sqrt{\epsilon} \left(\frac{|x-y|}{r_0} \right)^\alpha.$$

Next we can find finitely many balls $B_{|x|/2}(x_0), \ldots, B_{|x|/2^k}(x_k)$ such that $x_0 = x$, $x_k = z$ and $x_{i+1} \in B_{|x_i/2|}(x_i)$. Then applying Lemma 7.2, we get

$$d(f(z), f(x)) \leq \left(1 + \tfrac{1}{2} + \ldots + \tfrac{1}{2^{k-1}} \right) c_\alpha \sqrt{\epsilon} \left(\frac{|x|}{r_0} \right)^\alpha \leq 4c_\alpha \sqrt{\epsilon} \left(\frac{|x-y|}{r_0} \right)^\alpha.$$

Hence,

$$d(f(x), f(y)) \leq 28c_\alpha \sqrt{\epsilon} \left(\frac{|x-y|}{r_0} \right)^\alpha.$$

It follows that f extends to a Hölder continuous map from $\Delta_{r_0/2}$ to M.

7.4 Proof of the Theorem

Now we are in position to prove Theorem 7.1.

First we observe: There is a uniform constant c depending only on g and $[\omega](f_{i*}[\Sigma_i])$ such that for any f_i,

$$\int_{\Sigma_i} |Df_i|_g^2 dz \leq c.$$

Therefore, the L^2-norm of Df_i is uniformly bounded.

Next we observe: If ϵ in Lemma 7.2 is sufficiently small, say $c_{1/2}\sqrt{\epsilon} \leq \delta_{1/2}$, then for each i, either f_i is a constant map or

$$\int_{\Sigma_i} |Df_i|_g^2 dz \geq \epsilon > 0.$$

This can be seen as follows: If the above inequality is reversed, Lemma 7.2 implies that the image $f_i(\Sigma_i)$ lies in a Euclidean ball; on a ball the symplectic form ω is exact and so the energy is zero, that is, f_i is constant.

Consider the following class of metrics h_i on the regular part of Σ_i. The metrics h_i have uniformly bounded geometry, namely, for each $p \in \Sigma_i$ there is a local conformal coordinate chart (U, z) of Σ_α containing p such that U is identified with the unit ball Δ and

$$h_i|_U = e^\varphi dz d\bar{z}$$

for some $\varphi(z)$ satisfying:

$$\|\varphi\|_{\mathscr{C}^k(U)} \le c_k, \quad \text{for any } k > 0,$$

where c_k are uniform constants independent of i. We also require that there are finitely many cylinder-like necks $N_{i,a} \subset \Sigma_i$ ($a = 1, \ldots, n_i$) satisfying:
(1) n_i are uniformly bounded independent of α.
(2) The complement $\Sigma_i \backslash \bigcup_a N_{i,a}$ is covered by finitely many geodesic balls $B_R(p_{i,j})$ ($1 \le j \le m_i$) of h_i in Σ_i, where R and m_i are uniformly bounded.
(3) Each $N_{i,a}$ is diffeomorphic to a cylinder of the form $S^1 \times (\alpha, \beta)$ (α and β may be $\pm\infty$) satisfying: If s, t denote the standard coordinates of $S^1 \times [0, \beta)$ or $S^1 \times (\alpha, 0]$, then

$$h_i|_{N_{i,a}} = e^\varphi (ds^2 + dt^2),$$

where φ is a smooth function satisfying uniform bounds as stated above.

We will say that such a h_i is admissible. We will call $\{h_i\}$ uniformly admissible if all h_i are admissible with uniform constants R, c_k, etc.

Admissible metrics always exist on any sequence Σ_i of Riemann surfaces of the same genus. We will start with a fixed sequence of uniformly admissible metrics h_i on Σ_i. We will introduce a new sequence of uniformly admissible metrics $\tilde{h}_i$ on Σ_i such that there is a uniform bound on the gradient of f_i. Once this is done, the theorem follows easily.

We will define $\tilde{h}_i$ by induction.

Set

$$r_i = \inf \left\{ r \,\Big|\, \int_{B_r(x, h_i)} |Df_i|^2_{h_i, g} \, dz \ge \epsilon \text{ for some } x \in \Sigma_i \right\}.$$

Here $|\cdot|_{h,g}$ denotes the norm induced by g on M and h on the domain. If r_i is uniformly bounded from below, the induction stops and we just take $\tilde{h}_i = h_i$. Then our main theorem follows from Lemma 7.2 and standard convergence theory.

Now assume that r_i tends to zero as i goes to infinity. By going over to a subsequence we may assume $r_i \le 1/2$ for all i. Let p_i^1 be the point where r_i is attained. Let z be a local complex coordinate on Σ_i centered at p_i^1 and with values containing $2\Delta \subset \mathbb{C}$. Define $h_i^1 = h_i$ outside the region where $|z| < 1$ and

$$h_i^1 = \frac{r_i^{-2}}{\chi_i(r_i^{-2}|z|^2)} h_i \quad \text{for } |z| < 1.$$

Here $\chi_i : \mathbb{R} \to \mathbb{R}$ is a cut-off function satisfying: $\chi(t) = 1$ for $t \le 1$, $\chi_i(t) = t - 1/2$ for $t \in [2, r_i^{-2}]$, and $\chi_i(t) = r_i^{-2}$ for $t \ge r_i^{-2} + 1$; we may also assume that $0 \le \chi_i'(t) \le 1$. Clearly, we have $h_i^1 \ge h_i$ and it holds $h_i(z) = r_i^{-2}h_i$ for $|z| \le r_i^2$ and $h_i(z) = h_i$ for $|z| \ge r_i^{-2} + 1$. It is easy to check that the sequence h_i^1 is uniformly admissible. Moreover, we have

$$\int_{B_1(p_i^1, h_i^1)} |Df_i|^2_{h_i^1, g} \, dz = \epsilon$$

where $B_1(p_i^1, h_i^1)$ denotes the geodesic ball of radius 1 and centered at p_i with respect to the metric h_i^1.

Next we define

$$r_i^1 = \inf\left\{ r \,\middle|\, \int_{B_r(x,h_i^1)} |Df_i|^2_{h_i^1,g} dz \geq \epsilon \text{ for some } x \in \Sigma_i \right\}.$$

If r_i^1 is uniformly bounded from below, the induction stops and we just take $\tilde{h}_i = h_i^1$. Then our main theorem again follows from Lemma 7.2 and standard convergence theory. Otherwise, by taking a subsequence if necessary, we may assume that $r_i^1 \to 0$ as $i \to \infty$ and $r_i^1 \leq 1/2$ for all i. Let p_i^2 be the point where r_i^1 is attained. Then for i sufficiently large, $p_i^2 \in \Sigma \backslash B_2(p_i^1, h_i^1)$. For simplicity, we assume that this is true for all i. Now we can get h_i^2 by repeating the above construction with h_i replaced by h_i^1. Clearly, h_i^2 coincides with h_i^1 on $B_1(p_i, h_i^1)$, so

$$B_1(p_i^1, h_i^2) = B_1(p_i^1, h_i^1).$$

We also have $B_1(p_i^2, h_i^2) \cap B_1(p_i^1, h_i^1) = \emptyset$ and

$$\int_{B_1(p_i^2, h_i^2)} |Df_i|^2_{h_i^2,g} dz = \epsilon > 0.$$

We can continue this process to construct metrics h_i^L $(L \geq 2)$ and find points p_i^α $(\alpha = 1, \ldots, L)$ such that $B_1(p_i^\alpha, h_i^L) \cap B_1(p_i^\beta, h_i^L) = \emptyset$ for any $\alpha \neq \beta$ and

$$\int_{B_1(p_i^\alpha, h_i^L)} |Df_i|^2_{h_i^L,g} dz = \epsilon > 0.$$

It follows that

$$c \geq \int_{\Sigma_i} |Df_i|^2_{h_i^L,g} dz \geq L\epsilon.$$

Hence the process has to stop at some L. We obtain $\tilde{h}_i = h_i^L$ and a uniform $r_0 > 0$ such that for any $x \in \Sigma_i$,

$$\int_{B_{r_0}(x,\tilde{h}_i)} |Df_i|^2_{\tilde{h}_i,g} dz < \epsilon.$$

By uniform admissibility of $\tilde{h}_i$, we may choose m and R such that there are finitely many cylinder-like necks $N_{i,\alpha} \subset \Sigma_i$ $(\alpha = 1, \ldots, l)$ satisfying:

(1) $\Sigma_i \backslash \bigcup_\alpha N_{i,\alpha}$ is covered by geodesic balls $B_R(q_{ij}, \tilde{h}_i)$ $(1 \leq j \leq m)$ in Σ_i

(2) Each $N_{i,\alpha}$ is diffeomorphic to a cylinder of the form $S^1 \times (a_{i,\alpha}, b_{i,\alpha})$ $(a_{i,\alpha}$ and $b_{i,\alpha}$ may be $\pm\infty)$

Now by taking a subsequence if necessary, we may assume that for each j, the sequence $(\Sigma_i, \tilde{h}_i, q_{ij})$ of pointed metric spaces converges to a Riemann surface $\Sigma^0_{\infty,j}$. Such a limit $\Sigma^0_{\infty,j}$ is of the form

$$\Sigma_{\infty,j} \backslash \{p_{j1}, \ldots, p_{j\gamma_j}\},$$

where $\Sigma_{\infty,j}$ is a compact Riemann surface. More precisely, there is a natural admissible metric $\tilde{h}_{\infty,j}$ on each $\Sigma^0_{\infty,j}$ and a point $q_{\infty j}$ in $\Sigma^0_{\infty,j}$, such that for any fixed $r > 0$, when i is sufficiently large, there is a diffeomorphism $\phi_{i,r}$ from $B_r(q_{\infty j}, \tilde{h}_{\infty,j})$ onto $B_r(q_{ij}, \tilde{h}_i)$ satisfying: $\phi_{i,r}(q_{\infty j}) = q_{ij}$ and the pull-backs $\phi^*_{i,r}\tilde{h}_i$ converge to $\tilde{h}_{\infty,j}$ uniformly in the $\mathscr{C}^\infty$-topology over $B_r(q_{\infty j}, \tilde{h}_{\infty,j})$. Note that such a convergence of $\tilde{h}_i$ is assured by uniform admissibility.

Next we put together all these $\Sigma_{\infty,j}$ to form a connected curve Σ'_∞ as follows: For any two components $\Sigma_{\infty,j}$ and $\Sigma_{\infty,j'}$, we identify punctures $p_{js} \in \Sigma_{\infty,j}$ and $p_{j's'} \in \Sigma_{\infty,j'}$ (j may be equal to j') if for any sufficiently large i and r, the boundaries of $B_r(q_{ij}, \tilde{h}_i)$ and $B_r(q_{ij'}, \tilde{h}_i)$ specified above are contained in a cylindrical neck $N_{i,\alpha}$. In this way, we get a connected curve Σ_∞ (not necessarily stable) since each Σ_i is connected.

Since we have

$$\int_{B_{r_0}(x,\tilde{h}_i)} |Df_i|^2_{\tilde{h}_i,g} dz < \epsilon,$$

by taking a subsequence if necessary, we may assume that f_i converge to a J-holomorphic map f_∞ from $\bigcup_j \Sigma^0_{\infty,j}$ into M. By Proposition 7.3, f_∞ extends to a Hölder continuous J-holomorphic map from Σ_∞ into M. There is clearly also a limiting metric $\tilde{h}_\infty$ on Σ_∞, and Σ_∞ has the same genus as Σ_i for large i.

It remains to show that the homology class of f_∞ is the same as that of f_i. By convergence we have

$$\int_{\Sigma_\infty} |Df_\infty|^2_{\tilde{h}_\infty,g} dz = \lim_{r\to\infty}\lim_{i\to\infty} \int_{\bigcup_j B_r(q_{ij},\tilde{h}_i)} |Df_i|^2_{\tilde{h}_i,g} dz.$$

In fact, since the complement of $\bigcup_j B_r(q_{ij},\tilde{h}_i)$ in Σ_i is contained in the union of cylindrical necks $N_{i,\alpha}$, it suffices to show that for each i, if $N_{i,\alpha} = S^1 \times (a,b)$, then

$$\lim_{r\to\infty}\lim_{i\to\infty} \int_{S^1\times(a+r,b-r)} |Df_i|^2_{\tilde{h}_i,g} dz = 0.$$

This can be seen as follows: By our choice of $\tilde{h}_i$, we know that for any $p \in N_{i,\alpha}$,

$$\int_{B_1(p,\tilde{h}_i)} |Df_i|^2_{\tilde{h}_i,g} dz \leq \epsilon.$$

It follows from Lemma 7.2 that

$$\mathrm{diam}(f_i(B_2(p,\tilde{h}_i))) \leq c\sqrt{\int_{B_4(p,\tilde{h}_i)} |Df_i|^2_{\tilde{h}_i,g} dz},$$

where c is a uniform constant. Since ϵ is small, both $f_i(S^1 \times \{a + r\})$ and $f_i(S^1 \times \{b - r\})$ are contained in geodesic balls of radius $c\sqrt{\epsilon}$. Moreover, by varying r slightly, we may assume that

$$\int_{S^1\times(a+r,b-r)}|Df_i|^2_{\bar{h}_i,g}dz \leq 10\int_{B_4(p,\bar{h}_i)}|Df_i|^2_{\bar{h}_i,g}dz.$$

It follows that there are two smooth maps $u_{ij}: \Delta_1 \to M$ ($j=1,2$) with

$$u_{i1}|_{\partial\Delta_1} = f_i|_{S^1\times\{a+r\}}, \quad u_{i2}|_{\partial\Delta_1} = f_i|_{S^1\times\{b-r\}}.$$

and such that

$$\int_{\Delta_1}|Du_{i1}|^2_{\bar{h}_i,g} \leq c\int_{S^1\times(a+r-2,a+r+2))}|Df_i|^2_{\bar{h}_i,g}dz$$

and

$$\int_{\Delta_1}|Du_{i2}|^2_{\bar{h}_i,g} \leq c\int_{S^1\times(b-r-2,b-r+2))}|Df_i|^2_{\bar{h}_i,g}dz,$$

where c is a uniform constant. The maps $f_i|_{N_{i,\alpha}}$ and u_{ij} can be easily glued together to form a continuous map from S^2 into M. Since each $f_i|_{S^1\times[d-1,d+1]}$, where $d \in (a+r, b-r)$, is contained in a small geodesic ball of M, this map must be null homologous. It follows

$$\int_{S^1\times(a+r,b-r)}|Df_i|^2_{\bar{h}_i,g}dz = \int_{S^1\times(a+r,b-r)}f_i^*\omega = \int_{\Delta_1}u_{i1}^*\omega - \int_{\Delta_1}u_{i2}^*\omega.$$

Therefore, we have

$$\int_{S^1\times(a+r,b-r)}|Df_i|^2_{\bar{h}_i,g}dz \leq c\int_{S^1\times(a+r-2,a+r+2)\cup(b-r-2,b-r+2)}|Df_i|^2_{\bar{h}_i,g}dz.$$

This implies that the homology classes of f_i converge to the homology class of f_∞. So Theorem 7.1 is proved. $\square$

Remark 7.4 If (Σ_i, f_i) are stable maps, we may construct a stable limit $(\Sigma_\infty, f_\infty)$. Observe that Σ'_∞ may have components $\Sigma_{\infty,j}$ where f_∞ restricts to a constant map and which are conformal to $\mathbb{CP}^1$ and contain fewer than three other components. There are two possibilities for such $\Sigma_{\infty,j}$'s. If a $\Sigma_{\infty,j}$ attaches to only one other component, we simply drop $\Sigma_{\infty,j}$ from the construction; if $\Sigma_{\infty,j}$ contains exactly two other components, then we contract $\Sigma_{\infty,j}$ and identify those points where $\Sigma_{\infty,j}$ intersects with the other two components. Carrying out this process inductively, we eventually obtain a connected curve Σ_∞ such that the induced $f_\infty: \Sigma_\infty \to M$ is a stable map.

8 Second Variation of the $\bar{\partial}_J$-Equation and Applications

In Sect. 2 we saw that one prime difficulty in proving the isotopy theorem is the existence of a smoothing of a singular pseudo-holomorphic curve. Under positivity assumptions nodal curves can always be smoothed according to

Proposition 3.5. So to solve this problem it remains to find criteria when a pseudo-holomorphic map $\varphi : \Sigma \to M$ can be deformed to an immersion with only transversal branches. This seems generally a difficult problem, but over generic paths of almost complex structures miracles happen.

The content of this section is the technical heart of Shevchishin's work in [Sh]. The purpose of our presentation is to make this work more accessible by specializing to what we actually need.

8.1 Comparisons of First and Second Variations

Any of our moduli spaces $\mathcal{M}$ of pseudo-holomorphic maps is the zero set of a transverse section s of a Banach bundle $\mathcal{E}$ over a Banach manifold $\mathcal{B}$. This ambient Banach manifold also comes with a submersion π to a Banach manifold of almost complex structures. The purpose of this paragraph is to relate the first and second variations of s with those of π.

After choosing a local trivialization of the Banach bundle we have the abstract situation of two submersions of Banach manifolds. For the first variation the following result holds.

Proposition 8.1 Let $\Phi : \mathcal{X} \to \mathcal{Y}$, $\Psi : \mathcal{X} \to \mathcal{Z}$ be locally split submersions of Banach manifolds. For $P \in \mathcal{X}$ let $\mathcal{M} = \Phi^{-1}(\Phi(P))$, $\mathcal{F} = \Psi^{-1}(\Psi(P))$ be the fibers through P and $\bar{\Phi} = \Phi_{|\mathcal{F}}$, $\bar{\Psi} = \Psi_{|\mathcal{M}}$ the restrictions.
  Then there exist canonical isomorphisms

$$\mathrm{kern}(D\bar{\Phi}_{|P}) \ = \ \mathrm{kern}(D\Phi_{|P}) \cap \mathrm{kern}(D\Psi_{|P}) \ = \ \mathrm{kern}(D\bar{\Psi}_{|P}),$$
$$\mathrm{coker}(D\bar{\Phi}_{|P}) = (T_{\mathcal{Y},\Phi(P)} \oplus T_{\mathcal{Z},\Psi(P)})/(D\Phi_{|P}, D\Psi_{|P})(T_{\mathcal{X},P}) = \mathrm{coker}(D\bar{\Psi}_{|P}).$$

Proof. Let $X = T_{\mathcal{X},P}$, $Y = T_{\mathcal{Y},\Phi(P)}$, $Z = T_{\mathcal{Z},\Psi(P)}$, $M = \mathrm{kern}\,D\Phi = T_{\mathcal{M},P}$, $F = \mathrm{kern}\,D\Psi = T_{\mathcal{F},P}$. The claim follows from the following commutative diagram with exact rows.

$$
\begin{array}{ccccccccc}
0 & \to & \mathrm{kern}(D\bar{\Phi}_{|P}) & \to & F & \xrightarrow{D\bar{\Phi}_{|P}} & Y & \to \mathrm{coker}(D\bar{\Phi}_{|P}) & \to & 0 \\
& & \| & & \downarrow & & \downarrow & \downarrow{\scriptstyle\simeq} & & \\
0 & \to & F \cap M & \to & X & \xrightarrow{(D\Phi_{|P}, D\Psi_{|P})} & Y \oplus Z \to & (Y \oplus Z)/X & \to & 0 \\
& & \| & & \uparrow & & \uparrow & \uparrow{\scriptstyle\simeq} & & \\
0 & \to & \mathrm{kern}(D\bar{\Psi}_{|P}) & \to & M & \xrightarrow{D\bar{\Psi}_{|P}} & Z & \to \mathrm{coker}(D\bar{\Psi}_{|P}) & \to & 0
\end{array}
$$

$\square$

As an application of this lemma we can detect critical points of the projection $\mathcal{M} \to \mathcal{J}$ by looking at critical points of s for fixed almost complex structure. Note also that the linearization of a section $\mathcal{B} \to \mathcal{E}$ of a Banach bundle at any point P of its zero set is a well-defined map $T_{\mathscr{B},P} \rightarrow \mathscr{E}_P$. In fact, if $Q \in \mathscr{E}_P$ lies on the zero section then $T_{\mathscr{E},Q} \simeq T_{\mathscr{B},P} \oplus \mathscr{E}_P$ canonically.

The intrinsic meaning of the second variation is less apparent. In the notation of the proposition we are interested in situations when $\Psi|_{\mathscr{M}}$ is totally degenerate at $\overline{P}$, that is, if $T_{\mathscr{M},P} = T_{\mathscr{M},P} \cap T_{\mathscr{F},P}$. Now in any case the second variations of $\overline{\Phi}$ and $\overline{\Psi}$ induce two bilinear maps

$$\beta_1 : (T_{\mathscr{M},P} \cap T_{\mathscr{F},P}) \times (T_{\mathscr{M},P} \cap T_{\mathscr{F},P}) \longrightarrow \operatorname{coker}(D\overline{\Phi}_{|P})$$

$$\beta_2 : (T_{\mathscr{M},P} \cap T_{\mathscr{F},P}) \times (T_{\mathscr{M},P} \cap T_{\mathscr{F},P}) \longrightarrow \operatorname{coker}(D\overline{\Psi}_{|P}),$$

as follows. For $v, w \in T_{\mathscr{M},P} \cap T_{\mathscr{F},P}$ let $\tilde{v}, \tilde{w}$ be local sections around P of $T_{\mathscr{F}}$ and $T_{\mathscr{M}}$, respectively, with $\tilde{v}(P) = v$, $\tilde{w}(P) = w$. Then $D\overline{\Phi} \cdot \tilde{v}$ is a section $\alpha_{\tilde{v}}$ of $\overline{\Phi}^*T_{\mathscr{Y}}$ with $\alpha_{\tilde{v}}(P) = 0$ since $\tilde{v}(P) = v$ lies in $T_{\mathscr{M},P} = \operatorname{kern} D\overline{\Phi}$. If

$$\operatorname{pr}_{\mathscr{Y}} : T_{\overline{\Phi}^*T_{\mathscr{Y}},P} = T_{\mathscr{F},P} \oplus T_{\mathscr{Y},\Phi(P)} \longrightarrow T_{\mathscr{Y},\Phi(P)}$$

denotes the projection define

$$\beta_1(v,w) = \operatorname{pr}_{\mathscr{Y}}(D\alpha_{\tilde{v}} \cdot w),$$

viewed modulo $D\overline{\Phi}(T_P\mathscr{F})$. This definition does not depend on the choice of extension $\tilde{v}$ by applying the following lemma with $\tilde{v}$ the difference of two extensions.

Lemma 8.2 *If $\tilde{v}(P) = 0$ then $\operatorname{pr}_{\mathscr{Y}}(D\alpha_{\tilde{v}} \cdot w) \in \operatorname{im}(D\overline{\Phi})$.*

Proof. In the local situation of open sets in Banach spaces $\mathscr{X} \subset X = T_{\mathscr{X},P}$, $\mathscr{Y} \subset Y = T_{\mathscr{Y},\Phi(P)}$, $\mathscr{Z} \subset Z = T_{\mathscr{Z},\Psi(P)}$ we have

$$\operatorname{pr}_{\mathscr{Y}}(D\alpha_{\tilde{v}} \cdot w) = \partial_w(\partial_{\tilde{v}}\Phi) = \partial^2_{w\tilde{v}(P)}\Phi + D\Phi \cdot \partial_w\tilde{v}.$$

The claim follows because $\tilde{v}(P) = 0$ and $\partial_w\tilde{v} \in T_P\mathscr{F}$. $\square$

The analogous definition with Φ and Ψ swapped defines β_2.

Proposition 8.3 *Let $\Phi : \mathscr{X} \rightarrow \mathscr{Y}$, $\Psi : \mathscr{X} \rightarrow \mathscr{Z}$ be submersions of Banach manifolds with splittable differentials. For $P \in \mathscr{X}$ let $\mathscr{M} = \Phi^{-1}(\Phi(P))$, $\mathscr{F} = \Psi^{-1}(\Psi(P))$ be the fibers through P and $\overline{\Phi} = \Phi|_{\mathscr{F}}$, $\overline{\Psi} = \Psi|_{\mathscr{M}}$ the restrictions. Let $\Lambda : \operatorname{coker}(D\overline{\Phi}_{|P}) \rightarrow \operatorname{coker}(D\overline{\Psi}_{|P})$ denote the canonical isomorphism of Proposition 8.1 and β_1, β_2 the bilinear maps introduced above. Then*

$$\beta_2 = \Lambda \circ \beta_1.$$

Proof. In the local situation of the proofs of Proposition 8.1 and Lemma 8.2, by the definition of $\tilde{v}$, $\tilde{w}$ it holds $\partial_{\tilde{v}}\Psi = 0$, $\partial_{\tilde{w}}\Phi = 0$. Hence

$$
\begin{aligned}
(D\Phi_{|P}, D\Psi_{|P})[\tilde{w}, \tilde{v}] &= (D\Phi_{|P}[\tilde{w}, \tilde{v}], D\Psi_{|P}[\tilde{w}, \tilde{v}]) \\
&= (\partial_w(\partial_{\tilde{v}}\Phi) - \partial_v(\partial_{\tilde{w}}\Phi), \partial_w(\partial_{\tilde{v}}\Psi) - \partial_v(\partial_{\tilde{w}}\Psi)) \\
&= (\beta_1(v,w), 0) - (0, \beta_2(v,w)).
\end{aligned}
$$

Hence $\beta_1(v,w)$ and $\beta_2(v,w)$ induce the same element in $(Y \oplus Z)/X$, that is, $\beta_2 = \Lambda \circ \beta_1$. $\square$

8.2 Moduli Spaces of Pseudo-Holomorphic Curves with Prescribed Singularities

Let $\varphi : \Sigma \to (M, J)$ be a pseudo-holomorphic map with $D\varphi$ vanishing at P of order $\mu - 1 > 0$. Then $C = \operatorname{im} \varphi$ has a singular point at $\varphi(0)$. The number μ is the *multiplicity* of the singularity, which agrees with the degree of the composition of φ with a *general* local projection $M \to \mathbb{C}$ with J-holomorphic fibers. (For non-general projections this mapping degree can be larger than the multiplicity.) Choosing charts we may assume $M = \mathbb{C}^2$, $\Sigma = \Delta$, $\varphi(0) = 0$ and J to agree with the standard complex structure I at 0. Let j be the complex structure on Δ. Writing $0 = D\varphi + J \circ D\varphi \circ j = (D\varphi + I \circ D\varphi \circ j) + (J - I) \circ D\varphi \circ j$ gives the estimate

$$\left| \partial_{\bar{t}} \varphi(t) \right| \leq c \cdot |t|^{\mu - 1} \cdot |\varphi|.$$

Thus the higher dimensional analogue of Proposition 2.13 shows that φ is polynomial in t up to order $2\mu - 1$. It is not hard to see that this defines a holomorphic $(2\mu - 1)$-jet on $T_{\Delta, 0}$ with values in $T_{M, \varphi(0)}$. Note that this jet generally does not determine the embedded topological type of C at $\varphi(0)$. In the integrable situation one needs twice the Milnor number of C at $\varphi(0)$ minus one coefficients to tell the topological type, and the Milnor number can be arbitrarily large for given multiplicity.

The induced jet with values in the normal bundle $N_{\varphi, 0}$ (see Sect. 3.2) vanishes either identically or to order $\mu + \nu$, $0 \leq \nu \leq \mu - 1$. Define the *cusp index* of φ at P to be μ in the former case and to equal $\nu \leq \mu - 1$ in the latter case. (In [Sh] the multiplicity and cusp index are called primary and secondary cusp indices, respectively.)

For example, let $\varphi : \Delta \to (\mathbb{C}^2, J)$ be a pseudo-holomorphic singularity of multiplicity 2. Then

$$\varphi(t) = \left(\alpha t^2 + \beta t^3 + a(t), \gamma t^2 + \delta t^3 + b(t) \right)$$

with one of α or β non-zero and $a(t) = o(|t|^3)$, $b(t) = o(|t|^3)$. A linear coordinate change transforms $\begin{pmatrix} \alpha & \beta \\ \gamma & \delta \end{pmatrix}$ to $\begin{pmatrix} 1 & 0 \\ 0 & \delta \end{pmatrix}$, and hence we may assume $\alpha = 1$, $\beta = \gamma = 0$ and $\delta = 0$ or 1. Then φ defines the 3-jet with values in $T_{M, \varphi(0)}$ represented by $t \mapsto t^2 \partial_z + \delta t^3 \partial_w$. Going over to N means reducing modulo ∂_z. This leads to the 3-jet represented by $t \mapsto \delta t^3$. Thus φ has cusp index 0 if $\delta \neq 0$ and cusp index 1 otherwise. In analogy with the integrable situation the former singularity is called an *ordinary cusp*. We have seen that in this case

$$\varphi(t) = (t^2, t^3) + o(|t|^3)$$

in appropriate complex coordinates. We will use this below.

We can now define moduli spaces $\mathscr{M}_{\boldsymbol{\mu}, \boldsymbol{\nu}}$ with prescribed multiplicities $(\mu_1, \ldots, \mu_m)$ and cusp indices $(\nu_1, \ldots, \nu_m)$, $0 \leq \nu_i \leq \mu_i$, in k marked points $P_1, \ldots, P_m \in \Sigma$, and an immersion everywhere else. A straightforward transversality argument shows that $\mathscr{M}_{\boldsymbol{\mu}, \boldsymbol{\nu}}$ is a submanifold of the total moduli space $\mathscr{M}$ (without marked points) of real codimension

$$2(|\boldsymbol{\mu}|n - m(n+1)) + 2(n-1)|\boldsymbol{\nu}|, \tag{8.1}$$

where $n = \dim_{\mathbb{C}} M$. For details see [Sh], Sects. 3.2 and 3.3.

8.3 The Locus of Constant Deficiency

By the implicit function theorem critical points of the projection $\pi : \mathcal{M} \to \mathcal{J}$ have the property that the $\bar{\partial}_J$-operator for fixed J is obstructed. In fact, according to Proposition 8.1 the cokernels of the respective linearizations

$$D_{(j,\varphi,J)} = D_{\varphi,J} + J \circ D\varphi \circ j'$$

and $D\pi_{|(j,\varphi,J)}$ are canonically isomorphic. It is therefore important to study the stratification of $\mathcal{M}$ into subsets

$$\mathcal{M}^{h_1} := \{(j,\varphi,J) \in \mathcal{M} \mid \dim \operatorname{coker}(D_{(j,\varphi,J)}) \geq h^1\}.$$

To obtain an analytic description note that the discussion in Sect. 3.2 implies the following:

$$\operatorname{coker} D_{(J,\varphi,j)} = \operatorname{coker} D_{\varphi,J}/H^1(T_\Sigma) = \operatorname{coker} D_{\varphi,J}^N.$$

So in studying $\operatorname{coker} D\pi$ we might as well study the cokernel of the normal $\bar{\partial}$-operator $D_{\varphi,J}^N$.

The bundles $N = N_\varphi = \varphi^* T_M/D\varphi(T_\Sigma[A])$ on Σ from Sect. 3.2 do not patch to a complex line bundle on $\mathcal{M} \times \Sigma$ because their degree decreases under the presence of critical points of φ. However, once we restrict to $\mathcal{M}_{\boldsymbol{\mu},\boldsymbol{\nu}}$ the holomorphic line bundles $\mathcal{O}([A])$ encoding the vanishing orders of $D\varphi$ vary differentiably with φ; hence for any $\boldsymbol{\mu}, \boldsymbol{\nu}$ there exists a complex line bundle N on $\mathcal{M}_{\boldsymbol{\mu},\boldsymbol{\nu}} \times \Sigma$ with fibers N_φ relative $\mathcal{M}_{\boldsymbol{\mu},\boldsymbol{\nu}}$. For the following discussion we therefore restrict to one such stratum $\mathcal{M}_{\boldsymbol{\mu},\boldsymbol{\nu}} \subset \mathcal{M}$. Denote by $\mathcal{N}, \mathcal{F}$ the Banach bundles over $\mathcal{M}_{\boldsymbol{\mu},\boldsymbol{\nu}}$ with fibers $W^{1,p}(N_\varphi)$ and $L^p(N_\varphi \otimes \Lambda^{0,1})$, respectively. The normal $\bar{\partial}$-operators define a family of Fredholm operators

$$\sigma : \mathcal{N} \longrightarrow \mathcal{F}$$

with the property that for any $x = (j,\varphi,J) \in \mathcal{M}_{\boldsymbol{\mu},\boldsymbol{\nu}}$ there is a canonical isomorphism

$$\operatorname{coker} \sigma_x = \operatorname{coker} D_{\varphi,J}^N = \operatorname{coker} D\pi_{|x}.$$

To understand the situation around some $x_0 = (j,\varphi,J) \in \mathcal{M}_{\boldsymbol{\mu},\boldsymbol{\nu}}$ choose a complement $Q \subset \mathcal{F}_{x_0}$ to $\operatorname{im} \sigma_{x_0}$ and let $V \subset \mathcal{N}_{x_0}$ be a complement to $\ker \sigma_{x_0}$. Extend these subspaces to subbundles $\mathcal{V} \subset \mathcal{N}$ and $\mathcal{Q} \subset \mathcal{F}$. (Here and in the following we suppress the necessary restrictions to an appropriate neighborhood of x_0.) Then

$$\mathcal{V} \oplus \mathcal{Q} \longrightarrow \mathcal{F}, \quad (v,q) \longmapsto \sigma(v) + q$$

is an isomorphism around x_0 since this is true at x_0. In particular, $\sigma(\mathcal{V}) \subset \mathcal{F}$ is a subbundle and there are canonical isomorphisms

$$\mathcal{Q} \xrightarrow{\simeq} \mathcal{F}/\sigma(\mathcal{V}), \quad \mathcal{V} \xrightarrow{\simeq} \mathcal{F}/\mathcal{Q}.$$

Having set up the bundles $\mathcal{Q}$ and $\mathcal{V}$ we fit $\mathrm{kern}\,\sigma_{x_0}$ into a vector bundle by setting

$$\mathcal{K} := \mathrm{kern}(\mathcal{N} \to \mathcal{F}/\mathcal{Q}).$$

By the Fredholm property of σ this is a bundle of finite rank; it contains $\mathrm{kern}\,\sigma_x$ for any x and it is complementary to $\mathcal{V}$:

$$\mathcal{K} \oplus \mathcal{V} = \mathcal{N}.$$

We claim that σ induces a section $\bar{\sigma}$ of the finite rank bundle $\mathrm{Hom}(\mathcal{K}, \mathcal{Q})$ with the property that there are canonical isomorphisms

$$\mathrm{kern}\,\sigma_x \simeq \mathrm{kern}\,\bar{\sigma}_x, \quad \mathrm{coker}\,\sigma_x \simeq \mathrm{coker}\,\bar{\sigma}_x,$$

for any $x \in \mathcal{M}_{\mu,\nu}$ in the domain of our construction. In fact, this follows readily from the Snake Lemma applied to the following commutative diagram with exact columns and rows.

$$
\begin{array}{ccccccccc}
& & \mathrm{kern}\,\bar{\sigma}_x & & \mathrm{kern}\,\sigma_x & & & & \\
& & \downarrow & & \downarrow & & & & \\
0 & \longrightarrow & \mathcal{K}_x & \longrightarrow & \mathcal{N}_x & \longrightarrow & \mathcal{N}_x/\mathcal{K}_x & \longrightarrow & 0 \\
& & \bar{\sigma}_x \downarrow & & \sigma_x \downarrow & & \downarrow \simeq & & \\
0 & \longrightarrow & \mathcal{Q}_x & \longrightarrow & \mathcal{F}_x & \longrightarrow & \mathcal{F}_x/\mathcal{Q}_x & \longrightarrow & 0 \\
& & \downarrow & & \downarrow & & & & \\
& & \mathrm{coker}\,\bar{\sigma}_x & & \mathrm{coker}\,\sigma_x & & & &
\end{array}
\qquad (8.2)
$$

Because σ_x maps $\mathcal{N}_x/\mathcal{K}_x$ isomorphically to $\mathcal{F}_x/\mathcal{Q}_x$ we also see that

$$\mathcal{M}_{\mu,\nu}^{h^1} = \left\{ (j, \varphi, J) \in \mathcal{M}_{\mu,\nu} \mid \dim \mathrm{coker}\, D_{(j,\varphi,J)} \geq h^1 \right\}, \quad h^1 = \dim \mathrm{coker}\,\sigma_x$$

equals the zero locus of $\bar{\sigma}$ viewed as section of $\mathrm{Hom}(\mathcal{K}, \mathcal{Q})$ locally.

Proposition 8.4 (*[Sh], Corollary 4.4.2.*) $\mathcal{M}_{\mu,\nu}^{h^1}$ *is a submanifold inside* $\mathcal{M}_{\mu,\nu}$ *of codimension* (index $+ h^1) \cdot h^1$.

Proof. This follows from the implicit function theorem once we prove that $\bar{\sigma}$ is a transverse section for $\mathrm{Hom}(\mathcal{K}, \mathcal{Q})$ since

$$\mathrm{rank}\, \mathrm{Hom}(\mathcal{K}, \mathcal{Q}) = (\mathrm{index} + h^1) \cdot h^1.$$

To this end we look at variations at $x \in \mathcal{M}$ with φ and j fixed and with the variation J_s of J constant along $\operatorname{im} \varphi$. Note that such a path stays inside $\mathcal{M}$. The pull-back to the path of the bundles $\mathcal{N}$ and $\mathcal{F}$ with fibers

$$W^{1,p}(N) = W^{1,p}(\varphi^*T_M)/D\varphi(W^{1,p}(T_\Sigma)[A])$$
$$L^p(N \otimes \Lambda^{0,1}) = L^p(\varphi^*T_M \otimes \Lambda^{0,1})/D\varphi(L^p(T_\Sigma)[A] \otimes \Lambda^{0,1})$$

are manifestly trivial. Now σ is fiberwise given by $D_{\varphi,J}^N$, which in turn can be computed by lifting a section of N to φ^*T_M, applying

$$D_{\varphi,J} = \nabla + J \circ \nabla_{j(.)} + \nabla J \circ D_{j(.)}\varphi$$

and reducing modulo $D\varphi(T_\Sigma[A])$. The result of our variation is thus

$$(\nabla_{J'}\sigma)(v) = (\nabla_v J') \circ D\varphi \circ j,$$

written in the form lifted to φ^*T_M. Following the discussion above we now want to look at the derivative of the induced section $\bar{\sigma}$ of $\operatorname{Hom}(\mathcal{K}, \mathcal{Q})$. Write $h^0 = \dim \operatorname{kern} \sigma_x$, in analogy to $h^1 = \dim \operatorname{coker} \sigma_x$. For the construction of $\mathcal{Q}$ let $W \subset M$ be an open set such that $\varphi^{-1}(W) \subset \Sigma$ is a unit disk and such that there are complex-valued coordinates z, w on M with

$$\varphi(t) = (t, 0) \quad \text{for } t \in \Delta$$

in these coordinates. Note that $D_\varphi(T_\Sigma[A]) = \langle \partial_z \rangle$, so for the induced section of N only the ∂_w-part matters. Let χ be the characteristic function of $\varphi^{-1}(W)$ in Σ, that is, $\chi|_{\varphi^{-1}(W)} = 1$ and $\operatorname{supp} \chi \subset \operatorname{cl} \varphi^{-1}(W)$. Then

$$\chi \partial_w \otimes d\bar{t} \in L^p(\Sigma, \varphi^*T_M \otimes \Lambda^{0,1}).$$

Because φ is injective away from finitely many points, open sets of the form $\varphi^{-1}(W)$ span a base for the topology of Σ away from finitely many points. Now characteristic functions span a dense subspace in L^p. We can therefore find pairwise disjoint $W_1, \ldots, W_{h^1} \subset M$ such that the corresponding $\chi_j \partial_w \otimes d\bar{t}$ span the desired complementary subspace Q of $\operatorname{im} \sigma_x$.

To compute $\nabla_{J'}\bar{\sigma} \in \operatorname{Hom}(\mathcal{K}_x, \mathcal{Q}_x) = \operatorname{Hom}(\operatorname{kern} \sigma_x, Q)$ it suffices to restrict $\nabla_{J'}\sigma$ to $\operatorname{kern} \sigma_x \subset \mathcal{N}_x$, and to compose with a projection

$$q : \mathcal{F}_x \longrightarrow \mathbb{R}^{h^1}$$

that induces an isomorphism $Q \to \mathbb{R}^{h^1}$. This follows from Diagram (8.2). For q we take the map

$$\mathcal{F}_x \ni \xi \longmapsto \left(\operatorname{Im} \int_{\varphi^{-1}(W_j)} dt \wedge \langle dw, \xi \rangle \right)_{j=1,\ldots,h^1}.$$

This maps $\chi_j \partial_w \otimes d\bar{t}$ to a non-zero multiple of the jth unit vector. Hence q is one-to-one on Q.

Then $q \circ \nabla_{J'} \sigma$ maps $v \in \ker n \, \sigma_x$ to

$$\left(\operatorname{Im} \int_{\varphi^{-1}(W_j)} dt \wedge \langle dw, \nabla_v J' \circ D\varphi \circ j \rangle \right)_{j=1,\ldots,h^1}.$$

Now consider variations of the form $J' = gw \, d\bar{z} \otimes \partial_w$ in coordinates (z, w), that is, $\begin{pmatrix} 0 & 0 \\ gw & 0 \end{pmatrix}$ in matrix notation. If v is locally represented by $f\partial_w$ then

$$\nabla_v J' \circ D\varphi \circ j = (fg d\bar{z} \otimes \partial_w) \circ D\varphi \circ j = \sqrt{-1} fg \, d\bar{t} \otimes \partial_w,$$

and

$$\left(q \circ \nabla_{J'} \sigma \right)(v) = \left(\operatorname{Im} \int_{\varphi^{-1}(W_j)} \sqrt{-1} fg \, dt \wedge d\bar{t} \right)_{j=1,\ldots,h^1}.$$

By the identity theorem for pseudo-analytic functions the restriction map $\ker n \, \sigma_x \to L^p(\varphi^{-1}(W_j), N)$ is injective. Thus for each j there exist $g_{j1}, \ldots, g_{jh^0}$ with support on W_j such that

$$\ker n \, \sigma_x \longrightarrow \mathbb{R}^{h^0}, \quad f\partial_w \longmapsto \left(\operatorname{Im} \int_{\varphi^{-1}(W_j)} \sqrt{-1} fg_{jk} \, dt \wedge d\bar{t} \right)_{k=1,\ldots,h^0}$$

is an isomorphism. The corresponding variations J'_{jk} of $\bar{\sigma}_x$ (with support on $\varphi^{-1}(W_j)$) span $\operatorname{Hom}(\ker n \, \bar{\sigma}_x, Q)$. $\square$

Remark 8.5 The proof of the proposition in [Sh] has a gap for $h^1 > 1$, as pointed out to us at the summer school by Jean-Yves Welschinger. In this reference Q_x is canonically embedded into $L^p(\Sigma, N^* \otimes \Lambda^{1,0})$ as kernel of the adjoint operator. The problem is that the proof of surjectivity of the relevant linear map

$$T_{\mathscr{M}_{\mu\nu},(j,\varphi,J)} \longrightarrow \operatorname{Hom}(\mathcal{K}_x, Q_x)$$

relies on the fact that $\langle \mathcal{K}_x, Q_x \rangle$ spans an $h^0 \cdot h^1$-dimensional subspace of $L^p(\Sigma, \Lambda^{1,0})$, where $\langle \, , \, \rangle$ is the dual pairing. Our proof shows that this is indeed the case.

Corollary 8.6 *For a general path $\{J_t\}_{t \in [0,1]}$ of almost complex structures any critical point (j, φ, t) of the projection $p : \mathscr{M}_{\{J_t\}} \to [0,1]$ is a pseudo-holomorphic map with only ordinary cusps and such that $\dim \operatorname{coker} D^N_{\varphi, J_t} = 1$.*

Proof. This is a standard transversality argument together with dimension counting. Note that each singular point of multiplicity μ causes $\dim(\ker n \, \sigma_{(j,\varphi,J_t)})$ to drop by $\mu - 1$. $\square$

8.4 Second Variation at Ordinary Cusps

Corollary 8.6 leaves us with the treatment of pseudo-holomorphic maps with only ordinary cusps and such that D^N_{φ,J_t} has one-dimensional cokernel. Then the projection $p : \mathcal{M}_{\{J_t\}} \to [0,1]$ is not a submersion. The maybe most intriguing aspect of Shevchishin's work is that one can see quite clearly how the presence of cusps causes these singularities. They turn out to be quadratic, non-zero and indefinite. In particular, such a pseudo-holomorphic map always possesses deformations *with fixed almost complex structure J_t* into non-critical points of π.

Let us fit this situation into the abstract framework of Sect. 8.1. In the notation employed there $\mathcal{Z} = (-\varepsilon, \varepsilon)$ is the local parameter space of the path $\{J_t\}$, $\mathcal{X}$ is a neighborhood of the critical point $P = (j, \varphi, t)$ in the pull-back (via $\mathcal{Z} = (-\varepsilon, \varepsilon) \to \mathcal{J}$) of the ambient Banach manifold

$$\mathcal{B} := \mathbb{T}_g \times W^{1,p}(\Sigma, M) \times \mathcal{J},$$

and $\mathcal{Y} = L^p(\Sigma, \varphi^* T_M \otimes \Lambda^{0,1})$. The map $\Phi : \mathcal{X} \to \mathcal{Y}$ is a local non-linear $\bar{\partial}_J$-operator obtained from s via a local trivialization of the Banach bundle $\mathcal{E}$, while $\Psi : \mathcal{X} \to \mathcal{Z}$ is the projection, and $\bar{\Psi} = p$. The fiber of Φ through P is (an open set in) $\mathcal{M}_{\{J_t\}}$, and the fiber of Ψ through P is the ambient Banach manifold for $\mathcal{M}_{J_t}$.

Because $\mathcal{Z}$ is one-dimensional and (j, φ, J) is a critical point of the projection $p : \mathcal{M}_{\{J_t\}} \to [0,1]$ it holds $T_{\mathcal{M}_{\{J_t\}},P} \subset T_{\mathcal{F},P}$. Proposition 8.3 now says that we can compute the second order approximation of p near P by looking at the second order approximation of Φ restricted to $\mathcal{F}$, composed with the projection to the cokernel of the linearization. In Sect. 8.1 this symmetric bilinear form was denoted β_1. We are going to compute the associated quadratic form. We denote tangent vectors of the relevant tangent space $T_{\mathcal{M}_{\{J_t\}},P} \cap T_{\mathcal{F},P} = T_{\mathcal{M}_{\{J_t\}},P}$ by pairs (j', v) with j' a tangent vector to the space of complex structures on Σ and $v \in W^{1,p}(\Sigma, \varphi^* T_M)$.

Recall that the linearization of the $\bar{\partial}_J$-operator for fixed almost complex structure J is

$$D\bar{\partial}_J : (j', v) \longmapsto D_{(j,\varphi,J)}(j', v) = D_{\varphi,J} v + J \circ D\varphi \circ j',$$

where

$$D_{\varphi,J} v = \nabla v + J \circ \nabla \circ j(v) + \nabla_v J \circ D\varphi \circ j = 2\bar{\partial}_{\varphi,J} + R.$$

Near a cusp choose local coordinates z, w on M and t on Σ such that $J_{|(0,0)}$ equals I, the standard complex structure on $\mathbb{C}^2$, and $\varphi(t) = (t^2, t^3) + o(|t|^3)$ in these coordinates. Let $0 < \varepsilon < 1$ and let $\rho : \Delta \to [0,1]$ be a smooth function with support in $|t| < 3\varepsilon/4$, identically 1 for $|t| < \varepsilon/4$ and with $|d\rho| < 3/\varepsilon$. Ultimately we will let ε tend to 0, but for the rest of the computation ε is fixed. We consider the variation of σ along (j', v) with

$$v = D\varphi(\rho t^{-1}\partial_t) = \rho t^{-1}\partial_t\varphi, \quad j' = j \circ \bar\partial(\rho t^{-1}\partial_t) = it^{-1}\partial_{\bar t}\rho\,\partial_t \otimes d\bar t.$$

Again we use complex notation for the complex vector bundle $\varphi^*T_M \otimes_{\mathbb{C}} T_\Sigma$. Taking the real part reverts to real notation. Note that j' is smooth and supported in the annulus $\varepsilon/4 < |t| < 3\varepsilon/4$. For any $\mu \in \mathbb{C}$ the multiple $\mu \cdot (j', v)$ is indeed a tangent vector to $\mathcal{M}_{J_t}$ because

$$D_{\varphi,J} \circ D\varphi = \bar\partial_{\varphi,J} \circ D\varphi = D\varphi \circ \bar\partial,$$

since $R \circ D\varphi = 0$ and by definition of the holomorphic structure on φ^*T_M; taking into account pseudo-holomorphicity $J \circ D\varphi \circ j = -D\varphi$ of φ this implies

$$D_{\varphi,J}(\mu v) + J \circ D\varphi \circ (\mu j') = (D\varphi \circ \bar\partial)(\mu\rho t^{-1}\partial_t) + (J \circ D\varphi \circ j)\big(\bar\partial(\mu\rho t^{-1}\partial_t)\big) = 0,$$

as needed.

At this point it is instructive to connect this variation to the discussion of the normal $\bar\partial$-operator $D^N_{\varphi,J}$ in Sect. 3.2. There we identified the tangent space of $\mathcal{M}$ relative $\mathcal{J}$ with kern $D^N_{\varphi,J} \oplus H^0(\mathcal{N}^{\mathrm{tor}})$, see (3.4). Away from the node the vector field v lies in $D\varphi(W^{1,p}(T_\Sigma))$, and indeed v is a local frame for the complex line bundle $D\varphi(T_\Sigma[A])$. Thus we are dealing with a variation whose part in kern $D^N_{\varphi,J}$ vanishes and which generates the skyscraper sheaf $\mathcal{N}^{\mathrm{tor}}$ locally at the cusp.

The variation (j', v) is concentrated in $\mathscr{B}_\varepsilon(0) \subset \Sigma$ and we can work in our local coordinates t, z, w. Then

$$v = \rho t^{-1}\partial_t\big((t^2, t^3) + o(|t|^3)\big) = \rho \cdot (2, 3t) + o(|t|),$$

which can be represented by the variation

$$\varphi_s = \varphi + s\rho t^{-1}\partial_t\varphi = \varphi + s\rho \cdot \big((2, 3t) + o(|t|)\big).$$

Similarly, we represent j' by a variation of holomorphic coordinate t_s of t with associated $\bar\partial$-operator

$$\partial_{\bar t_s} = \partial_{\bar t} + a_s\partial_t, \quad a_s = sit^{-1}\partial_{\bar t}\rho.$$

The derivative with respect to s yields $it^{-1}\partial_{\bar t}\rho\partial_t$, and hence $\bar\partial_s$ indeed represents j'.

The non-linear $\bar\partial$-operator for (j_s, φ_s) applied to φ_s yields

$$\bar\partial_s\varphi_s = \frac{1}{2}\Big(D\varphi_s + J_{|\varphi_s} \circ D\varphi_s \circ j_s\Big) = (\partial_{\bar t_s}\varphi_s)d\bar t_s + \frac{1}{2}K_s,$$

with

$$K_s = (J_{|\varphi_s} - I) \circ D\varphi_s \circ j_s.$$

Using $d\bar t_s$ to trivialize $\mathscr{E}$ along the path (j_s, φ_s) we now compute for the second variation

$$\frac{d^2}{ds^2}\Big|_{s=0}\partial_{\bar{t}_s}\varphi_s = \frac{d^2}{ds^2}\Big|_{s=0}(\partial_{\bar{t}} + a_s\partial_t)(\varphi + s\rho \cdot ((2, 3t) + o(|t|)))$$

$$= it^{-1}\partial_{\bar{t}}\rho \cdot (\rho \cdot 3\partial_w + o(1)) = \frac{3i}{2}t^{-1}\partial_{\bar{t}}\rho^2 \cdot \partial_w + t^{-1} \cdot o(1),$$

$$\text{(8.3)}$$

and

$$K'' := \frac{d^2}{ds^2}\Big|_{s=0}K_s = (J_{|\varphi} - I) \cdot \frac{d^2}{ds^2}\Big|_{s=0}(D\varphi_s \circ j_s) + \nabla_v J \circ \nabla v \circ j + \nabla_v J \circ D\varphi \circ j'.$$

This is non-zero only for $|t| < \varepsilon$, and the first two terms are bounded pointwise, uniformly in ε. Here ∇ denotes the flat connection with respect to the coordinates z, w. For the last term we have

$$|\nabla_v J \circ D\varphi \circ j'| \leq \text{const} \cdot |t||t^{-1}| \cdot |\partial_{\bar{t}}\rho| \leq \text{const} \cdot \varepsilon^{-1},$$

by the choice of ρ. Taken together this gives the pointwise bound

$$|K''| \leq \text{const} \cdot \varepsilon^{-1} \cdot \chi_\varepsilon,$$

where χ_ε is the characteristic function for $B_\varepsilon(0)$, that is, $\chi_\varepsilon(t) = 1$ for $|t| < \varepsilon$ and 0 otherwise.

Lemma 8.7 *Let* $\lambda dt \in \text{kern}(D^N_{\varphi,j})^* \subset W^{1,p}(N^* \otimes \Lambda^{1,0})$ *and denote by* $\Lambda :$ *coker* $D^N_{\varphi,j} \to \mathbb{R}$ *the associated homomorphism with kernel* $\text{im}\, D^N_{\varphi,j}$ *induced by*

$$L^p(\varphi^*T_M \otimes \Lambda^{0,1}) \to \mathbb{R}, \quad \gamma \longmapsto \text{Re}\int_\Sigma \langle \lambda dt, \gamma\rangle.$$

Then for $\mu \in \mathbb{C}$ *it holds*

$$(\Lambda \circ \beta_1)(\mu \cdot (j', v), \mu \cdot (j', v)) \xrightarrow{\varepsilon \to 0} -3\pi \text{Re}(\mu^2)\lambda(\partial_w)(0).$$

Proof. The formal adjoint of $D^N_{\varphi,j}$ is, just as the operator itself, the sum of a $\bar{\partial}$-operator and an operator of order zero:

$$(D\bar{\partial}_J)^* = \bar{\partial}_{N^* \otimes \Lambda^{1,0}} + R^*.$$

Thus

$$\bar{\partial}_{N^* \otimes \Lambda^{1,0}}(\lambda dt) = -R^*(\lambda dt) \qquad \text{(8.4)}$$

is uniformly bounded pointwise. By the definition of β_1 and the discussion above we need to compute

$$\int_\Sigma \langle \lambda dt, (\frac{d^2}{ds^2}\Big|_{s=0}\partial_{\bar{t}_s}\varphi_s)d\bar{t} + \tfrac{1}{2}K''\rangle = \frac{3i}{2}\int_{B_\varepsilon(0)} \lambda(\partial_w)\partial_{\bar{t}}\rho^2 t^{-1}dt \wedge d\bar{t}$$

$$+ \int_{B_\varepsilon(0)} \big(o(1)t^{-1} + \text{const} \cdot \varepsilon^{-1}\big)dt \wedge d\bar{t}.$$

The second integral tends to 0 with ε. The first integral can be rewritten as a sum of

$$\frac{3i}{2} \int_{B_\varepsilon(0)} t^{-1} \rho^2 \bar{\partial}\big(\lambda(\partial_w)dt\big),$$

again tending to zero with ε in view of (8.4), and

$$-\frac{3i}{2} \int_{B_\varepsilon(0)} t^{-1}\bar{\partial}\big(\rho^2 \lambda(\partial_w)dt\big) \;=\; \frac{3i}{2} \cdot (2\pi i)\big(\rho^2 \lambda(\partial_w)\big)(0) \;=\; -3\pi\lambda(\partial_w)(0).$$

Here the first equality follows from the Cauchy integral formula.

Our computation is clearly quadratic in rescaling (j', v) by μ. Thus replacing (j', v) by $\mu \cdot (j', v)$ and taking the real part gives the stated formula. $\square$

Proposition 8.8 *Let $\{J_t\}_{t\in[0,1]}$ be a general path of almost complex structures on a four-manifold M. Assume that $P = (j, \varphi, t)$ is a critical point of the projection $p : \mathcal{M}_{\{J_t\}} \to [0,1]$ with φ not an immersion. Then there exists a locally closed, two-dimensional submanifold $Z \subset \mathcal{M}_{\{J_t\}}$ through P with coordinates x, y such that*

$$p(x, y) = x^2 - y^2.$$

Moreover, Z can be chosen in such a way that the pseudo-holomorphic maps corresponding to $(x, y) \neq 0$ are immersions.

Proof. By Corollary 8.6 the critical points of φ are ordinary cusps and $\dim \operatorname{coker} D^N_{\varphi,j} = 1$. Assume first that there is exactly one cusp. Another transversality argument shows that for general paths $\{J_t\}$ a generator λdt of $\operatorname{kern}(D^N_{\varphi,j})^*$ does not have a zero at this cusp. Let (j', v) be as in the discussion above with $\varepsilon > 0$ so small that the quadratic form

$$\mathbb{C} \ni \mu \longmapsto (\Lambda \circ \beta_1)\big(\mu \cdot (j', v), \mu \cdot (j', v)\big)$$

is non-degenerate and indefinite. This is possible by Lemma 8.7 and by what we just said about generators of $\operatorname{kern}(D^N_{\varphi,j})^*$. Let $Z \subset \mathcal{M}_{\{J_t\}}$ be a locally closed submanifold through P with $T_{Z,P}$ spanned by $\operatorname{Re}(j', v)$ and $\operatorname{Im}(j', v)$. The result is then clear by the Morse Lemma because $\beta_1|_{T_{Z,P}}$ describes the second variation of the composition $Z \to \mathcal{M}_{J_t} \to [0,1]$.

In the general case of several cusps, for each cusp we have a tangent vector (j'_l, v_l) with support close to it. Now run the same argument as before but with $(j', v) = \sum_l(j'_l, v_l)$. For ε sufficiently small these variations are supported on disjoint neighborhoods, and hence the only difference to the previous argument is that the coefficient $\lambda(\partial_w)$ for the quadratic form gets replaced by the sum $\lambda\big(\sum_l \partial_{w_l}\big)$. Again, for general paths, this expression is non-zero. $\square$

Remark 8.9 We have chosen to use complex, local notation as much as possible and to neglect terms getting small with ε. This point of view clearly exhibits the holomorphic nature of the critical points in the moduli space near a cuspidal curve and is also computationally much simpler than the full-featured computations in [Sh]. In fact, in the integrable case coordinates can be chosen in such a way that all error terms $o(1)$ etc. disappear and the formula in Lemma 8.7 holds for $\varepsilon > 0$.

9 The Isotopy Theorem

9.1 Statement of Theorem and Discussion

In this section we discuss the central result of these lectures. It deals with the classification of symplectic submanifolds in certain rational surfaces. As a consequence the expected "stable range" for this problem indeed exists. In this range there are no new symplectic phenomena compared to complex geometry.

Theorem 9.1 *(1) Let M be a Hirzebruch surface and $\Sigma \subset M$ a connected surface symplectic with respect to a Kähler form. If $\deg(p|_\sigma) \leq 7$ then Σ is symplectically isotopic to a holomorphic curve in M, for some choice of complex structure on M.*

(2) Any symplectic surface in $\mathbb{CP}^2$ of degree $d \leq 17$ is symplectically isotopic to an algebraic curve.

A Hirzebruch surface M is a holomorphic $\mathbb{CP}^1$-bundle over $\mathbb{CP}^1$. They are projectivizations of holomorphic 2-bundles over $\mathbb{CP}^1$. The latter are all split, so $M = \mathbb{F}_k := \mathbb{P}(\mathcal{O} \oplus \mathcal{O}(k))$ for some $k \in \mathbb{N}$. The k is determined uniquely as minus the minimal self-intersection number of a section. If $k = 0$ we have $M = \mathbb{CP}^1 \times \mathbb{CP}^1$ and there is a whole $\mathbb{CP}^1$ worth of such sections; otherwise the section is holomorphically rigid and it is in fact unique. Topologically $\mathbb{F}_k$ is the non-trivial S^2-bundle over S^2 for k odd and $\mathbb{F}_k \simeq S^2 \times S^2$ for k even. It is also worthwhile to keep in mind that for any k, l with $2l \leq k$ there is a holomorphic one-parameter deformation with central fiber $\mathbb{F}_k$ and general fiber $\mathbb{F}_{k-2l}$, but not conversely. So in a sense, $\mathbb{CP}^1 \times \mathbb{CP}^1$ and $\mathbb{F}_1$ are the most basic Hirzebruch surfaces, those that are stable under small deformations of the complex structure. Note also that $\mathbb{F}_1$ is nothing but the blow-up of $\mathbb{CP}^2$ in one point.

The degree bounds in the theorem have to do with the method of proof and are certainly not sharp. For example, it should be possible to treat the case of degree 18 in $\mathbb{CP}^2$ with present technology. We even believe that the theorem should hold without any bounds on the degree.

In Sect. 6.4 we saw the importance of this result for genus-2 Lefschetz fibrations and for Hurwitz-equivalence of tuples of half-twists in the braid group $B(S^2, d)$ with $d \leq 7$.

9.2 Pseudo-Holomorphic Techniques for the Isotopy Problem

Besides the purely algebraic approach by looking at Hurwitz-equivalence for tuples of half-twists, there exists only one other approach to the isotopy problem for symplectic 2-manifolds inside a symplectic manifold (M, ω), namely by the technique of pseudo-holomorphic curves already discussed briefly in Sect. 2.4,III, as explained in Sect. 2.4,II. This works in three steps. (1) Classify pseudo-holomorphic curves up to isotopy for *one particular* almost complex structure I on M; typically I is integrable and the moduli space of holomorphic curves can be explicitly controlled by projective algebraic geometry; this step is quite simple. (2) Choose a general family of almost complex structures $\{J_t\}_{t \in [0,1]}$ with (a) B is J_0-holomorphic (b) $J_1 = I$. By the results on the space of tamed almost complex structures this works without problems as long as the symplectic form ω is isotopic to a Kähler form for I. (3) Try to deform B as pseudo-holomorphic curve with the almost complex structure, that is, find a smooth family $\{B_t\}$ of submanifolds such that B_t is J_t-holomorphic.

The last step (3) is the hardest and most substantial obstruction for isotopy results for two reasons. First, while for general paths of almost complex structures the space of pseudo-holomorphic curves $\mathcal{M}_{\{J_t\}}$ over the path is a manifold, the projection to the parameter interval $[0, 1]$ might have critical points. If $\{B_t\}_{t \leq t_0}$ happens to run into such a point it may not be possible to deform B_{t_0} to $t > t_0$ and we are stuck. To avoid this problem one needs an unobstructedness result for deformations of smooth pseudo-holomorphic curves. The known results on this require some positivity of M, such as in Proposition 3.4. And second, even if this is true, as in the cases of $\mathbb{CP}^2$ and $\mathbb{F}_k$ that we are interested in, there is no reason that $\lim_{t \to t_0} B_t$ is a submanifold at all. The Gromov compactness theorem rather tells us that such a limit makes only sense as a stable J_t-holomorphic map or as a J_t-holomorphic 2-cycle. In the sequel we prefer to use the embedded point of view and stick to pseudo-holomorphic cycles. In any case, these are singular objects that we are not allowed to use in the isotopy. So we need to be able to change the already constructed path $\{B_t\}$ to bypass such singular points. This is the central problem for the proof of isotopy theorems in general.

In view of unobstructedness of deformations of smooth curves it suffices to solve the following:

1. Find a J_{t_0}-holomorphic smoothing of $C = \lim_{t \to t_0} B_t$
2. Show that any two pseudo-holomorphic smoothings of C are isotopic

In our situation the smoothing problem (1) of a pseudo-holomorphic cycle $C = \sum_a m_a C_a = \lim_{t \to t_0} B_t$ has the following solution. For a general path $\{J_t\}$ we know by the results of Sect. 8 that each J_{t_0}-holomorphic curve has a deformation into a nodal curve. Now for each a take m_a copies $C_{a,1}, \ldots, C_{a,m_a}$ of C_a. Deform each $C_{a,j}$ slightly in such a way that $\sum_{a,j} C_{a,j}$ is a nodal curve. This is possible by positivity. Finally apply the smoothing result for nodal curves (Proposition 3.5) to obtain a smoothing of C.

Problem (2) concerns the isotopy of smoothings of singular pseudo-holomorphic objects, which boils down to a question about the local structure of the moduli space of pseudo-holomorphic cycles as follows:

Let $C = \sum_a m_a C_a$ be a J_{t_0}-holomorphic cycle. Looking at the space of pairs (C', t) where $t \in [0, 1]$ and C' is a J_t-holomorphic cycle, do the points parametrizing singular cycles locally disconnect it?

So we ask if any point has a neighborhood that stays connected once we remove the points parametrizing singular cycles. We believe this question has a positive answer under the positivity assumption that $c_1(M) \cdot C_a > 0$ for each irreducible component C_a of C. In the integrable case this follows from the unobstructedness results of Sect. 5, which say that there is a local complex parameter space for holomorphic deformations of C; the subset of singular cycles is a proper analytic subset, and hence its complement remains connected. However, as already discussed briefly in Sect. 5.1 no such parametrization is known for general almost complex structures except in the nodal case of Sect. 3.4.

9.3 The Isotopy Lemma

Instead of solving the parametrization problem for pseudo-holomorphic cycles we use a method to reduce the "badness" of singularities of $\lim_{t \to t_0} B_t$ by cleverly adding pointwise incidence conditions. This technique has been introduced into symplectic topology by Shevchishin in his proof of the local isotopy theorem for smoothings of a pseudo-holomorphic curve singularity [Sh].

How does this work? The pseudo-holomorphic cycle C consists of irreducible components C_a. Write C_a as image of a pseudo-holomorphic map $\varphi_a : \Sigma_a \to M$. Pseudo-holomorphic deformations of C_a keeping the genus (*equigeneric deformations*) can be realized by deforming φ_a and the complex structure on Σ_a. The moduli space of such maps has dimension

$$k_a := c_1(M) \cdot C_a + g(C_a) - 1 \geq 0.$$

Each imposing of an incidence with a point on M reduces this dimension by one, *provided the number of points added does not exceed $c_1(M) \cdot C_a$*. Thus choosing k_a general points on C_a implies that there are no non-trivial equigeneric deformations of C_a respecting the incidences. Doing this for all a we end up with a configuration of $\sum_a k_a$ points such that any J_{t_0}-holomorphic deformation of C containing all these points must somehow have better singularities. This can happen either by dropping $\sum_a (m_a - 1)$, which measures how multiple C is, or, if this entity stays the same, by the virtual number δ of double points of $|C|$. The latter is the sum over the maximal numbers of double points of local pseudo-holomorphic deformations of the map with image $|C|$ near the singular points. Pseudo-holomorphic deformations where the pair (m, δ) gets smaller in the way just described are exactly the deformations that can not be realized by deforming just the φ_a.

On the other hand, to have freedom to move B_t by keeping the incidence conditions there is an upper bound on the number of points we can add by the excess positivity that we have, which is $c_1(M) \cdot C$. In fact, each point condition decreases this number in the proof of Proposition 3.4 by one. Thus this method works as long as

$$\sum_a \left(c_1(M) \cdot C_a + g(C_a) - 1 \right) \; < \; c_1(M) \cdot C. \tag{9.1}$$

If C is reduced ($m_a = 1$ for all a) then this works only if all components have at most genus one and at least one has genus zero. So this is quite useless for application to the global isotopy problem.

The following idea comes to the rescue. Away from the multiple components and from the singularities of the reduced components not much happens in the convergence $B_t \to C$: In a tubular neighborhood B_t is the graph of a function for any t sufficiently close to t_0 and this convergence is just a convergence of functions. So we can safely replace this part by some other (part of a) pseudo-holomorphic curve, for any t, and prove the isotopy lemma with this replacement made. By this one can actually achieve that each reduced component is a sphere, see below. If C_a is a sphere it contributes one less to the left-hand side than to the right-hand side of (9.1). So reduced components do not matter! For multiple components the right-hand side receives an additional $(m_a - 1)c_1(M) \cdot C_a$ that has to be balanced with the genus contribution $g(C_a)$ on the left-hand side.

Here is the precise formulation of the Isotopy Lemma from [SiTi3].

Lemma 9.2 *[SiTi3] Let $p : (M, J) \to \mathbb{CP}^1$ be a pseudo-holomorphic S^2-bundle. Let $\{J_n\}$ be a sequence of almost complex structures making p pseudo-holomorphic. Suppose that $C_n \subset M$, $n \in \mathbb{N}$, is a smooth J_n-holomorphic curve and that*

$$C_n \overset{n \to \infty}{\longrightarrow} C_\infty = \sum_a m_a C_{\infty,a}$$

in the $\mathscr{C}^0$-topology, with $c_1(M) \cdot C_{\infty,a} > 0$ for every a and $J_n \to J$ in $\mathscr{C}_{loc}^{0,\alpha}$. We also assume:

(*) If $C' = \sum_a m'_a C'_a$ is a non-zero J'-holomorphic cycle $\mathscr{C}^0$-close to a sub-cycle of $\sum_{m_a > 1} m_a C_{\infty,a}$, with $J' \in \mathscr{J}$, then

$$\sum_{\{a \mid m'_a > 1\}} \left(c_1(M) \cdot C'_a + g(C'_a) - 1 \right) < c_1(M) \cdot C' - 1.$$

Then any J-holomorphic smoothing $C_\infty^\dagger$ of C_∞ is symplectically isotopic to some C_n. The isotopy from C_n to $C_\infty^\dagger$ can be chosen to stay arbitrarily close to C_∞ in the $\mathscr{C}^0$-topology, and to be pseudo-holomorphic for a path of

almost complex structures that stays arbitrarily close to J in $\mathscr{C}^0$ everywhere,
and in $\mathscr{C}^{0,\alpha}_{\mathrm{loc}}$ away from a finite set.

In the assumptions $\mathscr{C}^0$-convergence $C_n \to C_\infty$ is induced by $\mathscr{C}^0$-convergence inside the space of stable maps.

Using the genus formula one can show easily that the degree bounds in the theorem imply Assumption $(*)$ in the Isotopy Lemma, see [SiTi3], Lemma 9.1.

9.4 Sketch of Proof

We want to compare two different smoothings of the pseudo-holomorphic cycle C_∞, one given by C_n for large n and one given by some J-holomorphic smoothing, for example constructed via first deforming to a nodal curve and then smoothing the nodal curve, as suggested above. There is only one general case where we know how to do this, namely if J is integrable locally around the cycle, see Sect. 5. But J generally is not integrable and we seem stuck.

Step 1: Make J integrable around $|C|$. On the other hand, for the application of the Isotopy Lemma to symplectic geometry we are free to change our almost complex structures within a $\mathscr{C}^0$-neighborhood of J. This class of almost complex structures allows a lot of freedom! To understand why recall the local description of almost complex structures with fixed fiberwise complex structure along $w = \mathrm{const}$ via one complex-valued function b:

$$T^{0,1}_M = \mathbb{C} \cdot (\partial_{\bar{z}} + b\partial_w) + \mathbb{C} \cdot \partial_{\bar{w}}.$$

The graph $\Gamma_f = \{(z, f(z))\}$ of a function f is pseudo-holomorphic with respect to this almost complex structure iff

$$\partial_{\bar{z}} f(z) \;=\; b(z, f(z)).$$

Thus the space of almost complex structures making Γ_f pseudo-holomorphic is in one-to-one correspondence with functions b with prescribed values $\partial_{\bar{z}} f$ along Γ_f. On the other hand, the condition of integrability (vanishing of the Nijenhuis tensor) turns out to be equivalent to

$$\partial_{\bar{w}} b = 0.$$

Thus it is very simple to change an almost complex structure only slightly around a smooth pseudo-holomorphic curve to make it locally integrable; for example, one could take b constant in the w-direction locally around Γ_f.

It is then also clear that if we have a $\mathscr{C}^1$-convergence of smooth J_n-holomorphic curves $C_n \to C$ with $J_n \to J$ in $\mathscr{C}^0$ it is possible to find $\tilde{J}_n$ integrable in a fixed neighborhood of C such that C_n is $\tilde{J}_n$-holomorphic and $\tilde{J}_n \to \tilde{J}$ in $\mathscr{C}^1$.

For the convergence near multiple components of C a little more care shows that Hölder convergence $J_n \to J$ in $\mathscr{C}^{0,\alpha}$ is enough to assure sufficient convergence of the values of b on the various branches of C_n. Finally, near the singular points of $|C|$ one employs the local holomorphicity theorem of Micallef-White to derive:

Lemma 9.3 *([SiTi3], Lemma 5.4) Possibly after going over to a subsequence, there exists a finite set $A \subset M$, a $\mathscr{C}^1$-diffeomorphism Φ that is smooth away from A, and almost complex structures $\tilde{J}_n, \tilde{J}$ on M with the following properties:*

1. *p is $\tilde{J}_n$-holomorphic*
2. *$\Phi(C_n)$ is $\tilde{J}_n$-holomorphic*
3. *$\tilde{J}_n \to \tilde{J}$ in $\mathscr{C}^0$ on M and in $\mathscr{C}^{0,\alpha}_{\mathrm{loc}}$ on $M \setminus A$*
4. *$\tilde{J}$ is integrable in a neighborhood of $|C|$*

Thus we can now assume that J is integrable in a neighborhood of $|C|$, but the convergence $J_n \to J$ is only $\mathscr{C}^0$ at finitely many points.

Note that if the convergence $J_n \to J$ is still in $\mathscr{C}^{0,\alpha}$ everywhere we are done at this point! In fact, in the integrable situation we do have a smooth parametrization of deformations of holomorphic cycles when endowing the space of complex structures with the $\mathscr{C}^{0,\alpha}$-topology. So the whole difficulty in the Isotopy Problem stems from the fact that the theorem of Micallef-White only gives a $\mathscr{C}^1$-diffeomorphism rather than one in $\mathscr{C}^{1,\alpha}$ for some $\alpha > 0$.

Step 2: Replace reduced components by spheres. The next ingredient, already discussed in connection with (9.1), is to make all non-multiple components rational. To this end we use the fact, derived in Proposition 2.8, that any J-holomorphic curve singularity can be approximated by J-holomorphic spheres. Let $U \subset M$ be a small neighborhood of the multiple components of C union the singular set of $|C|$. Then from C_n keep only $C_n \cap U$, while the rest of the reduced part of C_n gets replaced by large open parts of J-holomorphic approximations by spheres of the reduced branches of C at the singular points. For this to be successful it is important that the convergence $J_n \to J$ is in $\mathscr{C}^{0,\alpha}$ rather than in $\mathscr{C}^0$, for the former implies $\mathscr{C}^{1,\alpha}$-convergence $C_n \to C_\infty$ near smooth, reduced points of C_∞. As this is true in our case it is indeed possible to extend $C_n \cap U$ outside of U by open sets inside J-holomorphic spheres.

There are two side-effects of this. First, the result $\tilde{C}_n$ of this process on C_n is not a submanifold anymore, for the various added parts of spheres will intersect each other, and they will also intersect C_n away from the interpolation region. There is, however, enough freedom in the construction to make these intersections transverse. Then the $\tilde{C}_n$ are nodal curves. Second, $\tilde{C}_n$ is neither J_n nor J-holomorphic. But in view of the large freedom in choosing the almost complex structures that we saw in Step 1 it is possible to perform the construction in such a way that $\tilde{C}_n$ is $\tilde{J}_n$-holomorphic with $\tilde{J}_n \to \tilde{J}$ in $\mathscr{C}^0$, and in $\mathscr{C}^{0,\alpha}_{\mathrm{loc}}$ on $M \setminus \tilde{A}$, and $\tilde{J}_n, \tilde{J}$ having the other properties formulated above.

Now assume the Isotopy Lemma holds for these modified curves and almost complex structures, so an isotopy exists between $\tilde{C}_n$ for large n and some smoothing of $\tilde{C}_\infty = \lim_{n\to\infty} C_n$. Here "isotopy" means an isotopy of nodal, pseudo-holomorphic curves, with the almost complex structure and the connecting family of pseudo-holomorphic curves staying close to J, in $\mathscr{C}^{0,\alpha}$ away from finitely many points where this is only true in $\mathscr{C}^0$. Then one can revert the process, thus replace the spherical parts by the original ones, and produce a similar isotopy of C_n with the given smoothing of C_∞.

Thus we can also suppose that the reduced parts of C are rational, at the expense of working with nodal curves rather than smooth ones in the isotopy. As we can mostly work with maps rather than subsets of M, the introduction of nodes is essentially a matter of inconvenience rather than a substantial complication. We therefore ignore this for the rest of the discussion and simply add the assumption that the reduced components of C are rational.

Step 3: Break it! Now comes the heart of the proof. We want to change C_n slightly, for sufficiently large n, such that we find a path of pseudo-holomorphic cycles connecting C_n with a J-holomorphic smoothing of C_∞. Recall the pair (m, δ) introduced above as a measure of how singular a pseudo-holomorphic cycle is. By induction we can assume that the Isotopy Lemma holds for every convergence of pseudo-holomorphic curves where the limit has smaller (m, δ). This implies that whenever we have a path of pseudo-holomorphic cycles with smaller (m, δ) then there is a close-by path of smooth curves, pseudo-holomorphic for the same almost complex structure at each time. Thus in trying to connect C_n with a J-holomorphic smoothing of C_n we have the luxury to work with pseudo-holomorphic cycles, as long as they are less singular than labelled by (m, δ). We achieve this by moving C_n along with appropriate point conditions that force an enhancement of singularities throughout the path.

We start with choosing $k \leq c_1(M) \cdot C_\infty - 1$ points $x_1, \ldots, x_k$ on $|C_\infty|$ such that $k_a = c_1(M) \cdot C_{\infty,a} + g(C_{\infty,a}) - 1$ of them are general points on the component $C_{\infty,a}$. Then there is no non-trivial *equigeneric* J-holomorphic deformation of $|C_\infty|$ incident to these points, provided J is general for this almost complex structure and the chosen points. One can show that one can achieve this within the class of almost complex structures that we took J from, e.g. integrable in a neighborhood of $|C_\infty|$. With such choices of points and of J any non-trivial J-holomorphic deformation of C_∞ decreases (m, δ). This enhancement of singularities even holds if we perturb J in a general one-parameter family. By applying an appropriate diffeomorphism for each n we may assume that the points also lie on C_n, for each n.

For the rest of the discussion in this step we now restrict to the most interesting case $m > 0$, that is, C_∞ does have multiple components. Then Condition $(*)$ in the statement of the Isotopy Lemma implies that there even exists a multiple component $C_{\infty,b}$ of C_∞ such that

$$c_1(M) \cdot C_{\infty,b} + g(C_{\infty,b}) - 1 < c_1(M) \cdot C_{\infty,b} - 1.$$

Thus we are free to ask for incidence with one more point x without spoiling genericity. Now the idea is to use incidence with a deformation $x(t)$ of x to move C_n *away* from C_∞, uniformly with n but keeping the incidence with the other k points. The resulting C_n' then converge to a J-holomorphic cycle $C_\infty' \neq C_\infty$ incident to the k chosen points and hence, by our choice of points, having smaller (m, δ) as wanted.

The process of deformation of C_n incident to $x(t)$ and to the k fixed points works well if we also allow a small change of almost complex structure along the path to make everything generic – as long as (1) we stay sufficiently close to $|C_\infty|$ and (2) the deformation of C_n does not produce a singular pseudo-holomorphic cycle with the $k + 1$ points unevenly distributed. This should be clear in view of what we already know by induction on (m, δ) about deformations of pseudo-holomorphic curves, smoothings and isotopy. A violation of (1) actually makes us happy because the sole purpose was to move C_n away from C_∞ slightly. If we meet problems with (2) we start all over with the process of choosing points etc. but only for one component $\hat{C}_n$ of the partial degeneration of C_n containing less than $c_1(M) \cdot \hat{C}_n - 1$ points. To keep the already constructed rest of the curve pseudo-holomorphic we also localize the small perturbation of J_n away from the other components. Because each time $c_1(M) \cdot \hat{C}_n$ decreases by an integral amount (2) can only be violated finitely many times, and the process of moving C_n away from C_∞ will eventually succeed.

This finishes the proof of the Isotopy Lemma under the presence of multiple components.

Step 4: The reduced case. In the reduced case we do not have the luxury to impose one more point constraint. But along a general path of almost complex structures incident to the chosen points non-immersions have codimension one and can hence be avoided. One can thus try to deform C_n along a general path $J_{n,t}$ of almost complex structures connecting J_n with J and integrable in a fixed neighborhood of $|C_\infty|$. This bridges the difference between $\mathscr{C}^0$-convergence and $\mathscr{C}^{0,\alpha}$-convergence of J_n to J. If successful it leads to a J'-holomorphic smoothing of C_∞ that falls within the smoothings we have a good parametrization for, and which hence are unique up to isotopy. The only problem is if for every n this process leads to pseudo-holomorphic curves moving too far away from C_∞. In this case we can again take the limit $n \to \infty$ and produce a J-holomorphic deformation of C_∞ with smaller (m, δ). As in the non-reduced case we are then done by induction.

References

[Ar]   N. Aronszajn: *A unique continuation theorem for elliptic differential equations or inequalities of the second order*, J. Math. Pures Appl. **36** (1957), 235–239.

[Au1]   D. Auroux: *Fiber sums of genus 2 Lefschetz fibrations*, Turkish J. Math. **27** (2003), 1–10.

[Au2]   D. Auroux: *A stable classification of Lefschetz fibrations*, Geom. Topol. **9** (2005) 203–217.

[AuSm]   D. Auroux, I. Smith: *Lefschetz pencils, branched covers and symplectic invariants*, this volume.

[Bl]   Barlet, D.: Espace analytique réduit des cycles analytiques complexes compacts d'un espace analytique complexes de dimension finie. Sem. F. Norguet 1974/75 (Lect. Notes Math. **482**), 1–158.

[BtPeVe]   W. Barth, C. Peters, A. van de Ven: *Compact complex surfaces*, Springer 1984.

[Bi]   J. Birman: *Braids, links and mapping class groups*, Princeton University Press 1974.

[Ch]   K. Chakiris: *The monodromy of genus two pencils*, Thesis, Columbia Univ. 1983.

[Do]   S.K. Donaldson: *Lefschetz pencils on symplectic manifolds*, J. Differential Geom. **53** (1999) 205–236

[DoSm]   S. Donaldson, I. Smith: *Lefschetz pencils and the canonical class for symplectic four-manifolds*, Topology **42** (2003) 743–785.

[Fu]   T. Fuller: *Hyperelliptic Lefschetz fibrations and branched covering spaces*, Pacific J. Math. **196** (2000), 369–393.

[GiTr]   D. Gilbarg and N. S. Trudinger: *Elliptic partial differential equations of second order*, Springer 1977

[Go]   R. Gompf: *A new construction of symplectic manifolds*, Ann. Math. **142** (1995), 527–595.

[GoSt]   R. Gompf, A. Stipsicz: *4-manifolds and Kirby calculus*, Amer. Math. Soc. 1999.

[GrRe]   H. Grauert, R. Remmert: *Theory of Stein spaces*, Springer 1979.

[GrHa]   P. Griffiths, J. Harris: *Principles of Algebraic Geometry*, Wiley 1978.

[Gv]   M. Gromov: *Pseudo-holomorphic curves in symplectic manifolds*, Inv. Math. **82** (1985), 307–347.

[HaMo]   J. Harris, I. Morrison: *Moduli of curves*, Springer 1998.

[IvSh1]   S. Ivashkovich, V. Shevchishin: *Pseudo-holomorphic curves and envelopes of meromorphy of two-spheres in* $\mathbb{CP}^2$, preprint `math.CV/9804014`.

[IvSh2]   S. Ivashkovich, V. Shevchishin: *Structure of the moduli space in a neighbourhood of a cusp curve and meromorphic hulls*, Inv. Math. **136** (1998), 571–602.

[Ka]   A. Kas: *On the handlebody decomposition associated to a Lefschetz fibration*, Pacific J. Math. **89** (1980), 89–104.

[KpKp]   B. Kaup, L. Kaup: *Holomorphic functions of several variables*, de Gruyter 1983.

[La]   F. Lárusson: *Holomorphic neighbourhood retractions of ample hypersurfaces*, Math. Ann. **307** (1997), 695–703.

[LiTi]   J. Li, G. Tian: *Virtual moduli cycles and Gromov-Witten invariants of general symplectic manifolds*, in: Topics in symplectic 4-manifolds (Irvine, 1996), R. Stern (ed.), Internat. Press 1998.

[MD]   D. McDuff: *The structure of rational and ruled symplectic four-manifolds*, J. Amer. Math. Soc. **3** (1990), 679–712.

[McSa1]   D. McDuff and D. Salamon, *Introduction to symplectic topology*, Oxford Univ. Press 1995.

[McSa2]   D. McDuff and D. Salamon, *J-holomorphic curves and symplectic topology*, Amer. Math. Soc. 2004.

[MiWh]   M. Micallef, B. White: *The structure of branch points in minimal surfaces and in pseudoholomorphic curves*, Ann. of Math. **141** (1995), 35–85.

[OzSt]   B. Ozbağçi, A. Stipsicz: *Noncomplex smooth 4-manifolds with genus-2 Lefschetz fibrations*, Proc. Amer. Math. Soc. **128** (2000), 3125–3128.

[Pn]   P. Pansu: *Compactness*, in: Holomorphic curves in symplectic geometry, M. Audin, J. Lafontaine (eds.), 233–249, Birkhäuser 1994.

[PrWo]   T. Parker and J. Wolfson, *Pseudo-holomorphic maps and bubble trees*, J. Geom. Anal. **3**(1993), 63–98.

[PsPi]   U. Persson, H. Pinkham: *Some examples of nonsmoothable varieties with normal crossings*, Duke Math. J. **50** (1983), 477–486.

[RuTi]   Y.B. Ruan and G. Tian, *A mathematical theory of quantum cohomology*, J. Diff. Geom. **42**(1995), 259–367.

[Sh]   V. Shevchishin: *Pseudoholomorphic curves and the symplectic isotopy problem*, preprint `math.SG/0010262`, 95 pp.

[Si]   B. Siebert: *Gromov-Witten invariants for general symplectic manifolds*, preprint `dg-ga 9608005`.

[SiTi1]   B. Siebert, G. Tian: *On hyperelliptic symplectic Lefschetz fibrations of four manifolds*, Communications Cont. Math. **1** (1999), 255–280

[SiTi2]   B. Siebert, G. Tian: *Weierstraß polynomials and plane pseudo-holomorphic curves*, Chinese Ann. Math. Ser. B **23** (2002), 1–10.

[SiTi3]   B. Siebert, G. Tian: *On the holomorphicity of genus-two Lefschetz fibrations*, Ann. of Math. **161** (2005), 955–1016.

[Sk]   J.-C. Sikorav: *The gluing construction for normally generic J-holomorphic curves*, in: Symplectic and contact topology: interactions and perspectives (Toronto, 2001), 175–199, Y. Eliashberg et al. (eds.), Amer. Math. Soc. 2003.

[Sm1]   I. Smith: *Symplectic geometry of Lefschetz fibrations*, Ph.D. thesis, Oxford University, 1998.

[Sm2]   I. Smith: *Lefschetz fibrations and divisors in moduli space*, Geom. Topol. **5** (2001), 579–608.

[SU]   J. Sacks and K. Uhlenbeck, *The existence of minimal immersions of 2 spheres*, Ann. of Math. **113** (1981), 1–24.

[Te]   M. Teicher: *Braid groups, algebraic surfaces and fundamental groups of complements of branch curves*, in: Algebraic geometry – Santa Cruz 1995, 127–150, J. Kollár et al. (eds.), Amer. Math. Soc. 1997.

[Ti]   G. Tian: *Quantum cohomology and its associativity*, in: Current developments in mathematics (Cambridge, 1995), 361–401, Internat. Press 1994.

[Ve]   I. N. Vekua: *Generalized analytic functions*, Pergamon 1962.

[Ye]   R. Ye, *Gromov's compactness theorem for pseudo-holomorphic curves*, Trans. Amer. Math. Soc. **342** (1994), 671–694.

List of Participants

1. Angulo Pablo
   University of Complutense
   Madrid, Spain
   pangulo@mat.ucm.es

2. Anjos Silvia
   Instituto Superior Tecnico
   Lisboa, Portugal
   sanjos@math.ist.utl.pt

3. Auroux Denis
   M.I.T., USA
   auroux@math.mit.edu
   (lecturer)

4. Battaglia Fiammetta
   University of Firenze, Italy
   fiamma@dma.unifi.it

5. Bauer Ingrid
   University of Bayreuth
   Germany
   ingrid.bauer@uni-bayreuth.de

6. Bauer Thomas
   University of Bayreuth
   Germany
   thomas.bauer@uni-bayreuth.de

7. Bedulli Lucio
   University of Firenze, Italy
   bedulli@math.unifi.it

8. Borrelli Giuseppe
   University of Roma Tre, Italy
   borelli@mat.uniroma3.it

9. Burakovsky Evgene
   University of Bar-Ilan
   Israel
   evgen_b@hotmail.com

10. Buse Olguta
    Michigan State University
    USA
    buse@math.msu.edu

11. Cannas da Silva Ana
    Instituto Superior Tecnico
    Lisboa, Portugal
    acannas@math.ist.utl.pt

12. Canonaco Alberto
    University of Roma
    "La Sapienza"
    Italy
    canonaco@mat.uniroma1.it

13. Castano-Bernard Ricardo
    I.C.T.P., Trieste, Italy
    rcastano@ictp.trieste.it

14. Catanese Fabrizio
    University of Bayreuth
    Germany
    catanese@uni-bayreuth.de
    (editor, lecturer)

15. Colombo Elisabetta
    University of Milano
    Italy
    colombo@mat.unimi.it

16. De Poi Pietro
    University of Bayreuth
    Germany
    depoi@uni-bayreuth.de

17. Franciosi Marco
    University of Pisa, Italy
    franciosi@dma.unipi.it

18. Francisco Sandra
    M.I.T., USA
    francisco@math.mit.edu

19. Frediani Paola
    University of Pavia, Italy
    frediani@dimat.unipv.it

20. Ghigi Alessandro
    University of Pavia, Italy
    ghigi@dimat.unipv.it

21. Girolimetti Valentina
    University of Trento, Italy
    girolim@science.unitn.it

22. Iacono Donatella
    University of Roma
    "La Sapienza"
    iacono@mat.uniroma1.it

23. Jahnke Priska
    University of Bayreuth
    Germany
    jahnke@uni-bayreuth.de

24. Kadykova Tatyana
    Institute of Mining Devision
    of Ekaterinburg, Russia
    tien_shan@mail.ru

25. Kaplan Shmuel
    University of Bar-Ilan, Israel
    kaplansh@macs.biu.ac.il

26. Leveque Fanny
    Ècole Normale Superieur
    de Lyon, France
    leveque@umpa.ens-lyon.fr

27. Libermann Eran
    University of Bar-Ilan, Israel
    manliber@netvision.net.il

28. Ljungmann Rune
    University of Arhus, Denmark
    runel@daimi.au.dk

29. Manetti Marco
    University of Roma
    "La Sapienza", Italy
    manetti@mat.uniroma1.it
    (lecturer)

30. Manfredini Sandro
    University of Pisa, Italy
    manfredi@dm.unifi.it

31. Martinez David
    University Carlos III de Madrid
    Spain
    dmtorres@math.uc3m.es

32. Melnikov Nikolay
    Moscow State University
    of Russia
    n_melnikov@mail.ru

33. Migliorini Luca
    University of Bologna
    Italy
    migliori@dm.unibo.it

34. Mouftakhov Artour
    University of Bar-Ilan
    Israel
    muftahov@yahoo.com

35. Oliverio Paolo Antonio
    University of Calabria
    Italy
    oliverio@unical.it

36. Pali Nefton
    Institute Fourier, Grenoble
    France
    nefton.pali@uif-grenoble.fr

37. Pandharipande Rahul
    Princeton University, USA
    rahulp@math.princeton.edu

38. Perelli Alberto
    University of Genova, Italy
    `perelli@dima.unige.it`

39. Perroni Fabio
    SISSA, Trieste, Italy
    `perroni@sissa.it`

40. Pignatelli Roberto
    University of Trento, Italy
    `pignatel@science.unitn.it`

41. Polizzi Francesco
    University of Roma
    "Tor Vergata", Italy
    `polizzi@mat.uniroma2.it`

42. Radloff Ivo
    University of Bayreuth
    Germany
    `ivo.radloff@uni-bayreuth.de`

43. Rapagnetta Antonio
    University of Roma
    "Tor Vergata", Italy
    `rapagnet@mat.uniroma2.it`

44. Sansonetto Nicola
    University of Padova
    Italy
    `sanson@math.unipd.it`

45. Seidel Paul
    University of Chicago, USA
    `seidel@math.uchicago.edu`
    (lecturer)

46. Sena Dias Rosa
    University of Harvard, USA
    `rdurao@math.harvard.edu`

47. Shadrin Sergei
    Indip. University of Moscow
    Russia
    `shadrin@mccme.ru`

48. Siebert Bernard
    University of Freiburg
    Germany
    `siebert@math.uni-freiburg.de`
    (lecturer)

49. Smith Ivan
    University of Cambridge, UK
    `i.smith@dpmms.cam.ac.uk`
    (lecturer)

50. Tian Gang
    M.I.T. and Princeton
    University, USA
    `tian@math.mit.edu`
    (editor, lecturer)

51. Tonoli Fabio
    University of Bayreuth
    Germany
    `fabio.tonoli@uni-bayreuth.de`

52. Tzachy Ben-Yitschak
    University of Bar-Ilan
    Israel
    `benitzi@macs.biu.ac.il`

53. Van Geemen Lambertus
    University of Milano
    Italy
    `geemen@mat.unimi.it`

54. Viola Carlo
    University of Pisa
    Italy
    `viola@dm.unipi.it`

55. Welschinger Jean-Yves
    École Normale Superieur
    de Lyon, France
    `jwelschi@umpa.ens-lyon.fr`

56. Zannier Umberto
    University of Venezia
    Italy
    `zannier@iuav.it`

LIST OF C.I.M.E. SEMINARS

Published by C.I.M.E

Published by Ed. Cremonese, Firenze

1966   39. Calculus of variations
         40. Economia matematica
         41. Classi caratteristiche e questioni connesse
         42. Some aspects of diffusion theory

1967   43. Modern questions of celestial mechanics
         44. Numerical analysis of partial differential equations
         45. Geometry of homogeneous bounded domains

1968   46. Controllability and observability
         47. Pseudo-differential operators
         48. Aspects of mathematical logic

1969   49. Potential theory
         50. Non-linear continuum theories in mechanics and physics and their applications
         51. Questions of algebraic varieties

1970   52. Relativistic fluid dynamics
         53. Theory of group representations and Fourier analysis
         54. Functional equations and inequalities
         55. Problems in non-linear analysis

1971   56. Stereodynamics
         57. Constructive aspects of functional analysis (2 vol.)
         58. Categories and commutative algebra

1972   59. Non-linear mechanics
         60. Finite geometric structures and their applications
         61. Geometric measure theory and minimal surfaces

1973   62. Complex analysis
         63. New variational techniques in mathematical physics
         64. Spectral analysis

1974   65. Stability problems
         66. Singularities of analytic spaces
         67. Eigenvalues of non linear problems

1975   68. Theoretical computer sciences
         69. Model theory and applications
         70. Differential operators and manifolds

Published by Ed. Liguori, Napoli

1976   71. Statistical Mechanics
         72. Hyperbolicity
         73. Differential topology

1977   74. Materials with memory
         75. Pseudodifferential operators with applications
         76. Algebraic surfaces

Published by Ed. Liguori, Napoli & Birkhäuser

1978   77. Stochastic differential equations
         78. Dynamical systems

1979   79. Recursion theory and computational complexity
         80. Mathematics of biology

1980   81. Wave propagation
         82. Harmonic analysis and group representations
         83. Matroid theory and its applications

Published by Springer-Verlag

Lecture Notes in Mathematics

For information about earlier volumes
please contact your bookseller or Springer
LNM Online archive: springerlink.com

Vol. 1803: G. Dolzmann, Variational Methods for Crystalline Microstructure – Analysis and Computation (2003)

Vol. 1804: I. Cherednik, Ya. Markov, R. Howe, G. Lusztig, Iwahori-Hecke Algebras and their Representation Theory. Martina Franca, Italy 1999. Editors: V. Baldoni, D. Barbasch (2003)

Vol. 1805: F. Cao, Geometric Curve Evolution and Image Processing (2003)

Vol. 1806: H. Broer, I. Hoveijn. G. Lunther, G. Vegter, Bifurcations in Hamiltonian Systems. Computing Singularities by Gröbner Bases (2003)

Vol. 1807: V. D. Milman, G. Schechtman (Eds.), Geometric Aspects of Functional Analysis. Israel Seminar 2000-2002 (2003)

Vol. 1808: W. Schindler, Measures with Symmetry Properties (2003)

Vol. 1809: O. Steinbach, Stability Estimates for Hybrid Coupled Domain Decomposition Methods (2003)

Vol. 1810: J. Wengenroth, Derived Functors in Functional Analysis (2003)

Vol. 1811: J. Stevens, Deformations of Singularities (2003)

Vol. 1812: L. Ambrosio, K. Deckelnick, G. Dziuk, M. Mimura, V. A. Solonnikov, H. M. Soner, Mathematical Aspects of Evolving Interfaces. Madeira, Funchal, Portugal 2000. Editors: P. Colli, J. F. Rodrigues (2003)

Vol. 1813: L. Ambrosio, L. A. Caffarelli, Y. Brenier, G. Buttazzo, C. Villani, Optimal Transportation and its Applications. Martina Franca, Italy 2001. Editors: L. A. Caffarelli, S. Salsa (2003)

Vol. 1814: P. Bank, F. Baudoin, H. Föllmer, L.C.G. Rogers, M. Soner, N. Touzi, Paris-Princeton Lectures on Mathematical Finance 2002 (2003)

Vol. 1815: A. M. Vershik (Ed.), Asymptotic Combinatorics with Applications to Mathematical Physics. St. Petersburg, Russia 2001 (2003)

Vol. 1816: S. Albeverio, W. Schachermayer, M. Talagrand, Lectures on Probability Theory and Statistics. Ecole d'Eté de Probabilités de Saint-Flour XXX-2000. Editor: P. Bernard (2003)

Vol. 1817: E. Koelink, W. Van Assche (Eds.), Orthogonal Polynomials and Special Functions. Leuven 2002 (2003)

Vol. 1818: M. Bildhauer, Convex Variational Problems with Linear, nearly Linear and/or Anisotropic Growth Conditions (2003)

Vol. 1819: D. Masser, Yu. V. Nesterenko, H. P. Schlickewei, W. M. Schmidt, M. Waldschmidt, Diophantine Approximation. Cetraro, Italy 2000. Editors: F. Amoroso, U. Zannier (2003)

Vol. 1820: F. Hiai, H. Kosaki, Means of Hilbert Space Operators (2003)

Vol. 1821: S. Teufel, Adiabatic Perturbation Theory in Quantum Dynamics (2003)

Vol. 1822: S.-N. Chow, R. Conti, R. Johnson, J. Mallet-Paret, R. Nussbaum, Dynamical Systems. Cetraro, Italy 2000. Editors: J. W. Macki, P. Zecca (2003)

Vol. 1823: A. M. Anile, W. Allegretto, C. Ringhofer, Mathematical Problems in Semiconductor Physics. Cetraro, Italy 1998. Editor: A. M. Anile (2003)

Vol. 1824: J. A. Navarro González, J. B. Sancho de Salas, $\mathscr{C}^\infty$ – Differentiable Spaces (2003)

Vol. 1825: J. H. Bramble, A. Cohen, W. Dahmen, Multiscale Problems and Methods in Numerical Simulations, Martina Franca, Italy 2001. Editor: C. Canuto (2003)

Vol. 1826: K. Dohmen, Improved Bonferroni Inequalities via Abstract Tubes. Inequalities and Identities of Inclusion-Exclusion Type. VIII, 113 p, 2003.

Vol. 1827: K. M. Pilgrim, Combinations of Complex Dynamical Systems. IX, 118 p, 2003.

Vol. 1828: D. J. Green, Gröbner Bases and the Computation of Group Cohomology. XII, 138 p, 2003.

Vol. 1829: E. Altman, B. Gaujal, A. Hordijk, Discrete-Event Control of Stochastic Networks: Multimodularity and Regularity. XIV, 313 p, 2003.

Vol. 1830: M. I. Gil', Operator Functions and Localization of Spectra. XIV, 256 p, 2003.

Vol. 1831: A. Connes, J. Cuntz, E. Guentner, N. Higson, J. E. Kaminker, Noncommutative Geometry, Martina Franca, Italy 2002. Editors: S. Doplicher, L. Longo (2004)

Vol. 1832: J. Azéma, M. Émery, M. Ledoux, M. Yor (Eds.), Séminaire de Probabilités XXXVII (2003)

Vol. 1833: D.-Q. Jiang, M. Qian, M.-P. Qian, Mathematical Theory of Nonequilibrium Steady States. On the Frontier of Probability and Dynamical Systems. IX, 280 p, 2004.

Vol. 1834: Yo. Yomdin, G. Comte, Tame Geometry with Application in Smooth Analysis. VIII, 186 p, 2004.

Vol. 1835: O.T. Izhboldin, B. Kahn, N.A. Karpenko, A. Vishik, Geometric Methods in the Algebraic Theory of Quadratic Forms. Summer School, Lens, 2000. Editor: J.-P. Tignol (2004)

Vol. 1836: C. Năstăsescu, F. Van Oystaeyen, Methods of Graded Rings. XIII, 304 p, 2004.

Vol. 1837: S. Tavaré, O. Zeitouni, Lectures on Probability Theory and Statistics. Ecole d'Eté de Probabilités de Saint-Flour XXXI-2001. Editor: J. Picard (2004)

Vol. 1838: A.J. Ganesh, N.W. O'Connell, D.J. Wischik, Big Queues. XII, 254 p, 2004.

Vol. 1839: R. Gohm, Noncommutative Stationary Processes. VIII, 170 p, 2004.

Vol. 1840: B. Tsirelson, W. Werner, Lectures on Probability Theory and Statistics. Ecole d'Eté de Probabilités de Saint-Flour XXXII-2002. Editor: J. Picard (2004)

Vol. 1841: W. Reichel, Uniqueness Theorems for Variational Problems by the Method of Transformation Groups (2004)

Vol. 1842: T. Johnsen, A. L. Knutsen, K_3 Projective Models in Scrolls (2004)

Vol. 1843: B. Jefferies, Spectral Properties of Noncommuting Operators (2004)

Vol. 1844: K.F. Siburg, The Principle of Least Action in Geometry and Dynamics (2004)

Vol. 1845: Min Ho Lee, Mixed Automorphic Forms, Torus Bundles, and Jacobi Forms (2004)

Vol. 1846: H. Ammari, H. Kang, Reconstruction of Small Inhomogeneities from Boundary Measurements (2004)

Vol. 1847: T.R. Bielecki, T. Björk, M. Jeanblanc, M. Rutkowski, J.A. Scheinkman, W. Xiong, Paris-Princeton Lectures on Mathematical Finance 2003 (2004)

Vol. 1848: M. Abate, J. E. Fornaess, X. Huang, J. P. Rosay, A. Tumanov, Real Methods in Complex and CR Geometry, Martina Franca, Italy 2002. Editors: D. Zaitsev, G. Zampieri (2004)

Vol. 1849: Martin L. Brown, Heegner Modules and Elliptic Curves (2004)

Vol. 1850: V. D. Milman, G. Schechtman (Eds.), Geometric Aspects of Functional Analysis. Israel Seminar 2002-2003 (2004)

Vol. 1851: O. Catoni, Statistical Learning Theory and Stochastic Optimization (2004)

Vol. 1852: A.S. Kechris, B.D. Miller, Topics in Orbit Equivalence (2004)

Vol. 1853: Ch. Favre, M. Jonsson, The Valuative Tree (2004)

Vol. 1905: C. Prévôt, M. Röckner, A Concise Course on Stochastic Partial Differential Equations (2007)

Vol. 1906: T. Schuster, The Method of Approximate Inverse: Theory and Applications (2007)

Vol. 1907: M. Rasmussen, Attractivity and Bifurcation for Nonautonomous Dynamical Systems (2007)

Vol. 1908: T.J. Lyons, M. Caruana, T. Lévy, Differential Equations Driven by Rough Paths, Ecole d'Été de Probabilités de Saint-Flour XXXIV-2004 (2007)

Vol. 1909: H. Akiyoshi, M. Sakuma, M. Wada, Y. Yamashita, Punctured Torus Groups and 2-Bridge Knot Groups (I) (2007)

Vol. 1910: V.D. Milman, G. Schechtman (Eds.), Geometric Aspects of Functional Analysis. Israel Seminar 2004-2005 (2007)

Vol. 1911: A. Bressan, D. Serre, M. Williams, K. Zumbrun, Hyperbolic Systems of Balance Laws. Cetraro, Italy 2003. Editor: P. Marcati (2007)

Vol. 1912: V. Berinde, Iterative Approximation of Fixed Points (2007)

Vol. 1913: J.E. Marsden, G. Misiołek, J.-P. Ortega, M. Perlmutter, T.S. Ratiu, Hamiltonian Reduction by Stages (2007)

Vol. 1914: G. Kutyniok, Affine Density in Wavelet Analysis (2007)

Vol. 1915: T. Bıyıkoğlu, J. Leydold, P.F. Stadler, Laplacian Eigenvectors of Graphs. Perron-Frobenius and Faber-Krahn Type Theorems (2007)

Vol. 1916: C. Villani, F. Rezakhanlou, Entropy Methods for the Boltzmann Equation. Editors: F. Golse, S. Olla (2008)

Vol. 1917: I. Veselić, Existence and Regularity Properties of the Integrated Density of States of Random Schrödinger (2008)

Vol. 1918: B. Roberts, R. Schmidt, Local Newforms for GSp(4) (2007)

Vol. 1919: R.A. Carmona, I. Ekeland, A. Kohatsu-Higa, J.-M. Lasry, P.-L. Lions, H. Pham, E. Taflin, Paris-Princeton Lectures on Mathematical Finance 2004. Editors: R.A. Carmona, E. Çinlar, I. Ekeland, E. Jouini, J.A. Scheinkman, N. Touzi (2007)

Vol. 1920: S.N. Evans, Probability and Real Trees. Ecole d'Été de Probabilités de Saint-Flour XXXV-2005 (2008)

Vol. 1921: J.P. Tian, Evolution Algebras and their Applications (2008)

Vol. 1922: A. Friedman (Ed.), Tutorials in Mathematical BioSciences IV. Evolution and Ecology (2008)

Vol. 1923: J.P.N. Bishwal, Parameter Estimation in Stochastic Differential Equations (2008)

Vol. 1924: M. Wilson, Littlewood-Paley Theory and Exponential-Square Integrability (2008)

Vol. 1925: M. du Sautoy, L. Woodward, Zeta Functions of Groups and Rings (2008)

Vol. 1926: L. Barreira, V. Claudia, Stability of Nonautonomous Differential Equations (2008)

Vol. 1927: L. Ambrosio, L. Caffarelli, M.G. Crandall, L.C. Evans, N. Fusco, Calculus of Variations and Non-Linear Partial Differential Equations. Cetraro, Italy 2005. Editors: B. Dacorogna, P. Marcellini (2008)

Vol. 1928: J. Jonsson, Simplicial Complexes of Graphs (2008)

Vol. 1929: Y. Mishura, Stochastic Calculus for Fractional Brownian Motion and Related Processes (2008)

Vol. 1930: J.M. Urbano, The Method of Intrinsic Scaling. A Systematic Approach to Regularity for Degenerate and Singular PDEs (2008)

Vol. 1931: M. Cowling, E. Frenkel, M. Kashiwara, A. Valette, D.A. Vogan, Jr., N.R. Wallach, Representation Theory and Complex Analysis. Venice, Italy 2004. Editors: E.C. Tarabusi, A. D'Agnolo, M. Picardello (2008)

Vol. 1932: A.A. Agrachev, A.S. Morse, E.D. Sontag, H.J. Sussmann, V.I. Utkin, Nonlinear and Optimal Control Theory. Cetraro, Italy 2004. Editors: P. Nistri, G. Stefani (2008)

Vol. 1933: M. Petkovic, Point Estimation of Root Finding Methods (2008)

Vol. 1934: C. Donati-Martin, M. Émery, A. Rouault, C. Stricker (Eds.), Séminaire de Probabilités XLI (2008)

Vol. 1935: A. Unterberger, Alternative Pseudodifferential Analysis (2008)

Vol. 1936: P. Magal, S. Ruan (Eds.), Structured Population Models in Biology and Epidemiology (2008)

Vol. 1937: G. Capriz, P. Giovine, P.M. Mariano (Eds.), Mathematical Models of Granular Matter (2008)

Vol. 1938: D. Auroux, F. Catanese, M. Manetti, P. Seidel, B. Siebert, I. Smith, G. Tian, Symplectic 4-Manifolds and Algebraic Surfaces. Cetraro, Italy 2003. Editors: F. Catanese, G. Tian (2008)

Vol. 1939: D. Boffi, F. Brezzi, L. Demkowicz, R.G. Durán, R.S. Falk, M. Fortin, Mixed Finite Elements, Compatibility Conditions, and Applications. Cetraro, Italy 2006. Editors: D. Boffi, L. Gastaldi (2008)

Vol. 1940: J. Banasiak, V. Capasso, M.A.J. Chaplain, M. Lachowicz, J. Miękisz, Multiscale Problems in the Life Sciences. From Microscopic to Macroscopic. Będlewo, Poland 2006. Editors: V. Capasso, M. Lachowicz (2008)

Vol. 1941: S.M.J. Haran, Arithmetical Investigations. Representation Theory, Orthogonal Polynomials, and Quantum Interpolations (2008)

Vol. 1942: S. Albeverio, F. Flandoli, Y.G. Sinai, SPDE in Hydrodynamic. Recent Progress and Prospects. Cetraro, Italy 2005. Editors: G. Da Prato, M. Röckner (2008)

Vol. 1943: L.L. Bonilla (Ed.), Inverse Problems and Imaging. Martina Franca, Italy 2002 (2008)

Vol. 1944: A. Di Bartolo, G. Falcone, P. Plaumann, K. Strambach, Algebraic Groups and Lie Groups with Few Factors (2008)

Recent Reprints and New Editions

Vol. 1702: J. Ma, J. Yong, Forward-Backward Stochastic Differential Equations and their Applications. 1999 – Corr. 3rd printing (2007)

Vol. 830: J.A. Green, Polynomial Representations of GL_n, with an Appendix on Schensted Correspondence and Littelmann Paths by K. Erdmann, J.A. Green and M. Schoker 1980 – 2nd corr. and augmented edition (2007)

Vol. 1693: S. Simons, From Hahn-Banach to Monotonicity (Minimax and Monotonicity 1998) – 2nd exp. edition (2008)

Vol. 470: R.E. Bowen, Equilibrium States and the Ergodic Theory of Anosov Diffeomorphisms. With a preface by D. Ruelle. Edited by J.-R. Chazottes. 1975 – 2nd rev. edition (2008)

Vol. 523: S.A. Albeverio, R.J. Høegh-Krohn, S. Mazzucchi, Mathematical Theory of Feynman Path Integral. 1976 – 2nd corr. and enlarged edition (2008)